SANTORINI AND ITS ERUPTIONS

FOUNDATIONS OF NATURAL HISTORY is a series from the Johns Hopkins University Press for the republication of classic scientific writings that are of enduring importance for the study of origins, properties, and relationships in the natural world.

Published in the Series

Materials for the Study of Variation: Treated with Especial Regard to Discontinuity in the Origin of Species, by William Bateson, with an introduction by Peter J. Bowler and an essay by Gerry Webster

Nicholas Copernicus: On the Revolutions, translated by Edward Rosen

Nicholas Copernicus: Minor Works, translated by Edward Rosen

Problems of Relative Growth, by Julian Huxley, with an introduction by Frederick B. Churchill and an essay by Richard E. Strauss

Fishes, Crayfishes, and Crabs: Louis Renard's "Natural History of the Rarest Curiosities of the Seas of the Indies," edited by Theodore W. Pietsch

Historical Portrait of the Progress of Ichthyology, from Its Origins to Our Own Time, by Georges Cuvier, edited by Theodore W. Pietsch, translated by Abby J. Simpson

Santorini and Its Eruptions, by Ferdinand A. Fouqué, translated and annotated by Alexander R. McBirney

SANTORINI

AND

ITS ERUPTIONS

Ferdinand A. Fouqué

TRANSLATED AND ANNOTATED BY
Alexander R. McBirney

The Johns Hopkins University Press, Baltimore and London

Originally published as *Santorin et ses Eruptions,* G. Masson, Editeur,
Libraire de L'Académie de Médecine, Paris, 1879

English translation ©1998 The Johns Hopkins University Press
All rights reserved. Published 1998
Printed in the United States of America on acid-free paper

9 8 7 6 5 4 3 2 1

The Johns Hopkins University Press
2715 North Charles Street
Baltimore, Maryland 21218-4363
The Johns Hopkins Press Ltd., London
www.press.jhu.edu

A catalog record for this book is available from the British Library.

This book has been published from camera-ready copy prepared
electronically by the author.

Library of Congress Cataloging-in-Publication Data

Fouqué, F. (Ferdinand), 1828–1904.
   [Santorin et ses eruptions. English]
   Santorini and its eruptions / by Ferdinand Fouqué ; translated and
annotated by Alexander R. McBirney.
       p.   cm. — (Foundations of natural history)
   Includes bibliographical references and index.
   ISBN 0-8018-5614-0 (alk. paper)
   1. Santorini Volcano (Greece)—Eruptions. 2. Geology—Greece—
Thera Island Region.   I. McBirney, Alexander R.   II. Title.
III. Series.
QE523.S27F6813   1998
551.21'09495'85—dc21   98-4674 CIP

Contents

Translator's Preface   vii   Biographical Note   ix

INTRODUCTION   1

CHAPTER ONE: Historical Accounts of the Formation of the Kameni Islands   3

CHAPTER TWO: The Eruption of 1866   36

CHAPTER THREE: Prehistoric Structures of Santorini   94

CHAPTER FOUR: Description of the Present State of the Kamenis and Two Underwater Cones in the Bay of Santorini   132

CHAPTER FIVE: Chemical Study of the Products of the Recent Eruption of Santorini   189

CHAPTER SIX: Descriptions of the Older Parts of the Santorini Archipelago   235

CHAPTER SEVEN: Petrographic Study of the Dikes in the Northern Part of Thera   298

CHAPTER EIGHT: Petrographic Study of the Rocks in the Southwestern Part of Thera   341

CHAPTER NINE: Considerations on the Origin of the Ancient Parts of Santorini (Thera, Therasia, and Aspronisi)   379

SUMMARY   434

Subsequent Geological Studies   453

Eruptive History of Santorini   459

Translator's Notes   465

References   481   Index   491

Translator's Preface

Fouqué's monograph on Santorini has long been a standard reference for geologists and archaeologists alike. For the former, it is best known for having given the coup de grâce to the theory of craters of elevation, replacing it with the modern interpretation of calderas. For archaeologists, the account of the ruins of Akrotiri, discovered by Fouqué in the course of his geological survey, provided the earliest evidence of an advanced civilization in the Aegean. In both respects, the book took on added importance when, in 1939, Spyridon Marinatos proposed that the eruption that buried Akrotiri was also responsible for the demise of the Minoan civilization on Crete. But like so many classic works, the book is more often cited than read. This is understandable, for of the eight hundred copies printed, few remain outside the largest libraries.

During the centenary of the 1883 eruption of Krakatau, I had occasion to review the topic of caldera, and this led me to resurrect some ancient notes left to me by Howel Williams who, while writing his well-known monograph, *Calderas and Their Origin* (1941), had reviewed most of the early literature. Included in his notes was an extended summary of *Santorin et Ses Eruptions* with comments, not only on Fouqué's recognition of the origin of calderas but on other volcanological insights as well. My interest thus aroused, I delved into the book and was astonished to discover the wealth of original work that had been done by this extraordinary but sadly neglected man.

I would not have had the courage to undertake this translation had I not had the encouragement and guidance of friends who have worked extensively on Santorini, especially Tim Druitt, Walter Friedrich, Jörg Kellor, Hans Pichler, and Dorothy and Charles Vitaliano. Professor Jack L. Davis and J. M. L. Newhard annotated most of chapter 3 and provided invaluable advice on the archeological sections of the book. Hatten Yoder, through his intimate knowledge of the history of petrology and mineralogy, helped in evaluating several of Fouqué's original contributions, and Robert Symonds' knowledge of volcanic gases rescued me from a serious error in my comments on Fouqué's gas analyses. Jacque-Marie Bardintzeff, Jean-Louis Cheminée, Jaroslav Lexa, Volker Lorenz and dedicated librarians in France, Germany, and Britain helped dig out obscure references. Hans Pichler generously furnished the geological map included with this text. Madame Claire Tissot opened to me a wealth of information on the personal life of her great-grandfather. Finally, I shall always be indebted to my special friend, Bruce Marsh, without whose timely aid this book might never have found a publisher. To these and many other friends who provided help and encouragement, I offer my sincerest thanks.

FERDINAND ANDRÉ FOUQUÉ
1828–1904

Biographical Note

Ferdinand André Fouqué was born on the 21st of June 1828 at Mortain, Normandy. After primary school, he left Mortain to attend the lycée Henri IV in Paris and, through competitive examinations, won admission to both the military academy of Saint Cyr and the equally prestigious École d'Administration. He elected instead to enter the École Normale Supérieure where, in 1848, he began his studies in natural history and soon became an assistant to Henri Sainte-Claire Deville. In 1853 he published, together with Sainte-Claire Deville, his first paper, a brief note on the volatile content of fluorine-bearing minerals. About this same time, Fouqué developed a small business that prepared and sold chemical reagents for industrial use. After a few years he was able to leave the day-to-day operations to a manager, and with the freedom that the income from this business provided, he returned to school to obtain a degree in medicine. Upon presenting a thesis on the temperature of the human body, he obtained his doctorate in 1858, but he seems to have had little enthusiasm for medicine, for he never took up an active practice. In 1861, he was invited to join Henri Sainte-Claire Deville's brother, Charles, as a voluntary assistant in the latter's studies of the chemical and geological relationships of volcanic gases and fumaroles at Vesuvius. Greatly impressed by this experience, Fouqué decided to devote himself entirely to volcanology.

In 1865, Fouqué made two trips to Mt. Etna to observe the eruptions then in progress. While there, he witnessed a variety of eruptive phenomena, including the slow propagation of a radial fissure from near the summit down the northeastern flank of the volcano. During the same year he revisited Vesuvius and made a short excursion to the Eolian islands. At both places, he collected samples of volcanic gases that he used as part of a thesis on chemical aspects of volcanism. For this work he was awarded a doctorate in physical sciences on 9 August 1866. The same year, he began his studies of Santorini, which, a few months earlier, had entered a new eruptive phase.

In intervals between trips to Santorini, he visited the Azores in 1867 and 1868 to study the unusual eruptions on San Jorge which, in 1808, had devastated much of the island. Drawing on eye-witness accounts of the eruptions of 1580 and 1808, he deduced the nature of pyroclastic flows, a phenomenon that until then had not been described by geologists. He adopted the name *nuée ardente* from the Portuguese term *ardente nuvem* coined by Joao Inacio da Silveira, a village priest who witnessed the eruption of 1808. Fouqué's student and son-in-law, Alfred Lacroix, would later ensure the place of this term in volcanological literature by applying it to the eruptions he witnessed at Mount Pelée (Lacroix, 1904, p. 170; Hooker, 1965).

Meanwhile, a fortuitous discovery diverted Fouqué's studies into the realms of archaeology. During construction of the Suez Canal, large amounts of pumice were excavated at Santorini, loaded onto ships, and sent to Egypt where it was used in the construction of concrete piers. Quarrying on the island of Therasia revealed the remains of prehistoric structures buried under the ash (Lenormant, 1866), and when Fouqué returned to Santorini in 1867, he extended these excavations at his own expense. His independent studies at another site near Akrotiri (Fouqué, 1869b) led to a major archaeological discovery. From the buildings and artifacts found there, he was able to deduce conditions on the island prior to the eruption, and from a combination of archaeological and geological evidence, he concluded that the eruption occurred between 2000 and 1300 BC. Modern techniques place the date near 1628 to 1645 BC. One of most important innovations introduced by his work was the use of the petrographic microscope to determine the mineral compositions and possible sources of ceramic clays (Fouqué, 1876d).

In 1870, Fouqué's scientific work was abruptly interrupted by the Franco-Prussian War. After taking his family to the safety of his parents' home in Normandy, he returned to Paris and volunteered his service as a medical assistant. In addition to caring for the sick and wounded in the besieged capital, he converted his small commercial laboratory to manufacturing gun powder. More than a hundred letters sent out of the city by balloon to his wife and sister-in-law give a graphic account of conditions in the capital during the long winter siege (Tissot, 1991). Even with all these activities, Fouqué found time during periods of relative quiet to translate into French Zirkel's book on petrography. Throughout the turbulent years of the war and Commune, he was publishing at least five papers a year. With the return of peace, he returned to the Azores in 1873 to study the geysers of San Miguel and visited Santorini a third time in 1875.

Santorin et Ses Eruptions, published in 1879, brought Fouqué wide recognition among both geologists and archaeologists. In addition to his account of the eruptions of 1866-67 and his detailed descriptions of the excavations on Therasia and Thera, the book brought together a wealth of innovative observations and deductions based on his extensive field work and careful analyses of rocks, minerals, and gases.

In a historical sense, Fouqué's most conspicuous contribution was his analysis of the origin of calderas. The debate over von Buch's theory of *Erhebungskrater* (craters of elevation) had been going on for more than half a century. Several earlier workers, including Charles Lyell, Poulett Scrope, Prévost, and Darwin, had rejected von Buch's theory, and a few geologist recognized collapse as a more plausible alternative. Indeed, Virlet d'Aoust (1838, 1866a) had concluded on the basis of his visits to Santorini in 1829 and 1830 that the uplift theory was inadequate and that Santorini was a "cratère d'enfoncement." But the debate was prolonged, largely through the influential work of Élie de Beaumont and Dufresnoy, who continued to offer impressive arguments in support of von Buch's ideas.

In 1868, Fouqué had pointed out that the cliffs surrounding the bay of Santorini are too steep to have been formed by explosions, but the deductions he drew from this simple observation were not generally accepted. Von Fritsch (1871), for example, maintained that Santorini was a huge explosion crater slightly modified by erosion and by only minor amounts of collapse due to internal melting of the volcano. In his 1879 monograph, Fouqué explained, by simple reasoning based on careful fieldwork, that uplift had not been a significant factor and that explosion alone could not account for the missing volume of the volcano. Pointing out the small proportion of older debris in the pumice relative to the volume of rock that had disappeared, he reasoned that evacuation of a large volume of magma had undermined the stability of the volcano. Had the huge depression been formed solely by explosion, the volume of older rocks among the ejecta should be comparable to that of the caldera. It obviously was not. He concluded that "collapse, preceded, accompanied, and followed by explosions," could account for the steepness of the walls, the size of the cavity, and the scarcity of older rocks among the ejecta. Although this idea was not entirely new, it was Fouqué's careful observations and thoughtful analysis that finally won general acceptance for the basic concept that calderas are the result of collapse following rapid evacuation of a shallow magma chamber (McBirney, 1990).

Fouqué's accounts of the 1866-67 eruptions of Santorini include the first detailed description of rise and growth of a dome from the sea. They emphasize the role of viscosity in governing the contrasting eruptive styles and morphologies of domes and more common types of volcanic cones. In keeping with his long-term interest in the role of volatiles, he analyzed emissions from the erupting lavas and offered explanations for their unusual character. Even by modern standards, these analyses are of excellent quality. In addition, Fouqué provided a wealth of analytical information on the individual mineral components of igneous rocks. He developed a method of magnetic separation that enabled him to analyze for the first time, not just phenocrysts but fine-grained components of the groundmass. This work revealed that, contrary to widely held views of the time, volcanic rocks can contain more than one feldspar and that groundmass plagioclase tends to be more evolved than phenocrysts. Equally important was his observation that the last remaining interstitial liquids consist essentially of silica and alkali feldspar. Fouqué was able to show this by carefully separating and analyzing minute amounts of glass from the interstices between crystals. He also separated and analyzed the inclusions in minerals to provide the first reliable analyses of both glasses and fluids.

Noting that olivine occurs only in silica-poor rocks, Fouqué proposed a two-fold division of volcanic rocks based chiefly on the mutually exclusive occurrence of olivine and the silica minerals. This fundamental distinction, which he was the first to bring to the attention of petrologists, is the basis of most modern petrologic classifications. He seems also to have been the first person to realize that silica-rich magmas have a lower temperature than their mafic counterparts (Harker, 1909, p. 186).

Fouqué's wide-ranging interests led him into remarkably diverse fields of research. Largely responsible for bringing to France the methods of microscopic petrography that were then being developed by Sorby, Vogelsang, Zirkel, and Rosenbusch, he showed how the polarizing microscope could be used in other types of research, such as archaeology. He developed new techniques of mineral separation and was the first to utilize electromagnetic methods. As part of the project to produce a geologic map of France, he mapped four quadrangles in the volcanic regions of the Cantal and compiled the first comprehensive volcanic history of that region. In later years, he became interested in seismology, meteorites, the origin of hydrocarbons, and experimental studies of the crystallization of silicate melts.

Awarded the Cuvier Prize in 1876, he was given the chair at the Collège de France that he was to occupy for the next twenty-seven years. In 1881 he was elected to the Academy of Sciences and served as president of that body in 1901. During much of this period, he collaborated with Michel-Lévy on a study of the crystallization of minerals from artificial melts. In addition to synthesizing a number of common minerals, they produced the first feldspars of barium and strontium, demonstrated the breakdown of hydrous minerals at low pressures, and showed that equal amounts of mica and microcline were transformed to leucite and olivine. They observed that this latter reaction could account for the mineral assemblages of certain kinds of lamprophyres. Much of this experimental work was devoted to reproducing the mineral assemblages and textures of mafic igneous rocks and meteorites. First published in a series of short papers, mainly between 1878 and 1882, the work was later compiled in the book, *Synthèse des minéraux et des roches*, published in 1882.

His close association with Michel-Lévy extended beyond their mineralogical experiments. Following the devastating earthquake in Andalousia in 1884, they undertook some of the first measurements of the velocity of seismic waves through rocks (Fouqué and Michel-Lévy, 1885a, b, c). A few years later, Fouqué (1889b) made another important contribution to archaeology by identifying the brilliant pigment used in "Egyptian blue."

Despite his excursions into different fields, Fouqué had two major projects to which he continued to return throughout his long career. One was a systematic study of the feldspars. For nearly three decades, he continued to analyze mineral separates of differing compositions and to correlate their compositions with optical properties. The other was the relations of hydrothermal emanations to the structure and eruptive behavior of volcanoes. Several of his last publications were syntheses of this work.

Fouqué was greatly admired by a wide circle of close friends. His students and colleagues knew him as a dedicated professor who took a keen interest in their work and professional development. His best known student, Alfred Lacroix (1863-1948), married his eldest daughter, Catherine. It is said that Fouqué refused to give the couple permission to marry until the young man had completed his doctoral thesis! With this added inducement, Lacoix obtained his degree in 1889 and went on to pursue a career that was remarkably parallel to that of his father-in-law. He is best known, of course, for his classic account of the 1902 eruption of Mt. Pelée and for his petrologic studies of igneous rocks from all corners of the globe.

Fouqué's life spanned what was certainly one of the most vital periods in modern science, art, and literature. Born during the reign of the last French king, Louis-Philippe, he witnessed the revolution of 1848, the second republic, the Empire of Louis Napoleon, the Franco-Prussian War, the Paris Commune, and, finally, the firm rooting of democracy during the Third Republic. Fouqué's political views were strongly republican, but he avoided overt involvement in political affairs. He detested Napoleon III and the Second Empire, and yet, as we have seen, he did not hesitate to join in the defense of Paris. Though a strong supporter of Dreyfus, he declined to join other scientists and intellectuals in their public support of Zola's incendiary article "J'Accuse".

Ferdinand Fouqué died suddenly on 7 March 1904 while still active in research and teaching. The following year, his long-time coworker, Michel-Lévy (1905), provided a full account of his scientific achievements, and a modest plaque was placed on the house of his birth on the rue du Bassin in Mortain. The house was destroyed during the fighting of August 1944, but in 1959, the municipality gave his name to a street in the newly rebuilt part of the town. This is the only public memorial to this extraordinary man who contributed so much to one of the most exuberant periods in the development of western science.

SANTORINI AND ITS ERUPTIONS

SANTORIN

ET

SES ÉRUPTIONS

PAR

F. FOUQUÉ
PROFESSEUR AU COLLÉGE DE FRANCE

> Santorin, une des îles les plus remarquables et les plus instructives de la terre.
> Élie de BEAUMONT.
> *Ann. des Sc. nat.*, 1re série, t. XIV. 1830.

PARIS
G. MASSON, ÉDITEUR
LIBRAIRE DE L'ACADÉMIE DE MÉDECINE
120, Boulevard Saint-Germain
—
1879

INTRODUCTION

Few volcanic centers have been examined with as much care as that of Santorini. The eruptions that have taken place there in historic times have been the subject of several detailed accounts.

Evidence found in the older rocks in which the record of this activity is preserved offers strong arguments for all those who support, as well as those who oppose, the famous theory of craters of elevation. Until now, however, none of the geologists who have dealt with Santorini have approached this question in a comprehensive way. I have attempted to achieve this aim. In order to do so, I have made use of the conventional types of observations and experiments and, in addition, have examined the related aspects of archeology and hydrology.

The recent progress in mineralogy has been of great benefit to me. This branch of science has recently entered a new phase. Instead of limiting itself, as in the past, to studies of large crystals found mainly in cavities and other atypical occurrences, it can now deal with the immense realm of massive rocks, so that one can now study all the constituent minerals of rocks, particularly those of microscopic scale. This has opened a limitless field of study. In undertaking to explore this new realm, I have made sought to contribute to its progress by focusing my attention on the forms, chemical compositions, and optical properties of the innumerable crystals that the microscope reveals in volcanic rocks. I have made a special effort to identify those small crystals known as microlites, the existence of which was not even suspected by earlier naturalists. For my chemical studies of these rocks, I have devised methods that bring new precision to the analyses of their minerals. This has enabled me to delve into one of the most serious problems of miner-alogy, namely the characterization of the feldspars. The summary given in the concluding pages contains a brief review of the most interesting results of this work.

The aim of this book is to trace the history of the rocks of the Santorini Archipelago, indicate the catastrophes the islands have undergone as a result of subsurface thermal phenomena, and show how they attained their present form. The sources drawn on for this are of three kinds: first, historical documents containing information based on traditions and on accounts of phenomena that have taken place recently in the bay; second, prehistoric records in the form of human artifacts; and third, geological evidence based partly on an examination of the rocks underlying Santorini and partly on studies of their structure and stratigraphic relations.

The islands in the outer parts of the archipelago were formed at a much earlier time than the Kamenis. The latter owe their origin to a series of volcanic eruptions, the first of which occurred about two centuries before the Christian era, and have continued almost to the present day when high-temperature emanations still take place. Before this phase of volcanic activity that is recorded in historic annals or in the most recent geologic record, there had been a period when thermal manifestations were dormant. For several centuries, a bay with essentially the same form as today had been entirely occupied by the sea; no islets rose in the interior, and there were no obstacles to the free circulation of currents. It had this form during the height of the brilliant Hellenic civilization. Thus, we can distinguish at Santorini the formations laid down prior to this period of calm from those of a later date and divide the following account of events into two parts. The most logical order would be to start by considering the most ancient parts of the bedrock series, but I prefer to follow the reverse approach, because it offers the advantage of starting with what is best known and most readily observed. A thorough study of these recent phenomena facilitates a better understanding of similar events for which the only historical record is the geological one. This is why I have chosen to present, first, the subject of the Kameni islands and, second, all that is known about Thera, Therasia, and Aspronisi.

On three occasions I have spent several months at Santorini. I was first sent there by the Academy of Sciences (1866) and later, in 1867 and 1875, by the Ministry of Public Instruction. I want to express my gratitude for the honor these missions bestowed on me. It is to the generosity of the Minister of Public Instruction and to the good will of the Director for Sciences and Letters that I owe the opportunities to publish this work.

Finally, I must express my gratitude to the scientists who have helped me with my work. I render special homage to the memory of the illustrious masters who have preceded me in the chair of inorganic natural history of the Collège de France. Consciously following the path they have opened, I have relied on their guidance and long experience. At the same time, however, I have not hesitated to reject and even to actively oppose certain theoretical ideas that were dear to them. They themselves constantly encouraged this method. At the moment of my departure on the mission entrusted to me, Mr. Élie de Beaumont[1] offered me his final advice, saying: "You may observe facts that are not in accord with established views. On your return do not be afraid to point them out, without concern for any theory they may upset." Thus, in discussing one of the doctrines that Mr. Élie de Beaumont defended so energetically, I do so in the spirit of his glorious memory. For the study of the volatile components, I have followed the path opened to me with such brilliance by my late predecessor and mentor, Charles Sainte-Claire Deville.[2]

CHAPTER ONE

HISTORICAL ACCOUNTS OF THE FORMATION OF THE KAMENI ISLANDS

We possess no published eyewitness account of the first appearance of these islands, but Strabo, Seneca, Plutarch, Pausanias, Justin, Eusebe, and Ammianus Marcellinus all speak of them in more or less detail, and all are in general agreement. Pliny also made mention of the event but in a different context from that of other writers. Although only Pausanias, Eusebe, and Ammianus Marcellinus named Hiera specifically as the first island formed in the interior of the bay, the information on the appearance of this island provided by the other authors I have just cited is precise enough to confirm that their accounts refer to the same event. Strabo* recounts what happened in such detail that he must have received his information from a contemporary authority or from a well-preserved local tradition. He states that "for four days, flames were seen to rise from the sea between Thera and Therasia, and the entire bay seemed to be on fire. These fires gradually lifted, as if by a giant lever, an island composed of incandescent material and having a circumference of 12 stades (2,220 meters). The people of Rhodes, who at that time ruled a maritime empire, were the first to dare to land on the new island after these events. They raised a temple in honor of Neptune Asphalios." According to Titus-Livy (Book XXXI, Chap. 15), Rhodes was a dominant maritime power in the period around 197 B.C., the probable date of the eruption. As we shall see later, several other writers also indicated this period as that of the height of the power of Rhodes.

Although Seneca† gives a lively account of the eruption, he indicates the time and place of the event in much vaguer terms. He writes "Posidonius tells us that in the time of our ancestors an island appeared in the Aegean Sea. For a long time before, the sea foamed, and smoke rose from its depths. Then the fire became visible but remained intermittent. From time to time, it expanded like flashes of lightning, while the heated layers of water became strong enough to overcome the pressure of the overlying rocks. Then stones

---

* Strabo, 48, 10. Didot edition
† Seneca, *Natural History*, book II, Chap 26.

were thrown out wildly. Some were propelled by the underground heat and remained unaltered; others were corroded and became as light as pumice. In the end, one could see protruding from the highest point a peak that had a burned appearance and continued to grow in height until it became an island."

There is no question that the accounts of Strabo and Seneca are consistent. It is likely that both drew on the writing of Posidonius. This geographer is referred to in another passage of Seneca as having been an eye witness of the volcanic phenomena that they discuss.

Plutarch*, in listing events that were forecast and subsequently took place, specified the date and place of the eruption. As an example of a successful prediction, he cites the rise of an island between Thera and Therasia. The statement of the oracle was interpreted in this way: "When the descendants of the people of Troy gain supremacy over their adversaries, incredible things will come about. The sea will burst into immense flames, lightning will flash in clouds of steam, while rocks will rise to the surface of the water, and a previously unknown island with appear and become permanent. Then," he says, "within a short lapse of time, the Romans defeated Hannibal and took him to Carthage, Philip was defeated by the Etolians and Romans, and finally, an island rose from the depths of the sea in the midst of violent boiling of the waves and ejections of burned debris. Who would dare to say that all this is coincidence? Does this not demonstrate the truth of the prophesies?"

According to Titus-Livy (Book XXXIII, Chap. 1-3), the treaty of peace between Philip and the Romans was signed after the battle of Cynocephale in 197 B.C. Therefore, this same date must be assigned to the appearance of Hiera.

It is true that the same author, in Book XXXIX, Chapters 46 and 47 of his history, speaks of a treaty concluded between Philip and the Romans and says that this took place under the consulate of Claudius Marcellus and Quintus Fabius Labeo, who administered the republic during the 570th year after the founding of Rome (184 B.C.). But the treaty in question here is evidently later than the one that followed the battle of Cynoscephalae and was concluded under the consulate of Quintus Flaminus. The date of the latter, 197 B.C., is closer than that of the second (184 B.C.) to the year 201 in which the second Punic war took place, and since Plutarch considered the conclusion of the treaty with Philip and the end of the second Punic war as having taken place within a very short time, it follows that the first of these two dates must be favored over the second as corresponding to the formation of Hiera as a phenomenon contemporaneous with the two events cited above.

---

* Plutarch, *De Pytheæ oraculis*, vol. VII, p. 570, Reiske edition.

Pausanias* also mentions the appearance of Hiera in his description of Greece. He says: "Fate has given us the opportunity to contemplate in our time a more moving spectacle than the joys or griefs of city life. An invasion of the sea has plunged the island of Chryse beneath the waves, while another island, named Hiera, has risen from the depths of the waters."

One must not take the words "in our time" too literally, for they would no longer apply in this case. Pausanias was alive around the year A.D. 170, or about three hundred and seventy years after the formation of Hiera. We can conclude, therefore, that he meant to say that the phenomena in question had taken place in the familiar historical times of Greece.

The account left by Justin in Book XXX, Chapters III and IV of his *Historiae Philippicae* is very close to that of Seneca and dates from nearly the same period. "In the year 555 after the founding of Rome (198 B.C.), Philip, the king of Macedonia, asked for a truce with the Romans. In the same year, an upheaval of the ground took place in the area enclosed by the coasts of Thera and Therasia. Then, to the great astonishment of navigators, the waters became hot and an island emerged."

Eusebe reported in his chronicle that an island named Hiera was formed in the third or fourth year of the 145th olympiad, which again gives a date of 197 B.C.

And, finally, Ammianus Marcellinus (*Res gestae*, XXVII) only noted that Hiera was formed near Thera without giving any date or other commentary on the matter.

Pliny reports the appearance of Hiera in the year 107 B.C., but other than that, he provides no information on the eruption. The date he gives differs from that of the other authors and cannot be correct. It appears in the author's *Natural History* along with other information that is clearly wrong. According to this work, the islands of Thera and Therasia would have been formed one hundred and thirty years earlier, in the fourth year of the hundred and thirty-fifth olympiad, or not until 237 B.C. Geological studies of these islands are in accord with the account of Heroditus in showing that they originated much earlier. Hence, there is no reason to accept Pliny's date. This tends to confirm the opinions of modern critics who have concluded that a copier's error slipped into the translation of these Roman naturalists and that the CXXXVth olympiad should be read as the CXXXXth. This hypothesis in no way eliminates the inconsistency, however; it only makes matters worse, because it places the event in the year 67 B.C.

---

* Pausanias, *Description of Greece*, VIII, 33, 4. Didot edition.

In view of these accounts, it seems likely that the island called Hiera was formed in the year 197 B.C.

Several authors agree in reporting an eruption during the reign of Claudius; some fix the location in the bay of Santorini and add that it resulted in another island being formed. Dion Cassius, Aurelius Victor, and Eusebe all give exactly the same date for the event, which they say took place in the 799th year since the founding of Rome, or A. D. 46, during the consulate of Valerius Asiaticus. Aurelius Victor says that a very large island was formed. He reports that there was also an eclipse of the moon at the time of the volcanic eruption. Labbe de Bourges* (*Histoire chronologique*, vol. I, p. 152) indicates two eclipses of the moon in the year 46, one on the second of January and the other during the night of 31 December 46 and 1 January 47, and thought that on that night an island 30 stades (5,555 meters) in circumference rose from the Aegean Sea between Thera and Therasia.

Pliny informs us that this island was called Thia, but he also gives inaccurate information on the date of formation of Hiera, just as he does about the date of birth of Thia and on it location. According to him, this island would have appeared the 8th day after the ides of July (7 July) in the year A.D. 3 and at a distance of 2 stades (370 meters) from Hiera. He adds that this happened during the consulate of Marcus Junius Silanus and Lucius Balbus. It is known from other historical sources that this consulate took place, not in the year 3, but in A.D. 19. Pliny is therefore in disagreement not only with the authors just cited, but even with himself. To reconcile his testimony, it has been assumed, as we noted earlier, that a copier's error put Pliny's dates forty years earlier. In that case, the eruption that led to Thia would have occurred in the year A.D. 43. But the conflict still remains, because Marcus Junius Silanus and Lucius Balbus were not consuls in that year but twenty-four years earlier.

It is quite extraordinary that Pliny makes no mention of an eruption during the year 46, for he was alive at that time. The error he committed regarding the names of the consuls shows that his account was clearly faulty. One need only read any random page of his natural history to see that he accepted without question any information that came to him; so one should attach little credence to what he says. We should therefore hold to the dates of 197 B.C. and A.D. 46 for the eruptions responsible for Hiera and Thia.

Where do these two islands fit into the Kameni group? Of what did they consist? Writers who have concerned themselves either with the history

---

* Labbe de Bourges, a distinguished commentator, lived from 1607 to 1667.

or geography of the Santorini archipelago are unanimous in considering Hiera equivalent to the main part of Palaea Kameni. It seems at first glance that this is the only acceptable hypothesis. As we shall see later, the dates are known for the relatively recent formation of Micra Kameni and Nea Kameni, and because these, along with Palaea Kameni, are the only islands that presently reach the surface within the limits of the bay of Santorini, it was entirely natural to consider the earliest known eruptions in this region as having produced the oldest of these islands, namely, Palaea Kameni. But if so, at what point did Thia appear? If Hiera is nothing more than part of Palaea Kameni, and if the eruption of Thia took place only two stades (370 meters) from Hiera, it is clear that Thia appeared at the present site of Nea Kameni, of which it presently constitutes the base for the modern lavas that concealed it after the eruption we shall discuss later. In fact, within a radius of 400 meters around Palaea Kameni there is no submarine feature that could be considered the remains of an early cone with a summit that once rose above the water. Within the bay of Santorini, there are only two submarine cones, both situated between Micra Kameni to the west and the cliffs of Thera to the east. The first, which presently serves as an anchorage for ships and is known as Banco, is situated about 2,500 meters (13 stades) from Palaea Kameni; the second, even closer to Thera, is about 3,100 meters (15 stades) from Palaea Kameni. Thus, if one takes Pliny's statements literally, neither of these two cones fulfills the conditions necessary for the location of Thia.

No detailed soundings had been made of the bay before the eruption that produced Nea Kameni at the beginning of the previous century; so it is impossible to say anything about the state of the seafloor before the deposition of lavas that took place at that time. One can only say that it is not unlikely that a submarine feature already stood at this spot as the result of an eruption like that of Thia, for it is common to see in volcanic regions products of successive eruptions accumulate and pile up in the same place. One can also add that this is rendered even more likely by the emergence at the beginning of the eruption that produced Nea Kameni in 1707 of a bank of pumice covered with living mollusks.

Nevertheless, two objections can be raised against this explanation:

1. The submarine feature that existed before 1707 at the present site of Nea Kameni rose at the time of that eruption in the form of a mound covered with white pumice; it seems, therefore, that one must consider it a random high place on the general floor of the bay, which had the same sort of carapace. The submarine cones that can be observed at the present time are not covered with pumice; they consist entirely of gray cinders and dark lapilli.

2. It is difficult to believe that Nea Kameni could have covered Thia completely. According to the ancient authors, Thia was quite large; Pliny and, after him, Eusebe, Cassiodorus, Orosius, and Cedrenus give its circumference as 30 stades (5,550 meters); whereas that of the present Nea Kameni does not exceed 4,200 meters. Recent soundings around the island should have revealed shallow water where talus from Thia was covered by water. Since nothing of the kind has been noted, some who have studied the problem have retreated before the difficulties raised by the earlier explanation and have devised another that allows them to take Thia as one of the two submarine cones mentioned above. Ross, for example, thought that a copier's error had slipped into Pliny's account and that instead of two stades, *duobus stadiis* (370 meters), it should read twelve stades, *duodecim stadiis* (2,200 meters) for the distance between Hiera and Thia. Since the distance between Banco and Palaea Kameni is about 2,500 meters, we see that Banco could reasonably be considered within about the same range as that part of Thia that has survived the erosion of waves. Whether one accepts Ross's hypothesis or not, it is certain that the eruption of A.D. 46 can be placed much more reasonably at Banco than at a submarine cone hidden beneath the lavas of Nea Kameni; the measurements given by Pliny are seldom precise enough to justify putting greater weight on them in this case and abandoning an interpretation on the basis of the distance he indicates between the locations of the two phenomena.

The eruption of 197 B.C., as well as that of A.D. 46, can therefore be considered with great probability as having produced the principal parts of Palaea Kameni and of Banco. But is it Hiera that corresponds to Palaea Kameni and Thia to Banco? I am of the opinion that this is not correct and that the opposite is more reasonable. In fact, although it is not necessary to accept literally the numbers given by Pliny nor to believe that Thia had a circumference of only 30 stades (5,500 meters), it is nevertheless certain that Thia was relatively large with dimensions greater than those of Hiera, which had a diameter of only 12 stades (2,220 meters). However great the erosion one thinks that Banco has suffered as a result of the action of waves, it is easy to see from the extent of its base that its dimensions have always been less than those of Palaea Kameni, for even if waves quickly destroyed those parts of the volcanic edifice affected by near-surface currents of the sea, their erosive action is restricted to such shallow depths that it is unlikely that the cone of Banco ever had a basal diameter much greater than it has at present. That being so, Palaea Kameni, being the larger of the two volcanic bodies in question, can logically be considered the island that represents Thia.

Moreover, the present dimensions of Palaea Kameni exceed even those that ancient authors attributed to Hiera, even though they have been reduced in the course of centuries by the action of waves and earthquakes. It is impossible, therefore, to take Palaea Kameni as Hiera.

Finally, the manifestations so clearly described by Strabo and Seneca as having been characteristic of the eruption of 197 B.C. seem to indicate a predominance of explosive activity, which is not in accord with the present state of Palaea Kameni, which is composed almost entirely of solid lava and has only minor amounts of pyroclastic material. On the other hand, judging from the soundings across it, Banco seems to consist in large part of scoria and lapilli. In short, the eruption of 197 B.C., which was marked principally by release of gas and ejected material, corresponds more closely to what we know about Banco, while the eruption of A.D. 46, which is notable for the size of the island it produced, cannot have taken place anywhere but at Palaea Kameni. Thus Banco would be the remains of Hiera and the main mass of Palaea Kameni an important part of Thia.[1]

The cone formed by the eruption of 197 B.C., which was originally only slightly above sea level, seems to have been quickly truncated by waves, so that the name "Hiera" would later have disappeared had it not been shifted to the products of the eruption of A.D. 46. As a result, it is the name of Thia that has been lost, and during the first millennium of the Christian era, the island formed in A.D. 46 was designated Hiera instead of Thia, the name it had previously borne. The idea that the true 197 B.C. cone of Hiera disappeared and that the names were changed seems to be confirmed by the fact that Pomponius Mela, who lived in the first century of the Christian era, mentions only Thia as one of the islands in the bay of Santorini, while later the only name we find in the chronicles is Hiera.*

ERUPTIONS OF 726

These eruptions took place during the reign of Leon the Isaurian. The circumstances that accompanied them have come down to us through the work of Nicephore and Theophanes, both of whom lived in the eighth century. In the early summer of 726, the sea was seen to boil as if it had been heated by an incandescent furnace. Heavy steam rose from it, and the smoke became

---

* Despite what has just been said about Palaea Kameni being Thia, one can invoke an archaeological argument that does not seem very strong to me but should nevertheless be mentioned. It is known from the writing of Strabo that the people of Rhodes had constructed a temple to Neptune on Hiera; we also know that in Greece most ancient temples of Neptune have been replaced by churches dedicated to Saint Nicolas. There is, in fact, a church dedicated to this saint on the eastern coast of Palaea Kameni.[2]

denser and denser until it was joined by incandescent pyroclastic ejecta. The jet was so hot that it seemed to be completely on fire. Large blocks of pumice were ejected in such quantity that they covered the surface of the sea over an immense area, and wind carried them to the coasts of Asia Minor and Macedonia. They formed rocky masses on and around the island that at that time was called Hiera but is most likely that which is now called Palaea Kameni. The lavas added a small promontory on the eastern shore of the island. They border a shallow inlet of the shoreline now called Saint Nicolas Cove. Their scoriaceous surface, dark black color, and fresh appearance distinguish them from the main mass of Palaea Kameni.³

It is worth noting that at this time there is nothing on Palaea Kameni that would indicate a source for the ejecta produced by this eruption. According to the historians already cited, however, the ejecta were exceptionally abundant.* It is impossible to imagine that some trace of this great mass of loose material would not remain. We should attempt to find within the bay of Santorini an elevated feature composed largely of pyroclastic material that could be related to the 726 eruption. Among the subaerial and submarine cones in the bay, there is only one to which we have not yet assigned a date of formation. This is the submarine cone closest to Thira. Because of its distance from Palaea Kameni (3,100 meters), it has never been regarded as a contemporary of the material forming the promontory of Saint Nicolas Cove, and its formation cannot be attributed to any known volcanic eruption. (I intentionally ignore the eruptions of the years 3 and 19 that have been listed on the basis of Pliny's account, because to me they seem completely apocryphal.) If one does not accept the cone as having been formed by the eruption of 726, one must consider it a result of a prehistoric volcanic event for which we have no record. That is an improbable hypothesis, for after the immense cataclysm that was responsible for the bay, it is reasonable that the subterranean forces remained quiet for a long while. It would indeed be extraordinary for an eruption to take place after a relatively short lapse of time; on the contrary, a longer period of repose would have followed this isolated event.

The distance of Palaea Kameni from the submarine cone in question is not really very great. The outbreak of an eruption is always signaled by opening of more or less straight linear fissures that give off gas, steam, and molten magma, sometimes at the same place, sometimes at different places

---

* The considerable interval of time that elapsed between this eruption and those that immediately preceded and followed it demonstrates that it must have been one of great importance.⁴

along their length. These fissures, ordinarily very narrow, are commonly several kilometers long. The one formed on Vesuvius in 1861 started at the center of the mountain, crossed the town of Torre del Greco, and extended into the sea where it gave off emissions of gas. There is no reason to suppose that a similar thing could not have happened in the bay of Santorini in 726. Micra Kameni did not exist at that time, and Banco is south of the straight line joining the submarine cone closest to Thera with the promontory of Saint Nicolas on Palaea Kameni. Consequently, nothing rules out a fissure opening between these two points and feeding a flow of lava at its western extremity while emitting gas and ejecting cinders and pumice from its eastern end.

Philostratus mentions in his biography of Apollonius a volcanic eruption between Crete and Thera which resulted in the formation of a new island, but this seems to have nothing to do with eruptions in the bay of Santorini, and I shall not discuss it here.[5]

EVENTS OF 1457

An interval of several centuries had passed without any significant changes in the configuration of the islands when, on the 25th of November 1457, a part of the cliffs of Palaea Kameni broke away with a great crash and was almost entirely submerged. The detached rocks created a new reef. A Latin inscription in verse, written on the walls of the Jesuit church in the ancient fortress of Skaro in honor of Franciscus Crispus II, Duke of Naxos, relates the details of the catastrophe. I reproduce it here in its entirety:[6]

> Magnanime Francisce, heroum certissima proles,
> (Crispe) vides oculis (nobis) clades quae mira dedere
> Mille quadringentis Christi labentibus annis
> Quinquies undenos, istis jungendo duobus
> Septima Calendas decembris, murmure vasto
> Vastus Theresinus immaniasaxa Kamenae
> Cum gemitu avulsit, scopulusque ex fluctibus imis
> Apparet, magnum gignit memorabile monstrum.

We see that, according to this account, it is not a question of either fire, an explosion, or formation of a new island. The phenomenon took place in a single day and was not prolonged in the manner of true volcanic events. Thus, it is impossible to see in it the beginning of a new eruption. Besides, this description agrees with the evidence found in the topography of Palaea

Kameni, for as we shall note later, the central part of the islands has a steep southeastern cliff more than 100 meters high, which could have been formed only by a violent avalanche of rocks. The formation of the reef mentioned in the inscription was nothing but the emergence of the detached debris above the water level.

ERUPTION OF 1570 OR 1573

This eruption is the one that produced Micra Kameni. No contemporary account has survived - only a note written by the Jesuit missionary, Father Richard, eighty years later. Father Richard spent part of his life on Santorini, so his testimony can probably be considered an exact expression of the tradition that had been maintained by the inhabitants of the island. According to Father Kircher (*Mundus subterraneus*, Books I, IV, Chap VI), Richard recounted it to him in Rome in the following way: "We know that the second island neighboring Palaea Kameni was formed in 1570. The terror at Santorini was great. Several of the older residents who had been eye witnesses assured me that the volcanic fires lasted a full year. In the middle of the modest-size island called Micra Kameni, one still can see today a large, deep cavity with a rounded funnel-like form. It is from there that the mixture of stones, rocky debris, and cinders came as though from a furnace. These fires, which were fed by sulfur and bituminous material, never went out; they flared up from time to time with great strength."

Pègues (*Histoire et Phénomènes du Volcan et des iles volcaniques de Santorin*, p. 143) borrowed from Coronelli another account attributed to Father Richard.

Another Jesuit, Father Gorée, recounted the appearance of Micra Kameni in almost the same terms (*Phil. Trans.*, vol. V, xxvii, 1711), but he gives the date as 1573. It seems to me that the date 1570 should be preferred, for one finds it in a popular poem in modern Greek about the eruption of 1650. It has also been adopted by two scientists, Girardin and Ordinaire, but it is likely that the eruption lasted several years, so it must have been going on during both of these years. There is no need to go into the various accounts published later by different authors, all of whom appear to have copied or repeated more or less exactly the story contained in the work of Kircher and in a number of incomplete and inaccurate commentaries, including even the book of Coronelli drawn on by Pègues.

ERUPTION OF 1650

The site of this phenomenon is outside the bay of Santorini, three and a half miles north of Thera. Today the sea covers the volcanic mass that was born in this cataclysm and is now referred to as the cone of Kolombos, the name of the cape on the nearest part of Thera.

If one considers the form of the bay and the islands within it, one sees that they represent an imposing mountain, the base of which has plunged beneath the water and the crater of which has been invaded by the sea. The submarine cone of Kolombos must be regarded as a parasitic cone on the flank of the main mass of Santorini, much like the lateral cones on the outer slopes of almost all large volcanoes.

We have several accounts of the eruption of 1650 that appear to have been written by eyewitnesses or, at least, were written under the influence of well-preserved traditions. The principal ones are, first, a poem contained in a collection of curiosities printed in modern Greek verse; second, an anonymous account in Italian; third, an ancient manuscript, printed in 1837 in modern Greek (the publication has several missing parts owing to the illegibility of the original text); and, fourth, a report by Father Richard printed in Paris in 1656 and written at Santorini where the author lived from 1644 to 1675. Such are the main sources that Pègues drew on for his detailed history of the eruption of 1650. At the beginning of his description, however, he says that he also drew on traditions that were "still widespread throughout the island and had been transmitted by word of mouth." He states that he restricted himself to arranging and coordinating the accounts of old people by taking details from one to fill in gaps in another.

In this extract, the essential points of his account are taken almost verbatim, while omitting the philosophical reflections and descriptions of religious processions and other ceremonies that abound throughout his book.

In the year 1649, Santorini was shaken by such violent earthquakes that for a long while the inhabitants longed to leave the island and escape the dangers with which they seemed to be threatened. During the first days of March 1650, the shaking began again with greater violence than ever. Houses were cracked open, rocks split, and blocks of stone detached from the coastal parts of the island rolled down to the sea. These earthquakes were followed by a great drought and a calm so complete and prolonged that old people could recall nothing like it. It marked the beginning of a famine.

Again, on the 14th of September, frequent violent shocks were felt until some time the next day, not only on Santorini but on all islands of the archipelago; they were accompanied by frightful subterranean roaring. They continued to grow in intensity throughout the month. One of the earthquakes was so strong that the inhabitants of the island thought they were going to be buried under the falling debris. On the 27th, in particular, the uproar reached a climax with such furious shaking that the houses rocked like cradles or leaves in the wind.

On three different occasions after this terrifying earthquake, clouds of dense smoke and flame were seen rising about four miles northeast of Santorini in the direction of the reef of Anhydros near the island of Amorgos. At the same time, a noxious odor spread from that area toward the closest region around Thera and soon extended over the entire bay of Santorini and the islands that enclose it. Six days later, green seawater was seen at the spot where this phenomenon occurred, indicating that foreign material was dissolved in the water.

While the inhabitants of Thera watched these events in amazement, they noticed among the waves at the place where smoke had risen a body of snow-white ground resembling a ledge where birds roost. They then saw a towering column of smoke rise and quickly disappear. A second column that rose an hour later was even denser and taller than the first one. Earthquakes continued all day but with less violence. The sea was covered with pumice.

On the morning of the 28th, the volcano burst forth again. A cloud of smoke shot from the mound formed the evening before, rising in waves to great heights and then disappearing an hour later. In the afternoon, the smoke rose again, even higher than before, blotting out the entire sky. This condition lasted until the following morning. One of the explosions made a horrible uproar. Flames burst like flashes of lightning while incandescent rocks shot from the volcano with a sound like thunder.

On the 29th, a dense cloud of smoke was still rising to prodigious heights when incandescent material suddenly appeared. The newly emerged islet formed a crater from which enormous, glowing rocks were ejected. Terrifying detonations accompanied the explosions, the earth shook, the air seemed to be in flames, flashes of lightning rent the sky, and thunder rumbled. The sound reached the Dardinelles about 400 km from Santorini. On the island of Chio about 200 km away, it was thought that a naval battle was going on in the neighboring waters, for the detonations were taken to be cannon fire. The ground was in a state of continual unrest, and Laugier, in his history of Venice, says that earthquakes were felt on the island of Crete, situated 25 to 30 leagues (100 to 120 km) away. The ash erupted from the volcano was carried as far as Anatolia and Platia, where it covered the grapes with a white blanket that looked like chalk. Large rocks were thrown to distances of more than two leagues (8 km). Father Richard states that he saw some in a field near Thira that were so large that fifty men could not move them. The falling volcanic debris and earthquakes caused the sea to be as rough as it would be in a violent storm; it invaded the lower parts of Thera, carried away animals grazing in the fields, and submerged forever about 500 acres of land along the eastern shore of the island. It tore away fig and olive trees, knocked down five churches, revealed two ancient villages that had been buried by similar catastrophes or simple earthquakes in the past, and carried away the main route of communication along the coast at the foot of the mountains of Profitis Ilias and Mesa Vouno.[7]

On the island of Ios, a short distance away, the sea rose about 20 meters on the rocky shore, uprooting vegetation and depositing pumice. On Sikino, an island not far from there, it advanced more than 350 yards inland. On Zea (formerly Céos), the sea also overran the coast and was so rough that a ship of the Turkish fleet that was moored there was thrown violently against the shore of the same port and badly damaged. Two other ships experienced the same fate in the port of Canea on Crete.

All the while, the air was charged with noxious fumes that spread a sickening miasma everywhere.

One of the seismic tremors was so strong and prolonged that it was thought that the entire island was going to be engulfed by the sea. Then the sky was obscured and brilliant flashes of lightning illuminated the clouds of smoke. The bolts struck the island of Thera twice and split large rocks not far from a place where part of the population was assembled.

Throughout the archipelago, great quantities of pumice thrown out by the volcano were seen floating on the sea. On the island of Thera, even buildings constructed entirely of volcanic tuff cemented with lime suffered greatly. Father Richard states, "I saw them sway like ships and then come upright again." The roofs of more than 200 houses collapsed, and more than fifty buildings were overturned. The cliffs of Merovigli on the island of Thera split, and for days blocks of rock fell away into the sea.

The sun did not appear at all during the afternoon of the 29th of September; it remained veiled by a dense cloud. A frightful darkness spread over all the region. In the evening, around eleven o'clock, the volcanic tempest seemed to relent, but at midnight the explosions began again. A bad odor returned and the same flames and lightning were seen. The underground rumbling and explosions were the same as before. Great amounts of acidic gas and hydrogen sulfide seem to have been released at that time. Pègues says that "the inhabitants of Santorini found that their gold (which was probably debased with other metals) and their silver, had taken on the colors of bronze and iron. Panels were covered with a moss-like substance, and walls of dwellings that had originally been white took on tints of green and yellow that resembled rust. Silver coins in purses, and sacred vases in their cases were tarnished, and paintings that had not been varnished looked as though they have been rubbed off. In time they were restored to their proper colors by cleaning them with oil and hot ash."

These emanations also had an adverse effect on the general health of the inhabitants of Santorini. The following day, the 30th of September, they experienced an indescribable pain in their eyes that brought continual tears. Many were reduced to complete blindness and let out mournful cries. Very few escaped the bad effects, and almost all were blinded for three days. Several succumbed to violent eye pains and cerebral congestion. On the islands closest to the eruption, a great number died of asphyxiation. Father Richard estimates that fifty persons died from the harmful effects of the volcanic gas and that more than a thousand animals died in the same way. Pègues reports that when they could see again, several inhabitants of Santorini who were driven by curiosity went to the shores closest to the site of the eruption, but they soon regretted this, for several of them, especially the most enthusiastic who had arrived first, were suffocated by the foul-smelling gases, just as Pliny had been at Vesuvius, and, like him, fell in place, victims of their curiosity. Those who did not come as close experienced only fainting, but they would probably have suffered the same fate if those who followed had not rescued them. Those who had seen the danger and retreated to watch from a high point noticed that cattle and other animals had been suffocated by the noxious fumes.

On the 2nd of October, two local boats were returning from the neighboring island of Amorgos when, in the obscurity of the night, one of them passed close to the site of the eruption and was enveloped in poisonous gas. The next day the inhabitants of Ios noticed a boat drifting aimlessly with the wind a short distance offshore and suspected that some misfortune had befallen the crew. They immediately set out to rescue them and found the boat with its sails deployed caught on an islet of pumice. Onboard were nine bodies that were unusually inflated. Their eyes were inflamed, their tongues were hanging out, and they were still in positions that indicated the action each person was engaged in at the time he was suddenly brought to his death. They also bore marks of burns. The bodies were brought to the island of Ios for burial.[8]

The other boat escaped this misfortune and arrived on the 3rd safe and sound at Santorini. The crew, however, had suffered while in the vicinity of the volcano, where almost all the sailors had fainted. One who held up better than the others had saved his companions by telling them to rub their nostrils with wine.

Several days later the intensity of the eruption began to diminish. The volcano was only active at intervals that at times were almost periodic. Flames were no longer visible except for irregular bursts, some of which were small and others of great height. Earthquakes became less violent and less prolonged. But a great quantity of pumice still came from the volcano and spread over the sea throughout the archipelago. It gave off a strong odor of sulfur. At last, calm returned and continued through the rest of the month of October.

On the morning of the 4th of November, a dense, black cloud rose from the crater, immediately blotting out the sky. People working in the fields in the part of Thera facing the volcano were surprised by the noxious cloud. About twenty of them lost consciousness and did not regain their senses until the end of the day when their funerals were being prepared.*

The next day more flames were seen, but they were weaker than before. The poisonous fumes were less noticeable, the earthquakes were almost imperceptible, and the smoke was scarcely visible.

Nevertheless, during the first days of December flames reappeared, earthquakes became stronger, and the sea was more restless than usual.

Finally, on the 6th of December, everything almost ceased. The water, which had been green, yellow, and red, regained its natural color. A few signs of high temperatures and very weak shocks continued to be detected for several years.

The island that had formed so violently disappeared after a short while leaving no part visible above the water. In its place was a steep-sided underwater cone with a minimum depth of 10 fathoms (18.2 meters).

These are the main facts concerning the eruption of 1650. Some of the matters related in this account have recently been the subject of public discussions, which at times were marked by a note of skepticism. The flames, lightning, and the formation of a crater, as well as the projection of large blocks as far as Thera and splitting of the cliffs of Merovigli, have all been questioned; the movements of the sea have been attributed to a simple retreat

---

* At Santorini the dead are not buried in an open cemetery. Almost immediately after death, a funeral is held and the body is taken down into a vast cavern dug into the pumice.

of the water due to the release of gas from the volcano rather than to earthquakes.

Let us consider for a moment how much confidence one can attach to the account of Father Richard and his commentator, Pègues. First of all, were flames produced by the eruption? Denials of this are based on the fact that, in all the ancient descriptions of the volcano, the luminous trails of incandescent ejecta are described as flames; critics have evidently been under the influence of the idea that has recently been current in science to the effect that the illumination in volcanic eruptions does not come from combustion of gas. There were reliable witnesses to this during a certain phase of the Santorini eruption of 1866, but their stay on Thera coincided with a period when the amount of combustible gas released from the active vent was relatively small. If there had been witnesses of the first phase of the eruption in which these gases had a very prominent role, they would probably have taken another view of Father Richard's account, because, it is said, the eruption of Kolombos took place far from the nearest shores of Thera, and observers at that point could not tell whether they were seeing true flames produced by combustion of gas or simply red-hot ejecta that gave that impression. There can be no doubt that no such distinction was made by those who observed the eruption of 1650, but, judging from what happened at Santorini in 1866 and from what I observed in 1861 on Vesuvius and in 1867 on the Azores, it seems probable that there were actual flames. In the eruption of 1650 which, like that of 1866, began under the sea, the gas was released mainly through the thin layer of water and came into sudden contact with the air. That is no doubt the reason flames from their combustion were observed, whereas later when the volcano rose above the sea, the flames disappeared, because the combustion of gas took place within the interstices of blocks on the summit of the volcano, where it was out of sight. Flames could still have been clearly observed from Thera during the eruption of 1650 provided, as is likely, the release of combustible gases was on a scale comparable to that seen in 1866. In fact, every night for several weeks in 1866 observers at the village of Manolas, situated on the crest of the cliffs of Therasia, could see flames burning on the surface of the newly formed mounds of volcanic material. The distance from these places is slightly less than that separating Thera from the bank of Kolombos. I have no hesitation, therefore, in affirming that one could have seen the release and burning of combustible gas during the eruption of 1650.

The origin of the lightning is also questioned by these critics. The projection of incandescent rocks at night has, according to them, the same appearance as flashes of lightning. They also argue that the lightning striking rocks would have broken them into pieces as it was observed to do near the church of Agia-Marina. But these two arguments are baseless; there is no similarity between the parabolic trajectory of an incandescent block projected from a crater and the zigzag tracks so common in electric discharges in the atmosphere. In fact, anyone who has witnessed any great eruption knows that spectacular electrical manifestations are constantly being produced within volcanic fume clouds. So why should anyone be surprised that lightning has split a rock into fragments? There can be no doubt, therefore, that flashes of lightning were observable phenomena during the eruption of Kolombos, just as flames were.

Should we doubt that a crater was produced by this eruption? Critics say that observations of what happened at Santorini in May 1866 cast doubt on this. During their stay at Santorini in 1866, they did not want to recognize a crater as the source of ejecta. This crater existed, however, even though the cavity was congested with piles of debris. Between this pile and the edge of the cone an open interval still remained, and it was from this trough that the ejecta came. Shortly thereafter, the crater was cleared by a violent explosion of the material filling it, and its typical form at that time was perfect evidence to refute the doubting critics. Moreover, one need only read the account of Pègues to see the important role played by ejecta during the eruption of 1650, and I do not think I am exaggerating if I suggest that, as a general rule, all volcanoes with abundant explosive ejecta have such a crater composed mainly of loose material.

How should one regard the fissure in the cliff of Thera? Is the report describing it just an invention designed to arouse astonishment? Is it simply, as the critics claim, the work of someone who likes to spread tall tales? The author of the account, they say, relies only on very superficial observations. This reproach is unfair, for examination of the part of the cliff of Thera indicated by the author shows quite clear evidence of a significant fault. Moreover, the fissure still exists. It is only superficially filled at this time, and many people say that they saw it quite open only a short time ago. The Lazarist fathers and the Sisters of Charity who presently live on Santorini assure me that they had seen it throwing out stones and that the sound of them falling could be heard only after several seconds had elapsed. The mason who closed the opening showed me the signs of his work in the garden of the Lazarists, so I have seen with my own eyes the clear evidence in the cellars of the house of the Sisters of Charity. The extension of this crack under the house of Mr. Pierre Rubin has also been attested to me by several persons who saw the opening there and noticed the considerable depth of the crevice. And, finally, Mr. Moskakis informed me that he had observed it in an excavation made by Chondo Chorio quite close to the church.

Thus the doubt expressed about this point is no more justified than it is in the preceding cases.

We must still resolve the question of whether the low parts of Santorini and of the neighboring islands, such as Ios and Sikino, were due to ground movement or simply to some kind of tidal wave produced when the sea was driven back by the turbulent explosive action of gases during the eruption. During the explosion of 1650, there were genuine earthquake shocks, as demonstrated by the details cited above with regard to the houses that were cracked or knocked down on the island of Thera and the avalanches that were set off along the edges of the cliff. There were, therefore, ground movements; beyond that it is reasonable to conclude that these movements were strong enough to send a long wave that could have momentarily risen above the normal shore level and could have led to flooding of the low coastline and nearby islands.[9]

A tidal wave caused by the sudden release of an enormous volume of gas during the eruption of 1650 would not be an attractive explanation of the large invasions of the sea on the coasts of Thera, Ios, and Sikino. The first of these islands is 7 kilometers from the cone of Kolombos, the second is 21, and the third is 29. A discharge of gas capable of causing a 10 meters rise of sea level would be unreasonably large.

What other evidence is needed to confirm this hypothetical tidal wave? Father Richard says nothing whatsoever about flooding of the part of the coast of Thera that faces the bay. The debarkation place on this side of Thera would have been completely inundated by the sea if the ground had momentarily been lowered by several meters at this point, and it would have resulted in disasters that the contemporary accounts would not have failed to mention. Critics forget that all the houses on the landing dock of Thera were constructed quite recently. There is not a single one that was built before the beginning of this century. Fear of pillage prevented the inhabitants of Thera from building far from the places of refuge, such as Skaro, Apano Meria, Pyrgos, and Akrotiri. The town of Thira did not exist at that time, and if a storage place was maintained at the foot of the cliff, it was at most a primitive shed or, even more likely, a cave dug into the conglomerate at the base of the coastal escarpment. It is not surprising, therefore, that damage on this coast was insignificant and, therefore, was not mentioned in the account of Father Richard.

Thus, careful consideration confirms rather than discredits the accuracy of the reports we possess on the eruption of 1650.

At the site of this eruption there is now a submarine cone the summit of which is about 300 meters above the local bottom of the sea and 18 meters below the surface of the water. The island formed in 1650, now represented by the bank mentioned previously, had been referred to by the name Kolombos. It is likely that it was dismantled by waves and that it subsequently disappeared under the water shortly after its formation. In any case, Father Richard says that four months after this event the summit of the cone was already submerged and covered by the sea to a depth of eight cubits.[10]

Until very recent times, various volcanic manifestations that seem to be related to the eruption of 1650 were seen at Cape Kolombos. Warm springs flowing out along this stretch of the shore were accompanied by escaping gas. At times the seawater there was colored with reddish tints due to decomposition of dissolved salts of iron. These emanations have diminished considerably in the last thirty years or so, but what makes them more difficult to observe is that the sea has severely encroached on that part of the shoreline. The thermal springs are now found at a distance of about 20 meters from the shore and are covered by a layer of loose rock and about 3 meters of water, so that one can now only distinguish their location when the sea is exceptionally calm.

Around eleven o'clock on the evening of the 17th of October 1848, at the same moment when Lieutenant Leycester of the English navy was in the course of making soundings along this coast, he heard a subterranean noise resembling thunder, even though there had been no true eruptions in that area since 1650.

ERUPTION OF 1707

The Appearance of Nea Kameni

The history of this important volcanic event is laid out in detail by Pègues in the work he published on Santorini. The information he provides is drawn from three main sources. First, he studied a report written by a Jesuit, Father Tarillon, on what was communicated to him by two missionaries of the same organization who had been eyewitnesses to the phenomena. This report, included in the new memoirs of the missions in Levant, was printed in Paris in 1715. The second document used by Pègues is a manuscript found in the archives of the Catholic mission of Santorini. It was written by another Jesuit (probably Father Gorée), who was also an eye witness to the eruption. Finally, the third is a manuscript written by an inhabitant of the island, Jean Delenda, who also lived at that time.

"These different accounts," says Pègues, "do not all contain the same information, but all are in agreement on what they have in common, and they complement one another. This is why, by taking from one what is missing from another, I have combined all three while adding to them what the still-fresh traditions have preserved in the form of reliable information about this event."

We, in turn, can lay out the principal fact by extracting as objectively as possible the essential parts of this account.

On the 18th of May 1707, two light earthquake tremors were felt on Santorini; on the 21st of the same month, a third shock, like the others, almost passed unnoticed.

At sunrise on the 23rd, a mass that seemed to be floating on the water was seen about 200 meters west of Micra Kameni at a spot where the sea had been only eight fathoms deep and fishermen had formerly cast their nets. It was at first taken for a ship wrecked on the reefs of Micra Kameni but soon was recognized as a new bank that had just been formed of black rocks with white ground in the center. No audible sound or violent shaking had accompanied the appearance of the island. For several days it was possible to visit it without undue risk. Those who landed there found that the white material was nothing but blocks of very porous pumice with the color and texture of loaves of bread. From the surface of the rocks, they collected large numbers of live oysters and sea urchins. But suddenly the visitors felt the ground shake and move under their feet as the reef began to tilt. The sea became agitated and turned yellowish around the shore; it gave off suffocating sulfurous odors, and dead fish floated to the surface of the water. The island grew before their eyes; in a few moments it rose in height about 7 meters and spread laterally to twice its former diameter. From the 23rd of May until the 13th or 14th of June, it was seen to grow in both height and width. This came about almost imperceptibly, without violence, sound, or shaking, until the elevation reached about 70 or 80 meters and the diameter was five to six hundred meters. The sea became increasingly restless; currents carried discolored water for distances of up to twenty miles and spread a fetid, sulfurous odor.

In the beginning, the growth of the island was very irregular; often one side was seen to sink or become smaller while the elevation and lateral extent of another became larger. On one day, a rock that was remarkable for its shape and size appeared at the surface of the sea at a distance of 40 or 50 paces from the island, then, after four days, it disappeared beneath the waves. Several other blocks, after appearing and disappearing several times, reappeared once more and remained stable.

A long crevice crossed the summit of Micra Kameni. From the 14th of June until the end of that month, the sea continued to be colored with varying shades of green, red, and yellow, while a foul-smelling odor spread over great distances.

On the 30th of June, the sea became very agitated around the island, and the temperature of the nearby water became so hot that it became difficult to approach the island. The tar caulking of boats melted as if it were exposed to a strong fire. The stench of gaseous emanations was so unpleasant that at Skaro, on the island of Thera, it caused considerable discomfort.

After that day, the rate of growth of the island became even greater than before, and a distinct peak began to mount near its center. On the 2nd of July, rocks were seen again, just as they had been at the beginning, seeming to float on the water like the debris of a ship wreck. On the 5th, for the first time, great flames were seen to emerge. During the following days, the rocky carapace that seemed to cover the volcanic vent continued to develop, and an opening was seen in it.

At sundown on the 16th, a great chain of separate black rocks, 17 or 18 in number, was seen about 200 yards from the new islet in the direction of Micra Kameni, in a narrow place where the depth of the water had never been measured. At the same time, dense white fumes with no disagreeable odor rose from the same site. The rocks soon coalesced on the north side to form a small island that was referred to as "the black island" because of the color of the rocks forming it. Later, on the 9th of September, these two islands joined to form a single island to which the name Nea Kameni has since been given. They did not join suddenly but very gradually as they slowly grew progressively larger. Beginning with the appearance of the black island, however, the volcanic activity seemed to be focused on that side. The white island, which bounded it on the southeast, changed little if at all. Until 1866, it was still visible between the cone of Nea Kameni and a small inlet of that island called Voulcano Cove.

On the 17th, the rocks, the rise of which have just been described, could be clearly seen; and those whose emergent points previously could scarcely be seen rose in a great mass that joined itself to the earlier ones of the black island. On the 18th, dense fumes and flames similar to those of a roaring furnace burst from the middle of these rocks. Underground rumbling like the sound of distant thunder appeared to be coming from the center of the island.

On the 19th, new tongues of flame were noticed, but they were weaker than at first and had such a pale color that many persons doubted that they were really fire. The illumination came from a single spot a short distance from the black island and were not visible during daylight. As for the white island, no one ever saw there either flames or smoke. It did not cease to grow, however; even though the black island grew much faster. Every day one could see large rocks emerging and adding to its length and width in a manner so perceptible that one could notice changes from one moment to the next. Some of these rocks came out of the water next to the island; others appeared farther away, so that in less than a month there were four small black islets. But as they rose and grew, they promptly joined one another until there was only a single island that soon would become united with the main black island.

It was noticed that the fumes had increased greatly, and as no wind blew at that time, they rose so high that the cloud could be seen from Naxos, Crete, and several other islands over a distance of 70 miles. Wherever it was carried, it tarnished silver and copper. During the night, the fumes appeared to be mostly flames to a height of 6 to 8 meters. The sea was covered with foam that was reddish in some places but more yellow in others.

The new island grew each day, spreading on all sides but mainly toward the north. The sea was as agitated as before. The water was laden with sulfur and acids, the boiling was stronger, fumes denser and more voluminous, and the flames stronger and more extensive. The odor of sulfur that spread over Thera was so strong and noxious that it frequently caused fainting, violent headaches, and vomiting. These effects varied over the island depending on the direction of the wind. Attempts were made to alleviate them by igniting large fires in the streets but with little success. For thirty-six hours, the fumes were quite unbearable, but a fresh southwest wind finally dissipated them. As the emanations from the volcano continued to pass over Thera, they were mixed with a heavy fog. The fumes (probably charged with hydrochloric and sulfurous acid) caused great damage to the vineyards with which the island is covered. Leaves were desiccated and burned as though they had been exposed to high temperatures.

On the 31st of July, it was noticed that the sea threw up fumes and boiled in two different places, one thirty and the other sixty yards from the black island. In these two areas, each of which formed a perfect circle, the water looked like oil on fire; this lasted for more than a month. During this time, quantities of dead fish were found along the shore. The following night, a dull sound was heard like that of several cannons discharging in the distance, and almost immediately two long tongues of flame were seen to rise quite high, as from a furnace, before quickly going out.

On the 1st of August, a fume cloud, not white as before but bluish black, rose as a column to a prodigious height despite a strong northerly wind. The chronicler adds that if it had been night the entire cloud would have appeared to be on fire.

On the 7th of August, the sound that could be heard the previous days was no longer as muffled. It was like that produced by several large masses of stone falling at the same time into a large, deep pit. The extremities of the island were in continual motion, and the rocks that formed them were shaking back and forth, disappearing and reappearing in different positions. As the sound continued to mount in intensity, it finally became like the sound of thunder.

On the 21st of August, the fume cloud and flames diminished noticeably; nothing was visible, even at night. But at daybreak, they returned with even greater vigor than before. The fumes were red (doubtless because of the effect of refraction) and very dense, and the flames that rose were so intense that the sea around the black island steamed and boiled in an extraordinary way. Around the large vent that glowed at the summit of the island, one could count during the night up to sixty secondary, but still quite lively, incandescent spots that were visible from Thera, and it is likely that an equal number would have been visible on the opposite side.

On the morning of the 22nd, the island was much higher than it had been the evening before, and a chain of rocks, about 16 meters long, had risen during the night, greatly increasing the island's length. The sea was still covered with a reddish foam that spread an unbearable stench everywhere.

On the 5th of September, a new vent opened at the extremity of the black island toward Therasia, but it did not persist more than a few days, during which time the incandescence in the central vent was less intense.

During the night, there rose from the principal vent of the volcano what looked like three brilliant rockets. The following nights, the situation was quite different. After the usual underground rumbling, one could hear detonations like those of a large cannon volley. From the crater came a jet of incandescent bombs, and at the same time swarms of sparks glowed from a million particles that followed one another, rising very high and then falling back in a shower of stars that illuminated the island. In the middle of these sparkling jets, one could trace a separate line that cut through the sky above the fortress on Skaro and seemed to remain motionless for some time before suddenly disappearing. The superstitious inhabitants of Thera saw in this phenomenon a forecast of impending evil.

The discharge of the volcano was so furious that it rattled doors and windows two or three miles away, and it was necessary to leave them open to prevent them from being broken. On several occasions stones weighing up to a ton rose like rockets and then fell back again a league or more from where they started. While these discharges were going on, a great burst of flames was followed by a powerful jet of fumes mixed with cinders that fell back in a rain of dust. The sound of the detonations was comparable to that of several large-calibre cannons firing at once. With each explosion, the opening from which it came was enlarged. Incandescent stones were often thrown out in such great quantity that, after their fall, the entire island of Micra Kameni was covered with a brilliant illumination.

On the 9th of September, the white and black islands were completely united. After their junction, the form of the southeast end of the island underwent no further change, but the growth continued toward the north. Of all the vents formed by the volcano, only four ejected incandescent material, but at times strong fume columns came from all four and at other times from only one or two, sometimes with noise, sometimes silently, but almost always with whistling and prolonged reverberations like the sound of a pipe organ mixed with the roaring of ferocious beasts.

On the 12th of September, explosions were numerous and very violent. Ten or eleven times in twenty-four hours, jets of huge stones were seen to rise from the crater. Dense fumes released at the same time rose in waves and spread a wide rain of cinders. The sound of the detonations was again as strong as that of a barrage of heavy artillery.

On the 18th, a weak earthquake was felt on Santorini but did no damage. The new island grew noticeably. Flames and fumes began anew. The intensity of the flames and the strength of the explosions were greater than any time in the past. Their violence was so great that the houses of Skaro were shaken to their foundations.

On the 21st, each explosion brought a rain of incandescent stones falling on Micra Kameni. In one of these explosions, three great flashes suddenly cut through the atmosphere, and at the same instant the new island shook so much that half of the large vent fell in while throwing out great masses of incandescent rock to distances of more than two miles.

It was thought that this violent convulsion had exhausted the store of energy, and when, during four days of calm, neither flames nor fumes were seen, this belief was confirmed in the minds of the people of Santorini. On the 25th of September, however, the volcano became more terrifying than ever. The detonations were continuous and the sound so loud that on Thera two persons could scarcely hear each other speak. A formidable explosion made everything shake at Skaro, and the frightened populace fled to the churches.

In the month of October, the detonations were even more frequent. At least one or two explosions came from the main volcanic vent each day, and at times there were as many as five or six.

The explosions continued, but the sound was weaker and the ejected blocks were not so large or numerous. The sea began to regain its natural color, and for a month and a half one could no longer smell the bad odor that, in the beginning, had been so unbearable. Nevertheless, every day the fume cloud became thicker, blacker, and larger. Flames, greater than ever, shot into the skies. Although the explosions had ceased, the subterranean sounds were more continual and violent, like rolling thunder. The rain of cinders was incessant and so abundant that there was fear for the crops. Fortunately, this fear proved to be unfounded.

Although the activity of the volcano had diminished, the new island now took on a frightful aspect, mainly toward the southwest side of the part that had formed the white island. It was thought that the way the lava flows had progressed would create a port capable of sheltering ships of all sizes, but it only resulted in the indentation that has since been known as Port Saint Georges.

Around eight o'clock on the morning on the 10th of February 1708, a rather strong earthquake hit Santorini. The previous night there had been another, but it was much weaker. These tremors were the prelude to new volcanic explosions. The flames, fumes, and detonations were terrifying. Great blocks that until then had been awash at the surface of the water rose well above sea level. The boiling water around the new island increased to the point that the inhabitants of Thera were stricken with terror. The underground groaning could be heard night and day without interruption. In a quarter of an hour, the large vent broke up into five or six smaller ones. In this way, the volcanic manifestations that had signaled the eruption during autumn of the previous year, after having momentarily lost their intensity, took on a new burst of energy. The day of the 15th of April was particularly remarkable for the number and violence of detonations. The flames, eruptive column, and great blocks of lava filled the air. Everyone believed that the new island would be destroyed by this explosion. It did not happen, however. Only half of the large vent was destroyed. The part of the island consisting of lava that disappeared in this paroxysm had already suffered a similar destruction, but this soon recovered its elevation, and it grew even more by accumulating cinders and ejected blocks.

From the 15th to the 23rd of May, conditions remained more or less unchanged. The island grew mainly in its height and little in width, so that the main active vent became quite high. The molten material that piled up gave it the form of a cone with a summit about 100 meters high. It can still be seen today on western Micra Kameni, almost in the form it had at the time of the eruption.

Thereafter, everything declined imperceptibly. The flames and fumes diminished; the underground rumblings became weaker and less frightening, and they were less frequent. Of all the vents that had been active during the eruption, only four of the shallow depressions on the cone and upper flanks remained.

On the 15th of July, Father Tarillon visited the new island in the company of François Crispo and other ecclesiastics. The weather was fine, the sea calm, and the activity mild. They had taken care to choose a boat well-caulked with a double layer of oakum in all the cracks. At first they landed at a place where the sea was not boiling but where the fumes were strong. They had scarcely entered the cloud of vapor when they felt a suffocating heat. On putting their hands in the water, they found it very hot, even though they were still five hundred yards from their destination. They then tried to land at another place farther from the main vent of the volcano, but in vain. At the northern end of the island, the sea was throwing up great boiling jets, and the heat was intense, so they had to make a wide circuit to land on Palaea Kameni. From there, they could observe at their leisure all the western side of the new island, which was not visible from Skaro. They estimated the height of the cone as only 200 feet, the length of the island at 1,000 feet, and its circumference as 5,000. After leaving Palaea Kameni, they tried again to land on the new island, this time on the southern coast on the part corresponding to the white island. But even at two hundred yards from the shore, the seawater was very hot, and the temperature rose as they came closer to the coast. A sounding line was dropped. The entire length of 95 fathoms was let out without reaching the bottom. Then, frightened by two violent explosions, one after the other, and enveloped in a thick ash cloud thrown out by the volcano, they chose to distance themselves as quickly as possible. When they landed on Thera, the sailors noticed that the heat of the water had melted and carried away all the pitch from their boat.

Until the 15th of August, when Tarillon left Santorini, the island continued to produce flames and fumes and ejected incandescent bombs. The noise, though less than during the preceding months, was still intense.

Tarillon relates in his report that, according to letters and verbal communications he received, in June of 1710 the island was again in flames and the water was still boiling in the surrounding sea. It does not appear, he says, that this activity was likely to end soon. At the end of this account, he gives details furnished by another priest who was on Santorini on the 14th of September 1711, and on that and several following days made a tour of the island, taking care to avoid the places where the water was hot. He estimated the circumference of the island to be five or six miles and the height of the cone more than 400 feet. This cone was very irregular, and its surface was covered with grayish ash. A crater occupied the summit, and the flanks were encrusted with material resembling a welded mixture of sulfur and vitriol. The principal vent had a diameter of about 40 feet; three other window-like openings a little lower measured 6 or 7 feet across. The author remarks that on the western side the cone was very steep and the talus rose so abruptly that "only a cat could climb it." But on the other sides of the island one can go up almost to the vent by climbing over the large boulders resting one upon the other.

On the afternoon of the day when the priest visited Nea Kameni for the first time, the volcano was very active. In a period of two hours there were seven quite strong explosions. The sound was like thunder, and incandescent blocks more than 20 feet in length were thrown as far as two miles. The fumes were dense and white. During this time the three lower openings discharged streams of sparkling incandescent material that was violet or yellowish-red. Long after each detonation, one could hear the echoes of shattering sounds. From that time on, the volcano has not discharged molten material, and the noise has come to an end. From time to time, the three lower orifices give off puffs of dense fumes that come out quietly and in such small quantities that they cannot rise above the crater. The author remarks, however, that during heavy rains the main mass of the volcano continues to steam and give off a sound like that of hot metal when it is plunged into water. According to another account, these phenomena were still continuing near the summit seven years after the end of the eruption.

After the 14th of September 1711, the volcano was no longer the scene of violent events. In all, the eruption had lasted five and a half years. Some signs of activity persisted at this place until the reappearance of intense volcanism in 1866. At the southern foot of the cone of Nea Kameni, near the same place where the white island appeared close to the present indentation of the coast known as Voulcano Cove, abundant emanations continued to be emitted for a century and a half. The mineral components were dissolved in the seawater, tinting it yellow, green, and red. Depending on the direction of the wind, this coloration was carried to more remote places. On some occasions, the colored water formed a broad layer; at other times, it extended in an irregular trail, 50 meters wide and 4 or 5 miles long. When, for some unknown reason, these streaks did not appear, as was often the case during calm weather or when the wind was from the south or southwest, the inhabitants of Santorini feared that there would be another earthquake.

The temperature of the sea was warmer in the small cove than in the rest of the bay. In March 1836, Pègues found it to be 20 degrees centigrade, while that of the bay was 14.4 degrees and the air was 16.

It was noticed that at this same place the steady release of countless small bubbles of gas caused continual turbulence of the water. These exhalations killed fish in the surrounding area. Pègues reported that in 1836 he found 37 fish and at another time 7 or 8, most of which were dead while the rest were half-asphyxiated and wandered about in a drunken state. All the shoreline was covered by a rust-colored stain that covered the rocks to a height of 40 centimeters above the water level.

Comparing the eruption of 1707 with that of 1650, one sees that the later eruption was much less of a disaster for the inhabitants of Santorini than was the earlier one. It occurred at a place that was more centrally located and closer to the islands; it broke out on an island of considerable size that still survives today, while the cone of 1650 was soon completely truncated by waves. The long duration and the sequence of events that preceded it were also quite different from those of 1650. The prolongation of the period of volcanic activity seems to have attenuated the violence; in particular, the ground tremors were much weaker than those of the century before, and the emanations of noxious gases were also much less severe. But one must not attribute these differences only to the differences of abundance and composition of the products of the eruption. The location of the vent in 1650 could have played an important role in making the effects more disastrous for the inhabitants of Thera. Because of its form, the entire area of the island was exposed to the corrosive action of fumes, while, in contrast, the volcanic gases and steam of 1707 could spread to the island only by crossing the elevated barrier of the cliffs. Thus, according to Tarillon, the gases from the volcano mixed with those of the atmosphere to fill the bay with wispy clouds, the upper surface of which scarcely reached the crest of the cliffs. The small part of this cloud that managed to surmount the crest caused great damage to the vineyards. What would have happened if it had spread over the entire surface of the island as it did in 1650?

As for the ash, it initially did considerable damage to the vegetation, but, as is always the case, the loss was made up in the following years by the excellent crop that it produced.[11]

In addition to the documents on the eruption of 1707 that I have analyzed here, I should also mention an account taken from the *Journal de Voyage* of Aubry de La Mottraye, a French gentleman who visited Santorini, first in August 1707 and a second time in 1710.*

The description he gives of the beginning of the eruption of Santorini was based on information he received from a Greek priest and agrees perfectly with the account of Tarillon and confirms the latter in all respects.

LA MOTTRAYE'S DESCRIPTION OF THE ERUPTION OF 1707

First Visit to Santorini

We set sail on the 21st of August from Smyrna and on the 25th reached a place southeast of Naxos, where we were stopped by a great calm that left us immobile until the evening of the 26th, when we began to hear what sounded rather like cannon fire but at times like the roar of a storm. This noise was heard repeatedly until well into the night. The wind came up enabling us to get closer to the source of the sound, and we saw flames mixed in the middle with large fragments of incandescent material and with smaller ones that were carried almost out of view over the surrounding region, where they were extinguished in the waves where they fell as soon as they lost the

---

* The two quarto volumes containing the *Histoire des Voyages* of La Mottraye were published in the Hague by Johnson and van Duren in 1727 by public subscriptions. Among the subscribers were: the Duke of Orléans, regent of France; the king of England; the king and queen of Sweden; the king and queen of Prussia; princesses and princes of the English kingdom; Prince Eugène of Savoy; the Prince of Orange; the chancellor of Aguesseau; Cardinal Dubois; the Archbishop of Canterbury; the Cardinal of Noailles; the Archbishop of Paris; Lord Newton, president of the Royal Society of London; and Mr. de Pontcarré, first president of the parliament of Normandy. Although almost all the other subscribers were English, the list also includes a few Parisian bookshops - Coignard de Lespine, Mariette, and Rollin - which together subscribed for eighty-one copies in the large format and one hundred and thirty-two in the small one.

From 1696 to 1722, La Mottraye traveled through Italy, England, Turkey, Syria, Tunisia, Tripli, etc.

force with which they had been ejected. The light given off by the flames revealed two rocks, one black and the other white; this differences of color allowed us to distinguish between the two. The black material that came from the interior of the black island was much more extensive and higher than that of the white one. On coming closer, we observed that one of the vents, marked 3 on the point shown on the lower part of my map B, where the harbor of Santorini is indicated, was disgorging a torrent of lava that marked this point as a hot, fiery mountain. My nose and throat were suddenly seized by a sulfurous and highly disagreeable odor that took away my voice and almost kept us from breathing; it caused a fit of sneezing among the crew of the boat. In addition, I had a strong headache, pain in my heart, and nausea, and had I not been accustomed to the sea, I would have thought that I was seasick. Meanwhile, the pilot and seamen who had spent time there some months earlier were very surprised to see rocks in places where before there had been at least thirty to forty fathoms of water. The seamen added that when they were sounding the areas where the rocks are now they could find no place to anchor their ship. So they told the captain to head for the cape and wait there until daylight rather than approach this dangerous place during the darkness; to do otherwise could be fatal. Some of the more timid ones who made up a large part of the crew, wanted to leave Santorini completely and go directly to Candia, disregarding the agreement to let me land at Santorini. I firmly protested their plan and threatened to register a complaint with the authorities of Candia or Canea if they took me to one of those place against my will. Thereupon, the captain, being more responsible than most of the Greek masters of these ships, assured me that he would not fail to drop me, if possible at Skaro,* and if not at some other place on the island. He gave these orders to the pilot, who carried them out seven or eight hours later, as I shall explain below. It was fortunate that the wind was weak, for if it had not been, it would have been impossible to remain at the cape and keep his promise, since the form of these boats prevents them from tacking against the wind and leaves one at the mercy of the weather.

A little after eleven in the evening the noise declined markedly and remained less frequent until between two and three in the morning; the only sounds one heard from time to time were those that resembled the whistling of stormy winds. The flames became brighter and contained fewer or smaller fragments of opaque material. Small flashes of flames with all the colors of the rainbow shot into the air and rose as fast as lightning before disappearing

---

* Skaro is the capital of Santorini and is sometimes referred to by the latter name, which is also that of the islands. (Note in the work of La Motteraye.)

in the dense cloud of smoke that completely enveloped our boat. The sea around us was very restless and, in the vicinity of the rocks, was colored green and white.

At six o'clock on the 27th, we advanced as far as the small Kameni island, which the local people normally call the "burned island," marked no. 4 on the same map. We encountered no boats from Santorini. The Caravokery at that place lent me his own so I could go to Skaro. I asked him at the same time to give me the address of someone he knew, since I had no acquaintances there. He told me that he was a stranger there, but advised me to go straight to the house of the head Greek priest who, according to the custom of his order, would receive me. The scribe who was there said that he had a relative in this town and offered to take me to him. I accepted this offer, and the two of us set off in the boat.

The port - if one can give this name to a place that has no good anchorage but is protected from the winds by mountains - resembles that of Tripoli in its horseshoe shape. It is shown on the same map. Upon setting foot on land I was surprised to find that Skaro, which had been described to me as a well-populated place, had taken on the appearance of a desert, for when we entered it, there was no one in the streets. We concluded that fear had caused the inhabitants to flee and gave up hope of finding the person we were looking for. But it was my good fortune that this person was less timid and, unlike the others who had retreated to the far end of this and other neighboring islands, remained in his house, which was one of the last and most distant from the land side. I was well received by my host, who did not seem surprised to see me so hoarse that I could scarcely speak or make myself understood. He told me that this was a general problem among all the inhabitants. He said that the sulfurous fumes that were the cause of this trouble attacked everything, especially silver; he showed me a spoon of this metal that had been cleaned the evening before and was now already tarnished to a degree that was scarcely believable.

A Greek priest who lived nearby invited me to dine with him the next day. I accepted his offer quite willingly in the hope of learning more from him about the matters that had brought me there than I could from the secular residents. And since he seemed more enlightened, I put to him every question I could think of. The following are more or less the responses he gave me.

The white islands marked no. 3 on my map, which had almost become joined with the black island by the prodigious growth of the latter, had been the first to rise from the bottom of the sea. There was no visible fire or sound except for two or three small earthquakes felt at Skaro and in the immediate vicinity. But on the evening of the 7th of May (the 18th of May by the Gregorian calendar), a terrible shock was felt throughout the island, and on the 8th (the 19th of May by the Gregorian calendar) there was another even larger one, after which all remained peaceful until the 12th (the Gregorian 23rd) when fishermen noticed that the sea was boiling and churning, especially around the small Kameni island, numbered 2 on the map. They believed that this boiling and churning was caused by great numbers of fish, because there not the slightest wind. With this thought in mind, they went to the shallowest place but instead of fish they found to their astonishment quantities of white rock of the kind called pumice rising from the seafloor and floating on the surface of the water, where it collected and formed a single mass in the way a swarm of bees comes together and attaches itself to a tree. Their astonishment was even greater because several of them had passed by this place in their boats the day before without having seen anything. Three days later they discovered, to their surprise, numerous dead fish floating on the agitated water. The same spectacle had alarmed other inhabitants of the town when it spread a stifling odor of decay. After it had grown and taken on the form of a rather high pyramid, the mass formed of pumice fell back into the sea and disappeared. But it reappeared a few days later at three different places, as points rising above the water. These continued to grow and expand until they were united by new pumice into a single pyramidal peak higher and larger than the first. Around the beginning of June, this peak ceased to grow and spread, but there was a large earthquake that was stronger and louder than had ever been felt or heard before. This was followed by the appearance of several pieces of black material that rose from the bottom of the sea in places where before it was not possible to reach the bottom. This material floated and stirred the water, then collected into a single mass in the same manner as the pumice. These rocks and black material disappeared into the boiling and churning seawater as it did when the first island was formed. They disappeared and reappeared more quickly to form a larger mass that grew visibly each day.

Pliny reports that the large and small Kamenis rose in this manner from the bottom of the sea during the one hundred thirty-fifth olympiad. The former supported a few small patches of thin, short grass, but the other was entirely barren.

The same priest added that some weeks later, three large vents opened and vomited fire, like three great furnaces, throwing out hot stones with black, sulfurous torrents that were accompanied by a much greater sound than what we were hearing at that time. The main changes that were noticed were that for some time the fume clouds, flames, and hot rocks came only from a single vent and were discharged with less force, in smaller quantities, and less frequently.

The fumes that preceded, accompanied, and followed these eruptions covered almost the entire horizon of Santorini and not only bothered the inhabitants but did considerable damage to the vineyards and fruit trees.

All remained rather peaceful until the 2nd of September, when there was an earthquake and explosion, the sound of which surpassed any that I had heard before. The extraordinary, natural furnace ejected incandescent fragments larger than any ever seen before. So terrified were the inhabitants who had been courageous enough to remain in the town that almost everyone fled, especially those who had enough food to survive in the countryside.

La Mottraye's Second Voyage to Santorini (1710)

We arrived on the 20th of September at seven in the morning. We found that the first island, no. 3, which one might call the "Combined Island," since it was no longer the same island that had been covered with white and was no longer distinguished by its color, had not only continued to grow but was so close to the small Kameni (no. 4 on my map B) that a vessel could no longer pass between them without dangerous risk, and the passage was nothing more than a landing place for small boats.

Only the black part of this combined island continued to grow and increase in length and width; it was now about four miles in circumference. The vent that had previously been open no longer ejected more than small flames with some fused black material. The wind being favorable, we prepared to continue our journey without landing and passed close enough to the combined island to notice the different green, black, and yellow colors. I bought some fish in order to question two fishermen who came to the area. They told me that having braved the heat, which still continued but was now somewhat less intense, they had collected some large pieces of sulfur that was so fine and naturally pure that even art could not match its perfection.

CHAPTER TWO

THE ERUPTION OF 1866

Few eruptions have been observed as carefully as the last eruption of Santorini. Its opening phases were followed with especially close attention. As soon as the first activity appeared, Doctor de Cigalla[1], a distinguished medical doctor who lived in Thira, the principal town on the island of Thera, frequently went to Nea Kameni and recorded what was happening there. Enthusiastically noting what he saw, he was able to prepare a detailed journal of the volcanic phenomena that occurred at Santorini during 1866. The first part of this journal covers the period from the 30th of March to the 13th of April, the second from the 7th of May to the 15th of June, and the third from the 15th of July to the 22nd of September of the same year. Thanks to the kindness of Dr. de Cigalla, I had these documents at my disposal and had them translated. In the account that follows, I have also used a series of letters from the same person that were written in French recounting the same events. The documents furnished by Dr. de Cigalla on the beginning of the eruption are of the greatest importance, because, with his special knowledge, he was better prepared than any of his countrymen to recognize the importance of the exact information he would be able to offer science. Nevertheless, I have also benefitted from a journal kept by two gentlemen, Gauzente and Hypert, Lazarist missionaries living in Thira. This collection relates the phases of the eruption from the 30th of January to the 2nd of May, 1866. Finally, I have obtained valuable information from the oral accounts that were obtained for me by Nicolas Sfoscoti, who served as my boatman during the second half of my first visit to Santorini and during my two other visits in 1867 and 1875. The information he obtained for me about the beginning of the eruption is all the more valuable because it came from eye witnesses of the first manifestations of the volcano. Even when he no longer lived on Nea Kameni, he returned there more often than any other person of Santorini, because his work as a boatman led him to bring there various persons who wanted to visit the site of the eruption. Verbal accounts were also given to me by some of these visitors who were on Nea Kameni at the end of January or the beginning of February 1866. Each time that new documentation was transmitted to me, I recorded it immediately. I have tried to keep track of these accounts and almost always succeeded in reducing to their true value the exaggerations that were naturally inspired by seeing such phenomena. Reiss and Stübel[2] have published very important work on the history of the volcano during the period between the 23rd of April and the 31st of May 1866, and I have not hesitated to borrow from them a great number of their observations. Nevertheless, I must mention a minor deficiency of the work of these scientists. Being suspicious of some of the statements regarding events to which they had not personally been witnesses, they have often unknowingly been overly skeptical, if not annoyed, rather than too credulous. As a result, they have rejected as uncertain a number of observations that I can verify as perfectly true.

Professor von Seebach, who was sent there by the government of Hanover, was at Santorini from the 26th of March until the 10th of April and reported his observations during that period.

Professor J. Schmidt, director of the Athens observatory, was a member of a commission sent to Santorini by the Greek government and made up of Messrs. Palasca, Bouyouka, Chistomanos, Mitzopoulos, and Schmidt. He published in 1874 a volume on the complete history of the eruption. The main part of this work deals with the observations of the commission from the 10th of February until the 26th of March and with those of the author himself from the 4th until the 10th of January 1868.

Mr. Christomanos also published a note on the same subject.

Finally, reports have been sent to the Academy of Sciences of Paris by Messrs. Lenormant, da Coragna, Delenda, de Cigalla, Hypert, and Gorceix. Others have been sent by Mr. Schmidt to Petermann's journal in 1866 and 1868. A letter from the same author was sent to Haidinger, another to the English ambassador, Mr. Erskine, and communicated to the Geographical Society of London by Sir R. Murchison.

Several ships of various nations were stationed at Santorini during the course of the eruption. Their commanding officers carried out soundings and made sketches of the locality at the times of their visits. A map that was carefully prepared in 1870 by the officers of the Austrian gunboat *Reka*, commanded by Captain Germouny, deserves special mention because of the precision with which it was made and the many soundings that it shows.

I have made use, not only of these sources, but of other evidence drawn from the letters of Mr. Gorceix and the information that was regularly furnished to me during the entire eruption by the Lazarist missionaries, with whom I continue to maintain a lively correspondence.

Having been an eye witness of the eruption in 1866 and 1867, I have used my own personal observations in the account that follows.

During the winter, the island of Nea Kameni was inhabited only by Nicolas Sfascoti. This man, owner of the house situated at the northern end of the pier that borders the canal between Nea Kameni and Micra Kameni, watched out for all the houses on the island and, at the same time, maintained a small canteen for the sailors of ships anchored in the vicinity. On the morning of the 26th of January 1866, he noticed blocks rolling down the slopes of the cone and cracks in the walls of his house. These signs of ground movement, which became more pronounced on the 27th and 28th, began to frighten him. He notified the authorities on Thera on the 29th. When he awoke on the 30th of January, he noticed that these symptoms had become much worse. Until then, he had attributed them to the weak earthquakes, but soon, without feeling any shaking of the ground, he saw the cracks in the houses grow rapidly. By the end of the day, the houses closest to the cove of Voulcano* were threatening to fall into ruin. They were tilted in the direction of the cove. Large blocks broke away from the edge of the crater of Nea Kameni and rolled noisily to the base of the slope. Meanwhile, the dull sounds and shocks that seemed to come from the depths of the earth proved to be the terrifying premonitions of what was to be the first eruptive outbreak.

The following day, all these phenomena redoubled their intensity, the sounds becoming louder and the shocks more violent. Not only did the walls of the houses continue to open but deep crevices appeared along the length of the pier. All over the waters of Voulcano Cove, countless bubbles of gas were being released; the temperature of the water rose, white vapor appeared over its shores, and a strong odor of hydrogen sulfide was given off. In the afternoon, the ground began to sink at the southern point of Nea Kameni and particularly along the shore of the cove. The houses that had been constructed there were tilted more and more in the same direction as the evening before, and a few walls had crumbled.

On the first of February at five in the morning, flames appeared over the western side of Voulcano Cove and at the surface of the nearby water. A conical pile was formed measuring 4 or 5 meters high and 10 to 15 square meters in area at its base. The beach between the base of the cove and the adjacent muddy pond had sunk until a large part of it was submerged. Frightened by all this, Sfoscoti left Nea Kameni and took his family to Thira.

---

* See page 147.

The same day, however, he returned and brought Dr. de Cigalla, the prefect of Santorini, and several other persons, who devoted part of the day to a detailed examination of the southern part of the island. They noted that the southwestern part of Nea Kameni was covered with crevices. A crack began at the harbor of Saint Georges on the west and continued eastward, splitting the cone in two parts along a straight line. A great number of other cracks were seen in the part of the island that ends at Cape Phleva; some ran east-west, others were aligned perpendicular to that direction. Around the cove of Voulcano, the ground shook continually under their feet while it slowly sank. The depression of the ground was more marked on the western side than on the opposite shore; the center of the sinking movement was at the northwestern end of the inner part of the cove. The western shore had been depressed six meters, the opposite side only about three meters. In four hours the visitors witnessed a decline of 60 centimeters. The depression of the ground caused the sand bar separating the muddy lake from the cove to disappear completely. At the same time, in the southeastern part of Nea Kameni,* it led to the formation of four small ponds, the largest of which had an area of 12 square meters. According to Dr. de Cigalla, the water of these ponds was clear and not salty. Although this latter observation may seem unreasonable, there is no reason to doubt it. It is certain that in all cases ponds of water have been produced by a local depression of the ground below sea level. Their surface was at a slightly higher level, and the water that filled them was clear, while the nearby sea was strongly colored. Natural filtration through the volcanic material making up this part of Nea Kameni could well have removed from the water the ferruginous precipitate it held in suspension, but the dissolved salts of iron it contained would still remain, and soon the decomposition caused originally by contact with the air would resume, and the water of the small ponds would take on a yellowish color. So the small puddles of water described by Dr. de Cigalla do not owe their origin to a simple penetration of seawater into the depressed area, but at the same time, it is difficult to say where the fresh water came from. Could it have been from currents of fresh water circulating at a shallow depth in the ground, or was it from distillation of seawater that penetrated the hot crevices? This question is impossible to answer.

In four hours, the level of the sea in these puddles rose five centimeters. During the afternoon, the release of gas in the cove of Voulcano

---

* There can be no doubt about the location of these small bodies of water. I have positive information on this point.

never stopped increasing, and the water there was like a boiling cauldron. Nevertheless, according to Dr. de Cigalla, the sea elsewhere in the bay showed no noticeable increase of temperature; he noticed only that the water seemed to have a bitter taste. In the cove, however, the water was quite warm, at least locally, for it gave off white mist indicating that the water was at a higher temperature than the atmosphere. The odor of hydrogen sulfide was suffocating. Almost everywhere, the troubled waters of the cove were dark red.[3] From time to time, a few green patches broke the uniformity of the red stain. Since then, I have had more than one occasion to observe this same phenomenon around the new lavas, especially in the passage between Nea Kameni and Micra Kameni. The surface of the sea is converted to a warm, red layer by decomposition of the salts of iron, while the underlying water remains cool and clear. Under these conditions the least motion is enough to break the continuity of the surface layer. A large, sudden release of gas from the surface, for example, disrupts the reddish layer so that the normal greenish color of the underlying water can be seen. This is certainly the explanation for what Dr. de Cigalla reported seeing.

At five o'clock on the evening of the same day, a light earthquake was felt at Santorini. Dr. de Cigalla's mention of this is confirmed by other accounts I was able to collect.

During the following night, the sea appeared to have a whitish color around Nea Kameni. On several occasions red flames lasting several minutes at a time were seen in Voulcano Cove.

On the 2nd of February, the movement of water in the cove became much more turbulent than the evening before. It was augmented by a strong wind from the south. All around Nea Kameni the sea was colored with various shades of red, green, and brown. In the southeastern part of Nea Kameni the four fresh-water ponds grew continually, while similar puddles of water formed in the same area. According to Dr. de Cigalla, the water there was still fresh and clear. The sinking of the east bank of the cove continued during the day at a rate of about 10 centimeters per hour. One could now enter the houses with a boat, because their door sills were now two or three meters under water. Cracks in the ground and in the houses grew wider, but only in the southwestern part of the island. Until then, no ground motion was noticed in the area north of the harbor of Saint Georges. One could hear an incessant, dull underground rumbling, and feel a light agitation of the ground. The water in Voulcano Cove was warm and was giving off much steam. The odor of hydrogen sulfide could be smelled as far as Thera. The suffocating effect of the gas and steam was so strong that it drove away the sea gulls and other sea birds that, the day before, had been attracted in great numbers to the area around Nea Kameni by the dead or half-asphyxiated fish floating on the water.

During the following night, flames rose again from the western side* of Voulcano Cove.

At daybreak on the 3rd of February, flames and a thick column of steam were escaping with a whistling sound. The ground continued to sink, and the fissures grew larger. The southernmost of the cracks that crossed the cone of 1707 became particularly prominent. The water of the ponds on the southeastern side of Nea Kameni took on a salty, bitter taste. That of Voulcano Cove had become very hot, and elevated temperatures were noticed all along the adjacent western bank.

The reddish color of the surface of the sea, which previously had been limited to the southern coast of Nea Kameni, had now spread over much of the bay. During the night, the release of steam and sulfur-rich gases became more pronounced and on several occasions faint feathery lights were seen.

For an hour and a half around three in the morning of the 4th of February, a reddish, flame-like light was seen where the activity was strongest. This light, varying in strength, was surrounded by thick clouds of dark vapor. At this same place, at four-thirty, a reef was seen to appear and visibly increase in size. It was composed of a jumble of irregular dark blocks. The site of its appearance was on the western end of the sand bar that formerly separated Voulcano Cove from the muddy pond.

Toward eleven in the morning, Dr. de Cigalla, who by then had already landed on Nea Kameni, found that the new reef had already been transformed into a small island of considerable dimensions. He had earlier tried to reach Voulcano Cove in a boat but had been prevented from doing so by the agitation and heat of the water. No earthquake tremors or subterranean noises accompanied the formation of the island. At the time Dr. de Cigalla arrived, the island was separated by a distance of about ten yards from the closest shore of the cove. It was about ten meters high, 20 to 25 meters long, and 8 to 10 meters wide. It was elongated in the north-south direction and approached the southern edge of the cove. Its general form was that of a thick wall surrounded by boiling water, and it had the shape of half an ellipsoid that was narrowest in the east-west direction. On the surface of the

---

* In a letter to Ch. Sainte-Claire Deville on the 26th of March 1866, I wrote that the flames rose mainly on the eastern side. This was an error.

blocks one could see debris from the seafloor, such as rounded pieces of pumice, rounded pebbles of lava, organic material, and fragments of worn and rounded bivalve shells - not living mollusks as in 1707. But among the objects that the emerging rocks lifted in this way the most curious without doubt was a piece of wood, 15 to 20 meters long, which was nothing other than the keel of a boat that had been wrecked about 15 years earlier and had been buried in the mud on the bottom of Voulcano Cove. Beside it were seen a few planks from the same ship and some pieces of rope that had probably gone to the bottom more recently.

The growth of the island continued silently without shocks or explosions but with such rapidity that Dr. de Cigalla compared it to the development of a soap bubble. It expanded from inside outward with blocks seeming to come from the center and moving from there to the margins. One had difficulty following all these blocks by eye as they were in constant motion. During this growth, small objects torn from the seafloor quickly disappeared, so that the remains of the boat were soon the only visible debris left on the surface. The blocks were mostly dark black. Some, however, were reddish or gray. No flames were visible, only dense fumes that were very hot and suffocating. The temperature of the water was very high throughout the cove, particularly around the island and western shore. Part of what looked like lively boiling was no doubt due to the great amounts of gas being given off, but in a few places it was probably true boiling. In the early evening, the continual growth of the island had already doubled the dimensions it had at midday. At three in the afternoon, the muddy pond was invaded and soon disappeared under the rocks.

Sailors who were stationed around Nea Kameni during the following night told Dr. de Cigalla that in places the rocks were incandescent and resembled great blocks of burning coal. A thick cloud of fumes covered everything. Phosphorescent lights were seen at the surface of small ponds on the southeastern part of Nea Kameni, and with time red flames appeared intermittently in the large crack that split the cone of 1707. The volcanic manifestations were not limited to Voulcano Cove. Between that place and Cape Phleva a dull sound could be heard, and white fumes were escaping noisily from the ground.

On the morning of the 5th of February, it was noticed that the body of lava continued to grow but less rapidly than the evening before. The sinking of the ground seemed to have ceased, but the sea was still very agitated and strongly colored. Toward the middle of the day, Dr. de Cigalla landed on

Nea Kameni and noted that the sinking of the ground had essentially stopped. Around Cape Phleva the temperature of the sea ranged from 17.5 to 50 degrees depending to the distance from the coast and the direction of the currents. Near the volcano the depth of the sea had diminished considerably. Dr. de Cigalla continued his careful observations of the cone, to which he gave the name Giorgios. Contrary to what had been seen the evening before and the commonly held view that the new pile of lava grew exclusively from the center toward the periphery, on that day the reverse seemed to be true. It expanded mainly by continual additions of banks of compact lava. A scarcely perceptible movement raised these masses and caused them to emerge. The rocks that appeared in this way had temperatures that differed little from that of the surrounding water. Nearby, steam and gas were no more abundant, nor was the boiling of the water more intense.

By now the new lavas were not just an islet but a sizable mound joined by reefs on the west and north to the older ground of Nea Kameni. It was now 70 meters long, 30 wide, and about 20 high. The entire surface gave off thick white vapor that was not suffocating, even when one breathed it for a long time close to the source. Except in a few localized areas, the rocks were not particularly hot. One of the people of Santorini drawn to the place, Mr. Vambaris, who at that time was the head of the post office at Thira, climbed the new knoll and was soon followed by several of his assistants. They noticed that the summit was not especially hot and that the hottest places were mostly around the margins and on the slopes. A few local places were incandescent. From time to time, small, lively tongues of red flame could easily be seen in the interior of the vapor. The large piece of wood that had been raised on the surface of the lava was in flames, either from the surrounding gas or because it was ignited by contact with the incandescent blocks. The carbonaceous odor that Dr. de Cigalla attributed to the vapor given off in the eruption may simply have come from combustion of this piece of wood, which continued to burn throughout the afternoon of the 5th of February.

The temperature of the water around Giorgios was so high that one could cook an egg in it quite quickly. According to Dr. de Cigalla, however, the most intense activity was observed outside the cove in the part of Nea Kameni extending from Voulcano to Cape Phleva, especially along the edge of the coast. The cracks in the ground that had been seen there since the 30th of January had grown considerably, and the fumaroles in the same area were more active. Along the entire shoreline, bubbles of gas were rising in enormous quantities. It was there that sounds had first been heard coming from deep in the ground. All these phenomena continued with redoubled intensity. The abundance of steam with a strong odor of sulfur was particularly noticeable, as were underground sounds like volleys of rifle fire.

During the following night, the flames were very visible at the surface of Giorgios, and steam continued to rise in a high plume. On two or three occasions, a dull rumbling sound was heard, and toward the end of the night there was an explosion.

The cloud of vapor that rose above Santorini had attracted the attention of the inhabitants of the neighboring islands. Deducing the cause, the people of Anafi sent a rescue boat, which arrived at Santorini on the 6th of February.

Toward noon of that day, Dr. de Cigalla returned to Nea Kameni. The sea was troubled and colored reddish-yellow over a less extensive area than the previous days. A dirty green streak about 500 meters long stretched from the entrance of the cove toward the southeast. Near the shore, the temperature varied from 17.5 to 56.3 degrees. New soundings carried out near the southern coast of the island indicated the depths were considerably reduced. A depth of 30 fathoms was found at a point where the English chart indicated 100, and at another point toward the southeast where the chart indicated 17 fathoms, they found only 3.

The area around Cape Phleva showed the same phenomena as the days before. The sinking of the ground had begun again on the margins of Voulcano Cove, particularly on the western shore. Giorgios had grown until it covered the entire bottom of the cove and had already protruded outward on the south side as a narrow promontory. The progression of lavas toward the south was advancing rapidly, for Dr. de Cigalla estimated that the lengthening in that direction amounted to 95 meters in six or seven hours. Again, several persons climbed the dome that day and, as before, found debris from the seafloor among the blocks on the surface. They were not seriously bothered by either the heat or the fumes.

On the 7th of February, Dr. de Cigalla went to Nea Kameni and observed that the length of Giorgios was about 150 meters, its width about 60, and its height about 30. The temperature of the sea and rocks seemed to have fallen; at the base of the dome the temperature was found to be 50 to 75 degrees, while higher on the slope it was 25.5°. The subterranean sounds continued. Some of the escaping gas made a whistling sound.*

---

* I believe I can state that the temperature at this point must have been at least 100 degrees and was probably even higher.

On the night of 7 to 8 February, a great increase of flames was seen on Giorgios, and incandescent blocks were noticed at several places. The gas given off had become unusually violent, and on two occasions produced small explosions. The fumes carried by the wind toward Thera formed a thick white cloud with a strong odor of hydrogen sulfide. In the morning, a statue painted with white lead and placed in front of the Lazarist cloister was found blackened by the emanations of the volcano, and on the following day copper and silver objects belonging to the inhabitants of Thira were equally tarnished. Toward midday, the seawater all around the bay of Santorini was agitated and colored with various tints. Between noon and one o'clock, the sea began to boil at a point between Nea Kameni and a place on Palaea Kameni known as Diapori. The diameter of the surface area stirred by this disturbance was about five meters; the turbulence moved from east to south and then to the west, depressing the surface of the water where one could see a small jet of pumice that rose to a height of about two meters. The sea at this place had a strong greenish-yellow color, and the escaping gas was very turbulent.

The sinking of the ground had continued in the southeastern part of Nea Kameni and especially near the entrance to Voulcano Cove. All the coastal part of this area, which before had been slightly above sea level, was now submerged. The other shore of the cove, which had been much higher, still remained dry, even though it too had been considerably depressed.

Dr. de Cigalla points out that during the evening of the 8th of February the sea became rough and colored around Kolumbos, which is situated outside the bay at the foot of the outer slopes on the northeastern side of Thera.

On the 9th of February, these phenomena remained unchanged. In the middle of the day there was a small ejection of pumice lapilli between Nea Kameni and Palaea Kameni at the same place where a similar event had occurred the evening before. Toward four o'clock in the afternoon, the fumes became more abundant. At ten o'clock in the evening, weak earthquake shocks were felt, and then, around midnight, a great tempest broke loose. The heavy darkness was broken by brilliant flames rising from Giorgios.

At six o'clock on the morning of the 10th of February, flickering red flames were seen on the surface of Giorgios. The lavas were progressing actively toward the south, almost filling Voulcano Cove.

At five in the morning on the 11th of February, flames and great quantities of gas came from the northern shore of the dome. At nine fifteen in the morning and at one in the afternoon there were dull rumblings. Soundings where the eruptions had occurred on the 8th and 9th of February indicated a depth of 6 fathoms where there had previously been 30.

The members of the scientific commission sent by the Greek government arrived at Santorini at three o'clock in the afternoon and immediately began their investigation. At ten thirty in the morning, when calm weather had left their ship still forty miles north of the site of the eruption, they could see the rising fume cloud, and later at a distance of 16 miles, they could distinguish the main features of its form. Around four o'clock their ship reached the channel between Micra Kameni and Nea Kameni and was tied up by means of cables attached to posts set in each of the two banks. The volcano was very calm and wrapped in a thick cloud of white vapor. When they passed by the landing place of Thira, the seawater in that area was found to have a temperature of 16.6 degrees, while around the southern coast of Nea Kameni it was notably hotter. At a distance of 350 meters the temperature was between 25 and 30 degrees. At the place where the islet called Aphroessa was later to appear, they observed a large discharge of gas and measured a depth of 20 fathoms. At this point the temperature of the sea was 24 degrees, while it was notably warmer in the surrounding area. This effect must have been due to the movement of the seawater and release of gas, which mixed layers of deep water with that of the surface and in that way lowered its temperature. The green color of the turbulent area was due to the same effect. Immediately south of Giorgios, the temperature of the sea varied from 24 to 60 degrees.

The dome had no crater. At the summit there was a confused pile of large gray blocks, and during daylight hours no incandescence could be seen, but in the evening the entire summit appeared to be on fire and, the fumes that emanated from it were illuminated by radiation from red-hot rocks. Cooled rocks covered the lower slopes, but when some of them occasionally rolled down, a bright incandescence appeared in the partly molten material that could then be seen in the interior of the blocks.

Around the point of Phleva, the steam from fumaroles was frequently streaked with what looked like flames but in reality was the illumination caused by radiation from incandescent spots. The temperature of the escaping vapor produced in the middle of the older lavas of Nea Kameni was found to be around 50 degrees, while at the extremity of Giorgios, the temperature already reached 104 degrees.

The members of the Greek commission explored the coast of Giorgios again on the 12th of February. They found that the muddy pond and Lophiscos (the white island of the 1707 eruption) were entirely covered with new lava. The dome was still growing and invading new terrain on all sides, especially toward the south. Between the new lava and the older cone of Nea

Kameni there was an opening through which one could easily pass from the southeastern to the southwestern part of Nea Kameni.

The paved road that formerly led from the pier at the eastern end of the island to Voulcano Cove was still intact up to a point about 120 meters beyond the Greek church. From there on, it disappeared under water that had a temperature of 60 to 70 degrees. The eastern shore of the cove and all the houses that had been built there were almost completely hidden by water. Of all the dwellings on this part of the island, only three remained standing. They were located near a small inlet a short distance east of the entrance to Voulcano Cove. But as a result of the subsidence they were invaded by the sea and immersed in water up to their roofs. A little farther to the north one could still see in the middle of the blocky lava the remains of a house built at the northeastern angle of the cove.

On the western coast, Giorgios was in direct contact with the southwestern shore of Nea Kameni. The abrupt slope of older talus that formerly bordered the shore of the cove was now covered at its base by the new lavas, but its upper part remained uncovered and separated from Giorgios by a small declivity, one side of which was formed by the new lavas. The fumes of the volcano had not prevented a few plants from growing on this slope.

At ten in the evening, there was a strong explosion. A dark red shower shot from the summit of Giorgios and rose to a height of about 100 meters. It was followed by a rain of cinders and lapilli.

The 13th of February was a day that heralded the formation of a new eruptive center. The members of the Greek commission had for some time been at the summit of Nea Kameni observing the activity on Giorgios when one of them, Professor Christomanos, glanced toward Cape Phleva and saw a black reef breaking the surface of the water. The place where this happened was exactly where a circular motion of the water and a release of gas had been noticed earlier. The water churned again, as it had the evening before, and the reef rose near the southern edge of this disturbance. After about six minutes, the newly emergent rock foundered and disappeared. A quarter of an hour later the same thing happened again. The second reef remained above the water level for only four minutes. In the afternoon, rocks again reappeared out of the sea. At the edges of the disturbed area one could see from afar two long, dark-colored streaks that were nothing more than submarine lava just below the surface. On the evening of the same day, Mr. Christomanos says that the emergent rocks showed signs of incandescence; he also noticed that between the new reef and the coast of Nea Kameni a bright yellow flame was rising to a height of one meter. During the day Giorgios produced two or three rather strong explosions. In the evening its entire summit was incandescent, and the members of the Greek commission noticed weak blue flames in the large fissure crossing the dome. Incandescence was noted between blocks on its flanks. Flames were also noticed on the lower part of the southern slope of the cone of 1707 at a point where fumaroles were giving off large amounts of steam and hydrogen sulfide.

On the 14th of February, new rocks appeared about 160 meters from Cape Phleva* at the same location as the reefs of the previous evening but instead of disappearing as they had before, they remained above water. The highest rose about 2 meters above the water. All around, the water was so warm that it was giving off steam.

During this same period, Giorgios continued to evolve. Its explosions were weak, with only one of any notable strength being noticed all day, but gases rich in hydrogen sulfide were given off. During the morning, the mound of lava had continued its advance and covered the westernmost of the three houses that remained standing at its eastern base; by four o'clock in the afternoon it had buried one of the remaining two.

The ground continued to sink in the southern part of Nea Kameni. In twenty-four hours the sea had advanced 7 or 8 meters along the western part of the road that passed between the Orthodox and Catholic churches. The reefs, which the English map shows at the entrance to Voulcano Cove, were still partly visible, but the lowest one had emerged, as had the nearby shore.

By the morning of the 15th of February, the new reefs in front of Cape Phleva had become quite large; at nine-fifteen in the morning there were six. The largest of these came up a few meters west of the place where the first had appeared during the preceding two days. They did not all have the same stability; five remained emergent, but the sixth, situated west of the others, disappeared after a few minutes. The members of the Greek commission who witnessed these phenomena gave the new eruptive center the name *Aphroessa* after the ship that had brought them to Santorini. Judging from everything that was seen, it is evident that a discharge of lava at this point was causing a progressive growth from the southeast toward the northwest[†]. Around six in the evening, the wind having momentarily swept aside the fumarolic gases,

---

* The location was determined by Messrs. Schmidt and Palasca by angular measurements.
† On the basis of incorrect information, I wrote in my letter of the 26th of March to Mr. Ch. Sainte-Claire Deville that the blocks of Aphroessa carried oysters and mollusks on their surface. That is now known to have been false.

the scientists of the Greek commission who were stationed on the cone of Nea Kameni noticed a bright glow at the northern base of Aphroessa. The luminous part formed a band 10 meters long and 50 centimeters high starting at sea level. As I have since had several occasions to observe this glow and identify it as due to flames, I believe that the illumination seen then had the same origin.

At about the same hour, a series of small explosions like pistol shots occurred in the middle of the rocks of Aphroessa. They were not nearly as strong as those that had come from Giorgios. Real explosions had taken place there during the day, and one of these was accompanied by projections of incandescent rocks. The dome was very steep-sided at the northern and southern ends of its western flank. Red-hot blocks were constantly rolling down that side, making much noise and falling into the 75-degree water that lapped up on its base.

On the opposite, southern side of the cone of Nea Kameni, fumaroles had been active for several days in a crudely defined area, but little by little they had extended up to the middle of the slope and released great quantities of gas. New fumaroles were also seen at the southeastern foot of the cone; these had broken out in the middle of the rubble that now covered the older surface.

By the 16th of February, Aphroessa formed a substantial island that was continually growing. The manner in which it grew resembled that of Giorgios during its first days in that there were no earth tremors or violent explosions. The shores of the island spread as a result of both the continual emergence of new rocks and the accumulation of blocks pushed from the center of the mass toward its periphery.

The configuration of the surrounding seafloor underwent great modifications. Soundings carried out between Nea Kameni and Palaea Kameni by Mr. Palasca indicated depths of 40 to 70 fathoms where the English map showed former depths of 103 fathoms.

Giorgios continued to grow steadily. Just as in the first days of the eruption, no crater could be seen there, but large curved fissures on its upper surface outlined a crude circle. The fumaroles had taken on a new aspect indicating that they were much more intense than on the preceding days. The emanations of white steam were joined by transparent bluish gas and by ejections of cinders and lapilli. The white vapor appeared mainly on the southern part of the dome, and reddish fumes came from the middle of the summit, while solid material was ejected at the northern extremity. It was not uncommon to see three separate types of fumaroles active at the same time.

At a distance of 6 to 7 meters from the eastern foot of Giorgios the temperature was 85 degrees.

On the 17th of February, Aphroessa was 100 meters long, 60 meters wide, and 10 meters high. A thick cloud of steam enveloped it completely. No reverberating sound could be heard, only the noise of escaping gas and water coming in contact with the hot rocks.

There were no explosions on Giorgios on that day. The vapor it gave off from the summit was immediately carried away by a strong wind blowing all day from the north. To the west the narrow declivity between the dome and the older shore of Voulcano Cove was still partly covered with green vegetation. To the north, the fumaroles had spread and were seen to continue up two-thirds of the height of the cone of Nea Kameni. To the east, the base of Giorgios was awash in a layer of water 9 meters wide and with a temperature varying between 55 and 85 degrees. The ground had continued to sink perceptibly.

During the day of the 18th of February, the growth of Aphroessa and Giorgios continued quietly; the progressive displacement of blocks went on without shocks or violent explosive ejections of debris. At nightfall, the summit of Giorgios showed only a bright incandescence; the glow of the red heat of the lavas illuminated the white fume cloud that rose from the middle of the dome to a height of 300 meters.

The next day the same calm continued at the two eruptive vents and the incandescence of the lavas was even brighter. After sundown, the fume cloud from Giorgios was brightly lighted and colored red to a height of at least 1000 meters.

In the evening, Mr. Palasca made soundings between Aphroessa and Cape Phleva and found a depth of 31 meters. He also noted real flames on the margins of Aphroessa as well as elevated temperatures in the surrounding water.

At eight o'clock on the morning of the 20th of February the air was quiet and clear, the wind weak, the temperature of the air 4.2 degrees, and the barometer at 759.3 mm (slightly less than the evening before when it was 758.3 mm, as it was the day before);* in other words, there was nothing about the atmospheric conditions that would have led one to expect a catastrophe.

---

* On the 10th of February, however, there was a sudden change of the atmospheric pressure when, within 24 hours, the barometer dropped from 750 to 740 mm and then rose again by the same amount in the same period of time.

The same was true of the volcano; it did not seem to indicate anything unusual, but we have noted that the incandescence of the two eruptive centers had increased, the temperature of the surrounding water had risen, and the nearby fumaroles had increased in number. Several explosions took place during the early morning hours, but they were rather weak and caused no alarm among the members of the Greek commission. Around eight in the morning they had approached the eastern base of Giorgios, and four of them had then climbed to the summit of the cone of Nea Kameni to carry on with their observations.

Suddenly, at nine thirty-six, there was a terrifying explosion. The first thing that was noticed was a rumbling like the sound of thunder, which was followed immediately by masses of rock rising from the summit of Giorgios. In an instant a thick black cloud of ejected material rose to great heights blotting out the sky. A few seconds later, thousands of incandescent bombs fell over a wide area. The members of the commission, caught in this rain of burning ejecta, instinctively sought the shelter of some large rocks. Two of them, Mr. Christomanos and Mr. Palasca, were injured by falling debris. The incandescent lapilli set fire to their clothing. The brush that covered the crater of Nea Kameni caught fire, and the same thing happened on Micra Kameni. I consider it unlikely that hot gas came from the fissures on the latter island, even though the great quantities of fumes coming from these cracks might indicate otherwise. A steamboat that was tied up close to the eastern pier of Micra Kameni had been badly damaged by the explosion. An incandescent bomb made a hole in the bridge, and fire had broken out in one of the cabins. A rowboat, pierced in the same way, sank to the bottom. A quartermaster and several sailors had injuries of greater or less severity.

A nearby ship that was loading pozzolana caught fire and soon went down. Captain Valianos, who commanded it, was mortally injured by a falling rock. Mr. Schmidt estimates that blocks were projected to a horizontal distance of 625 meters and that smaller ejecta went much farther with cinders reaching the island of Thera.

All these events caused much alarm among the population of Thera and caused the commercial ships that ordinarily called at the island with wheat and other supplies to stay away. Observers that had studied the volcano up until then found no boatmen willing to take them there and had to be content with distant observations.

This disastrous eruption lasted only a few minutes. Almost immediately, Giorgios returned to its normal behavior. A small eruption that took place around eleven o'clock on the same day was of moderate strength like those seen early in the morning. In the afternoon, however, stronger explosions were already beginning to take place filling the air with groaning noises.

The form of Giorgios was changed very little by the explosion. The summit appeared to have been flattened and slightly lowered, as if it had been truncated. The total length of the dome from north to south was about 380 meters.

During the day of the 21st, there were several very violent eruptions. One of these, which came a little before one in the afternoon, caused a rain of cinders to fall as far away as the island of Thera. It made a sound like thunder, and, according to a measurement by Mr. Palasca, the eruptive column reached a height of about 2000 meters. The sky was darkened for a long while. Another explosion took place around two-thirty in the afternoon. According to the observations made by Mr. Schmidt, the fume cloud took 57 seconds to rise to a height four times that of the cone of Nea Kameni, which would mean it had a velocity of about 7 meters per second.

On the 22nd of February, there were more eruptions of great intensity. The strongest one, which occurred around three o'clock, was even greater than the one that had occurred on the previous evening. Explosions followed at short intervals, giving rise to clouds that merged into a single black mass. This is something that is often observed in all large volcanic eruptions. The rise of the eruptive cloud lasted 80 seconds and reached an elevation of 2200 meters at an average velocity of 25.5 meters per second. A rain of stones fell all around. Blocks falling into the sea made violent waves on the water. According to Mr. Palasca, the projected material reached a distance of about a kilometer around Giorgios. A great number of these blocks were large and incandescent. Those that fell on the flanks of the cones of Nea and Micra Kameni continued to glow for some time after they fell and, seen from Thera, they looked like sparkling points of red light.

It was on the 22nd of February that cyclones were noted for the first time in the atmosphere over the bay. Mr. Schmidt, who has made a special study of these phenomena, described them as follows. They were seen to appear first on the south side of the volcanic edifices. They reached heights of 100 to 300 meters in the fume clouds coming from Giorgios and Aphroessa and then descended toward the sea in the form of a contorted, vertical cylinder that spread out as it came down. They often shot up in a vertical spiral. The direction of their drift was governed by that of the wind. Where their diameter was greatest they were transparent. Mr. Schmidt added that he saw several hundred of these eruption clouds, which he observed and sketched with the aid of a telescope. None of them resembled a water spout. Instead, he compared them to spiral dust columns that form around Athens and on the roads of Attica, even when there is not a breath of wind that might cause them. On the first day he noticed them, the sky was cloudy, and a few drops of rain were falling. All the largest cyclones were produced in the same place and always followed the same track.

The 23rd of February opened with a heavy rain of cinders that spread over all the eastern part of Thera. The main explosion, which took place around nine o'clock in the morning, was accompanied by a loud noise. A great amount of incandescent material was ejected. Part of the cinders and other ejecta appeared to come from Aphroessa. The eruption column rose to a height of 1000 meters.

In the early morning of the 24th, explosions and a heavier rain of cinders burst out with much noise. A strong odor of hydrogen sulfide could be smelled on Thera.

On the 25th, the water of the bay was strongly colored around the Kamenis, but the volcano remained quiet. The sea had the same color on the 26th. The rain of cinders, sulfurous odor, and occasional explosions continued.

Nothing unusual happened on the 27th. Hydrogen sulfide continued to fill the air. On the 28th, around eight in the morning, there was a terrible explosion. The entire summit of Giorgios seemed to rise as a single, large, incandescent mass. Countless bright-red blocks were projected from a brilliant eruption column and, for several moments after their fall, illuminated the flanks of the cones of Nea Kameni and Micra Kameni. The sound of the explosions was like strokes of thunder mixed with shrill whistling. The foul smell of hydrogen sulfide reached all the islands of Santorini. In the evening of the same day there was a rain of cinders.

On the first of March, the Greek commission, having returned to Santorini from an excursion to Milo, noticed that the glow of an eruptive vent could be seen a mile north of Nea Kameni. They also observed a large area of greenish flames on the northern side of Aphroessa. During the day there were several eruptions of modest size. Toward evening the Turkish frigate, *Sinope*, which had anchored at Banco, fired a salvo of seven rounds from large cannons. Mr. Schmidt compared the sound to that of the explosions from the volcano, and estimated that at the same distance the artillery fire produced a sound four times stronger than that of the loudest volcanic detonations. This statement may be accurate for the explosions that the witness had heard, but I am sure that it would not be true in all cases. The sounds of large explosions from Giorgios in 1866 and 1867 were much greater than that of any artillery barrage.

On the 2nd of March, the rain of cinders and the frequent explosions continued. Mr. La Motte, the gunnery officer of the Austrian ship *Reka*, twice tried in vain to climb Giorgios.

On the 8th of March, I arrived in Santorini in company with Mr. de Verneuil. The next day, thanks to the enthusiasm of the officers of the *Reka*, we visited the site of the eruption. (Local boatmen had refused to take us.) Here is what we observed.

Giorgios had the appearance of a blocky mound 50 meters high, 350 long, and 100 wide. From the southern side, a gentle slope extended smoothly from the summit situated near the northern end to the southern point, which was only slightly above sea level. The margins on the northern, eastern, and western sides of Giorgios were more abrupt and were broken by huge blocks projecting above the surface. From time to time, some of these blocks broke away and rolled noisily down the slope. Many were very unstable. A small rock thrown against one could be enough to cause it to fall. Going up Giorgios by any route was therefore quite dangerous. To climb the dome Mr. de Verneuil and I were obliged to start near the southern point. We could detect no regularity in the distribution of blocks. The crest that crowned the margins reached a height of 20 to 25 meters. Higher up, the dome was more rounded. No crater-like depression could be seen; only fissures oriented north-south. Blocks with very irregular dimensions were piled up in great disorder and with no connection between them. We turned over several by hand, even though they had volumes of several cubic meters. The rock is rough and scoriaceous or, more commonly, compact and vitreous. All around us we heard the incessant noise of cracks opening as the rock contracted during cooling. This sound was accompanied by a tinkling somewhat like that of broken china. It was produced by falling fragments detached from fractures. The temperature varied greatly from place to place. We explored a rather extensive stretch along the eastern crest of the dome and noticed that the temperature there did not exceed 30 or 40 degrees. As we approached Giorgios, however, and especially when we advanced toward the north, we found higher temperatures as well as many fumaroles between the blocks. White vapor came from the cooler places, but a great many of the fumaroles produced hydrochloric and sulfurous acids where the thermometer indicated temperatures of 100 to 300°. Finally, we passed by some colorless, transparent emissions in which we could fuse pieces of zinc wire in a few moments.[4] The rocks surrounding the channels through which the gas escaped were covered with a coating of sodium chloride. In some of the cracks we could see incandescent lava.

During this excursion, we witnessed several explosions at the summit of Giorgios, but none of these threw out solid ejecta, only a violent release of steam and gas. At the same time, the hottest fumaroles seemed to redouble their activity. The correlation was not always perfect, however; some fumaroles remained unchanged, while others suddenly increased during the period of the explosion.

We descended the eastern slope of Giorgios after first taking the precaution of knocking loose all the unstable rocks that covered the slope where we planned to go down.

Dr. de Cigalla, who had accompanied us to Nea Kameni and had been willing to serve as our guide, waited for us there. He pointed out to us the small ponds that at the beginning of the eruption had been filled with fresh water. Originally isolated from the sea, they were now connected to it and were filled with clear water that had essentially the same taste as sea water. The hottest place had a temperature of 76 degrees. Many bubbles of gas rose here and there. Their connection to the sea was through three openings the largest of which had a width of 1.20 meters and an average depth of 15 centimeters. We could see water flowing out at a rate of 12 centimeters per second. Its clarity contrasted with the yellowish color and turbulence of the seawater it was entering. The ponds were shallow, the maximum depth of the largest one being only 30 centimeters. The ground around them was almost at sea level. They had irregular shapes and, taken together, an overall diameter of about 15 meters. They were located near the middle of the southeastern point of Nea Kameni about 50 meters southeast of the two chapels.*

The subsidence of the ground led to flooding of all the stretch of land bordering the southeastern point of Nea Kameni. From the southern side of this point, the width of the submerged zone was about 25 meters. Near Giorgios, one could no longer see more than the roof of a single house that remained above the water.

---

* The topography of this part of Nea Kameni has since been considerably modified; the place where the ponds were is now in the middle of a small cove that borders the passage between Nea Kameni and Micra Kameni and reaches the base of Giorgios. A small part of their original location may now be covered by lavas.

The same was true of the two houses at the southeastern point of the island. The southern part of the pier bordering the canal of Micra Kameni was submerged, but the northern part, which had subsided less, was still dry.

This part of Nea Kameni had scattered blocks that had been ejected during the previous days. Most consisted of light gray, very porous lava covered with a compact glassy crust with numerous cracks. Almost all had broken when they fell; a few, however, remained intact. Some had volumes of several cubic meters. One that had fallen on the roof of the Orthodox chapel punched a hole and came to rest at the foot of the altar, which it equaled in size.

After visiting the southeastern part of Nea Kameni, we turned our attention to the opposite coast of Giorgios, where we observed that the distance between the orthodox chapel and the base of Giorgios had been reduced to 15 meters. We passed between Giorgios and the foot of the cone of Nea Kameni. The original ground surface, which could be seen in this space, was only 30 meters wide. Many fumaroles in this same area gave off a great deal of water vapor, carbon dioxide[5], hydrogen sulfide, and combustible gases. Most of the fumaroles had a temperature near 100 degrees. Their orifices were surrounded by deposits of crystalline sulfur. The western slope of Giorgios was still separated from the older ground of Nea Kameni by a narrow depression. The adjacent talus was totally devoid of vegetation; the abundant gases escaping in this ravine had stained and desiccated the plants that had covered the slope on the western side. The ground was covered with yellowish-white crusts of sulfur, calcium sulfate, and very small amounts of ammonium sulfate and chlorohydrate. The numerous fumaroles that had formed these deposits were still very active. The hottest ones had temperatures approaching 200 degrees. They extended along the shore and followed the edge of the water for about 100 meters to the west.

On returning, we got into row boats of the *Reka*, and, with the guidance of Lieutenant La Motte, headed for Aphroessa. All along the southern shore of Nea Kameni we found the water warm, agitated, and colored with various tints. Ten meters from the shore, the temperature was between 50 and 60 degrees. The water gave off thick fumes and smelled of hydrogen sulfide. At the entrance to the canal between Aphroessa and Cape Phleva,* the fumes were so thick that the shore could not be made out at a distance

---

* The prominence of this cape is exaggerated on the English map of Captain Graves. Despite the vertical ground movements, the coast was steep and abrupt at this point, and this is why the form of the coast was little changed by the eruption until it was covered by the new lava.

of 5 meters. According to Mr. La Motte's soundings, the depth of the water in the middle of the passage was only 10 meters, while three days earlier it was still 17 meters. The width of the channel was only about 10 meters, and the entrance had been narrowed by the reefs, making access to it difficult. One could cross it only with great caution. Bubbles of gas came up in great numbers around Aphroessa, producing an effect like boiling. Those that hit the bottom of the boat made a sound like the rolling of a drum. While we made a tour of the islet we were constantly enveloped in thick fog that cut off our view. We saw neither flames nor incandescent lava.

On reaching a point about 100 meters to the west, we found ourselves out of the fog and could see the form of the island more clearly. It was almost circular with a diameter of about 400 meters and a height of 30. Aphroessa produced no explosions or loud, sudden sounds, but from time to time we heard a dull rumbling that normally did not correspond to the eruptive pulses of Giorgios. The emanations from the summit of the islet had a peculiar character. They were not like the white vapor given off by Giorgios or from the area around Aphroessa but were transparent reddish fumes that appeared to have a very high temperature. We attributed the color to volatilized iron chloride.

On the next day, the 10th of March, we made another excursion to the southern part of Nea Kameni and to the area around Aphroessa. A rather strong wind swept the vapors from around the latter place and opened our view. We noticed that, like Giorgios, it was composed of incoherent blocks. A certain number of these are black, semivitreous, very compact, and speckled with small, white crystals of feldspar; others are brown, scoriaceous, and similarly spotted with white crystals. Compared to the rocks of Giorgios, those of Aphroessa are distinguished by relatively greater amounts of various types of porous lava than of compact varieties. The dome of Aphroessa had the form of a rather regular mound with no visible fissures. I set foot on the shore of the small island and tried to get closer to its center, but after a few steps the heat of the fumes forced me to retrace my steps. I tried again from other parts of the coast but always without success. In the area around Aphroessa the sea was yellowish-white and very warm. Many bubbles of gas came from it with turbulent churning, particularly at the northern point of the island. I collected these gases, not without difficulty, and filled several tubes which I then sealed with a lamp. The water of the sea close to the shore of Aphroessa was so hot that one could not hold one's hand in it. Fear that the heat might melt the mastic of my stopcock led me to search for more distant places that were not so hot in order to collect my samples of gas. I made a quick analysis of these emanations on the spot and saw that they were combustible. I also separated the solids from a notable quantity of the seawater at a place where it was especially agitated. When the deposit collected on the filter was dried and heated on a platinum sheet, it burned with a strong order of sulfurous acid and left a ferruginous residue. Thus, the material stirring the water was a mixture of sulfur and hydrated iron oxide.

Continuing to explore the area around Aphroessa, we saw that a new islet had developed on the western side; I proposed that we name it *Reka*. According to the information we obtained, the first signs of the island had appeared around eight o'clock that morning. By three o'clock in the afternoon, it was 30 to 40 meters in diameter. Its height above sea level was a meter and a half, and it was separated from Aphroessa by a channel 10 meters wide and 10 meters deep. The blocks composing it were dark brown, rough, and similar to some of the porous lavas of Giorgios and especially to those of Aphroessa. The interior of the blocks was somewhat glassy with scattered crystals of feldspar in a seemingly homogeneous groundmass. The rocks of this new islet had the same temperature as the surrounding water or even less. One could handle them and walk over the entire islet without danger. The steady growth was by a continual elevation of the rocks. We made a reference mark on a large block with the point of a hammer and found that it rose 12 centimeters in a period of one hour. It appeared to us that the displacement of this block, which was situated on the western shore of Reka, was not entirely vertical. The horizontal line we had drawn on one of its faces had become inclined during the period we watched it.

The growth of Reka was much quieter than that of Giorgios or Aphroessa. One could hear none of the incessant cracking or grinding of the other two islands but only the occasional sound of a falling block.

Near the northern point of Reka at the entrance to the channel separating the islet from Aphroessa the temperature of the sea was about 60 degrees, and much gas rose from the water. I took several samples and found that these gases were not combustible but differed in composition from those produced a little farther to the north at the northern end of Aphroessa.

As darkness extended over the bay at nightfall of the same day, we saw the rocks of Aphroessa gradually become illuminated. A great many of the blocks around the margins of the island were incandescent, and the entire center of the dome was composed of bright-red rocks resembling an immense brazier. The reddish fumes at the summit came from undulating flames tens of meters high, sometimes shooting out violently, sometimes quietly carried by the wind. They had the same bright-yellow color as the high-temperature emissions from sodium vapor.

Flames also came from places between many of the blocks covering Aphroessa; they were fed by gases coming from narrow, incandescent crevices. Bubbles breaking at the surface of the sea also gave rise to flames. When the air was calm, these gases were ignited on contact with the rocks of Aphroessa, and the flames then spread over the entire area where the gas was released. These flames were often extinguished by the gusty wind, but they were quickly reignited and spread again over the surface of the water. The heat was intense enough to ignite paper 15 meters from the shores of Aphroessa.

The gas coming from the water around Reka was not combustible; during the evening we saw no flames coming from that area.

Flames were seen at the summit and on the flanks of Giorgios. Along the slopes they were bluish and flickering, and they jumped from place to place, going out and igniting again with each change in the strength of the wind. At the summit of Giorgios they were less conspicuous than on Aphroessa, because they came mainly from spaces between the blocks of lava. For that reason, the only visible fumes were those illuminated by strong radiation from incandescent material below. Nevertheless, when the wind blew the flames in a favorable direction and when a pulse of activity caused a new surge, one could hear a violent rushing sound and see jets of flame several meters high rising turbulently from the black mass of the dome.

On the 13th of March, the space between Reka and Aphroessa was reduced to a narrow channel with emergent reefs in the middle. The following day the two islets were united. Aphroessa had grown toward the west as well as eastward, and although the canal separating it from Cape Phleva bristled here and there with blocks of lava, a small boat could still pass through.

We visited Aphroessa and the area around Giorgios and then went on to Palaea Kameni. I collected gases at the northern point of Aphroessa and in the harbor of Saint Nicolas on Palaea Kameni. Returning to Nea Kameni, we climbed the cone of 1707, where we found on the summit many blocks thrown there during the eruptions of the previous days, especially on the 12th of February. The crater of 1707 was crossed by several fissures. One with a trend of north 26° east extended from the eastern shore to the middle of the cone; it had a depth of only 2 or 3 meters and about the same width. It did not give off appreciable amounts of gas. A second fissure, more important than the first, followed a contour all around the southern half of the crater. The other fissure formed a cord across the arc of this second one. The depth of the latter at places reached 4 meters. It was the focus of numerous fumaroles that gave off water vapor, and sulfuric and carbon dioxide. The temperature seemed to be between 30 and 50 degrees. We could not descend to the bottom of this fissure, because the cinders and lapilli of the walls threatened to fall in on us, and this is the reason why my temperature estimate is so imprecise.

Our special attention was drawn to the original ground surface of Nea Kameni lying between Giorgios and Cape Phleva. Already, two evenings earlier, we had noticed the deep cracks that split the ground on this part of the island. The main ones were four in number, and they were about 150 meters long, approximately parallel to a line of very hot and active sulfurous fumaroles aligned 20 degrees east of north and about 40 meters farther north. The temperatures of several of these fumaroles were as high as 300 degrees, but none had a higher temperature. They gave off sulfurous acid, carbon dioxide, and water vapor. The orientation of the fissures and fumaroles conforms more or less to that of a line extending between the summits of Giorgios and Aphroessa. The fissures are 8 to 10 meters deep and 3 to 4 meters wide. They are pointed and separated from one another by spaces 15 to 20 meters wide. In a few places the cracks can be traced for 40 of 50 meters cutting ledges of dense lavas. Thick flows erupted in 1707 are cut as though they had been split in two by a large cleaver. We were so impressed by one particular section that was sharply cut into exceptionally compact lava that, on the advice of my companion and fellow scientist, Mr. de Verneuil, I mentioned this in my letter to Mr. Charles Sainte-Claire Deville on the 26th of March, 1861.

At the bottom of the fissures, rapid currents of salt water flowed freely from Giorgios toward Aphroessa and came out in the channel between Aphroessa and Nea Kameni. This water had a temperature of between 60 and 70 degrees. Along its entire course this water had combustible gas rising through it. We amused ourselves by igniting the bubbles with a match. A rough on-the-spot analysis I made of the gases indicated that they were a mixture of small amounts of oxygen and hydrogen sulfide along with much greater amounts of carbon dioxide and nitrogen, together with notable quantities of hydrogen.

These fissures were connected to each other by transverse cracks at their eastern end.

Several times during the day, explosions of mild intensity occurred at the summit of Giorgios, but they were too weak to bother us during our tour. We saw no eruptions of cinders. We estimated the height of the volcano to be 45 meters.

On the 15th of March, there were several explosions during the course of the day. The volcano continually gave off large amounts of steam and hydrogen sulfide. On Thera the odor of the gas was unpleasant, and the people of the island dreaded the damage it could do to the vines. At a quarter after two in the afternoon, a wind from the northwest carried a light rain of cinders over Thera. Two rather strong eruptions occurred at daybreak on the 16th.

On the 17th of March, we went to Nea Kameni with Mr. Lenormand and Mr. Da Corogna, who had joined our group the evening before. Conditions had not changed much since the 14th. We noticed only that the fumaroles on the southwestern part of Nea Kameni had become more active. They produced a great deal of hydrogen sulfide and sulfurous acid. Those with the lowest temperatures deposited thin crusts of sulfur, calcium sulfate, and iron compounds. These deposits took the form of tufts of prismatic crystals, mainly in small cracks in the ground and under blocks that only touched the ground at irregular parts of their lower surface. The crusts and the ground immediately in contact with them turned litmus paper very red; they owed their acidity to free sulfuric acid. These various deposits were evidently the products of incomplete combustion of hydrogen sulfide. They did not occur where the temperature was high and contact with the air was unrestricted.

I collected samples of gas from the bottoms of fissures at the places where I had already made an in-place analysis on the 14th. The currents of dirty water from which the escaping gas was rapidly rising were on average 10 degrees hotter than three days earlier.

We went around the western side of Aphroessa in a boat. The channel between this island and Cape Phleva was so congested and cluttered with reefs that our boatmen did not dare to enter it. Reka was completely joined to Aphroessa, but a small depression in the mass of rock still marked the place where they had come together. We could distinguish no crater on Aphroessa nor any notable fissures in its upper surface. On the flanks, however, particularly at the northern point, we saw crevices up to 20 or 30 centimeters wide between the blocks. At the bottom of these cracks we could see in full daylight the incandescent, viscous lava. The heat it radiated was intense. We approached more closely, however, and after taking the precaution of first wrapping my face and hands in damp cloth, I thrust the point of an iron-tipped stick into the lava. After ten minutes the material was indented by the iron.[6]

We then climbed the cone of Nea Kameni and from there we witnessed the spectacle of several eruptions from Giorgios and Aphroessa. The explosions of Aphroessa were weaker than those of Giorgios, but the temperature of the gases given off there was higher, and vapor did not condense as quickly. The gases produced by Aphroessa always had a conspicuous reddish color; flames from their combustion are indicated in daylight by their transparency, by their low refractivity relative to air, and by the velocity and undulating motion with which they rise.

On the 18th, a whim of our boatmen prevented us from landing on Nea Kameni, so we were forced to retrace our path. Remaining in Thira the rest of the day, we heard the repeated complaints of the inhabitants. The eruption had so frightened the crews of the commercial ships that supplies had become very expensive. Bread sold for one franc twenty a kilo. Salt was scarce. The Lazarists with whom we were staying awaited impatiently the arrival of a boat that was due from Syra with wheat and rice. I decided to write a letter to the bishop designed to reassure the merchants who normally supplied the islands of Thera and Therasia.

On the 19th and 20th, we returned to the Kamenis. The junction of Aphroessa with the point of Phleva began during the 19th, and by the 20th was complete enough for one to walk on dry land from Nea Kameni to Aphroessa. Mr. Verneuil, despite his advanced age and poor eyesight, made this crossing with me twice, the first time around two in the afternoon and the second after night fall. The flames appeared to us the same as on the night of the 10th of March. In the evening, we approached an incandescent crack that was open on the eastern side of Aphroessa and was giving off flames. Mr. Verneuil repeated my experiment of the 17th by picking up a pasty mass of lava with the metal tip of his stick.

The same day we noticed that the crevices on the southwestern part of Nea Kameni had grown noticeably. They had become wider and deeper, while the flow of water through them was faster. A large, flat block on which I had kneeled to collect a sample of gas on the morning of the 17th and which at that time was scarcely 10 centimeters above the water, on the 20th of March was 1 meter 46 centimeters above the same level. On this block, there were still the remains of a tube I had accidentally broken on the 17th. We left these fragments in place in order to be able to recognize the block and use it to follow ground movements in the future. We also made marks on it with our hammers.

The flames were very well developed on Giorgios and had the same appearance as before. They were particularly evident toward the southwestern end of the dome.

On the 24th of March, we made another excursion to Nea Kameni and devoted part of the day to a detailed examination of the southwestern part of the island. The cracks had grown considerably. The block which, on the 20th, had led us to notice the inflation of the ground, was now 5.5 meters above the level of the water, showing that the ground had been elevated about 5.4 meters. At the eastern end of the fissures, the elevation appeared to have been even more. The crevices at this point were 15 to 20 meters deep and 7 to 8 meters wide; their bottoms remained slightly below sea level. It must be concluded from this that the cracks had opened more in recent days and their sides were higher. Thus, since the 10th of March, the original ground of Nea Kameni had risen several meters in the middle of the area between Giorgios and Aphroessa. This ground movement was local; it was much less along the former shore of Cape Phleva, which was still visible along its entire length. Uplift could also be detected there as well, but it did not appear to be more than 1.5 to 2.0 meters at the western end of the fissures. At the bottom of these fissures there were still currents of dirty water flowing toward Aphroessa, but they moved very slowly. They still gave off combustible gases that seemed to be richer in hydrogen sulfide than they were before. The highest temperature observed was 80 degrees. The average temperature in that area seemed to be a bit higher than a few days earlier. In this same period the channels had grow in depth and width, but their bottoms were choked with blocks that had fallen from their walls. The latter were now less abrupt, because piles of rock covered their sides. Instead of the streams of freely circulating hot water we had seen before, more of the water was in small pools between pieces of lava.

While the fissures in the southwestern part of Nea Kameni were undergoing these changes, the line of sulfurous fumaroles that ran parallel to them about 40 meters to the north had redoubled their activity. On the 12th of March, none of these had temperatures in excess of 400 degrees, but on the 24th of March they were hot enough at several place to melt zinc[4]. There was a strong smell of hydrochloric acid. The condensed vapor, when treated with silver nitrate, produced an abundant grayish precipitate of silver chloride and a small amount of silver sulfide. It also gave a precipitate in acidic barium nitrate.

Soundings carried out between Aphroessa and Palaea Kameni by the officers of the Italian frigate *Principe Carignano* showed that the maximum depth, which was 296 meters at the beginning of the eruption, had been reduced to 108 meters. This decrease was especially evident between Reka and the southern point of Palaea Kameni.*

On the 24th of March we could still detect subsidence at the southeastern point of Nea Kameni which seemed to have stopped for several days but now renewed its gradual motion. The pier subsided 40 centimeters below its previous level. Flames continued to burn each evening at the summits of Giorgios and Aphroessa. They were seen on the uppermost part of the small mound corresponding to Reka, but they were no longer produced around the base of any of these features.

During the evening of the 25th of March, we made another tour around Giorgios and Aphroessa. Flames were abundant at the two summits, and they were also well developed on the eastern flank of Aphroessa. On that side, the viscous incandescent lava moved slowly at the bottom of cracks between blocks, and flames one or two meters high were continually being given off. The Italian officers that accompanied us took pleasure in igniting the gas bubbles that came to the surface of the water.

On the 26th of March, we left Santorini for a while. One could see from what was happening that the volcano was in a stage of relative calm. Giorgios had attained a height of 50 meters, a length from north to south of about 500 meters, and an east-west diameter of 400 meters.

From the 29th of March until the 12th of April, the eruptive phenomena of Santorini were observed by Mr. von Seebach. During this interval, the volcano continued to develop without notable changes in the intensity of its activity. Several things, however, indicated that its continued growth should still be monitored.

Toward the end of March and the first week of April, it was no longer

---

* In my letter of 26 March 1866 to Mr. Ch. Sainte-Claire Deville, I attributed a decrease of the depth of the channel between Aphroessa and Palaea Kameni to this slow rise of the seafloor. I have always recognized, however, that the submarine lavas from this eruption also contributed to a change in the bathymetry. In general, I have attached too much importance in my earlier publications to modifications due to movements of the original ground. In the course of the last eruption of Santorini, the effects of these movements must be considered almost negligible compared to the changes due to lava flows.

impossible to climb Giorgios. Von Seebach succeeded in doing so three times, once on the 30th of March, again on the 4th of April, and a third time on the 11th of April. The form and structure of the dome's summit was the same as we described it earlier. When he climbed the dome on the 4th of April, von Seebach was able to get within 100 meters of the highest point. Most of the blocks were cold, but here and there it was necessary to cross cracks giving off gas and high-temperature steam. These cracks were randomly spaced and tended to be oriented radially about the summit. It was possible to cross them at only a few places where they happened to be obstructed by debris, and having crossed them one hastened to gain some distance from them, because as soon as the activity at the central vent increased, floods of burning incandescent gas came from them. One could smell the odor of hydrochloric acid and hydrogen sulfide, but it was not unduly troublesome.

Some of the blocks had unusually high temperatures, although during the daylight their color showed no incandescence. Near the sites of the most active fumaroles, the rocks had a reddish-brown color that was probably due to oxidation.

There was no true crater but only an irregular central opening about 7 meters in diameter formed by the intersection of several fissures.*

In the central part on Giorgios, one did not hear the constant cracking sound caused by contraction and breaking of the blocks.

According to Mr. von Seebach, the height of Giorgios during the first half of April was about 55 meters. The growth of its diameter continued in all directions. The lavas on the east side of Giorgios were still 8 meters from the Orthodox chapel on the 29th of March, but on the 12th of April they were in immediate contact with that small building.

One of the most interesting events during this period was the shifting of the original position of Giorgios toward the northwest.

---

* In exploring Giorgios, von Seebach found no high-temperature deposits of fumarolic salts, although they undoubtedly were there, for I had previously collected samples of them. He observed only acidic emissions (from the second type of fumarole according to the classification of Charles Sainte-Claire Deville[7]) which are characterized, as I have shown elsewhere, by just such a mixture of hydrochloric acid and hydrogen sulfide when the temperature is rather low or contact with the air is so restricted that the latter does not break down by combustion. Finally, while making a quick examination of the fumaroles that deposited sulfur at the southern base of Giorgios, he was able to observe vapors that contained no hydrochloric acid (Sainte-Claire Deville's third type of fumarole). Thus, there were at Santorini, as in other active volcanoes, regular variations in the distribution of fumaroles.

Observing Giorgios at night from the height of the cone of Nea Kameni, von Seebach clearly saw flames coming from the summit and flanks. He estimated their height at about 5 meters. In the cracks at the margins of the dome he also distinguished incandescent, semifluid lava.

At the end of the first week of April, the explosions grew in intensity and began to erupt cinders. The first took place on the 6th and 7th of April. At five thirty on the morning of the 8th, another dense column of cinders rose with the form of a bulbous, dark cloud on a pedestal. The people of Thera called a mass of cinders and gas with this form a κουνουπίδιον, or cauliflower. According to a measurement by Mr. von Seebach, the fume cloud produced by the eruption on the 8th of April rose 815 meters. From the 6th to the 11th of April, the same thing was observed seven times. Each time, it was preceded by prolonged underground rumbling like the sound of a railroad train going through a tunnel.

The fumes that caused the cauliflower clouds marked the different stages of the last eruption of Santorini. Though rare at the beginning of 1866, they became frequent in April and May, and after that time, accompanied every notable explosion until the end in September 1870. All observers who visited Santorini during the years the volcano was active were able to see eruption clouds of this kind composed of great mottled clouds surmounting a cylindrical stem of the same material. The length of the supporting column appeared to be proportional to the strength of the explosion. The size of the fume clouds varied with the number of explosions within a given pulse of eruptions. But an extraordinary event was witnessed by von Seebach at 6:40 on the evening of the 8th of April. A violent explosion produced a dense column of smoke that rose in the form of an enormous twisted rope to a height of 581 meters. Using a telescope of low magnification to make observations from the heights of Thera, he distinguished dark and light spiral bands within the column. It is evident that the darkness of this cloud must have been due to the cinders carried by the eruptive column. The gases of fumaroles or of the air moving around Giorgios often produced analogous phenomena; what von Seebach observed is remarkable in that it reveals exceptionally well the entrainment of ash in the swirling movement of the whirlwind motion.

According to notes in the journal of Mr. Hypert, the explosions of Giorgios increased in intensity during the second week of April but were not especially frightening. Like von Seebach, he was able to get near the volcano on two occasions, once on the 1st and a second time on the 6th of April, and managed to climb the rocks of the southern side. On the 11th, however, von Seebach attempted to return but was stopped by high-temperature fumaroles.

During the daylight hours of the 22nd of April, the explosions became unusually violent. They produced a louder and more prolonged noise. On several occasions, high cauliflower clouds rose to great heights. According to the observations of von Fritsch, Reiss, and Stübel, one of these jets reached a height of 1700 meters.

On the 24th of April, the scientists visited the volcano together with Mr. da Corogna. They noted that at the southern end of Giorgios high-temperature lavas formed a promontory that was advancing toward the south. Blocks were continually rolling down the front into the water, where they caused mild explosions and sent up large amounts of steam.

At the foot of Giorgios, close to the Orthodox church, there was a pond of salty water fed by springs coming out here and there between the blocks. At their source, the waters were acidic and clear with a temperature of 75 degrees. The basin communicated with the sea. Its bottom was covered with a light yellow deposit, possibly sulfur; the yellowish water was agitated and rich in iron compounds that were in the process of breaking down. Everywhere among the rocks but particularly at the sources of the springs, great numbers of gas bubbles were escaping. They consisted of a mixture of carbon dioxide and nitrogen. On the surface of Giorgios, the air in contact with the rocks and the fumarolic gases was very hot, as could be seen from its turbulent motion. On the eastern side of the dome, the rocks had no deposits, but in the rest of the space toward the north, between the base of Giorgios and the foot of the 1707 cone, there were numerous fumaroles with an unpleasant odor of hydrogen sulfide, as well as steam, deposits of sulfur, and traces of chlorides and sulfates. In places, the damp ground was so warm that it was difficult to stay long in one place. It was covered by a reddish yellow coat composed of half-decomposed salts of iron. The blocks on this side of Giorgios had a blackish color that contrasted with the lively colors of the neighboring ground.

From time to time the volcano rumbled and discharged columns of dense smoke and incandescent bombs.

An ascent of Giorgios was possible on the northern side, but only with difficulty. The surface formed a plateau which was gently inclined from north to south and chaotically littered with irregular blocks. Most of this debris was compact and angular. A mound or crude craterless cone stood at the extreme northern end of the lava field. Between blocks, hot air was given off along with a mixture of steam containing traces of hydrogen sulfide and sulfurous acid. The scientists who attempted to examine Giorgios at that time were forced to retreat by the unusually high-temperatures they encountered in the fumaroles.

At the western foot of the cone they observed that the original ground of Nea Kameni had four fissures oriented toward Aphroessa. These were the same fissures from which I had collected mixtures of combustible gases in March, but now they were not discharging gas. To the north of these crevices near the eastern extremity of the line of fumaroles between Giorgios and Aphroessa, the scientists discovered a small explosion crater at precisely the point where a few weeks earlier I had found fumaroles hot enough to melt zinc. The crater was a circular pit about 24 meters in diameter. The upper edge was at the level of the surrounding ground and had no rim of ejected debris, but all the older lavas on this side of Nea Kameni were covered with a thin layer of very fine grayish ash at least part of which came from the explosion that created the small crater. The inner walls of the pit revealed an almost vertical upper section between 1 and 2.5 meters thick and a lower, flat, circular part in the center of which was a crater 16 meters in diameter. The flat part of the floor was 3 to 4 meters wide, while the small central crater had walls inclined at 50 to 60 degrees. The bottom, which was at a depth of about 15 to 20 meters, was covered with soft, light-gray mud. Dried material of the same origin was plastered on the walls. All the area surrounding the edge of the crater was covered with brilliant yellow and red deposits. It gave off steam, hydrogen sulfide and sulfurous acid at temperatures ranging from 60 to 85 degrees. The walls of the pit gave off steam and hydrogen sulfide and were covered with a layer of reddish-yellow iron compounds that were in the process of decomposing. No volatile emissions came from the bottom.

The form of this small crater and the area over which it had ejected material reveal the manner in which it had been formed. It is likely that there were two successive violent explosions. What is most extraordinary is that after this happened, the temperature dropped markedly where the eruptions had taken place. At the same time the uplift of the southwestern part of Nea Kameni had stopped. The gases that had been discharged in such quantities from cracks in the ground were so reduced that they escaped the notice of the investigators.

Precisely when was the small pit-crater formed? It must have been during the period between the 26th of March, the day of my departure from

Santorini, and the 25th of April, when it was discovered by the German scientists. Moreover, I have reason to believe that it was on the 15th of April (not the 27th as I wrote earlier, owing to confusions arising from the difference between the Greek and Gregorian calendars). I have read in the journal of Mr. Hypert that on the 15th of April, Giorgios was peaceful and produced no explosions, but suddenly, at nine o'clock, a strong roar was heard and a thick cloud of white vapor was seen to rise between its base and that of the cone of Nea Kameni.

During the month of April, Aphroessa had developed much more than Giorgios. At the end of the month, lavas from the former had spread over a wider area than those of the latter. The height of Aphroessa had increased more rapidly than that of Giorgios and seemed destined to overtake it. Aphroessa was growing mainly toward the north, so that its summit, instead of being in the center, was about a third of the distance from its southern end. Its form was that of a very subdued cone sloping gently toward the north and more abruptly on its other sides, particularly the east and west, and surmounted by a small conical peak ending in slopes of 30 to 35 degrees. A long strip of lava of very unusual appearance extended between the summit and the northern end. It was not the usual jumbled mass of angular and compact blocks but a coherent trail of lava with parallel groves and curved ridges. The wrinkles were convex toward the north. The rock spread over this area was porous and scoriaceous. Each of the tongues was enclosed between two lateral ridges one or two meters high. At the points where several of the longitudinal depressions came together, the space between them was reduced to irregular, jagged points. The elongation of the tongues, the shape of their borders, and the conformity of their surface attest to the greater fluidity of the lava at these places. At the beginning of the month of May, these flows were still moving. When observed at close range, one at first had the impression that they were completely solidified, but after watching the same point for a short time one soon saw the shifting of the rocks relative to each other, and standing on one of the flows, particularly near its point of emergence toward the sea, the motion of the ground could be felt through one's feet. These observations, which I noted after my return to Santorini on the 2nd of May, had already been made by Reiss, von Fritsch, and Stübel, all of whom had managed to cross Aphroessa from one side to the other despite the high temperatures of the rocks and gases and, above all, the suffocating effects of the acidic gases that surrounded them. Hydrochloric acid was given off in strong concentrations, especially near the highest part of Aphroessa. In places, weak acidic fumaroles had temperatures of no more than 50 degrees, but more commonly they were above 100 degrees and so hot that hydrochloric acid seemed to predominate over water vapor. Very hot, dry fumaroles were common, and all around them the surfaces of the rocks had a very thin white coating. This insoluble material appeared to be a faint deposit of silica. On lifting these same blocks, I found that in nearly all cases their underside had a saline deposit, one or two millimeters thick, composed almost entirely of sodium chloride.

The main flow came from the central part of the dome, where the platy lava had the appearance of ground that had been furrowed by a giant plow. The crest was especially well developed on the western side. According to the account of the German scientists, they found on approaching the area that the blocks were cold or slightly warm to depths of three meters below the surface. But in the very center the heat was intense. In the spaces between blocks and in the bottoms of cracks the rocks were incandescent, even during daylight. At night, the whole central pinnacle became luminous.

The fumes coming from the central part of Aphroessa had, as in the beginning of the eruption, a remarkable reddish color due, no doubt, to iron chlorides, for all around in the interstices of the blocks there were fine crystals of hematite resulting from decomposition of these products.

The former site of Reka was no longer visible. In its place there rose a mound of blocky lava that by now had cooled completely. Reka and Aphroessa were closely joined with their bases welded solidly together and only a shallow linear depression marking the channel that had once separated them.

From the eastern side, Aphroessa had advanced until it reached Nea Kameni and bordered the former shore of that island for a length of three or four hundred meters without ever crossing it. The invasion of lavas was more marked at the southern end of the line of junction than at the opposite end. Between Aphroessa and Nea Kameni, there was a kind of ditch bounded on the western side by the talus of rubbly lava from Aphroessa, 15 to 20 meters high, while on the eastern side the blocky former shore of Nea Kameni did not rise more than four or five meters above the water. Here and there, the depression had blocks of debris that fell into it from the sides. The sea was visible there only in a few ponds at the northwestern end near the harbor of Saint Georges. At the other end, however, the former shore of Nea Kameni was still distinct. It is worth noting here that when the eruption began the point of Cape Phleva was not as pronounced as is shown on the English chart. At the beginning of May, this point was not yet covered by new lavas. At that time, I followed the former coast of Nea Kameni and could recognize everything I had seen the month before when the channel between Aphroessa and Nea Kameni had not been obstructed. When I brought this to the attention of my boatman, Nicolas Sfoscioti, he confirmed my observations. I never saw the southwestern end of Nea Kameni terminated in any way but by an obtuse cape formed by the two coast lines that converged almost at a right angle.

In the remaining ponds of water left along the former channel between Aphroessa and Nea Kameni at a place where there had formerly been copious discharges of combustible gas, I collected on the 4th of May a gas composed only of carbon dioxide and nitrogen. The water there had a temperature of 74 degrees; it was milky, blackened lead acetate paper,[8] and produced an acidic reaction. The seawater inside the harbor of Saint Georges had the same chemical characteristics but had a temperature of only 40 degrees.

The growth of Aphroessa continued toward the north-northwest until the middle of May. By the 17th of the month, the flows had extended beyond the harbor of Saint Georges and had formed a natural jetty part way across its opening, so that the entrance, far from being obstructed, now had a narrow outer opening toward the northwest. The people of Santorini greatly dreaded a renewed advance of the lavas on this side, for closure of the harbor would deprive them of one of the best shelters for ships anchoring in the bay, but, fortunately, the progress of the flows in this direction came to an end. From that time on, the harbor of Saint Georges has not changed its form.

None of the lavas of Aphroessa extends very far to the south, at least on the surface. Nevertheless, one could see on the surface of the newly formed mound currents directed toward the south and with the same form as those that flowed in the other direction. Extension in this direction had also been preceded by a continual emergence of new reefs and by elevated temperatures of the surrounding seawater. On the 12th of May about five meters from the end of the middle flow, the thermometer read 70 degrees. In the small cove on the southern coast at the junction of Aphroessa and Nea Kameni, it showed only 40 to 50 degrees.

The encroachment of rocky debris from Aphroessa advanced more slowly toward the east, owing to the obstacle formed by the older ground of Nea Kameni, and was stopped completely after the middle of the month by the talus of the cape protruding from the northern side of the entrance to the harbor of Saint Georges. It was expected that more of this material would thereafter escape in another direction. On the 19th, new islets were seen rising between Aphroessa and Palaea Kameni. The German scientists gave them the name *Islands of May* and did not hesitate to consider them the emergent parts of submarine flows coming from Aphroessa. I had left Santorini on the 17th of May, so my description of the appearance of these curious islets is based on the account of these scientists as well as the spoken and written testimony of witnesses passed on to me by Dr. de Cigalla and Mr. Hypert.

The rocks making up these islets barely rose above sea level; they were black and glassy, and when examined closely were seen to be full of small cracks that split them into innumerable little blocks. The lava of one of the islands was very compact, while that of the other had a few vesicles. Nowhere was there any sign of scoriaceous surfaces. The rocks were cool and only in a few spaces between blocks could one feel a slight warmth.

On the surface of the new islands there was no worn debris from the seafloor but only bits of marine plants and animals. This organic material was rare and appeared to have been dropped on the surface only a few days before.*

The emergence of the islets usually went on imperceptibly, but on certain days it was rapid enough to be visible. Their gentle slopes were prolonged under the water toward the north in a way that made a double crest the southern extremities of which were visible only beneath the water. The southern side culminated in a peak. We followed their growth during the second half of May and all of June. Each day they grew a little toward the north, while their southern side was eroded by waves and cut back by failure of the slopes. At the same time, they were lifted slightly but so slowly that it was difficult to judge the magnitude of the movement, because the blocks that were constantly coming down the slope provided no point of reference.

The two crests forming the islets that appeared on the 19th of May did not form a continuous surface. They produced new reefs that were elongated in the direction of the chain of emergent rocks. At the end of May, we could count four of them; on the first of June they were joined by a fifth and, on the 4th of June, by three more. Four of these were the most easterly points along the double ridge crest; the four others corresponded to the most elevated points on the western crest. The total length of these two rows of islets was 80 meters.

---

* Messrs. Reiss and Stübel, to whom I am indebted for this information, did not identify positively the species to which these remains belonged, nor their state of preservation, position on the surface, or state of development.

The rock making up the part of the two crests that was in the deepest water was much less cracked than the emergent part. This has surprised most observers, but, having made a careful study of the matter, I can state this quite positively. In May and June of 1866, when we examined the surface that was still emergent but covered by a thick layer of water, we saw that this surface was almost smooth. Dr. de Cigalla even concluded from this that each of the islets could be described as a bulge of lava with a molten interior that broke up about a meter from the surface of the water. What is certain is that the reefs that appeared on the 5th of June had been seen by several persons at the time of their emergence, and although they were completely coherent at that time, they were crossed by innumerable fractures two hours later.

The Isles of May were not produced by a simple uplift of the seafloor, for if they were one would have seen on their surface or in the spaces between the blocks remains of the original material from the bottom of the bay. One would have found fragments of pumice, the marine organisms that normally encrust these fragments, and, above all, some trace of the volcanic sand mixed with bits of pumice that cover the bottom in the interior of the bay.

Another explanation was offered by Dr. de Cigalla. In his opinion, the lavas came out in the form of a bulb and remained fluid inside a solid envelope and would have broken out just below the surface from the fractures produced in this way. But such an explanation is contradicted by all the observed facts. The form of the Isles of May, the regularity of the fracture pattern of their rocks, and the absence of any deposits filling the cracks are all inconsistent with the ideas of the indefatigable scientist of Santorini. We should also note that a discharge of molten lava through the opening of fractures very close to the surface of the sea would certainly have caused considerable heating of the water near the fractures and would probably have released gas and steam.

Messrs. Reiss and Stübel have proposed, as we have already noted, another explanation that seems much more plausible. They considered the Isles of May to be the extremities of a great flow of lava. "If one considers," they say, "the form of the seafloor and the apparent motion of the subaerial part of Aphroessa, one cannot doubt that the Isles of May are part of a great flow of lava coming from Aphroessa, for during the month of May the movement of material making up Aphroessa was mainly toward the northwest and west; the emergent parts have already reached the western point of Nea Kameni. The advance of these lavas was arrested by an elongated submarine indentation extending from Palaea Kameni (the part formed in 726) toward the western point of Nea Kameni. They must have had to press against this promontory until the crust forming the lateral wall of the flow finally gave way. The break took place toward the west and southwest, in an area near the deepest part of the channel. The lava then extended over the sea floor. It followed the declivity of the surface that turned it back toward the southeast. Little by little it has filled the depression of more than 100 fathoms (183 meters) between Palaea Kameni and Micra Kameni, so that the highest points of the extrusion appeared at the surface as small islands. Soundings during the month of March had already shown that the bottom of the channel was being filled since the beginning of the eruption, and one should not be surprised to see submarine plants and animals on the surface of the lava. This thick crust enveloped more or less completely the fluid part of the lava, so that it gave off only enough heat to cause a slight increase of the temperature in that part of the gulf where the water is circulated by currents."

This theory is based on two assumptions; first, that the Isles of May were formed by a slow discharge of lava, and, second, that these lavas came from a vent corresponding to the center of Aphroessa.

The nature of the material forming the Isles of May is enough to establish the validity of the first assumption and rule out the possibility that the seafloor was raised. The second, however, is purely hypothetical. The flow of lava from the emergent part of Aphroessa was mainly toward the north-northwest, not toward the northwest, the direction of the Isles of May. Between these islands and the closest lavas coming from Aphroessa, soundings indicate no continuous, well-defined protuberance that could be the result of a submarine flow. Islands such as Reka, that on several occasions were formed in the area around Aphroessa by the same mechanism proposed by Reiss and Stübel, had a character quite different from that of the Isles of May. They appeared in the form of isolated, scoriaceous reefs, and on each occasion were composed of masses of irregular blocks rather than by a distinctive kind of glassy material.

It is not the superficial lava flows of Aphroessa that could have produced the Isles of May; deep flows issuing from the central vent would have penetrated the mass of solid lava encasing them and piled up around the vent. Is it not more simple to suppose that the lava came from a submarine vent located in the space between Aphroessa and Palaea Kameni? In all volcanoes, the opening of the ground at the moment of eruption takes place on several fractures, usually more or less linear, along which activity breaks out. Formation of this kind of fissure was demonstrated in 1866 by the remarkable crevices between Giorgios and Aphroessa we have already mentioned. In both of the eruptions I have witnessed, that of Vesuvius in 1861 and Etna in 1865, I saw the main fissure extending from each side of the active vent. Moreover, I have observed that the same was true of several former eruptions on Mount Etna and in the Azores where the traces were even more clearly visible.

Prolongation of fissures of this kind seems to have been common; even at Santorini it is well shown by the crevices east of Giorgios where they are marked by the cone of Nea Kameni and, especially, where they cross Micra Kameni. It is quite reasonable, therefore, to conclude that the same thing happened west of Aphroessa, and one can therefore say that the lavas it would have discharged were responsible for partly filling the channel between Aphroessa and Palaea Kameni as well as the Isles of May. Slow spreading of underwater lava under these conditions seems to have produced precisely the same phenomena that characterize the Isles of May.

A remarkable occurrence supports this interpretation. The lavas making up these small islands contain many nodules of anorthite-bearing rock torn from the older lavas underlying the bay. Inclusions of this kind are rare in the lavas of Giorgios and Aphroessa, so their unusual character makes this lava very notable. They seem to come from primitive igneous material discharged through a fracture somewhere in the subsurface and not from the still molten part inside the long carapace of a submarine flow coming from Aphroessa.

While Aphroessa was extending itself, important changes were taking place in the channel between it and Palaea Kameni. Giorgios also underwent notable modifications. When I returned to Santorini at the beginning of May after an absence of five weeks, I noticed that all its dimensions had increased. Its form was that of a truncated cone about 60 meters high and a base with a diameter of 100 meters. From its flat top rose a small spire composed of irregular scoriaceous blocks, some of which had volumes of several cubic meters and outer surfaces that were strongly altered by steam and acidic emanations. This feature dominated a plateau of about 5 meters that was surrounded on all sides by rather gentle slopes of 10 to 12 degrees. A circular moat about 2 meters wide extended around all sides but was particularly conspicuous on the western side. Detached blocks covered most of the eastern and southern sides. According to the account of Reiss and Stübel, this moat did not exist on the 29th of April. It could have developed in a single day on the 30th or during the following night after explosions of great violence.* The existing central spine continued to grow and was almost isolated from the rest of the cone by the depression mentioned above.

Two days later, when I had the occasion to observe this curious situation, I had no doubt that the depression around the central body was the result of strong explosions. This was all the more likely, because powerful jets of gas were coming from it almost constantly, and from time to time explosions were accompanied by dense cauliflower clouds of ash and steam. I conclude, therefore, that the summit of the cone was a large crater, the interior of which has been filled by voluminous lavas. Events were not long in verifying this point of view, for at the time of my departure from Santorini on the 17th of May, the explosions had already considerably enlarged the circular, moat-like depression, and a month later more and more of the central mass of lava was uncovered and thrown out by the action of the gas, leaving in its place a large crater-like depression. In the course of the eruptions, however, lavas accumulated on several occasions inside the crater, particularly during the first year, thereby restoring its original form, but each time they were quickly thrown out by violent explosions leaving no visible trace.

Around the beginning of May, I realized that the summit area of Giorgios had moved during the preceding month. On the 2nd of May, the central spine was a short distance southwest of the position it had formerly occupied at the highest point of the volcano. The displacement amounted to about 50 meters. The focus of most of the activity of Giorgios had also moved. Although explosions occurred all around the circular depression, they were concentrated mainly in its western part.

Minor explosions had taken place in various places, but the strongest ones were accompanied by violent discharges of gas and steam all along the annular vent at the upper plateau of the cone and even from the spaces in the pile of blocks at the summit.

---

* Reiss and Stübel (*Geschichte*, p. 139) state that they observed these events, and one must trust their assertion. But when it comes to the reasons they call on to support their account they offer nothing of value. The blocks ejected by the explosions up to the first of May had been very abundant, and cinders had on several occasions covered the terraces of Thira, 4 kilometers away, with a layer a centimeter or so thick. In short, the amount of material ejected by the volcano was quite sufficient to account for a shallow crater 100 meters in diameter.

The argument that the western flank of the cone had a slope of 50 degrees is also of no importance, for at the end of the eruption after formation of the crater that remains today, slopes of 50 degrees are still seen on the western flank of Giorgios.

At times, the main mass of rising fumes had the effect of drawing the steam and gases given off by fumaroles into a single column. When this happened, the fumes rose as a giant cauliflower-shaped cloud.

Explosions took place, on average, every four or five minutes. They differed in frequency as well as in intensity. In general, they were much more numerous and more violent than those observed during the month of March. The ejected bombs were incandescent, and at night they rose in luminous sheaves that were more beautiful than a display of fireworks. They were often so abundant that they showered the area around Giorgios like hailstones during a storm. On certain days the explosions were relatively rare. On the 12th of March, for example, I was able to take a position in the narrow passage between Giorgios and the cone of Nea Kameni where I could collect gases and condensed steam from the fumaroles. This operation was not entirely without danger, however. On one occasion, one of my instruments was broken by falling rocks, and I was forced to seek shelter behind a large block that happened to be nearby. My companion on this excursion, Mr. da Corogna, miraculously escaped the rain of ejecta that was coming down all around us.

In many places the lavas of Giorgios were still very hot. The central part of the flow was incandescent at night, and the interstices of the part moving toward the south were also very luminous. At this time, the height of Giorgios was about 70 meters.

During my second visit to Santorini, between the 1st and 17th of May, 1866, I saw no flames above either Aphroessa or Giorgios nor at any place on the older ground of Nea Kameni. All the combustible gas seemed to have disappeared, but this disappearance was less complete than I had thought, for Reiss and Stübel collected mixtures of gases with small amounts of hydrogen from fumaroles near the foot of the 1707 cone.

The flow of lava from Giorgios during April and May of 1866 was mainly toward the south. In the middle of May its front was about 300 meters from the former southern shore of Nea Kameni. At the end of that same month it extended almost 500 meters on that side and had a width of 150 meters. The surrounding sea had temperatures of 50 to 80 degrees. At a distance of 30 meters the temperature was only about 40 degrees and remained rather warm out to a distance of several hundreds of meters in the direction of the currents. Minute amounts of sulfuric acid and iron compounds dissolved in the warm water gave it a greenish or reddish color, depending on the degree to which the iron was oxidized. In places, a deposit of limonite from decomposition of these salts floated on the surface, while the seawater had a very marked acidic effect on reactive paper.

In May of 1866, I made several soundings near the leading edges of the flows from Giorgios and along the western coast of Aphroessa in the direction of Palaea Kameni. These measurements showed that in both cases the lavas ended with very abrupt slopes. For a distance of 20 meters the fronts had slopes of 45 degrees then steepened so rapidly that they were essentially vertical. The lava forming the underwater front was composed of jagged blocks between which the sounding weight often became caught and could not be recovered.

The original ground surface of Nea Kameni underwent constant changes during the month of April and the first half of May.

By May there were many more fissures than the four fissures that crossed the southwestern part of the island at the end of March. They had the same orientation and sharply pointed forms. They were also very narrow and had vertical dimensions of 15 to 20 meters. The block that had previously been my reference point was still visible, but an avalanche of debris had partly buried it, making it impossible for me to use it without a leveling instrument. I was able to note, however, that the upheaval of this part of the island had ceased and may even have given place to a slight amount of subsidence. Currents of saline water with temperatures between 70 and 75 degrees circulated throughout the small channels. Gases were still given off in these places but in much smaller amounts, and they no longer burned, even after eliminating the great quantities of carbon dioxide that were in them.

The surface that remained uncovered between Giorgios and Aphroessa was now no more than 120 meters wide. It was furrowed by so many fractures that crossing it was very difficult.

Nearby could be seen the small explosion crater which, as already mentioned, was formed on the 15th of April.

The pier on the Micra Kameni side of Nea Kameni was still depressed by 30 centimeters. From one day till the next it seemed to rise and fall but long-term movements were not apparent. The apparent variations actually resulted from changes of the water level due to tides and variations of wind-driven currents.

At the southeast point of Nea Kameni, ponds of salty water fed by springs of water at 73 degrees were depositing abundant ferruginous sediments and releasing a little carbon dioxide.

The summit of the cone of 1707 had undergone notable changes. By May, the fissure, which in March had extended radially from the center out toward the eastern edge, now crossed all the way from one side to the other, and the circumferential fracture on the southern side of the cone had doubled its width.

As for the harbor of Saint George on the western shore of Nea Kameni, by the end of May 1866 it had already taken on the form it has today. A few days before my departure from Santorini, I made numerous measurements of temperature in the harbor and obtained a wide range of readings, most of which were above 30 degrees. The maximum I found was 80 degrees in the southeastern corner of the harbor near the juncture between the original ground of Nea Kameni and the lavas of Aphroessa. At this place the water was agitated and had a whitish color due to hydrogen sulfide that decomposed in the air and then dissolved in the water. The fetid odor was very strong, and lead acetate paper was quickly blackened, both by the air and in the seawater.

Before closing my account of the series of observations I was able to make at Santorini in 1866, I must add a few notes on the material ejected by the volcano during this period of activity. Volcanic bombs, both intact and broken, as well as fragments of lava, were strewn all over the cones of Nea Kameni and Micra Kameni, but they were most plentiful on the southeastern side of Nea Kameni. They completely covered the ground in that area. The road across this part of the island is so cluttered with them that it is now difficult to get through. The bombs range in size from that of walnuts up to several cubic meters. They have two distinct parts: first, a black crust speckled with a few crystals of feldspar and separated into thin concentric shells, and, second, a vesiculated core. The thickness of the crust increases with the size of the bomb, but in the smallest ones it is rarely less than seven or eight millimeters thick, while in the very large ones it can be as much as 15 centimeters. The thickness of the layers making up the crust is also varied; most are very thin, commonly not more than half a millimeter, but those closer to the interior may have thicknesses of as much as 4 centimeters. The outer crust is crossed by irregular cracks that are oriented more or less perpendicular to the surface. Some of these are simple cracks into which one can insert a piece of paper, but many have pronounced offsets. In general, the larger the bomb, the more the cracks penetrate directly through to the vesiculated core. It is not uncommon to see large bombs with cracks up to

15 centimeters wide. In small bombs the displacement of pieces of the crust averages about a centimeter, but one finds bombs with a diameter of less than 10 centimeters in which the crust is displaced as much as 3 centimeters. The cracks in large bombs commonly extend to depths of several decimeters, and in smaller ones they cross in the center of the core. This causes them to be so fragile that they fall to pieces with the slightest touch of a hammer. Of the numerous samples I brought back to France, very few arrived intact, despite the great care taken in packing them. On the older ground surface of Nea Kameni one finds a certain number of bombs that are still intact, even though they have deep cracks running through them. From that one must conclude that they were still more or less molten when they landed or at least that they were so hot that they were not brittle enough to crack; otherwise they could not have failed to break on impact. They were not very fluid, however, because most of the bombs are spherical or spindle shaped. In falling they were not flattened into thin, cake-like forms, such as those thrown out of the craters on the flanks of Etna in 1865 or those that come from most volcanoes that erupt basic lavas.

Further evidence that they solidified after falling comes from the fact that in almost all cases in which a bomb is found broken into pieces the fragments still remain almost in contact with one another. If they had been broken by a strong impact, such as that caused by striking the ground, they would be scattered over a wider area.

The form of these bombs is explained by the rapid solidification of their surface layer and the greater thermal contraction that occurs in that zone than in the center.

The ash that had been ejected on many occasions from Giorgios had accumulated on the terraces near Thira; all of it has the same characteristics. It is very fine-grained, with a light gray color and a gritty feel. It is strongly hygroscopic, and on absorbing moisture from the air becomes darker and agglutinated. When washed in distilled water, it yields a clear liquid that produces an acidic reaction on litmus paper but has no apparent effect on lead acetate paper. Silver nitrate and barium nitrate reveal the presence of substantial amounts of chlorides and sulfates. Microscopic examination of the residue of evaporation reveals sodium chloride and sodium sulfate. In addition, tests with standard reagents indicate traces of compounds of magnesium, calcium, and potassium.

A microscopic examination showed that the ash is nothing but the lava of the latest eruption reduced to a fragmented state. It is not at all pumiceous, as were the products of earlier eruptions of Santorini, and this is particularly true of the thin stratified layers resting on pumice in the upper parts of the cliffs of Thera.

Mr. da Coronga has carefully studied the effects of the ash and fumes on the plants and animals of Santorini. The main finding of his study was that the ash ejected during dry weather was essentially harmless, but that it had very serious effects during periods of rain or when it was accompanied by great quantities of saturated steam. One of the more fatal days for the vegetation of Santorini during 1866 was the 3rd of June. The vines had the appearance of having been burned everywhere that the fumes from Giorgios had been blown over them by the northwest wind. The quantity of ash that was ejected that day was not great, but a fine rain that was falling greatly aggravated its harmful effects. It could also be that part of this disaster can be attributed to the direct or indirect action of the great quantities of hydrogen sulfide that the volcano gave off that day.

The month of June was also marked by the cessation of all eruptive activity at Aphroessa. The extinction of this vent came about gradually. On the 15th of June all incandescent activity at the summit had ceased, although a few reddish fumes continued to escape. On the morning of the 17th, these fumes seemed to be notably denser. During a light rain that lasted several hours, they became black and looked like a storm cloud hanging over the bay. This was the last notable activity seen at Aphroessa. After that, all it gave off were a few light emissions of steam, indicating that heat still remained in the interior of the lavas.

During this period, the explosions from Giorgios continued with the same frequency and strength as before. The central mound disappeared on the 20th of June. From Thira the summit of the cone appeared to be obliquely truncated and surrounded on its upper flanks by a small crest indented on its southern side. At the end of the month the mound had reappeared, and, although its shape and dimensions changed, it persisted during the month of July. The ejecta were so abundant that they completely covered the flanks of Giorgios, and the blocks that made up most of its mass were buried under this mantle of pulverized ash. It lost its rocky appearance, and the irregularities of the surface disappeared, giving the cone of Giorgios the same form as that of 1707.

We saw in an earlier section that the lavas from Giorgios had extended mainly toward the south and that they formed a long triangular promontory extending to the base of the cone. During the following month, this promontory grew slowly, and the lavas from Giorgios took a new direction. At the beginning of June, an island made up of irregular blocks and elongated in a north-south direction appeared in the interval between the coast south of Aphroessa and the promontory just mentioned. The first blocks to reach the surface were cold, but the temperature soon rose in the middle of the island. Very shortly, this island merged with Aphroessa on one side and with the western promontory of Giorgios on the other. After that, it became a ridge that continued to grow rapidly. By the end of June, it formed a rocky hill three hundred meters long, about one hundred meters wide, and thirty to forty meters above sea level. During the month of July, this mass did not increase perceptibly, but it remained incandescent, particularly in the upper part, and during all this time it had many fumaroles. Mr. Hypert, to whom I am indebted for this information, gave this feature the name *Ascension Ridge*, because the first rocks to appear rose from the sea on that feast day (by the old calendar still used in Greece).

During the month of July, there were very violent explosions many of which were accompanied by ash. On the 20th of July, a particularly large amount of ash fell on the island of Thera and dried out the vines and figs around the locality known as Kamari. The ash reached the island of Anafi, where it killed large numbers of bees.

The same phenomena were repeated during the day of the 21st. The vineyards of Messaria on the island of Thera were particularly hard hit. On that same day, the emissions of sulfur gases caused much discomfort to the inhabitants of the island.

The same fumes could be smelled on the 25th at the same time that ash was falling. Ash fell in even greater amounts on Thera during the night of 29 - 30 July and during the following day, but, since it did not rain, the grapes remained intact. The leaves of fig trees, however, were more sensitive and were desiccated wherever the fumes reached them. It is noteworthy that the vines of Santorini, which for two years had been attacked by *Dium tuckeri*, continued to be ravaged by the infestation, and the emanations from Giorgios seemed to have had no effect on the spread of this malady.

The month of August was marked by the two strong explosions that we have already noted. The first, which took place at eight o'clock on the evening of the 3rd of August, covered Micra Kameni and Nea Kameni with incandescent ejecta. A luminous block of very large size fell near the northern point of the island. The eruptive column soon fell back and enveloped the Kamenis, completely hiding them from view. Meanwhile, the eruption continued to make a terrifying roar. Ejected blocks were no longer visible, but the sound of them falling was like a fusillade. The explosions continued for more than ten minutes.

A second violent eruption took place at ten thirty on the morning of the 19th of August. Its ejecta fell over a radius of three kilometers around the volcano.

At the beginning of August, flames had returned to the summit of Giorgios. They were so strong that their undulating bluish light could be seen quite clearly from Thira.

At this time a flow of lava coming from a low place in the crater rim of Giorgios seemed like a ribbon of fire at night. The center of the crater was still occupied by a pile of blocks, and explosions were coming from vents around its base. Mr. Hypert compared the summit at this time to a sieve the central part of which was plugged. Both Mr. Hypert and Dr. de Cigalla report that the crater underwent curious modifications during August, September, and October of 1866. On the 18th of August, a violent explosion reduced the cone to fragments and threw out the major part of the mass of blocks that had occupied the center of the crater. The central pinnacle quickly returned, however. During the first two weeks of September it was tens of meters high and, as in May, was composed of irregular blocks of lava. On the night of 20 - 21 September, the major part of this plug was carried away by a discharge of gas. During the 22nd, new explosions finally emptied the crater, but the next day it began to fill again, and soon one could see in its center a mass similar to but lower than the one that had just been destroyed. On the 2nd of October, this new mound was largely destroyed, only to form again almost immediately. This sort of behavior is not uncommon in erupting volcanoes; it is observed when the vent from which lavas are discharged coincides with the vent of volatiles, as was the case at Santorini in 1866.

Another event worth noting is the rains that were produced at Santorini during September 1866. The inhabitants have no doubt that they were caused by the volcano, for they were quite abnormal. This was the only time in ten years that rain had fallen on Thera during September, and it had never been significant at that time of year. But at seven in the morning of the 14th of September, there was a heavy rain. The rain water was salty and acidic. By the next morning, the small amount of vegetation that still remained on Thera had been completely wiped out. Leaves seemed to be burned and were reduced to powder by the slightest touch. Another short but very heavy rain then followed during the day of the 15th. The rainwater had no taste. Several ash falls also occurred during September and October; samples that were collected at that time consisted of powdery, light-gray material strongly impregnated with sodium chloride.

Although flames were still visible at night at the summit of Giorgios, there were no emissions of combustible gases elsewhere around the volcano. Professor Christomanos, who was at Santorini during the first fifteen days of September, collected gas samples, which he found to be mixtures of only carbon dioxide and nitrogen with a little hydrogen sulfide. He also observed that Aphroessa was completely cold.

For the last three months of 1866, the eruption continued without any notable incidents. The explosions, ejections of pyroclastic material, flows of lava, and coloration of the sea had the same intensity as they had during the summer. Until the month of October, the lavas coming from Giorgios took various directions toward the southwest and southeast. The southeastern point of Nea Kameni, which had been partly submerged by subsidence, was never covered by lava. It was not until the 21st of October that the flows began to go toward the east. These new flows were unusual in that, instead of taking a straight course, they followed a curved path. Emerging from the southern side of the cone and moving first toward the south, they then followed the southeastern base of Giorgios, with some moving toward the southeast while others went east. Still others made an even greater curve and were directed northward toward the southern tip of Micra Kameni.

Ash falls were less frequent than in the previous months. The odor of hydrogen sulfide reached Thera from time to time, but it was weaker and caused the inhabitants no discomfort.

I find among the notes of observers that I draw on for information on this period remarks that an ash fall on the 15th of November was followed by a heavy rain and caused concern among the farmers for their grape crop of the coming year. Another heavy rain on the 19th of November did not coincide with an ash fall.

I spent four weeks on Santorini, from the 23rd of February to the 21st of March, 1867, and carefully followed the evolution of the volcano. The configuration of the newly formed dome had undergone substantial changes.

The cone of Giorgios had greatly increased in width and height. Its highest point was now well above that of the cone of 1707. A measurement I made with the aid of a sextant on the 5th of March gave a height of 108 meters. Despite the frequency of explosions, the dimensions of the cone changed very little during my second stay at Santorini. The culminating peak was on the north side of the crater rim. The opposite side, which was about ten meters lower, was crossed by a large flow of lava descending toward the south. Lava occupied the entire center of the crater without rising above the surrounding rim. The explosions were coming mainly from a cross-shaped cavity between the northern rim and the central mass of lava.

The outer flanks of the cone had a slope of 35 degrees and were entirely covered with cinders. The western side was more irregular and somewhat steeper, in places having a slope angle of as much as 40 to 55 degrees. Great blocks of lava showed through the surface layer of loose material. The northern flank had an intermediate slope and its base encroached on that of the cone of 1707. The lavas of Giorgios had covered all the southwestern part of the original surface of Nea Kameni and even a small part of Aphroessa. Those that had spread toward the south and southwest formed an elongated mass of considerable thickness. Its extremities reached 40 to 50 meters in thickness and ended in talus slopes of 30 to 35 degrees. Although it appeared motionless, most of this flow was actually moving. From time to time, blocks broke away from the slopes at its front and came crashing down to the base. High-temperature lava that was still able to move could be seen moving visibly at five different places. On the southwest, one of these flows could be distinguished at a distance by a line of fumaroles; to the southeast, a very active flow ended at the sea where it produced great amounts of steam and gas. A third, which has already been mentioned, went around the base of the cone of Giorgios and moved toward Micra Kameni. Each day, the space between the flow and the southern base of the cone of Micra Kameni steadily diminished. On the 25th of February, when we had easily passed through it in a boat, it was about 30 meters wide and 15 to 20 meters deep. On the 21st of March, its width had decreased to 6.5 meters, and its depth did not exceed 3 meters. The blocks that were constantly rolling down the front of the flow made passage through the remaining space very dangerous. At the time I left Santorini, I expected that the channel would soon be closed completely.

For the most part, the explosions from Giorgios had been very violent. On the 1st of March, from six in the morning until mid-day, five explosions ejected large numbers of blocks and a rain of ash from a cauliflower-shaped eruption column (κουνουπίδιον). Four other explosions during this period were smaller but still threw out rocks and ash, and eight explosions produced only steam.

On the 5th of March, during the same period of time, I observed a single explosion with a cauliflower eruption column, three with moderately strong ejections of rock and ash, and six with only steam.

From the 23rd of February to the 21st of March, no significant amount of ash reached the island of Thera. The heights of the cauliflower fume clouds varied between 1500 and 2000 meters. The average height to which the clusters of incandescent bombs rose was 150 meters above the summit of the cone. In one of the explosions on the 1st of March, blocks ejected from the volcano fell into the sea east of Micra Kameni, a distance of 2 kilometers from Giorgios.

At night, flames were still visible but only at the summit of Giorgios. They could be seen only at the time of an explosion, when they accompanied ejections of gas, blocks, and ash. They had the form of jets 5 or 6 meters long, the largest ones being those that accompanied the strongest explosions. They had the same yellowish tint that characterizes emissions from sodium. During eruptions with a succession of explosions, the flames continued to be emitted but rose and fell in size. They were undulating or flickering only during weak explosions.

The pier of Nea Kameni continued to subside. Since the 15th of May 1866, the ground near the middle of the pier had gone down a total of 1.3 meters. The southeastern edge of the base of the cone on Micra Kameni subsided 15 centimeters. In March of 1867, the small masonry column that had served as a post for anchoring boats was only five centimeters out of the water. The two chapels built on the southeastern side of Nea Kameni were still visible, but the southernmost one was half buried by the eastern talus of the cone of Giorgios. The pond of water that spread there in 1866 was considerably enlarged but still separated from the canal between Nea Kameni and Micra Kameni by some ruined houses and a small remnant of the surrounding ground.

The cone of Giorgios also expanded toward the north where the small explosion crater was formed in April of 1866. It almost completely hid the fissures that had formerly extended toward Aphroessa at the southwestern point of Nea Kameni. The northern part of the passage way between Aphroessa and Nea Kameni could still be seen, but it was largely obstructed by boulders.

Gas could be seen escaping in several places, but it was mainly only carbon dioxide and nitrogen. This was the dominant composition of all the samples I collected on the 6th of March.

One of the emissions of mixed gases given off near Giorgios, however, still contained free hydrogen. This was the one near the source of the lava flows from Giorgios that had been identified earlier by Dr. Nicolas de Cigalla. The mixture was notable for the large amount of free oxygen it contained. I am tempted to attribute this to dissociation of water in contact with incandescent lavas, and yet I recognize that other flows that are equally hot enter the sea without producing any such gases.[9]

I explored the entire surface area of Aphroessa as well as that of the rocks that Mr. Hypert designated by the name *Ascension*. In the interstices between blocks in the upper parts of these two masses I found a considerable amount of saline material composed mainly of sodium chloride and having an alkaline effect on reactive paper.

I also visited the Isles of May and collected a great quantity of crystalline sublimates. These islands were now only four in number; three form a line to the east of the fourth one. The two small ridge crests of which they are the emergent points can be seen at a shallow depth beneath the surface of the water.

From the 21st of March until the 19th of April, Mr. Janssen carried out very interesting physical measurements on the volcano. With the aid of a clinometer and compass, he determined the variations of the magnetic field around Giorgios and found that the greatest effect coincided with the direction of the fissure on which Giorgios and Aphroessa are the two main features.[10] In addition, he carried out spectral analyses of the flames coming from the summit of Giorgios. This was the only method capable of yielding information on the nature of combustible gases at the instant they come from the volcano, for at that time the high temperatures and frequency of falling ejecta made it impossible to climb Giorgios. Chemical analyses of the gases released at the summit of the cone were therefore impossible. Mr. Janssen found that free hydrogen was the combustible component of the mixture of gases. His spectral analysis also revealed the presence in the flames of sodium, chlorine, and copper.[11] The presence of the latter among the products of eruptions had not previously been suspected.

During Mr. Janssen's stay at Santorini, the explosions were very frequent and quite intense. They came from numerous openings irregularly distributed around the interior of the crater. Several times each day, the mass of lava that occupied this space was lifted, torn apart by the explosions, and scattered in all directions. It constantly renewed itself, changing shape and size with each explosion.

The strongest of these explosions occurred around three in the morning on the 2nd of April. So much ejecta was thrown out that terror spread among the crews of boats anchored in the coves of Nea Kameni. Several moved to new positions.

The many letters I received from Mr. Hypert during the remainder of 1867 showed that during that period of eruptions the volcano followed a regular pattern, even though the explosions varied in frequency and average intensity. Among the many details provided by these letters I shall comment on only the following.

During the summer and autumn of 1867, flames were rarely seen at the summit of Giorgios; they appeared only in the strongest explosions. Among the largest eruptions, Mr. Hypert cites in particular those of the 18th of May and the 6th of June. After the first of these, the crater was enlarged by about 50 meters on its southern side, and the highest point of the rim was about 10 meters lower, so that, for a while, Giorgios was not as high as the cone of 1707. It was not until the month of July that the cone again regained its former height, and, thereafter, it remained higher than the older cone of Nea Kameni.

In the same month, the volcano seems to have reached its maximum level of activity, and while strong explosions were common during August, September, and October, none of these equaled in violence those of the 2nd of April, the 18th of May, and the 6th of June.

Throughout almost all of 1869, Mr. Hypert maintained records of the weather and strong explosions of the volcano. In comparing the two, he concluded that there was no relationship between the energy of eruptions and barometric pressure. He told me in his letters that although he at first expected such a relation, quite often he was disconcerted to find that his expectations were not fulfilled. "All that I was able to find was that the fume cloud given off by the volcano was denser when there was a westerly wind." This second observation is not surprising, since the air coming from that direction has more moisture and is therefore closer to saturation. Any additional contribution of water vapor, such as that given off by the volcano, must cause more condensation than under drier conditions.

Mr. Hypert also thought he could detect a greater frequency of explosions during the night than the day, but the data on which this is based are very incomplete, and we can only consider them suggestive. The cone of Giorgios reached its highest elevation for 1867 during the month of September; it began to decline at the beginning of October as a result of continuous explosions.*

---

* When I departed from Santorini in March, I left with Mr. Hypert a drawing of the volcano made from the terrace of the Lazarist monastery. On several occasions, he sent me this same drawing, modified according to the current state of the volcano. The upper edge of the cliff of

At the same time, the source of the explosions moved from the northern edge of the crater toward its southern side. At the end of September, all the explosions came from the southern part of the crater. We also noted at that time that ferruginous deposits began to appear in great quantities on the upper eastern flank of Giorgios. This brilliantly colored coating frequently disappeared under the ash that fell on it, but each time it quickly spread again over the layer of pulverized material that had briefly covered it.

The small crater now visible southwest of the large one began to form in 1867. Ejections from this point were first seen on the 26th of April. The cavity from which they came was very inconspicuous and did not become pronounced until a strong explosion occurred on the 6th of June. Since then, the crater was never filled again. In the autumn of 1867, it gave off white vapor, and incandescence could often be seen there at night. During the afternoon of the 29th of November, a cauliflower eruption column rose from the crater.

In summary, by the end of 1867, the summit of Giorgios already had nearly the same form it has today.

The volcano does not seem to have dropped great amounts of ash on Thera in 1867, nor did the vineyards suffer extensive damage, even though the fumes were heavily charged with hydrogen sulfide, sulfurous and hydrochloric acid. Rain that fell on the night of the first and second of May was found to be acidic and salty; it caused damage to the vegetation in the districts of Akrotiri and Megalokhorion in the southern part of Thera. On the night of 12 - 13 May, a damp fog mixed with fumes from the volcano had a very bad effect on vines in the area between Merovigli and Cape Kolombos in the northern part of Thera.

Only twice, on the 15th and 19th of November, was the odor of hydro-gen sulfide strong enough to remind the people of Thera of the stifling fumes they experienced during the earlier part of the eruption.

The flow of lava advanced very irregularly and in different directions during 1867. After April, the flows no longer moved toward the east and northeast; the channel between them and the shore of Micra Kameni remained open and could still be used by boats.

During the summer, the flows moved mainly toward the south and southwest. In October, the discharge of igneous material was very great, and

---

Palaea Kameni, shown in these drawings, provided an excellent reference to which variations of the height of Giorgios could be related. I think I can conclude from Mr. Hypert's observations that the highest elevation of Giorgios was about 120 meters on September 29th, while it was scarcely 110 meters a month later.

at that time it took a new direction. The main part of the flow went toward the west in front of the large mass of Palaea Kameni. Between the first of October and the first of December, it formed an immense current descending in a single flow from the plateau of Giorgios and spreading at its base over the entire coast in front of Palaea Kameni.

The lavas also advanced toward the southeast. During the summer, they had ceased to flow in that direction, but on the 29th of October, a streak of fire, 40 to 50 meters long, broke out for several hours from the flow front on that side, and for the rest of the year the flow continued to advance in that direction. It formed two points that grew rapidly and steadily.

The 1868 period is poorly documented, but the records are consistent in indicating a steady continuation of the eruptive activity of the previous year.

From the 4th to the 10th of January 1868, Mr. Schmidt was at Santorini and made very detailed notes for several hours each day on the number and intensity of explosions. These observations, made mainly from a ship anchored at Banco about a kilometer east of Giorgios, are not at all comparable to those he made in 1866. In the first place, they would lead one to believe that the output of ash in January 1868 was much more frequent than before, but this cannot be correct. All the information I have been able to collect from other sources contradicts this conclusion. Mr. Schmidt furnishes only vague comments on the state of the cone and the distribution of flows. The only thing of interest brought out by his observations is the appearance of flames 3 to 6 meters long from the middle of some large blocks visible near the eastern edge of the crater.*

Dr. de Cigalla states that flames came from the summit of Giorgios throughout the first three months of 1868. He also indicates that a few small explosions came from Giorgios but outside the crater. He notes explosions in the seawater at the head of the active flows and even says that flames were seen there. Mr. Hypert agrees with Dr. de Cigalla about this phenomenon, which had already been seen from time to time in 1866 and 1867.

According to Dr. de Cigalla,† the explosions that produced ash and incandescent blocks took place every four or five minutes. During the first four months of 1868, cauliflower eruption clouds rose to an average height of 1,200 to 1,500 meters. They did not affect vegetation except on rainy days.

---

* Julius Schmidt, *Vulcanstudien*, pages 20 to 85.
† De Cigalla, Sur la continuation des phenomenènes éruptifs à Santorin. *Comptes rendus de l'Acad. des Sci. de Paris*, LXVI: p. 553, 1868.

The lavas emerged slowly and in small quantities, although a few flows heading toward Plaka and Thermia (on the southern side of Thera) progressed more rapidly.

A photograph taken from the east by Stillmann on the first of July, 1869, shows the cone of Giorgios covered by ash. The talus slope of the cone had not advanced appreciably for fifteen months, but the nearby pond of water had grown larger, and its connection with the sea was more direct owing to subsidence of the strip of ground that separated it from the shoreline.

Mr. Leyer, a lieutenant on the French frigate *Thémis*, observed the eruptions on the 21st of July and on several days following. He stated that incandescent lava was still moving toward the southeast. The seawater was so hot, even at fifteen meters from the shore, that one could not put a hand in it without being burned. He gave the height of the cone as 167 meters (an estimate that was probably too low). Only three of the Isles of May rose from the bank that extended 3 meters below the water from the northern coast. Lieutenant Leyer counted fifteen eruptions in a period of an hour.

Mr. Schmidt remarked that this observer should have seen signs of incandescence under the water at the extremity of one of the flows.* Although it is very unusual, this would not be impossible, for if the lava was hot enough, a thin layer of steam would be maintained between it and the adjacent water. Nevertheless, in view of the possible sources of error, a phenomenon as remarkable as this should be reported by different observers before being accepted without question.[12]

The eruption continued in 1869 but with diminishing violence. Numerous explosions were seen, but only a few had the same energy. The output of lava was small and the area affected by the eruption was limited.

A note from Dr. de Cigalla to the Academy of Sciences of Paris on the first of March reported that at that time flames still were seen at the summit of Giorgios and when there were explosions the volcano gave off enormous amounts of steam charged with hydrogen sulfide and hydrochloric acid. The eruptions sometimes followed one another after only two or three minutes, but at other times they were separated by intervals of half an hour. Thus they were more irregular than during the two preceding years.

---

* Mr. Schmidt, in his work on the volcanic manifestations of Santorini (p. 99), complained that the Academy of Sciences of Paris had not included *in extenso* the notes of Dr. de Cigalla and Delenda. It is indeed regrettable that, for practical reasons, the Academy cannot publish in their entirety the written communications addressed to it. But why did Mr. Schmidt not publish these notes in his own work?

In December of 1869, Mr. Gorceix, a member of the School of Athens, visited Santorini. I take the following from two letters by him, one to Mr. Charles Sainte-Claire Deville,* the other to me on the 6th of January 1870.

On the 15th of December 1969 the cone had a height of 123 meters. At the summit there was a crude dome of porous lava. These rocks had spaces between them like those from which ash and fumes were discharged during eruptions. Few blocks were ejected.† The eruptions were sometimes continuous, while at other times scarcely anything would be produced for three or four hours. The steam emitted was accompanied by small amounts of sulfurous acid. A plateau that extended above the summit of the cone was covered by ash and ended in a talus slope of lava. Everywhere on this plateau the temperature of the ground was 100° at depths of a few centimeters.

The southeastern flows continued to advance slowly. Temperatures were quite high. Boats could still pass easily through the channel between the northeastern flow of Giorgios and the southern foot of Micra Kameni.

Near the former pier of Nea Kameni, two springs of salty ferruginous water gave off gases. One had a temperature of 21°, the other 50°. The gases were mixtures of nitrogen with much greater amounts of carbon dioxide.

One could barely see fumes on the walls of the cone; rare fumaroles rose here and there from the lava. All of these were mainly water vapor with smaller amounts of sulfurous and hydrochloric acid.

In his letter, Mr. Gorceix says he climbed to the summit of the cone but makes no mention of a crater-like depression. In a second letter, dated 13 April 1870, he wrote that he was again at Santorini for a few days and that he could climb to the summit of the cone several times without danger. The eruptions had about the same character as in December of 1869; they came at intervals of about 25 minutes. The activity of fumaroles had greatly declined and the flow toward the southeastern point seemed to have stopped.

The cone was 118.5 meters high. There was no crater, only a pile of lava that was broken up and ejected, mainly by the eruption of April of the preceding year. Shortly thereafter, this body was quickly replaced by another.

On the 18th of April around nine in the morning, there was a terrifying explosion. The summit of the cone blew up with a great roar. Enormous blocks were thrown to distances of up to 500 meters, and somewhat smaller

---

* *Comptes Rendus*, t. LXX, p. 274, 1870, session of 7 February.
† Mr. Gorceix is incorrect when he speaks of discharges of pumice from Giorgios. The last eruption of Santorini produced no true pumice.

debris was thrown as far as Micra Kameni. Incandescent bombs fell on a ship anchored in the cove of Nea Kameni at the eastern foot of the cone of 1707. The captain was killed by a falling block; the ship took fire and soon foundered. Incandescent stones also reached a second ship anchored farther away in a cove on the northeastern coast of Nea Kameni. One of these went through the bridge and penetrated barrels of wine that the ship was carrying, thus saving the ship. A bark sank and another was badly damaged.*

According to Mr. Gorceix, violent explosions continued for another fifteen days. On the 30th of May, a very strong explosion produced a fume column so dense that the Kamenis were completely obscured. A heavy rain of ash fell on Thera, and for an hour the sun was blotted out.

On the 16th of June, Captain Germounig, commandant of the Austrian corvette *Reka*, returned to Santorini.[†] He could see the eruption column 40 miles away. He saw no change in the outer form of Giorgios since his previous visit in 1868. Between the 16th and 20th of June an average of two hundred explosions a day were counted; half of these were strong. Several times flames were seen above the edge of the crater, and the fume cloud was illuminated by incandescent lavas. Strong explosions occurred again during the first days of June and again in July. An eruption on the 29th of June threw ejecta almost as far as the Banco.

In September, there were no more explosions, but discharges of ash and gas were frequent, and there was a little scoria. The fume clouds were strongly charged with electricity. Dark clouds above the crater were crossed by streaks of lightning that gave off sounds of thunder. These phenomena were particularly notable during the 3rd, 5th, 6th, and 9th of September. On the latter day, the noise was so loud it caused a moment of terror among the inhabitants of Thera. It is remarkable that these electrical phenomena, which are so common in volcanic eruptions, occurred only during the very last phase of the eruption.

A final explosion occurred on the 15th of October, then the volcano became quiet. It was not entirely extinct, however. At several places the temperature remained very high, and gas and steam continued to be given off. I found that they still persisted in 1875, and some probably continue today.

---

* *Augsburg all. Zeitung*, 1870, p. 2273, taken from the *Merimna* and the report of Commandant Germounig. *Wiener geogr. Mitheilungen*, June 1870.
† *Wiener geogr. Gesellschaft*, 1870.

CHAPTER THREE

PREHISTORIC STRUCTURES OF SANTORINI

The bay of Santorini was formed by a frightful cataclysm. Compared to the eruptions that created the Kameni islands during the historic period, this eruption was immense. The people that inhabited Santorini before the bay was formed perished completely at the time of this terrible catastrophe. They have been buried in the depths of the volcano or crushed under the great mass of material thrown out by its explosions.[1] Today, a layer of pumice, which in places reaches thicknesses of up to 30 meters, covers what remains of the former island. Houses and the relics of an advanced culture have recently been found beneath this shroud. The excavations have shed an unexpected light on the nature of the cataclysm and have revealed curious insights into the customs and level of civilization of the primitive population of the island.

The circumstances that led to these discoveries are interesting enough to warrant brief mention.

A few years ago, the Suez Canal Company, having undertaken the great masonry structures for the port and buildings of Port Saïd, had the idea of using pozzolana[2] to make high-quality cement. The pumice of Santorini, mixed with half or even a third as much lime, quickly turns exceptionally hard while at the same time becoming perfectly resistant to seawater. The available amounts of this material and the ease with which it can be mined and transported at low cost to Egypt induced the company to procure it in great quantities. From that time on, the pumice has been exploited on a large scale from the coasts of Thera and Therasia. Arriving ships are stationed at certain places close to the shore, where they load the pozzolana as it is broken away from the escarpments above and allowed to fall or slide down the tortuous slopes of the cliff. Substantial parts of the pumice layer of Thera and Therasia have been removed in this way. On Thera, the excavations were mainly confined to the shore of the bay near the village of Akrotiri; on Therasia, they extended along almost the entire length of the eastern cliffs and the southern shore facing the small island of Aspronisi. On the latter coast, about midway between the two capes of Tripiti and Kiminon, several different proprietors have excavated vast, open quarries quite close to one another. In all these excavations the pumice was removed down to its basal layer, which contains numerous lithic blocks that hindered the work and reduced the value of the pozzolana. Some of these blocks were found to follow regular lines that turned out to be the tops of walls. The workers and owners realized this perfectly well and knew they were dealing with ancient buildings, but ruins of this kind are encountered so often on Thera and Therasia that it raised no particular interest. The attention of the scientific world was first drawn to the ruins on Therasia by Mr. Christomanos, professor of chemistry at the University of Athens and a member of the scientific commission sent to Santorini by the Greek government to follow the progress of the recent eruption.

By chance, Mr. Christomanos visited this locality and happened to notice the sections of walls along the floors of the pozzolana quarries. He immediately concluded from this clear evidence that the structures he was looking at were older than the pumice. There were, however, many questions that could be raised, and it was necessary to confirm his deduction by digging into the debris. One might have thought, for example, that the place was an old burial ground and that these structures were tombs that could easily have been dug like catacombs into the pumiceous tuff, in which case they would be more recent than deposition of the pumice. We should note in passing that several funeral monuments of the Hellenic epoch have been found in similar situations at the base of tuff layers, such as the one on Thera, and one such tomb just below the tuffaceous unit on Santorini had even been seen and described by Bory Saint-Vincenet.[3] It could be, therefore, that he had come upon something similar. So it was necessary, first of all, to determine whether the remnants of walls were from burial caves or buildings.

Second, even if this question were resolved and the ruins were shown to be from houses built above ground, it would be necessary to find out whether the pumiceous tuff under which they were buried still occupied the same position it had when laid down by the volcano or whether it was a landslide or mudflow that had accidentally covered the buildings, as happens today to buildings constructed at the bases of certain cliffs, such as the one at the landing place of Thira,[4] which is now half covered by debris.

All the excavations completed on Therasia up to now have been on the property of one of the local proprietors, Mr. Alafouzos; only one of numerous structures indicated by the outlines in the floors of quarries has been completely uncovered, but one can already see from this work enough evidence to resolve the two questions just raised. The buildings were clearly built on what was then the ground surface and later filled with tuff that has not since been reworked. They are older than Santorini's great eruption of pumice. This is in accord with conclusions reached from other lines of evidence.

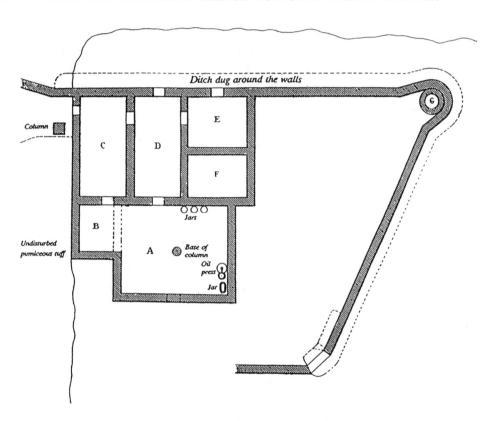

Structure above the southern cliff of Therasia explored in 1867 by F. Fouqué.
(Plan is drawn to a scale of 1 to 100)

The excavations started by Dr. Nomicos, the distinguished medical doctor of Santorini, were pursued by the owner of the land, Mr. Alafouzos. The building they revealed consists of several rooms of unequal size.

The largest, A, situated on the southern side, is 6 meters long and 5 wide; it has an additional extension toward the west, B, forming a small, square chamber 2.5 meters on a side. Room A and its annex B are separated from the rest of the building by a transverse wall.

From west to east on the other side of this wall are found the following:
1. room C, 6 meters long and 2.5 meters wide
2. room D, the same size as C
3. two rooms, E and F, separated by a partition. E measures 3.8 by 3 meters and F, 3.9 by 2.5

Messrs. Nomicos and Alafouzos at first tried to uncover the exterior on the western side of the building but were stopped by fear of causing a failure of the slope and confined themselves to clearing away the debris inside. They completely emptied the material that filled rooms D, E, and F, half of room C, and three-quarters of room A. Then, wanting to see what they would find outside the ruins, they followed an exterior wall that starts as an 8 meters prolongation of a fairly regular northern wall of the building then turns and follows a curving line toward the southwest and is interrupted 18 meters from the corner to form an opening on that side. The northeastern angle of this enclosing wall has a cylindrical masonry structure with an interior opening of about 80 centimeters and a height of one meter above the ground.[5] Finally, 24 meters from the main building they found a smaller one-room building, and near that a wall that passed under the pumice deposits.

The construction of the building is completely different from that used today on Therasia and Thera. Viewed from the inner side of the rooms, the walls appear to be composed entirely of irregular blocks of untrimmed lava piled up haphazardly and held together by reddish soil mixed with organic material. There is no sign of the use of lime, either by itself or mixed with pozzolana. Lengthwise between the stones on all sides are pieces of wood of various diameters taken from wild olive trees and still covered by bark. Most of the wood inside the thick walls is in a very advanced state of decay; on breaking down chemically, it has become as black and friable as charcoal.[6] In most instances, the slightest touch reduces it to powder. The interior sides of the walls do not seem to have been covered by any sort of plaster but simply finished with a layer of the same earthy material that was placed between the blocks of masonry. The roof was formed by a layer of soil and stones about 30 centimeters thick and held up by numerous wooden beams. In all the parts excavated by Messrs. Nomicos and Alafouzos, the roof had fallen in, and the debris was found mixed with strongly agglomerated tuff. Despite that, it was easy to see from the upper parts of the walls that the roof over each room must have had a simple low arch.[7] Room A was constructed in a way that differed from the others; in the middle of that room there was a stone cut in a rounded form. The upper face had a flat surface with a diameter of about 30 centimeters and seems to have been the base of a column that supported the roof.[8] This interpretation is confirmed by the fact that all the pieces of wood inserted into the walls of room A rise uniformly toward the center of the room as though they would meet in the middle and be supported there by the inferred column.

Room C is also distinctive. Halfway up from the floor, pieces of wood were placed in the wall where they must have formed a floor. Thus, it is very probable that this room, unlike any of the others, had another story above the ground level.

All the inside corners of the walls are rounded. This is especially clear in the small separate building 24 meters from the main one. Although this feature can be seen in the larger building, it is less evident. This peculiarity can certainly be attributed to the crude nature of the original construction, but an even more important factor is the settling resulting from crushing of the pieces of decomposed wood.

The alteration of the wood also had another effect. Wherever there were doors or windows, the wooden frames have given way under the weight of the walls, and the openings are obstructed by debris, so that it is difficult to see where they were located. Between rooms C and D, however, Mr. Nomicos discovered a doorway with two steps descending from C into D, and in the western wall of the latter close to the roof there is a small window made of roughly shaped stone.

At the base of all the rooms one finds the scoriaceous lava that immediately underlies the tuff horizon.

Aside from the peculiar features of the layout and construction of this old building, the excavations have revealed in the interior a great number of objects made of stone and pottery that are remarkable, not only for their forms but also for the material from which they were made. Piles of barley and other types of grain are found either on the ground near the base of the partitions between the rooms or, more often, inside jars of various styles and ornamentation.[9] The largest of these have a capacity of at least one hundred liters. It is noteworthy that no metal objects were found in any of the rooms of this building, and no nail has been found in the many pieces of wood making up the debris from the roof. Instead, one finds stone weights, millstones, troughs, and other stone objects, the point of a lance about eight centimeters long, and a small saw of flint 5 centimeters long.[10] In short, all the tools that have been found are made of stone.

Finally, in room A near the entry to the smaller room B, a human skeleton was found, but unfortunately only the lower jaw and a few fragments of the back and pelvis were preserved. The rest had been crushed by the fall of the roof. Despite the careless way in which the workmen had cleared the building, one could determine the position of the skeleton when it was found and deduce that the person seemed to have been bent down and not extended as he would be in a burial position. The lower jaw from which Mr. de Hahn, scientist and consul of Austria at Syra, had a cast made, shows no particular ethnological feature, with the possible exception of a slight vertical flattening of the two lateral branches. The teeth were worn from chewing; the tubercles are entirely gone but the teeth have no cavities and were not crooked as one might expect. The skeleton was certainly that of an old man of medium size.

When I first visited the site of these important discoveries in March of 1867, I immediately saw the importance of continuing the excavations undertaken by Messrs. Nomicos and Alafouzos. I began by having a deep trench dug all along the northern wall of the building, in order to uncover the foundations over a considerable length, then I had this excavation extended along the exterior wall that encloses the adjacent courtyard.[11] The main purpose of this operation was to verify whether the foundation really rested on lava. It would also make it possible to determine if there were any openings on this side that would go through to the interior of the building. I showed in this way that the foundations of the building had their base at a very shallow depth below the ground inside the rooms, so that the walls rest directly on the lava and not on a foundation. Nowhere is there any pumice between the base of the walls and the underlying lava.

In the parts of the walls corresponding to rooms D and E, two windows are located about a meter above ground level and much lower than the window found in room C by Mr. Nomicos. The masonry of these openings has no cut stone. They are crudely arched*; they were about 60 centimeters high and 50 wide.[12]

The northern wall of the building, seen from the outside, seemed better constructed than one would have expected from the appearance of the interior. Its whole length actually consists of irregular pieces of lava randomly

---

* Their upper part was supported by a horizontal piece of wood, so the arch was probably due to sagging of the stones in the middle.

piled on one another, but they are more regular than in the interior face of the wall. Moreover, at the angles at its two ends, the wall is formed by perfectly shaped stones, many of them quite large, that are seated in a flat position.[13] The highest block at the northeastern angle has a cylindrical hole about 5 centimeters deep, and on its northern side it even has marks that may have been letters or numerals.[14]

The two windows on this side show that this was a house and not a tomb as one might have concluded from seeing only the opening in room C. They also show, as does the position of the wall on the lava, that the construction preceded the fall of pumice, for they look out on the hillside, and it is clear that if the dwelling had been built at the foot of a steep slope of pumice, the windows would have been placed on the side looking toward the sea and not toward this escarpment.

Finally, examination of the pumice shows that it has not been reworked and that it was not brought in by water after first being deposited elsewhere. The pumice that fills the interiors of rooms is composed of angular fragments that still have their original sharp edges. The same is true of the pumice around the outside of the building, where a faint stratification is horizontal or slightly inclined parallel to the slopes it covers. These traces of bedding can be followed for great distances along the cliff where they pass over buildings without any deflection. One of these layers that lies about two meters about the tops of the walls is so distinct that one might take it for a layer of organic soil, but closer examination shows that this is not the case. The yellowish color of this thin band of tuff is due to a change of composition that is so slight that it can scarcely be detected by chemical analysis. The pumice deposits above this level are also in place; the entire mass is part of a single deposit laid down during a single volcanic eruption that may have been prolonged but had no important interruptions.

In the interior of the building and in the courtyard next to it, fragments of scoriaceous lava from the underlying horizon can be found within the pumice. There are also pieces of reddish cinders that come from weathering of the lavas and are identical to the material used in construction of the walls. Some of this scoriaceous material contains organic matter that covered the ground at the time the buildings that are now beneath the pumice were constructed. It is found just below the entire mass of pumice. Another observation that I will discuss later leads to the same conclusion that there was a distinctive layer of organic soil below the pumice layer on all the islands that now surround the bay.[15]

From all these facts we can certainly deduce that the structures brought to light by the excavations of Messrs. Nomicos and Alafouzos, as well as those that I have had made, are really dwellings dating from before the great pumice eruption and buried in the course of that eruption under a mass of pumice ejected by the volcano.

Everything that is to follow will only serve to reinforce that interpretation. I shall discuss the results of my studies without insisting on this further.

After following the northern side of the building, I had the digging extended along the western side where the pumice was still intact. In carrying out this work, we saw for the first time that the lower part of the northern wall of the building was connected with another wall that headed westward beneath the tephra, and after crossing it we entered a large arched chamber the wall of which was inclined toward the north. I had this space cleared for a distance of six meters without encountering any obstruction. To the left of the entrance I had made to get in at that point, there was a large stump of a column still standing but slightly inclined toward the southwest. Two blocks of square-shaped columnar lava, each about a meter high and 50 centimeters in diameter were perfectly cut and carefully mounted to form a kind of pedestal resting on a low, crudely rounded base.[16] Proceeding a little farther, we encountered the remains of another wall that was so badly damaged that I could not determine how it might have been built. All I can say is that the stump of the column is completely independent and isolated from it.

Fear that the steep cut might fail forced me to give up further excavations on this side and direct the work toward the southwestern angle of the building. Further digging outside the western wall uncovered the external part of the corner between annex B and the western wall of room A and exposed the southwestern corner of this room which, like the other corners of the building, was formed by regularly cut and superposed stones. The southern side that this digging revealed was more poorly preserved than the other parts of the exterior wall. I attribute this to a door that probably existed in the middle of the wall but had fallen among the chaotic debris. All this leads me to believe, however, that this was the main entrance to the house.

The wall of the courtyard, the foundation of which I uncovered earlier, stands, like all the rest, on the lava and shows nothing unusual, provided this is not its entry turned toward the southeast, to which one finds two steps down followed by two more narrower steps that rise toward the higher part of the ground. These rather crude steps, especially the latter two, were made mainly by taking advantage of natural irregularities in the underlying rock. The same is true of two other steps at the northwestern corner of the courtyard, where the wall joins the main building.

I attempted to have the interior of the cylindrical structure excavated, just as Mr. Nomicos had before me, but it was impossible to dig to a depth of more than about two meters. The narrowness of the opening and the lack of suitable tools prevented the workers from going deeper; so I have not been able to resolve the interesting question of what purpose it could have served. Despite its form, I do not think it was a well, for the fairly flat layer of lava at ground level is quite hard and thick, and to reach an aquifer it obviously would have been necessary to penetrate it. I doubt whether this could have been done with stone tools. Even if the hole were a natural crevice in the lava, there is no indication that the water table reached the conglomerate just below the lava. It seems even more unlikely that it was a cistern, for the ground level of the courtyard drops toward the west, and the dwelling occupies the lowest part. Between the ground under room C and that of the small structure, G, there is a vertical difference of about 5 meters. Rain water falling on the roofs of the house and on the courtyard could not, therefore, drain into a cistern dug at the very highest point. And, of course, one would still have to accept the unlikely possibility, as in the case of a well, that one could dig through the lava or that a large natural cavity had been used. I must conclude that the walls of the cylinder formed the opening of an underground structure, and I am inclined to believe that it was the base of something constructed above ground. In searching through the shallowest part, Mr. Nomicos found numerous pieces of pottery, while my own digging to deeper levels encountered less and less of this material, and when I stopped digging there were only small pieces of lava that differed little from the surrounding masonry.

I should add some additional comments that might throw some light on the role of this unusual structure:

1. When the exterior of the foundation is exposed, one can see that it is resting on the layer of lava at a shallow depth.

2. No wood was used in the masonry, and none of the stone has been worked.

3. Three low, very wide steps extend along the front; they average about 12 to 15 centimeters in height and about 80 centimeters in width.

After exploring the interior of the building, I had the workers continue the removal of the tuff started by Messrs. Nomicos and Alafouzos. In finishing room C, I found the remains of many jars similar to those found earlier. Some of these appeared to have been empty, but the majority contained carbonized barley and other substances less easy to identify.

In the small room R[17] that forms a corner of room A and had not been explored earlier, the roof had not fallen and was still intact. I could easily see the form of its low arch. Among the many objects found there were:

1. Earthen jars and fragments of pottery similar to those found in other parts of the house. Some that were made of crude pottery contained finely chopped straw that was probably feed for animals.

2. Two grinding stones of lava, each consisting of two hemispherical dome-like pieces superimposed on their flat faces and of unequal dimensions. The lower piece is 15 to 20 centimeters in diameter and 6 centimeters thick, while the upper one is only 12 to 15 centimeters in diameter and 4 thick.

3. Three weights, also of lava, weighing about 250 grams, 750 grams, and 3 kilograms.[18]

4. An elliptical disk of lava, 12 by 15 centimeters in diameter and 5 centimeters thick, pierced in its center by a hole the size of a finger. This hole must have been for a rope or strap by which the disk was suspended. One can see on the two faces grooves as thick as a finger, running parallel to each other from the center to the edge. These were certainly the result of rubbing of the rope on the surface of the stone along the side opposite to what would be its center of gravity when one suspends it from the middle.

5. A container made of lava, 30 centimeters long and 20 wide, with a cavity that is not more than 3 centimeters deep. The lower side has very short cylindrical knobs that serve as feet.[19]

6. Two other vessels made of lava in the form of small troughs. These have diameters of 20 to 25 centimeters and cavities 8 centimeters deep. Their bottoms, which are at least 20 centimeters thick, show that they must have been imbedded in the ground. A similar vessel, slightly larger than these, was found at the entrance to the courtyard.

7. Large numbers of bones of animals, such as sheep or goats. These remains belonged to three different individuals, because one commonly finds three of the same parts of the skeletons.

Many articles similar to those I had found were also discovered by Messrs. Nomicos and Alafouzos in other rooms of the building. Among the objects I have seen in their collection, I noted in particular a piece of wood, 10 centimeters in diameter, that was still wrapped in its bark. It had a crude hole, made at a right angle with a cutting tool, and two parallel mortises 5 centimeters long, 15 millimeters wide, and 3 centimeters deep, that are so regular and well cut that they could have been made with a steel chisel.

Finally, I should add a word about the easternmost part of room A. The most curious object I found there is a large vessel made from lava, hollowed out with a conical hole 40 centimeters in diameter and 30 deep. At the bottom a hole connects with a slot on the underside and comes out in a receptacle made of lava that seems to have been designed to receive liquid coming from the main vessel. The interior walls are worn by abrasion, and the local people did not hesitate to tell me that it was a mill for making olive oil.

These are the main results of the excavations I carried out on Therasia. They are far from revealing all that existed at that time and place. There is a prehistoric village there made up of a great number of dwellings whose walls can be seen reaching the surface of the ground. I do not doubt that future research will lead to very interesting discoveries.

Mention must also be made of my other excavations on the island of Thera that yielded most of the pottery and other objects that I brought back to France. Having been drawn by my geologic studies to the southern part of the island near the village of Akrotiri, I had the occasion to inquire whether traces of ancient structures had been found there. A peasant led me into a ravine, where he showed me sections of a wall that rose into the pumice. I would like to have opened these structures to find out whether they resembled those of Therasia and to determine exactly where their foundations are based, but a misunderstanding with the owner of the land prevented me from doing so. This raised my curiosity, however, and I decided to examine all the ravines in that area where erosion by running water has provided natural sections through the ground. In the same ravine where I had seen the structures just mentioned but under 3 or 4 meters of rounded gravel and organic debris, I discovered an extensive layer, about 30 centimeters thick, formed entirely of fragments of pottery. A few hours of digging in this layer yielded a considerable quantity of fragments of all kinds that I was later able to assemble into more or less complete jars.

In another ravine farther to the east, I was shown a rather compact layer of iron-rich cinders immediately below the pumice that had two arched tombs with narrow openings near the top. These tombs, about two meters in diameter and two meters high, had long since been opened. I found nothing there except a Byzantine coin, which would argue against great age if one did know that the same tombs were commonly used for later burials during very different periods.

A short distance away I noticed under the pumice a layer of fine-grained, reddish volcanic ash, only a few centimeters thick, that immediately reminded me of the material in the ancient soil at the base of the structures on Therasia. The calcification of a small amount of this soil clearly showed me that this was a layer of old organic soil.[20] After that, I made every effort to follow it along the ravine and find out what might be in it. In addition to a few fragments of pottery similar to those on Therasia, I found several small obsidian tools, and the local farmer who accompanied me then obtained others identical to the ones I had found, as well as two small gold rings found in the same bed.

The obsidian instruments have two very different forms. The most common ones look like small blades of elongated knives, 5 to 6 centimeters long and 1 centimeter wide. One side is more or less flat or slightly concave in the lengthwise direction, and the other has two faces inclined to each other like a roof with a truncated ridge line. The truncation is normally very narrow but in certain rare specimens it is quite wide and takes the place of almost all the two side faces.

The other type has the shape of a triangular arrow point. The length is almost one centimeter and the width a little less.[21]

The two small rings are of very pure gold. Boiling them in nitric acid showed no trace of silver or copper. They are about 3 millimeters thick. The metal shows no sign of welding, so they must have been made from a small piece of native gold that was hammered into a small disk then worked by a method known in industry as *repoussage*. The two edges of the small plate are brought close together but not in contact so they form a small channel. Each ring is hollowed out in the middle. At the two extremities of the same diameter the metal is pierced with small holes, as thick as a common sewing needle. One side has only a single hole, but the other side has three closely spaced holes. The slender size of the rings and the holes in them indicate that they were part of a necklace.[22]

The jars and fragments of pottery found on the island of Therasia and in the ravines of Akrotiri were made from the same types of clay. They were not made by hand but were turned on a potter's wheel.[23] Comparable jars from the two sites are so similar that they could well have been made by the same person.

We should note here that these jars can be grouped into different categories according to the nature of the clay used in them or the style of their ornamentation.

Those of the first group are made of fine, slightly yellowish clay that seems to have been very plastic. They are decorated with circular bands separated by regularly spaced vertical or slightly inclined lines. The dark brownish-red color was applied as a very clear liquid, which was no doubt prepared from a suspension of ferruginous clay in water.

The largest of these jars, which I was able to reconstruct almost entirely, is 24 centimeters high. It has an opening 17 centimeters in diameter, two handles, and a narrow neck.

Another with a very elegant form and very similar decoration is 12 centimeters high and has a mouth 7 centimeters in diameter.

The numerous fragments with similar ornamentation that Mr. Nomicos and I have collected show that this type was very common. Aside from the rather large pottery, others tend to be smaller and appear to have been made from a different material. They are covered with brownish designs in a style completely different from that of the first group. The designs are made up of dots and curved lines mixed with drawings, in some cases showing garlands of foliage. One jar in this group is unique in that the interior is covered with a brownish glaze. I do not have an unbroken example of this type.

A third type includes jars made with very fine, plastic white clay. These have very curious forms. The most common type, of which I have several unbroken examples, is an unusual imitation of the shape and appearance of a woman. These jars have a swollen part, a kind of stomach surmounted by a thin, turned-back neck; above the swollen part are two brown breasts surrounded by a circle of dots of the same color as that of the inside coating. Around the neck are two inclined circles of dots representing two necklaces. And finally, earrings have been drawn on each side of the handles in an elliptical band using the same brown as the rest of the ornamentation.[24] These jars are 18 to 25 centimeters high.

Another jar made from the same sort of clay has a very different form.

It is almost cylindrical with a height of 15 centimeters and a diameter of 8. Near the upper edge it has two handles with the shape of small rectangular plates attached at their edges and only slightly hollowed on their outer forms. Their surface has no ornamentation.

In a fourth group I place the pottery made of a red, very ferruginous clay. These have no designs on their surface, and their shapes and sizes vary widely. One that I brought back with me is a wide bowl with a single, very small handle near the edge. Others have the form of cups with or without handles. I also brought back a dozen examples of the shallow bowls that are found there in great numbers.

A fifth group includes all the common pottery made from a rose-colored clay in which I was able to distinguish partly decomposed fragments of feldspar. The most remarkable of these are the large jars with a capacity of at least a hundred liters. Three of these have been found intact on Therasia, and I found fragments of others at Akrotiri. These jars have a thick edge around their mouth and a little lower but above the neck a narrow design in relief that goes all around the jar and has a series of closely spaced indentations.

I should perhaps include in the first group the funnels made with yellowish clay and found in the first ravine at Akrotiri. I found two and a half broken ones and a third that is still perfectly preserved. These funnels have much thicker walls than the first type of jars and are covered with a very thin, uniform ferruginous coating not seen on the latter. They have a small handle and an edge that protrudes slightly around a large opening, 8 centimeters in diameter. The opening measures only a millimeter; the length is 15 centimeters. Mr. Nomicos found among the debris on Therasia a jar, several fragments of which he gave me, so I could relate them to the categories I have outlined. This pottery is made of very fine grayish clay and covered with a dark gray to black coating speckled with white designs representing leaves. It is the only jar I have seen with these two colored used one on the other.

Messrs. Gorceix and Mamet, both members of the School of Athens, have continued the excavations I started in 1867. With the aid of a subsidy from the French government they have been able not only to explore the localities that I had studied but to open other very rich ones on the edge of the cliff facing the bay near the village of Akrotiri.[25]

The first locality they explored was the ravine southeast of Akrotiri where I had worked earlier. An extensive area of the adjacent ground was cleared of pumice in order to expose the horizon at its base.

The ravine was cut with an earlier terrace, 3 or 4 meters wide, separated from the neighboring vineyards by a talus of about equal height. A path along the bottom has been cut into the lava that underlies the pumice in much of this part of the island. Above the lava is a dark organic layer then a pumice bed 3 to 5 meters thick. The latter is very irregular; it is surmounted by a layer of rounded pebbles, mentioned earlier, which has been left when the lighter material in the underlying bed has been carried away by erosion.

According to Messrs. Gorceix and Mamet, the pumice is no longer present along the sides of the road in the bottom of the ravine; it has been completely carried away by water, so that the layer of gravel and fragments of rounded pumice would have rested directly on the lava if it had not been separated from it by the old organic layer just mentioned. It was from this small strip of blackish soil that most of the pottery I have described was collected in 1867. I had thought at that time that there was a thin layer of pumice preserved between this soil and the lava, but more recent work has shown this to be incorrect; the organic soil is in direct contact with the lava. What I had taken in 1867 for a continuous strip of pumice was only a lens of reworked material resting on the lava at the foot of the talus slope. In other words, the bed of pumice, where it is still in place, is above the old organic soil and not below it as I stated in a note published in 1867 in the *Annales des missions scientifiques et litteraires.*

The first objective of the work undertaken by Messrs. Gorceix and Mamet was to expose the wall M that I had noticed projecting from the talus to the left of the ravine in the middle part of the gravel bed. In doing this, they soon began to uncover all the remains of a building composed of several rooms that are arranged as follows.

The walls, which rest directly on the lava, are 50 centimeters thick and made of rubbly stones. They are cemented with limeless mortar made from clay-poor earth containing bits of partly decomposed lava, trass, and pumice. The stones are similar to those employed in buildings throughout Santorini where they are known as "porites." In M the standing part of the wall has a height of only one meter, but around A it reaches two meters and the highest part almost reaches the level of the organic soil. The opening marked "O" seems to have been the entryway, but, because of the collapse of the wall in this area, I cannot be sure.

Beyond this wall one begins to see quite clearly that the bed of pumice is situated above the building and below the bed of rounded gravel. The contact between these two units is very sharp and devoid of pottery.

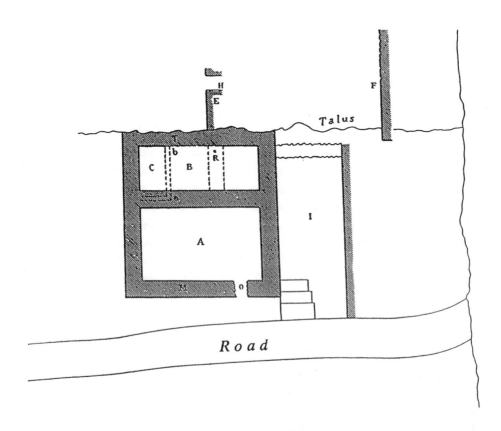

A large building excavated by Messrs. Gorceix and Mamet in the ravine of Akrotiri following earlier explorations by F. Fouqué
(Plan drawn by Mr. Gorceix at a scale of 1 to 100)

Behind this wall the excavations have revealed a large number of perfectly preserved jars buried under the pumice and filled with the same material. Their colors and shapes are quite varied. Most were probably meant to hold water; several have the same shape as those still used for this purpose in eastern countries.[26]

A considerable pile of stones from the fallen walls takes up much of the small room B.

The rear wall T stands under the talus, which it appears to be supporting.

Although all the walls were constructed on lava and are under a cover of various layers of pumiceous debris and gravel, one could well have taken them for modern structures that had been buried either by landslides or stream deposits, but a new discovery removed all doubt about their antiquity.

In trying to unearth wall R, which seemed to divide the small room B, an opening was made in wall T. As a result, a kind of corridor was found leading from the other structures well into the ground. Its arched ceiling was made of weakly consolidated pumice that fell at the slightest touch. The thinnest part was strewn over the ground, forming a layer one meter thick and leaving a void of about the same height between it and the ceiling. Wall E, extending to the right of this passageway, was found to be covered with a white coating several centimeters thick and decorated with bright colors.

Taking great care, it was possible to penetrate a few meters into the vaulted space. At 2.5 meters the wall was interrupted by a lateral opening, and the remains of boards that had formed a door could still be recognized. The width of this second room is 4.75 m, and the wall F opposite the door is covered with frescoes similar to those just mentioned.

Despite a strong temptation to pursue the excavations, which seemed to be yielding such important results, serious difficulties brought the work to a halt. Because the layer of pozzolana was not very thick (at most 2 to 2.5 m) or strong, and the gravel was even less resistant, the arch threatened to collapse, as it soon did, even though the work was suspended.

The only practical solution was to proceed with the work in an open excavation, but this would have required destroying an adjacent vineyard, which was in full production, and the compensation demanded by its owner exceeded the available financial resources.

Nevertheless, the discovery, though incomplete, is of great importance. The rooms A and B evidently belonged to the same dwelling as the walls E and F, and there can be no doubt that they were constructed prior to deposition of the pumice. The layout of the rooms alone was enough to dispel any idea that they had been part of a tomb, which, in any case, would be very difficult to excavate in such weak material.

As for the discovery of the coatings on the walls, they enable us to make two new deductions: first, the use of a considerable number of colors was well established at that time, and second, an unusual use was made of a lime-like substance.

The walls are decorated in a manner similar to that used today. A few centimeters of mortar made of beaten earth is applied to the surface of the stones, then this is covered with pure lime on which the artist draws designs in color. Four colors are used: bright blood red, pale yellow, a blue that is bright when first spread but is quickly affected by the air, and, finally, a blackish brown.

The lower part of the wall was covered with stripes in which these four colors were used alternately. The horizontal spaces between the stripes were traced with a pointed instrument that left a small scratch. Above these bands, red designs were drawn on a white base representing flowers in which the stamen is longer than the corolla.

The lime is pure, white, and resembles that manufactured today on the island of Thera. It could very well have been made at the site. It was used without any addition of sand, which would have reduced its whiteness but would have improved its adherence, while a mixture with pozzolana would have made it denser without affecting its whiteness. The inhabitants would certainly have used this mixture if at that time there were pozzolana on the island.

From the number of pieces of this plaster scattered on the ground one can conclude that the rooms of this house had painted ceilings made of the same material. Moreover, these fragments are thicker, as they would have to have been for this purpose.

A fifth room, I, is 5 meters long and 2 wide. The debris cleared out of it has the same characteristics as that from A and B. Long, thin plates of stones found in great number formed the roof. The narrow width of this corridor permitted this kind of roofing.

At the entrance, three stones 50 by 30 cm formed the steps of a small stairway descending into the room, the floor of which had been dug into the lava.

Rooms E and F also had walls decorated with frescoes but, unfortunately, they have not yet been entirely excavated.

While it is very likely that the opening O in wall M was the chance result of collapse, the room may have been a reservoir for collecting rain water. This is the explanation offered by Mr. Gorceix. The bottom is paved with well-placed flagstones, and the carefully constructed masonry walls formed a water-tight surface. If so, it is easy to understand that the large number of jars found in room B were used to draw, filter, and carry water. This also explains the trough of packed earth running along the top of the wall in C. The bottom of this trough is 140 centimeters long and 80 wide; it was inclined from b toward a. The pieces of pumice filling it were pressed into the soil showing that the latter was soft at the time the pumice fell, probably because water was in the trough and descending from the roof to the reservoir.

Room I, situated about a meter below the ground level, served as a hypogeum or cellar.

The quantities of stones that filled all the room and were piled up around them indicate the importance of the structures.

Knives, scrapers, and a saw, all made of obsidian, were found together with the jars. The saw is 5 cm long and is thinner at the ends so it can be held easily. One edge has irregular teeth that are pointed enough to permit making fairly deep cuts. Three knives seem to have been parts of larger instruments. They have two cutting edges, one with fine, closely spaced teeth. Numerous scrapers were collected; they have a rather irregular shape but resemble the same types of tools found in other countries.

The number of other objects found in the excavation was not great. They include, first, a disk of serpentine, 13 mm in diameter, with a central hole and several regularly spaced marks on one face. The surface around the hole shows no sign of wear as it would have had if it had been strung on a necklace. It was probably an earring or an amulet worn on the neck.

Second, a small ball of black earthy material with designs cut into it and pierced by a central hole. It may have been part of a necklace.

Utensils made of lava include: hand grinders, mortars and pestles, small troughs, and two elliptical disks, each with a hole near one end.

The number of earthen jars that Messrs. Gorceix and Mamet collected, either broken or intact, number about a hundred. Such a number of utensils should not be surprising when one considers that jars of this kind were used, as they still are, for all the domestic needs of Greek homes. Metal or wooden vessels are almost unknown there.

The jars can easily be separated into three types according to their shapes and the uses they were designed for.

1. water jars

2. containers for grain, oil, and the like

3. filters or jars used to drain coagulated material and pierced by a small hole at the bottom

Messrs. Gorceix and Mamet distinguished within the first group two types, each of which includes a variety of shapes. One is characterized by having two handles and an elliptical opening; the other, which is bent backwards, by a neck and a beak-shaped spout.[27]

Jars of the first type, which resemble the amphoras used to carry water, have no remarkable features. Their shapes have continued to be reproduced down to the present day. They were made from a red or yellow clay, and, although most have no trace of ornamentation, some have circular red bands between which are large, elliptical figures in the same color.

It is the second group that includes the most curious shapes and types of decoration. The most notable examples are those in which, as I have already described, the artist's models were women of different ages. The clay used to make this type of pottery and the colors used to decorate it are identical to those of the first group.

Because the pottery was not baked in a large fire, water filters easily through the pores. It seems that the jars served as decorative jars or as water coolers.

Apart from these types with such distinctive shapes, three others are notable for their remarkable decoration. The first is an almost cylindrical jar, slightly swollen at its base, 15 centimeters tall and 13 in diameter. It has a handle and two ear-like projections. It is coated in red with zones of black on which are drawn in white flowers and leaves resembling those of iris plants.

The second is made of a finer clay. It had an open handle, which unfortunately is broken, with one end near the middle and the other around the mouth. From the base up to about one-third of the height it is decorated with three circular red bands; above these are flowers, also in red, between which can be seen three galloping horses crudely drawn in black. The legs, tails, and easily recognizable heads with ears and eyes leave little doubt as to the kind of animal that has been portrayed. The design, though very crude, shows that the artist had a good feeling for the positions of the different parts of the body in motion. Although drawings of flowers are common, this is the only example yet found of an animal. The jar has another peculiarity: the spout is closed by a plate with six small holes that served as a sieve or filter.

The third jar has a rather elevated base with two handles above an orifice fitted with a cover. The only decoration is three circular black bands with vertical lines of the same color.

Numerous fragments of jars for grain and other types of food have been found here, but they are difficult to reconstruct. The largest, which are identical to those I found on Therasia in 1867, have a capacity of about 100 liters. Their outer surfaces have no trace of coloring but were decorated with rope-like designs in relief. Fingerprints are recognizable in the clay. In addition to the large jars, one finds others of smaller size fitted with a handle and a rather elegant neck and spout.

Messrs. Gorceix and Mamet found only a few fragments of funnels similar to those I found in great numbers in the talus of this same ravine in 1867, but they have collected two jars of different forms that were probably designed for the same purpose.

One is made of very fine material with a very regular, remarkable form. On two fragments broken along an old crack, six small holes are arranged in three groups of pairs. The broken jar had been repaired by a procedure still used in earthenware jars. A hole was made in the base after baking; the clay is brown and must have been carefully decanted.

Another smaller jar has a similar hole made after baking. The clay in it is cruder, and it is covered with a simple red coating.

Many cups of various shapes and saucers of very rough material have been found, particularly in lower levels of the excavations. Two of these saucers contain, along with the dirt that fills them, fragments of obsidian. Calcified wood and charcoal were mixed with this debris, but the completely decomposed wood amounted to only a small part of the mixture. The charcoal seems to have been made from pine. A few sea shells and many pieces of bone are spread throughout the debris with the fragments of pottery. The shells belong to all the edible species - limpets, pectens etc. - that one still finds on the shores of the island today.

Bones of horses, asses, dogs, cats, goats, and sheep have been found in rather great numbers, but most of those collected from debris have been reworked. Vertebrae and sides of goats and sheep are all that have been found where the pumice layer is still in place.

Sixty meters lower in the same ravine, two small cave-like diggings into the upper talus have revealed a layer of debris identical to the higher one. Stones from ruined structures can be seen on the ground and have even been used to build a retaining wall. The pumice immediately overlying the debris and enclosing the parts of the walls still standing was still more or less in place and was easy to remove. The loose talus above these newly uncovered structures cannot be approached without fear of it collapsing, so it is impossible to clear more than part of three of the rooms belonging to this same dwelling.

The first room, B, is only 1.4 meters wide; the walls extend under the talus, so the length has not been determined. The thickness of these walls is about 50 centimeters, and, like those of the first house, they were built of rough stone and mortar without lime.

An almost intact jar one meter high and 65 centimeters in diameter was found upright and filled with pumice in the angle a of this room. At its bottom it had a small amount of unrecognizable carbonized material. The shape and position of the jar resembled those in almost all the modern houses of Greece, where they are known as οἶδος. The walls are based on the same lava flow mentioned earlier. Room A, which has not been completely cleared, has a width of 2 meters. The floor is paved with slabs of lava embedded in mortar.

Near the middle of the wall at d a pile of ashes and charcoal indicates the location of a fireplace. One could even distinguish marks of smoke on the wall. No trace was found of a stove or chimney; the smoke must have escaped through a simple opening in the roof, just as it does in the houses of modern Greek villages. Jars, some broken and others still whole, tools made of obsidian and stone, hand grinders, etc., were spread over the middle of the floor and stones up to a meter high rested on the ground.

Beside A is a third room, G', that has not yet been completely cleared. Behind it is a fourth that does not seem to connect with any of the other three. It is similar to the one already described in the first house. In room G', which is much longer than wide (4.3 by 2 m), there may have been an opening at f that is now hidden by fallen debris, but it is clear that there was no cistern.

About a meters from the house, a separate wall is now almost completely ruined but can still be made out. It has been traced for 5 meters, and at C' it was one meter high. At H', some stones indicate its last traces. This wall belonged either to a second house or to a wall around an enclosure.

The nature of the different layers can be seen twenty-five meters lower, in a section cut in the same vineyard. The organic soil is mixed with a thin bed of gravel most of which has disappeared. Immediately below is a layer of pumice fragments with very thin beds of pozzolana, then a black soil 20 to 25 centimeters thick with fragments of pottery and obsidian and resting on the lava. Weathering of the latter has contributed most of the fragments in the basal layer.

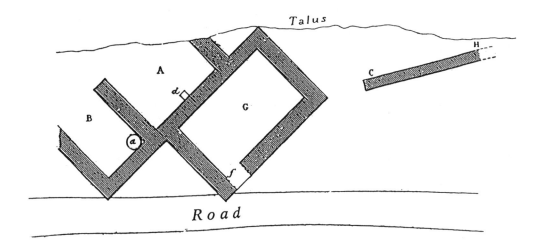

A small structure in the ravine at Akrotiri excavated by Messrs. Gorceix and Mamet.
(Plan drawn by Mr. Gorceix at a scale of 1:100)

This same bed, as has already been noted, is clearly exposed in a ravine toward the northeast. It has been followed by Messrs. Gorceix and Mamet for a distance of 500 meters. Ditches dug at various spacings have turned up many scrapers and obsidian knives, as well as a few fragments of pottery.

The geological level of the foundation of the second house is the same as that of the first. The walls are based on the lava, and if their position relative to the loose material seems unclear near the road, it must be due to reworking of the pumiceous tuff by water. In the direction of the talus, this tuff becomes thicker and more regular, and its position above the walls is very clear. The latter extend under the talus just as they do in the first house.

Here, as above, one sees no debris with artifacts between the gravel bed and the pumice. The tools, jars, and utensils are very similar to those already described. All are found under the pumice.

Numerous stones pierced with a hole have been found in various rooms of the second excavated building. Some are elliptical, others irregular. Some of the latter are made of rather heavy lava or limestone in which a natural cavity facilitated making the hole. They were probably used to stretch fishing nets. One of these stones with an elongated elliptical form has a hole near the center that narrows from the surface to two-thirds of its depth then increases again. A shallow groove about as wide as a finger extends across the surface of the stone from one hole to another and resembles a pulley. Two small balls of clay found in the same place have a narrow, round hole around which they are flattened and worn. They were probably part of a crude necklace.

The jars in this house had quite varied shapes and were much less elegant than those described earlier. A few jars made of rough clay to hold grain or carry water and some cups deformed in baking were the only household utensils in this house.

But two or three pieces of very fine pottery with beautiful designs were found among the cruder pieces. These isolated and unrelated fragments had an entirely different origin. In room C' of this poor house the excavators were quite surprised to find two columnar posts. The first, cut from a kind of friable millstone, is 30 centimeters high and has a diameter of 21 centimeters at the base and 12 in its upper part. It is only part of what must originally have been a much longer piece, for the remaining part is only a fragment.

The second is hollow and was made from brown local clay and covered with a brilliant black coating. It is 35 centimeters high, and has a diameter of 20 centimeters at the base and 11 at the top. The rings on the outside are separated by grooves of about the same size that diminish in size upward. The last is much thicker than the others; its surface has regularly spaced depressions that give it the appearance of a twisted cord. The top has a cylindrical part of small diameter and irregularly broken.

It is impossible to know the origin of this column, the use of which has not been determined. Is it a container of some strange shape used in the home? Could it represent some deity? Both of these hypotheses suggested by Mr. Gorceix seem equally unlikely.

Animal bones of the kinds already mentioned, a few marine shells, pieces of charcoal, and decomposed wood have also been collected from this house.

Many rounded pebbles were found in the different rooms at depths where they could not have fallen accidentally from the gravel layer above.

As I had already noted in the objects found on Therasia, these small cobbles have a simple relation in which their measured weights fall into groups that cannot be random: 105, 139, 175, 211, 320, 425, 535, 840, 956, 1167, and 1288 grams. Comparing these numbers, one sees that their ratios are very close to the following: 1, 4/3, 5/3, 2, 3, 4, 5, 8, 9, 11, 12. The small deviations are easily explained by the wear on the stones, which are much less resistant than the metal used to make weights today.

The deviations of these values are not much greater than those of weights in use in countries where they are not closely controlled. The rounded stones must therefore represent weights used by the inhabitants of the buildings where they are found. They have no marks on the surface. In addition to stones picked up on the beach and then abraded to give them the desired weight, one finds a certain number of balls the size of children's marbles and made of a different material. Most are of siliceous siltstone, but some are made of limestone; both appear to have been made by human hands. Their weights seem to have the same relations as the other stones, but they are in smaller fractions: 0.8, 0.3, and 0.15.[28]

The most interesting and fruitful excavations carried out by Messrs. Gorceix and Mamet are at the edge of the cliff northwest of the village of Akrotiri. For a period of about fifteen years, pumice was mined in the area above the small harbor of Balos. At this place the cliff is quite steep and one must descend along a crude trail to the place where several workmen mining the pumice uncovered a small platform. The workers pointed out to Messrs. Gorceix and Mamet an area in the pumice where they found a mortar made of lava, together with its pestle. Diggings were undertaken immediately. Thanks to the conditions that made it possible to throw almost all the excavated material over the cliff, the thickness of the pumice posed little problem, and from the first day they began to uncover the tops of walls belonging to buildings constructed on the lava bed. The pumice cover was 22 meters thick, and in its upper part was made up almost entirely of angular blocks. At lower levels it was a pure, very compact pozzolana. This thick layer of pumice was responsible for the perfect preservation of the houses which, in some cases, lacked only their roof.[29]

The geologic section of the cliff in these two places was very similar;

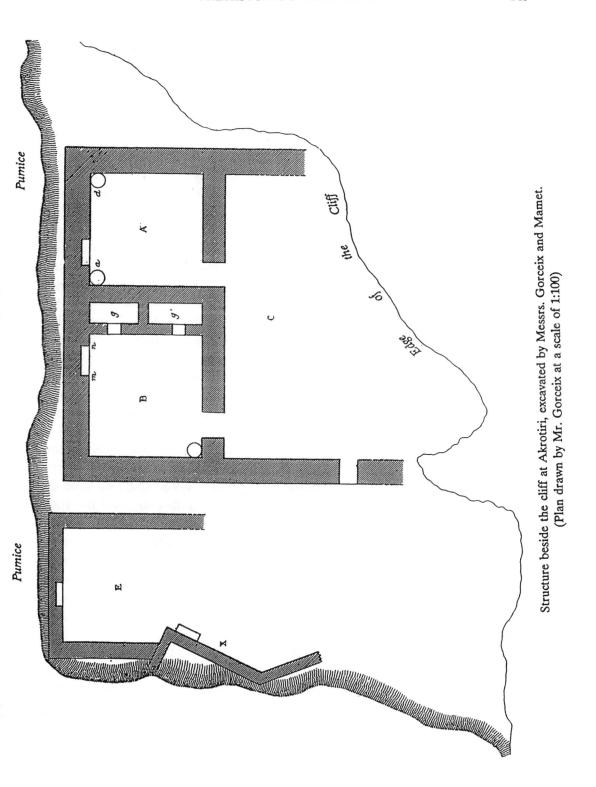

Structure beside the cliff at Akrotiri, excavated by Messrs. Gorceix and Mamet. (Plan drawn by Mr. Gorceix at a scale of 1:100)

the bed of pumice makes up the uppermost part; below that come the lava flows of various thicknesses up to 16 meters, then various beds of ash, pumice, lapilli, and other volcanic products extending almost to sea level for a total thickness of 62 meters.

The walls, 45 to 50 centimeters thick, were built with rough stone and small amounts of mortar; the interior was covered with rough plaster that still has traces of a yellowish finish. Each of the two main rooms, A and B, is almost a square, 4 meters on a side. The ground is covered with a layer several tens of centimeters thick and consisting of black, decomposed organic material in which bits of straw and large numbers of pieces of wood are still recognizable. Much pottery, almost all of it in good condition, and many pieces of obsidian were found in the detritus.

In the two corners opposite the door of room A, two enormous jars, each with a capacity of more than 100 liters, had been set into the wall. One was filled with chopped straw, the other with barley; everything else was completely carbonized.

In room B, the number of these jars was even greater. Three, which were smaller but fixed to the wall like those just mentioned, were filled with barley, lentils, and a kind of pea called *arakas* that is still cultivated on the island.[30] These three jars had covers of clay, hollowed out, and pierced at the bottom. In one, the excavators thought they recognized a piece of rope passing through the hole in the bottom where it was held in the middle with a knot. The cord was used to lift the cover.

Two openings a meter above the floor in the right-hand wall of the same room connected with two small sheds *g* and *g'* 65 centimeters wide, separated by a thin partition, and containing barley in one and chopped straw in the other. The floor was paved, the walls carefully plastered, and the roof made of platy rock covered with a thick layer of mortar.

An opening had been made into the wall at *m n*, probably to serve as a storage space. The remains of its door were still recognizable. Two round holes had been made in the two sides of the door, and in one of these were fragments of a wooden bar that served to close the door. The system of closure is still used in almost all the houses of Greece and even in many villages in France.

The number of jars in room B was not great, but all contained grain or straw except for one or two in which bones of goats were found in a mixture of unrecognizable carbonized material. In the corner was the entire skeleton of one of these animals.

The barren nature of rooms A and B, the absence of windows, and the great amount of vegetables and straw in them suggest that they were used as storerooms for food.

The main part of the dwelling was in front of C. Unfortunately, its walls are almost totally ruined; only part of the one on the left remains. On that side, there was either the main entrance or a connection with the adjoining structures; in the latter case, the entrance would have been on the side facing the cliff. Some of the debris indicated that there may have been a partition dividing this space into separate parts.

The house was cut in two when the central part of the island collapsed. The wall on the right runs up to the edge of the cliff.

In digging a trench in the pumice on the left side of the main part of the house the workmen uncovered a series of walls forming a rather irregular contour. The ground at E was scattered with a layer of straw, and in the middle were numerous bones of goats and sheep with some basin-like jars that still contained barley. This was probably a pen for animals. In one corner, there was a trunk of an olive tree more than 2 meters long, with several branches.

The few obsidian tools found in this excavation tended to be larger than those in the ravine of Akrotiri. Near the tree trunk, one of the diggers found a copper saw. This was the only metal tool found in the prehistoric ruins of Santorini. Shaped like a sickle, it is 22 centimeters long and 15 wide. The point is broken, and the other end was forged in a way to take a wooden handle. The metal contains no trace of tin. It is entirely covered with a green layer of copper carbonate, but despite this oxidation and the poor condition of the teeth, one can still make deep cuts into pine or poplar.

According to those who excavated the ruins, it is impossible to believe that this saw was not contemporaneous with the houses. The thick layer of pumice that covered the ground had not been disturbed by the quarrying of pozzolana. The soil layer, like that of other parts of the cliff, was composed of successive beds of pumice or pozzolana in regular, horizontal layers and no mixture of foreign material. The possibility that it was carried to its position by slumping can safely be ruled out.

The farther one penetrates into the cliff the thicker the pumice layer becomes, but behind the wall marked X, the lava drops away sharply, so that this wall serves both to enclose the courtyard and to separate it from the precipice. Small galleries dug under the tuff led to no discoveries, and it was necessary to give up further digging, leaving the structures exposed and the large, intact jars still in place. But the bed of organic soil and fragments of pottery could be followed for several kilometers, and in many places a few strokes of the pick uncover small fragments of obsidian.

Similar pottery is often unearthed by workers digging in nearby fields where the soil layer is thin. These diggings have a special purpose; where the pumice is not mixed with other material it is very poor for planting grain, so the owners dig down to the underlying organic layer. This practice is particularly notable north of Arkhangelos. According to the owners of these fields, great amounts of pottery have been found at different times in this part of Thera. Messrs. Gorceix and Mamet dug a trench down through the pumice to the organic layer in the middle of one of these fields and found pieces of pottery and large numbers of stones showing vestiges of walls, but everything had been disturbed and completely destroyed.

Obsidian tools found in the houses excavated near the cliff at Akrotiri appear to have come from the same source as those found elsewhere. One with the shape of a knife blade has a well-sharpened cutting edge. Other pieces were probably scrapers.

The jars are much better preserved than in the other places already described. They have little decoration but quite varied shapes, including some that had not been found before. Many of the jars, however, are identical to those in the other buildings.

One set of five jars is worthy of special note, because they are like modern bottles and were probably used for similar purposes. They are without decoration but have four lengthwise ridges along their sides indicating that they were made in a mold rather than on a wheel as most others were. Two pieces attached to the sides like ears have holes by which they could be suspended, so that evaporation could cool the contents. They were found together, and a sixth found nearby differed from modern bottles only in the nature of the material with which it was made. All the jars are made with a very common type of whitish clay; only two are decorated in red and black.

Pieces of cone-shaped filters, which also served as funnels, had already been found in the ravine at Akrotiri and are represented at the cliff-side locality. Four of them are almost complete. One has circular and longitudinal lines, drawn in red, black, and white over almost their entire surface; three others have a simple coating of red varnish.[31]

Containers for grain and other perishable material are very numerous and quite large. Disks made from platy pieces of lava were used to cover their openings. Edible material has also been found in a great number of very crude clay bowls.

The only example of internal decoration is seen in an elliptical vessel that is remarkable for its shape and excellent preservation.[32] It is a large oval basin with a base 40 centimeters long and 20 wide and an upper part that measures 70 by 40 centimeters. A series of festoons covers the long sides, and the bottom is painted red.

A few elliptical stone disks, numerous hand grinders, and mortars complete the list of objects collected from these excavations. As in the case of obsidian instruments and jars, the forms, designs, and materials used to make them differ in no essential way from those found in the houses near the ravine of Akrotiri.

Having described in detail the structures on Therasia and the objects found there and at Akrotiri, we can now see what conclusions one must draw regarding the level of civilization that had been reached by the inhabitants of the southern part of the archipelago at the time of the pumice eruption.

First of all, all the structures explored in the excavations described above belong to the same period and are older than the pumice covering Thera and Therasia. This is undoubtedly true of the structures found on the southern coast of Therasia and on the cliff near Akrotiri. Their situation under the pumice is beyond question, and it follows that they were built prior to the event that produced the bay. In the case of the houses discovered on the sides of the ravine and explored by Messrs. Gorceix and Mamet, I had earlier thought their foundations were above the pumice layer, and this is why, in my first report, I had considered them later than the eruption responsible for this unit. But the excavations carried out by Messrs. Gorceix and Mamet have shown this to have been incorrect. The houses uncovered in the ravine of Akrotiri, like those cited earlier, were beneath the pumice.

Important differences can, however, be found in the construction of these buildings. Some are built without cement, with a roof supported by a timber frame and bare interior walls; their masonry is crude. Others have signs of cement and plaster. One of the latter, the highest one found in the ravine of Akrotiri, even has vaulted rooms and walls decorated with frescoes. Messrs. Gorceix and Mamet attribute these differences to the level of wealth of the owners. They point out that the most carefully constructed house is also the one in which they found the largest number of decorated jars that had no utilitarian purpose and could only have been objects of luxury. They add that, in the second house found under the pumice in this same ravine, the obvious poverty of the furnishings corresponds to the imperfect construction of the building; the inhabitant was not a wealthy person like the owner of the neighboring house but only a simple fisherman. Many stones used for fishing nets are found in his house. The pottery is crude and without decoration, and everything is utilitarian. The house on Therasia and the one on the cliff near Akrotiri belonged to workers; they, too, have differences that, according to these same gentlemen, reflect mainly the differences of wealth of the owners.

But in addition to these differences, I think there is one other. The differences in their construction could also be due to their different ages. Even though they belong to the same archaeological period, they could have been built centuries apart. In such a lapse of time we know that customs, construction methods, and procedures used in masonry can change completely. Hence, the buildings found under the pumice of Santorini need not be absolutely uniform.

Except for these differences, the buildings have much in common. And if one considers the pottery and tools found in them, the similarity is even greater. Even a quick glance at these products is enough to convince one that they belong to the same period. The discovery of the copper saw by Messrs. Gorceix and Mamet shows that this metal was already in use. But the use of metal was not yet widespread, because this instrument is the only object of copper found thus far in the excavations.* All the crude articles were made from pieces of lava, the finer ones of obsidian. Stone was still the most commonly used material for most of the purposes where iron would now be used. In any case, working with metals was still in its infancy, as can be seen from the purity of the gold and the workmanship of the two small rings I collected at Akrotiri.

None of the stone artifacts is polished, but all are delicately carved. The small tools of obsidian or chert have thin edges and regular shapes. They are as carefully made as instruments of the same kind that are so common in Mexico where they were still being made at the time of the Spanish conquest. The articles made of lava look as though they had been made with the aid of

---

* Being unique among the objects found there, one must consider the possibility that this piece could have been brought there for some chance reason.

a steel chisel. The workmanship testifies to a remarkable degree of manual skill.

The cut stone seen in the corners of the buildings on Therasia and the prismatic stone columns nearby indicate a similar skill on the part of the workers who built these structures, especially when one considers the tools they had to work with. This is even more true of the main buildings uncovered in the ravine of Akrotiri. But proof of advanced techniques and extraordinary taste is even more apparent in the painted frescoes found by Messrs. Gorceix and Mamet, as well as in the many diverse pieces of pottery found wherever excavations have been carried out. The jars were turned on a wheel, and most are remarkable for their beautiful shapes and decoration. Some are truly works of art.

In an earlier report I expressed the belief that all the pottery had been brought to Santorini from elsewhere. That opinion was based mainly on the fact that today one finds no deposits of suitable clay on Thera, Therasia, or Aspronisi. A chemical test seemed to confirm this deduction by showing that one of the pieces found there differed in its basic composition from any of the sandy clays visible today on islands of the Santorini archipelago. This has proved to be incorrect. Further work has enabled me to show that, on the contrary, all the pottery found under the pumice on Santorini had been made in the same region but at a place now covered by water.[33]

This conclusion is based on a microscopic study of thin sections made of each piece of pottery in the same way that rocks are studied. In these sections I have recognized:

1. bits of trass and amphibole andesite of submarine origin from the southwestern tip of Thera

2. fragments of the various types of lava with labradorite and pyroxene that are so common on the three islands of Santorini

3. numerous bits of pumice or obsidian identical to those seen in the ash beds in the southern part of Thera

4. fragments of marble and mica schist identical to those making up the massif of Mt. Profitis Ilias on the island of Thera

5. grains of quartz, crystals of monoclinic and triclinic feldspar, pyroxene, hornblende, and biotite in the primary material of jars made mainly of trass

6. foraminifera, diatoms, and very abundant sponge spicules of several different kinds

Among the most common forms of the latter that have been identified are the following:

Rotalia lepida	Coscinodiscus minor
- senaria	- radiolatus
Gallionella varians	Lithasthericus tuberculatus
- distans	Cocconema
Planulina globularis	Vaginulina
	Spongolithis acicularis

The interiors of the foraminifera are filled with opal, calcite, and limonite.

All these diverse components are found in differing proportions in the material used in the prehistoric jars of Santorini, both in the crude pottery as well as in the fine-grained, more compact clays. One finds no fragments of the anorthite-bearing lavas that are common in the northern part of Thera. The foraminifera, diatoms, and sponge spicules are far more abundant and better preserved than they are in the trass on the island of Thera.

I conclude from all this that:

1. Without exception, all the pottery was made on Santorini.

2. It was made from clay obtained from a low-lying region within reach of seawater and where fresh water from time to time brought in all sorts of detritus from rocks on the southern part of Thera. No weathering products from the anorthite-bearing lavas of the northern part of the island came into this area.

3. The place where the pottery was made must now be under the sea, for there is no longer any place on Santorini that has these conditions; it was probably situated in a valley between the cliffs of southern Thera and a large cone that occupied the area now covered by the waters of the central part of the bay. Purely geological considerations that I discuss in a later section indicate that this valley came down to the sea near the island of Aspronisi. For this reason it appears that the source of the ceramic clay was in that region.

The presence of pumiceous debris in the pottery does not contradict what I said before about the stratigraphic situation of the houses uncovered on Therasia and at Akrotiri, because, in addition to the thick bed of pumice mantling all the islands of the Santorini group, the underlying sections of these same islands have a lower white pumice that could very well have been the source of these components.

I should add that most of the pottery of Santorini was simply dried in the sun or baked at low temperature, because otherwise the fragments of marble in it would have been altered. Some, however, appear to have been fired. These are a crude, bright red type of pottery that is impermeable to water. They contain no calcite, and the iron oxide seems to be entirely in the form of hematite. Some that I found among the debris in the ravine at Akrotiri must have been used for cooking food, for one side was still covered with soot.

The mineralogical composition I have just described as distinctive of Santorini is so characteristic that the thinnest fragment is enough to identify any one of them under the microscope.* Thus, there can be no doubt that all the broken or unbroken jars under the Santorini pumice came from a local source.

The coating of lime that covers the walls of the large structure at the ravine of Akrotiri, when seen under the microscope, contains many fragments of marble and schist and a few crystals of pyroxene and feldspar. This leaves no doubt that it was made at Santorini not far from the place where this same product is made today.

The discovery of the two small gold rings at Akrotiri is important, because it shows there was commerce between Santorini and the neighboring islands and mainland, and probably with Asia Minor where certain rivers were famous in ancient times for the quantities of gold found in their beds. It is certain that there were no deposits of gold on Santorini or on any of the neighboring volcanic islands.

The jars I have described as having turned-over necks and a prominent swollen stomach also support this conclusion regarding the relations of Santorini with the outside world. We have jars that are similar (though lacking the breasts) that were found in excavations elsewhere, particularly on Milo. Funnels like those of Santorini have also been found at Jalissos on the island of Rhodes. Other jars identical to some of those at Akrotiri have been found on Cyprus. Moreover, an eminent archaeologist has told me that an Egyptian painting in the Louvre shows the king of Egypt receiving Greek envoys and

---

* I have recently received from Mr. Dumont, director of the French School in Athens, a small fragment of a jar found in the ruins of the city of Mycene. A microscopic examination clearly showed that it came from Santorini. Several other jars found at the same locality were made with calcareous or quartzose clay from a quite different origin.

that among the articles of tribute brought by these men are pottery with shapes and decoration identical to those of the jars of Santorini.[34] Thus, a wide commerce already extended across the seas at that time.

One might think, perhaps, that the tools of flint and obsidian were made on the large pre-collapse island, for one finds siliceous sandstones and obsidian now near the village of Akrotiri and in other parts of Thera, and on Therasia the lavas tend to look very much like obsidian. But closer examination shows that the objects of flint or obsidian that I and the four other excavators have found were brought in from elsewhere. All the siliceous sandstones of Santorini are imperfect. Although I have gone over all the region of Akrotiri step by step, I have never found a siliceous rock like the yellowish, translucent, and homogeneous one in the small saw found in the excavations carried out by Mr. Nomicos; moreover, the lava of Santorini, even when it appears somewhat translucent, never has the clarity of true obsidian, and it is normally speckled with small crystals of white feldspar that are not seen in the tools I have collected. One must go to Milos to find an obsidian like that used for the knives and arrow heads of Akrotiri. Thus, study of the stone tools, as well as the objects made of gold, demonstrates that there was marine commerce during the stone age.[35]

This is not all we know about the inhabitants and the kind of life they had on the large island. We know that they cultivated grain. I have personally collected large amounts of carbonized barley from the buildings on Therasia, and, according to Mr. de Cigalla, chick peas and seeds of coriander and anise have also been found there. I have found no wheat, and, so far as I know, the others have not found it either, so it seems that it was unknown. The grain crops were stored in piles or in large jars. The small grinders, several pairs of which have been found, show that flour was known and probably bread as well. Oil was extracted from olives, domestic animals were raised, and, if our identification of the pasty material found in a jar by Mr. Nomicos is correct, cheese may have been made. The use of measures of weight was known, and one can assume there was a system of numbering. I believe that the balance used was the instrument known today as a beam or Roman balance. What leads me to think this is the discovery in the excavations of Therasia and Akrotiri of the disks described earlier that were made of lava and had a hole through them. The shape of these objects is easily explained if one thinks of them as having served as counter-weights of a beam balance, and this hypothesis is not unreasonable, since one always finds them together with weights. The workmen I employed think that they were used to hold threads of cloth being woven on a loom, because similar stones are used for this purpose today by the weavers of Thera and Therasia. If so, one would expect to find evidence that cloth was being woven at a time when the use of metal was almost unknown.[36]

The abundance of pieces of olive wood and pitch found in the ruins of Therasia and Akrotiri prove that the island was at that time well wooded and that the inhabitants were farmers and shepherds rather than cultivators of vineyards as they are today.[37]

Having acquired this knowledge of the lives of the former inhabitants of the island, let us lay out what is known about the catastrophe that produced the bay of Santorini. First of all, we can show that the great eruption of pumice preceded the collapse of the center of the island, for the tuff that covers the present cliffs of Thera and Therasia is truncated in the same way as the underlying lavas. This can only be explained if it had been cut by the same collapse as everything else. But it is quite likely that there was no geological event between the eruption of pumice and the collapse, and that one followed very closely on the other, for it is difficult, if not impossible, to imagine that two phenomena of such magnitude could occur independently at the same place. As for the pumice eruption itself, it does not appear to have been preceded, as one might expect, by any violent shocks or earthquakes, for if it had, the buildings on Therasia would have been knocked down, and we would not find the walls still standing. This is all the more remarkable, for the methods of construction of the newly discovered buildings show that the ground had already been subjected to earlier earthquakes, because the pieces of wood inserted into the walls could only have served to prevent the disastrous effects of ground movements. (This custom is followed today on all the islands of the archipelago for precisely this reason.) If there were any strong shocks during the eruption, they could only have come after the buildings of Therasia and Akrotiri were already filled with pumice, and their inhabitants had been crushed under the ruins of their dwellings.[38]

I must now address very carefully the remaining question of when the collapse took place.

The formation of the volcanic part of the original island had scarcely begun when the Pliocene deposits were laid down. The duration of its growth, measured in terms of the time necessary to accumulate the lavas seen in the cliffs of Santorini, could scarcely have been less than the entire

Quaternary period. Thus one arrives at the conclusion that, in terms of geological time, the collapse was a modern event. But the present geological period spans thousands of years, and we must try to establish more precisely how recently the pumice eruption could have occurred. Geological processes subsequent to the collapse offer a way to make such a calculation.

The first evidence I use comes from the development of the islands in the middle of the bay. After the collapse and the terrible phenomena that preceded it, there was certainly a long period of repose; it was not until 196 B.C. that a new eruption produced Palaea Kameni. After that date, successive eruptions followed during the first centuries of the Christian era, and the newly formed island continued to grow. A second period of relative calm followed through all the Middle Ages, and it was only in the fifteenth century that eruptions regained their frequency and strength to form new islands. The second period of calm having lasted about ten centuries, one can confidently assume that the first one lasted at least twice that long, especially when one considers the great difference in the intensity of the activity that preceded the two periods of repose. By this reasoning, the bay would have been formed around two thousand B.C.

A second line of evidence brings me to the same conclusion. At the northern tip of Therasia and on the adjacent part of Thera, the pumiceous tuff is covered with a layer of reddish gravel around 15 to 20 meters thick and containing marine shells. My scientific traveling companion, Mr. de Verneuil, and I noted the same layer on the eastern coast of Thera near Kolombos. All of these places have been below sea level since the pumice was laid down and then were uplifted by what must have been a slow emergence. On the part of Therasia that was raised in this way there are ancient structures with inscriptions that Mr. F. Lenormand was able to assign to the fifteenth century B.C. These structures were constructed at a time when the elevation of the ground was even greater than now, for some of them are still below sea level today. The time required for deposition of the marine sediments and for the subsequent uplift, which must date from before the eighth century A.D., would require a period that I confidently estimate to have been at least ten or twelve centuries. Thus one is brought again to the same date I arrived at for the age of the pumiceous tuff.[39]

Finally, historical evidence supports this line of reasoning, for we have firm evidence that the Phoenicians invaded the southern islands of the archipelago in the fifteenth century B.C. The buildings, tools, and pottery of these people and of those that later replaced them in the same area are totally different from those discovered on Therasia and at Akrotiri. The populations at the time the bay of Santorini was formed were on the island before the Phoenicians and, therefore, before the fifteenth century B.C. And because the ruins of their buildings and the abundance and variety of pottery found over a wide area show that they were well established there, it is reasonable to think that they had been there for several centuries and could therefore have been living there in two thousand B.C.[40]

Some archaeologists are of the opinion, however, that the original population of Santorini belonged to the Pelagean race and that they would therefore have lived there at a later time.[41] I only mention this view, which I do not feel competent to judge one way or the other. The only certain thing is that Santorini was inhabited before the collapse that produced the bay.

Excavations have revealed only a bare minimum of the archeological treasure hidden in the ground on Thera and Therasia, but small as it is, it is of great interest to us because of the link it provides between geology and history.

CHAPTER FOUR

DESCRIPTION OF THE PRESENT STATE OF THE KAMENIS AND TWO UNDERWATER CONES IN THE BAY OF SANTORINI

The islands that define the outer limits of the Santorini archipelago were formed during a much earlier period than the Kameni islands.[1] The latter, as we saw earlier, owe their origin to the series of volcanic eruptions that began two centuries before the Christian era and have continued to the present day. Before this phase of activity, the various events of which are related in historical legends and recent geological accounts, there was a period when magmatic activity was in a state of repose. For several centuries, the bay had essentially the same form as today but was occupied only by the sea; no islands rose within its boundaries to form barriers to the waves. This was the configuration of the region during the most flourishing period of Hellenic civilization. Thus we can separate the geologic formations prior to this period of calm from those that came later and divide the exposed features of the region into two corresponding parts.

It would be more logical to start by considering the rocks making up the very oldest part of Santorini and then proceed to successively younger units, but I prefer to follow the reverse order, which has the advantage of greater clarity and allows us to begin with what we know best. A detailed study of more recent features makes it easier to understand the older ones for which the only historical record is the geological one. I shall therefore deal first with the Kamenis and then go on to what is known about the origin and composition of the islands of Thera, Therasia, and Aspronisi.

PALAEA KAMENI

Palaea Kameni lies to the west of Nea Kameni and forms a prolongation of the latter island, which was produced by the eruption of 1866. It is elongated toward the south-southeast, and its main mass is surrounded by promontories oriented in various directions. The longest dimension of the island is 1,450 meters and its average width 400 meters. The surface of the central part is a plateau elongated in the same direction as the island; it slopes gently from the southeast to the northwest at an average of 2 to 3 degrees. Abrupt escarpments bound it on all sides. The plateau is much more uniform in its central part than at its extremities. Its area is roughly a third of that of the entire island. The north-northwestern point is a very irregular mass of chaotic blocks. Toward the opposite end there is a large, shallow depression surrounded by banks of platy lava the surfaces of which are disaggregated into piles of incoherent debris. Beyond, the ground rises rather abruptly and attains almost the same prominence as the highest part of the island.

The maximum elevation is 98.8 meters above sea level. The summit is only about 15 meters above the adjacent plateau on the northwest. On all sides it is surrounded by precipitous cliffs. It can be climbed on the south and west, but the eastern side is virtually inaccessible, owing to the huge blocks that litter the upper surface. It resembles the central court yard of a ruined castle buried under its own debris.

There is no question that the main mass of Palaea Kameni was formed by lavas discharged from a source near the present summit in the southeastern part of the island. Whether one examines the surface of the plateau, the vertical sections visible in the cliffs that surround the island on all sides, or the walls of depressions in the interior, one sees that all the ground is composed of a succession of superimposed, blocky lava flows that slope away from the dome forming the southeastern end of the island at gentle angles corresponding to that of the present surface. Most of the irregularities of the surface that one sees are simply random variations from the main slope angle. Most can be explained as the natural form of the flows, which solidified while moving and had an uneven fluidity that had the effect of making their advance very discontinuous. A flow that stopped momentarily almost always became inflated at its extremity as new molten material continued to move into its interior beneath the thick, scoriaceous crust. After being temporarily obstructed, it resumed its advance as soon as the increased pressure of the fluid interior caused it to break through the weakest part of the crust. After complete solidification of the flow, a more or less pronounced bulge remains at the place where its advance was interrupted. Beyond, the flow regains its normal size and shape and takes on the overall appearance of a flattened, rocky arc with a knobby surface.

Many of the other irregularities come from depression of the central part of the flows and formation of piles of blocky debris. Some of the blocky zones in the Kameni lavas are deep and very well defined. The marginal ridges are strongly accentuated and separated from the central part of the flow by vertical shear faces. Striking examples will be pointed out in the course of these descriptions.

When several tongues of lava are superposed, the irregularities, instead of increasing, are usually reduced with the flows stacked one on the other in what has the appearance of regular stratification. What seem to be wide layers are actually narrow flows that are interdigitated one beside the other.

All these features are well displayed in the walls of a remarkable fissure that crosses Palaea Kameni. This crack is 500 meters long but only 2 or 3 meters wide; its apparent depth ranges from 10 to 20 meters. Throughout most of its length it is a simple fissure, but it bifurcates toward its end near the southern coast. Its bottom is a jumble of disoriented blocks. It must certainly extend to greater depths, for small stones dropped into the cracks can be heard falling for several seconds. The fissure is oriented N40°W. Its length, straightness, depth, and vertical walls show that it is due to a violent event, most likely the shock of an earthquake.

Other surface irregularities of the plateau on Palaea Kameni are the results of some of the minor phenomena just described. Variations of the atmospheric temperature, humidity, and wind direction no doubt caused the originally massive lava to break down into piles of rubble of all sizes.

The plateau of Palaea Kameni is covered with a thin layer of organic soil formed by decomposition of the volcanic material and surficial alteration of the rocks by atmospheric weathering. Cinders thrown out during the various eruptions of the Kamenis have spread loose debris over the island and have contributed to the formation of a small layer of humus. The plants that have gained a tenuous foot-hold in this surface layer are still very feeble and have a difficult struggle to survive the effects of the population of Santorini which, for centuries, has considered this meager carpet of vegetation a common pasture.

The lavas that make up the dominant part of Palaea Kameni have very diverse forms, but most varieties are compact or platy and fine-grained. One also finds, however, vesicular or glassy and obsidian-like varieties. Their chemical composition closely resembles that of the products of the eruption of 1866. The glassy groundmass contains myriads of feldspar microlites and many grains of iron oxide. They also have phenocrysts of labradorite, most of which are clearly visible to the naked eye, as well as crystals of pyroxene and iron oxides that have dimensions ranging from 0.05 up to 0.5 mm. The feldspars among the microlites have very small extinction angles; it is likely that their compositions are the same as the crystals of oligoclase or sodic oligoclase found in the lavas of the 1866 eruptions. Several of these lavas are noteworthy for the abundance of tridymite seen in their glassy groundmass or lining the walls of vesicles. In some samples this mineral reaches diameters of as much as one millimeter; the crystals are perfectly clear and euhedral.

Scoria, conglomerates, and tuff are almost entirely absent among the rocks of Palaea Kameni.

The coast of the island is very irregular. On the southeast below the highest part of the plateau, a high cliff is cut into a thick mass of lava. The vertical cross-section shown in the cliff resembles a huge rocky mushroom. Parallel veins and fractures in the rock spread out like a fan in the upper parts of the escarpment. What one sees is a vertical cut through an enormous dike that fed the lavas making up much of Palaea Kameni. The action of the sea has contributed greatly to erosion of this part of the island, but to reveal the interior of such a mass required a more powerful agent than waves; the great quantity of material carried away can only be explained by powerful ground movements.[2] Before the other Kamenis were formed, the island was surrounded on all sides by deep water, and violent earthquake shocks could easily have shaken parts of its mass into the sea. This is not just a possibility; we have already seen that it is almost certain and that even the date of this event can be established.

None of the other parts of the coast of Palaea Kameni have such imposing cliffs. Most of the shore is formed by the fronts of lava flows that spread in all directions from the margins of the central plateau of the island. These flows descend toward the sea with slopes that are very steep (20 to 30 degrees) near the edge of the plateau and commonly become almost flat at their opposite ends. In most cases these are clearly branches of the great outpourings that were responsible for the plateau. Some end as low promontories, while others continue into the sea to produce a line of reefs.

Elsewhere, the flow fronts are cut back sharply by the waves. The steep levees of several flows border deep, narrow inlets, and in some instances the eroded sides are almost vertical. Where the marginal levees have been worn away by the water, the more compact, central parts of the flows are very prominently exposed, owing to their greater resistance to erosion. Some of these form narrow promontories between two deep coves, and their crests rise abruptly from the waterline to narrow ridges extending back to the central plateau.

The courses of most of the flows are easy to follow on the surface of Palaea Kameni. They can be seen to have started at the highest part of the plateau and spread toward the northwest with some of them even reaching the northwestern extremity of the island. As soon as they reached the edge of the plateau, they dropped to one side or the other and produced lateral currents. Examples of this are seen most clearly on the western slope of the island.

Some of the lateral flows result from currents that accidentally broke through their crust and drained part of the fluid magma from the interior of the flow, but examples of this form of branching are rare and only weakly developed on the flanks of Palaea Kameni. Promontories formed in the other way are much more common.

It is noteworthy that flows formed by breaching and draining of the interior are uncommon on all the islands of the Santorini Archipelago, both young and old. This is no doubt due mainly to their chemical and mineralogical composition. At the moment they are discharged, most of the lavas were rather viscous. They contain many clots of crystalline material caught up in the less viscous matrix, which itself was neither abundant nor very fluid. In addition to this, they are not particularly thick. As a result, when they become consolidated it is as a more or less single unit. As soon as the crust becomes rigid, the molten material in the interior, though quite viscous, still has the properties of a fluid rather than a solid and can flow out through cracks in its shell.

Thus, Santorini is not the best place to come to in search of good examples of secondary flows formed by lava breaching its crust. Flows of this kind are not entirely absent there, as they are on volcanoes that have erupted lavas that are distinctly more siliceous, but, at the same time, they are not at all common nor are they as well developed as they are on volcanoes with more basic lavas.

To summarize, Palaea Kameni can be thought of as being composed of three parts:

1. an enormous dike exposed in the cliff on the southeastern coast below the highest part of the island

2. multiple lava flows piled one on the other to form the mass of the plateau

3. secondary flows that descended in all directions away from the plateau

Before discussing whether there is a crater on Palaea Kameni, it is necessary to comment on the form of the southern part of the island.

At the southeastern end of the island, an elevated ridge extends down the southern flank of the main summit and forms a promontory toward the south. This point is made up of superposed banks of lava stepped back on each other and dipping gently toward the southeast.

The southern extremity forms another promontory that is almost as sharp but somewhat lower. It has suffered less wave erosion than the other. One would scarcely notice that it consists of the lower extremities of lavas that descended from the plateau.

Between these two promontories a semi-circular embayment is surrounded by high walls with piles of talus at their base. This basin-like feature standing against the mass of lava making up the main summit of the island has the form of a crater breached on its southern side. When one views it from a point high on the ridge that borders it on the east, one is tempted to consider it the remains of a large volcanic depression that would have been for Palaea Kameni what the well-preserved craters were for the other islands in the bay of Santorini. But this idea is ruled out by the absence of pyroclastic beds dipping radially inward and especially by the form of the lavas in the southwestern promontory. It is not a crater but simply an inlet between two rows of debris from lava flows.

In a similar way, the rocks show no evidence of a crater on the eastern side of the southeastern promontory; the regular arrangement of lavas in this promontory has nothing in common with that seen around craters.

If there was ever a crater on Palaea Kameni, no trace of it remains. All that can be said is that if such a crater indeed existed, it was at the southeastern end of the island on the eastern side of that point and has since disappeared into the depths of the sea along with part of the large dike that cuts the cliff.

In the middle of the eastern shore of Palaea Kameni, a small inlet, the harbor of Saint Nicolas, is remarkable for the amount of gas that rises through the seawater. A flow that follows the southern side of the cove emerged from a vent near the summit, crossed the plateau, and cascaded over its upper edge. This flow consists of a light-colored lava, rich in tridymite and split by a vertical platy structure parallel to the direction of flow.

The northern side of the same cove is bounded by a large field of lava that has a quite different appearance. The black lava is a chaotic jumble of large, irregular blocks. Its glassy appearance, dark color, and scoriaceous texture contrast with the trachytic aspect and schistosity of other lavas on Palaea Kameni. The promontory that it forms is low and abuts against the foot of the cliff at the edge of the plateau. One can clearly see that it was produced by an unusual eruption unlike those responsible for the rest of the island. A small pond in the middle of this field of lava is a source of water for domestic animals that are brought here each year when there is a shortage of fodder on Thera and Therasia. This body of water existed before the eruption of 1866 and may even be older than the lavas around it; there is nothing to indicate that is was formed by later subsidence of the ground.

The innermost side of the cove has a small beach formed by a landslide. A chapel built on this spot and dedicated to Saint Nicolas gives its name to this inlet of the coastline.*

MICRA KAMENI

As stated earlier, Micra Kameni owes its origin to the eruption of 1573. This island is situated about 1.8 kilometers northeast of Palaea Kameni and 2.5 kilometers from the nearest coast of Thera. It has an oval shape elongated north-south. The coast forms a smooth curve on the western side, but the eastern shore is much more irregular. A small inlet on the eastern side is caused by a sharp angle between the NW-SE and N-S trends of the shoreline. The long axis of the island measures about 490 meters, and the greatest width is 320 meters. Its height was 70.9 meters in 1866, but it was 1.1 meters lower in 1867, and in 1875 I found it still lower but by a smaller amount (about 30 centimeters).

---

* This chapel has already been discussed in the note on page 9.

Micra Kameni has two distinct parts. On the south one sees a craterless cone with a circular form and a diameter of about 300 meters. The northern part of the island consists of a pile of lavas about 200 meters long and 200 meters wide. These two parts are joined without any boundary clearly visible on the western flank of the island, but the cove on the eastern side lies exactly at their junction and results from the differences of their east-west dimensions.

The slopes of the cone are rather uniform between 28 and 35 degrees. The steepest side is on the east where wave erosion has cut into the base. If one considered only the surface of the cone, it would appear that it is composed entirely of loose cinders, lapilli, and a few isolate blocks. On the south and southwest, the cone is bordered at sea level by a narrow strip of rocks that seem to come mainly from blocks that originally covered the slopes. Sparse vegetation has developed on the flanks of the summit, particularly on the south and east. Many tufts of grass were also growing on the western slope before the eruption of 1866, but the gases and rocky ejecta from the new volcano killed them for a few years. Later, in 1875, I noted that the vegetation had returned to its former state.

The rim of the crater has a north-south diameter of about 60 meters. It is a little narrower in the other direction. The bottom is 40.8 meters below the highest point on the rim and about 29 meters above sea level. The wall of the crater is quite sharp on the south side. The breach is wide and extends almost to the lowest part of the bottom.

The section of the cone exposed in the inner walls of the crater has attracted the attention of several geologists who have visited it in recent years. One sees none of the beds of ash or scoria that one would expect after seeing the outer slopes of the cone. Instead of regular layers of lava separated by scoria, there are massive flows truncated and piled up in complete disorder. Large blocks cut by irregular cracks jut out menacingly from the wall, and much loose debris has tumbled down to the floor of the crater. Almost all the rocks are dense and semi-vitreous; they are remarkably fresh. There are no signs of fumaroles, even though a crater like this could have been formed only by strong gaseous eruptions.

It is likely that at the beginning of the eruption of 1573, the new lavas formed a large mass similar to that marking the opening of the eruption of 1866 and that an explosive vent did not open until later. The later development of the crater explains the abundance and irregularity of blocks in its walls and accounts for the mantle of loose material on the outer slopes.

One should not think, however, that the Kameni cones are very unusual. It is true that in most cones fragmental ejecta are dominant, but it is also common for them to be formed of lava flows. For example, the outer flanks of the cone of Vesuvius are covered with a variety of ash, lapilli, and scoriaceous debris mainly of explosive origin, but voluminous, massive lavas are seen in the walls of the crater, and they are seen to make up much of the central part of the cone. In 1869, soon after the crater of Vesuvius had undergone profound changes, considerable amounts of lava were exposed along the inner escarpment of its rim.

One also sees ledges of more or less regular lava flows in the walls of the crater of the small central cone of Etna, particularly in the part that is no longer active, but all the outer surface is covered with a blanket of ash and lapilli.

The form of the cone of Micra Kameni cannot, therefore, be considered as anything anomalous.

The northern part of the island is much rockier than the southern part. The surface is irregular and strewn with many blocks, including some of great size. Many low ridges and shallow troughs are aligned north-south. The alignment of these features suggests that the surface is immediately underlain by a lava flow that came from the crater and spread north. This part of Micra Kameni is crossed by shallow cracks that, for the most part, cross the island in a north-northeasterly direction. One of the most conspicuous cracks runs north-south. Judging from what I have been able to learn about them, these cracks opened during the explosion of 20 February 1866; I have been able to determine that in the interval of time between the beginning of March 1866 and the middle of that same month, they had become longer and wider, but when I examined them in 1867 and again in 1875, they had not changed notably.

The easiest place to land on Micra Kameni is at the head of a small cove on the eastern side of the island. Stumps of fig trees can be seen growing there between large blocks of lava.

Pègues reported that at the time of the eruption of 1707, the ground surface of Micra Kameni as well as that of the opposite shore of Thera had dropped noticeably. In support of this opinion, he cites a tradition that was still strong among the surviving witnesses of the event. It states that "the depression of the coast of Thera could be seen where certain storage caves that were formerly at considerable height above the shoreline are now five or six feet below the water level, so that boats could easily enter them." Such is the testimony of the local inhabitants that had seen it. I myself found this eye-witness information in the notes of a respectable missionary of propaganda and native of the island, Don Giovanni Alby, who died in 1830 and could have learned about it from a great many persons who had witnessed the phenomenon. In addition, it is cited in the memoirs of Trévoux and in the annals of the Levantine Jesuits, where it is mentioned in the following way. "At the same time the island of Nouvelle Camène (Nea Kameni) was rising, the nearby island of Petite Brulée (Micra Kameni) was greatly depressed and continued to sink every day; even the coast of Santorini (Thera) that is across from it continued to go down. Several storage places that had previously been a good five feet above sea level are now under water and can be entered by boats without touching bottom."

There is no question that the information cited by Pègues is quite reasonable. Under similar conditions during the course of the eruptions of 1866, the ground at the southern tip of Nea Kameni dropped about 2 meters, even though the center of the new eruptions is a good bit farther from that island than it was in 1707. So it is reasonable to suppose that the depression could have been at least as great at that time as it was during the most recent activity.

It would certainly be better if one could demonstrate this with precise measurements made before and after the event, but that would have required knowing in advance where the sinking would take place. It would also have been necessary for the importance of precise measurements to be understood by observers who were accustomed to carrying out careful scientific studies. None of this was possible, of course, but that is no reason to discount the figures in the chronicles. Foreigners who occasionally visited the region and could have seen the same places only a few times could have seriously misjudged the amount of ground movement, but local inhabitants know all the details of places they visit frequently, and the slightest changes of the terrain do not escape their notice. They note these changes carefully enough for us to accept them as real.

The sinking of Micra Kameni is further demonstrated by the contemporaneous depression of the opposite coast of Thera. Close to the pier at Thira near the Lazarists' building, one sees structures that could not have been built where they now are relative to sea level. When the weather is calm, the foundations of these building are under water. According to tradition, the storage places dug into the conglomerate at the base of the cliff were invaded by the sea in 1707, and one can still see them north of the landing place of Thera a few meters beyond the pillar that has been cut from the conglomerate to tie up boats. Several of the boatmen that I questioned about this did not hesitate to attribute the sinking of the caves to the eruption of 1707.

A recent report was in error in stating that other storage places at the base of Micra Kameni had been invaded by the sea in 1707. Examination of the shores of this island shows that shelters of this kind have never been dug there.

Most of the rocks of Micra Kameni are dense, semi-vitreous, and dark brown. Scoriaceous samples are relatively rare. Outwardly, these rocks closely resemble certain varieties of lava on Nea Kameni, especially some of those erupted during the last eruption. They resemble them in their chemical and mineralogical composition and in their textures and component minerals. But a remarkable peculiarity of these lavas is their abundant olivine. This mineral is detectable under the hand lens and even with the naked eye, so that at first glance one is tempted to consider the rocks of Micra Kameni to be anorthite-bearing lavas; in the lavas of Santorini, olivine phenocrysts are found with anorthite more often than with labradorite. Chemical analyses and tests showing a great resistance of the feldspar phenocrysts to the action of acids were necessary to determine that these are crystals of labradorite, not anorthite.

Nevertheless, an *a priori* factor indicates that these rocks are related to the labradorite basalts. Like the latter, they have microlites with an extinction angle close to zero corresponding to oligoclase or a monoclinic sodic feldspar and not to the labradorite microlites in the anorthite lavas, which have a much greater extinction angle.

On the surface of Micra Kameni one sees a great number of light gray, porous blocks with angular fractures. These rocks are nothing more than pieces of volcanic bombs produced mainly by explosions during the eruption of 1866.

NEA KAMENI

The form and structure of Nea Kameni are fully consistent with the accounts that have come down to us on the eruption of 1707. The southern side of the island is the only part that has been covered by ejecta from the 1866 eruption. The cone has been modified only slightly during these last events, and the lava flows discharged toward the north still preserve almost all the details of their original form. Hence, one can still make out most of that part of the island that was produced during the course of the eruption of 1707.

The longest dimension of Nea Kameni, from the point of Staki to the northern end close to the foot of Giorgios, is about 1350 meters; The greatest width is between the most prominent point on the eastern side and the entrance to the harbor of Saint Georges on the west; it measures 660 meters. The maximum length was noticeably reduced by the eruption of 1866, because the new volcano encroaches on what had been the southern coast of Nea Kameni, especially along the middle part and its western side. It can be estimated that the island has lost about 300 meters of its former length.

Before the eruption of 1866, Nea Kameni had the form of a scalene triangle with the sharpest angle toward the north. According to the English map, the western coast was 1650 meters long, the eastern coast 1440 meters, and the southern coast 910 meters. The area of the island was 65 hectares.

The coasts are sharply truncated, particularly on the east. Staki Point, which juts out toward the NNE, forms a promontory 330 meters long and 200 wide. The eastern shore between this point and the channel that separates Micra Kameni and Nea Kameni has three saliants of which the middle is the most prominent, having a length of about 250 meters; it is remarkable for the narrowness and depth of the two coves that border it. The other two points beside it are less pronounced.

The size and shape of the three points have been only slightly modified by the last eruption; the northern part of Nea Kameni, however, has undergone a notable depression, although it has been less than that of the southern part of the island. The posts set up at various places along the shore to tie up boats still indicate the shape of the coast, but they are now almost submerged, and from their present relationship to sea level one can conclude that the general depression of this part of Nea Kameni has been about a meter. This subsidence has created reefs from some of the blocky rocks that formerly rose from the water close to the shore. It has also caused the sea to enter farther into the coves, but on the whole, the coast has such a steep coastline that the shape has changed very little, and one can still see from the present configuration what the northeastern shore of the island looked like before 1866.

The western coast has only a single cove with a narrow entrance that widens in a north-south direction and ends toward the northeast in a deep, tortuous enclosure with the shape of a cross. This place is frequently used as a shelter for ships. It takes its name of Port Saint Georges from a small chapel situated in one of the niches in its banks. The ground sank there during the 1866 eruption when posts that had been erected along its sides were submerged by about 1.5 meters. During the last eruption, the entrance to the harbor took on a quite different appearance from what it had before. It no longer opens directly to the sea but is connected to it by a kind of channel; but apart from this channel, the harbor of Saint Georges itself has changed very little from what it was before. Its shape has not been notably affected by the depression of the ground.

Before the eruptions of 1866, the southern coast of Nea Kameni had only a single indentation, to which I have referred several times as the cove of Voulcano. This inlet was sometimes referred to as Calangue[3] of Emissions after the gases that are given off there. This part of the shore is now entirely covered by the cone and lavas from the eruptions of 1866.

Before the 1866 eruption, the highest point of Nea Kameni had an elevation of about 160 meters above sea level. It was gradually lowered during the course of the eruption, most notably between 1866 and 1868, when it is estimated to have lost about 5 meters. In 1875, the height of the cone was about 101.2 meters, and it has probably been further modified since that time.

The cone of Nea Kameni is well defined on the eastern, southern, and western sides. It extends to the north where the lava flows make up the northern part of the island. Thus it resembles on a larger scale the shape of Micra Kameni. The cones differ, however, in that, until recently, Nea Kameni was bounded on the east and west and around the southern side by a low strip of volcanic terrain that today is almost entirely hidden under the layer of ejecta of the last eruption that now covers all but a small part of the eastern side, whereas the base of the cone of Micra Kameni descends directly down to the water on all sides except the north.

Another difference is that the slopes of Micra Kameni are steeper and more irregular; the ash and other loose, fine-grained material forms a much thinner mantle than does the one on the flanks of Nea Kameni. Weathering and wave erosion have no doubt been the main factors responsible for the differences in the superficial forms of the two cones. That of Micra Kameni has been exposed to their action much longer, and it has no protection from the wind and waves coming from the east. So the absence of a strip of ground around the southern base has allowed the sea to undermine the base of its slopes. It is not surprising, therefore, that climbing the slopes on the east, south, and west is more difficult than going up the corresponding slopes of Nea Kameni.

The crater of Nea Kameni is elliptical with its long axis oriented northeast-southwest and measuring about 110 meters; the smaller axis is 80 meters. Irregularities of the wall are minor, especially toward the north where one can see no demarcation between the cone and adjacent lavas. The surface of the cone slopes from south to north; the center of the shallow depression at the summit is about 15 meters below the highest part of the rim on the southern side opposite the crater.

Some of the fissures that cross the crater are oriented north-south; others, including the largest ones, trend east-northeast. All were opened during the first months of the 1866 eruption.

Very large blocks of lava are scattered over the surface of the crater. Most are from the eruption of 1707, but some are from the last eruption. Most of the latter are distinguished by their light gray color, angular shapes, and porosity. Almost all the older blocks are superficially altered, because they have been exposed to high-temperature acidic gases after they solidified.

I was unable to determine the positions of the three secondary vents that were active at the same time as the main vent of 1707. These three openings were visible from Skaro, so they must have been on the eastern side of the summit. A contemporary account states that "they resembled embrasures and, on the 14th of September 1811, discharged streams of sparkling molten material with a violet and reddish color that in places tended to be more yellow." No trace of these small flows can be seen today, but this is hardly surprising, because the description indicates that they may not have been lava flows but simply saline deposits of fumaroles. In 1875, the surface of the cone of Giorgio-Kameni had deposits of this kind that, from their remarkable white, red, and yellow colors and their distribution in strips along the upper edge of the cone corresponded perfectly to the description I have quoted. If this is reasonable, one can easily understand that rains would dissolve the deposits, leaving no trace.

The upper edge of the crater has large massive and scoriaceous blocks, particularly on the southern side and toward the southwestern corner. The flanks of the cone, however, are covered with an almost uniform mantle of ash, lapilli, and scattered blocks. The latter are most abundant around the base of the slopes.

A short distance from the southern edge of the crater on the southern slope there is a long fissure that averages about two meters in width and is concentric to the crater. The crack appears to have opened close to the edge of the crater early in the eruption of 1866 and to have been modified as the eruption went on. The fresh appearance of the walls in March 1866 indicated that it had been formed very recently.

Irregular bodies of rock at several places on the southwestern flank of the volcano may mark the sites of earlier vents.

As mentioned earlier, the northern part of Nea Kameni is made up of a series of lava flows that spread from south to north. Those that are still close to the surface are easy to trace. Longitudinal grooves and ridges indicate the direction they followed. Wrinkles that are convex toward the north show that the lava was still semi-fluid when it was moving, but their thickness shows that it was very viscous. The main current flowed from the northern flank of the cone directly northward and veered toward the north-northeast only when it was close to its toe, where it forms the point called Staki. Its slope toward the north is gentle and irregular. At its source the surface is more or less continuous with that of the cone, and at Staki the thickness of the escarpment formed by its toe is still about 10 meters. The mass of lava that accumulated on this side of the island is quite considerable.

Secondary flows have branched toward the right and left. On the eastern side they form the most prominent of the three points mentioned earlier. The flanks of these flows, like those of the main part, are very abrupt. The one that forms the principal promontory on the east is separated at its origin from the branch bordering it on the north and from the main current just to the west by a kind of narrow corridor open to the sky and bordered by high walls of lava. Other branches are on the west; the thickest ones enclose the harbor of Saint Georges on the north and east and form a narrow high-walled passageway at the northeastern corner of the harbor.

The 1707 lavas also spread toward the southwest and southeast, but in that area they did not accumulate in great amounts. Toward the southwest they formed an irregular platform sloping toward the sea, and at its highest point it is only 30 or 40 meters above the water. The promontory of Cape Phleva where it ends is slightly exaggerated on the English map; it ends in a small cliff about 5 or 6 meters high.

On the southeastern side there is another similar plateau, but it is smaller and even lower.

Between these two plateaus rises the mass of pumice that, when the eruption of 1707 began, formed White Island, which was later enveloped by the lavas of Black Island and thereafter called Lophiscos. The cove of Voulcano was also formed during this period.

Lophiscos was a mass of light gray pumice. It rose from the shoreline at the innermost part of the cove as a round mound about 10 meters high and 30 to 40 meters in diameter. The side facing south had a vertical face with two storage places, about 6 meters long and 3 wide, which had been dug into it. The entrances to these two caves were separated from the inner shore of the cove by a kind of walkway that was only 2 meters wide. Fragments of pumice can still be seen around the sides of the cove and on the surface of tiny islets near its entrance, but it is likely that these very thin deposits were

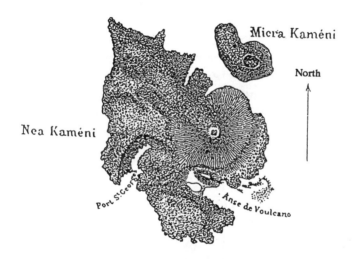

Map of Nea Kameni before the eruption of 1866. (After Bylandt Palstercamp)

brought in from Lophiscos and deposited by waves. Several fig trees grew at the top of the mound. The material that it was composed of was small, loose pieces of very porous pumice. To obtain the cohesion needed for this material to support a vaulted roof and vertical walls, it had to be covered with lime to form a surface layer of pumiceous cement, and from time to time this was renewed where the coating broke away.*

---

* The mound of Lophiscos and the muddy pond at the end of Voulcano Cove are clearly indicated on the map by Bylandt Palstercamp shown above. The contours of Nea Kameni are fairly correct, but those of Micra Kameni were so far off that I had to modify then. This latter island should have been shown about a centimeter farther toward the southeast.

Lophiscos is clearly distinguished from the surrounding lavas by its shape and color. This led Fiedler to state that "this isolated mass of pumice seems to have fallen from the sky." Judging from the sizable dimensions that the white island is said to have had (80 meters high and 1000 meters in circumference) on the 14th of June 1707, when it had not yet acquired its full development, it is evident that the Lophiscos cannot be considered equivalent to the entire island. It is likely that the neighboring parts of the area south of Nea Kameni were also part of White Island, even though they were composed of nothing but dark lavas similar to those of Black Island. It would be wrong, however, to consider all the southern part of Nea Kameni as having formerly been covered by pumice that was stripped by commercial exploitation. A notable part of the former southeastern point of Nea Kameni is still visible today, and one can confirm that it has no trace of pumice. The part that is no longer visible was equally lacking in pumice, as I was able to determine for almost all of this zone at the beginning of March 1866. I also found that there was no pumice whatever on the southeastern part of the island in the area between the cove of Voulcano and Cape Phleva. At the time of my arrival at Santorini, this area, which was soon to be buried beneath lavas, was still exposed. The surface was covered with scattered blocks of lava that were generally scoriaceous and concentrated in the middle of a thin bed of cinders and lapilli, but none of this material resembled that of Lophiscos.

One of the most curious things about this remnant of White Island was the abundance of marine seashells found on it. Lieutenant Leycester, in the memoir he published on Santorini in 1859, said that these shells were mixed with the pumice and badly abraded. But this assertion is contradicted by Edward Forbes, to whom we owe a very detailed description of well-preserved fossils he says he collected from Lophiscos. The conclusions he reached from a study of the species he found at this locality in 1841 was one of the finest known examples of an exact determination of the habitat of marine organisms. He had collected at this place remains of the following sixteen species: *Pectunculus pilosus, Arca tetragona, Cardita trapezia, Cytherea apicalis, Trochus zizyphinus, Trochus fanulum, Trochus exiguus, Trochus Couturii, Turbo rugosus, Turbo sanguineus, Phasianella pulla, Turritella tricostata, Rissoaciminoides, Cerithium lima, Pleurotoma gracilis*, a serpulid, and fragments of *Cellepora* and of *Millepora*.[4]

Forbes' extensive studies of marine flora and fauna from different depths near the Mediterranean coast have enabled him to distinguish eight different submarine levels, each characterized by a special assemblage of plant and animal life. The first level extends to a depth of 2 fathoms, the second is between 2 and 10 fathoms, the third between 10 and 20, the fourth between 20 and 35, the fifth between 35 and 55, the sixth between 55 and 79 (with some of the species of this level ranging up to 24 fathoms), the seventh from 80 to 105, and the eighth from 105 to 230. Accordingly, if the sea floor rises and is exposed, one can examine its organic debris and determine the depth at which the species lived and hence the amount of uplift.

Applying these principles to the case of Lophiscos, Forbes states that the species he found there came from the fourth and fifth levels or between 15 and 55 fathoms. He found, for example, that among the mollusks listed above *Trochus zizyphinus* would have been replaced by *Trochus millegranus* if the depth had exceeded 55 fathoms. If the depth had been less than 15 fathoms, one would have found *Arca tetragona*. All the species listed above are found between these two limiting depths, especially between 20 and 35 fathoms (36 and 63 meters). It is in that range that two of the species, *Pectunculus pilosus* and *Trochus zizyphinus* are most common and that several other species Forbes noted tend to be concentrated. As the depth of the bay of Santorini around Nea Kameni was until recently about 100 fathoms, except in the channel separating the island from Micra Kameni, one can see that the marine life that left these remains in the pumice of Lophiscos must have lived on a submerged peninsula that was relatively shallower than the surrounding seafloor. This conclusion is confirmed elsewhere by the information left by contemporary chronicles of the 1707 eruption, for we have seen that, according to Pègues, that part of the bay where White Island appeared was said to be shallow.

How can one explain this submarine mound in the middle of the bay of Santorini prior to the eruption of Nea Kameni and in exactly the spot where the island later appeared? I have already answered this question and cited the evidence that supports my interpretation, but the question is of sufficient interest to warrant a more detailed look at the proposed explanation.

An explanation I do not support but will nevertheless consider because it is favored by several geologists is that the mound was only a prolongation toward the southwest of rocks underlying Nea Kameni. According to this view, the bank was formed by the eruption responsible for the island; it was based on lavas similar to those making up the northern part of Micra Kameni and discharged from that source during the eruption of 1573. The material that forms its summit and appeared in 1707 as White Island would be nothing but part of the volcanic material that accompanied the eruptions of lavas from Micra Kameni. Or it could have been a mass of pumice that slumped from the cliffs of Thera or Therasia and was transported by the sea to be deposited on the underlying rocks dating from 1573. This latter interpretation was especially favored by the German scientists who visited Santorini in 1866.

The reasoning invoked in support of this hypothesis is as follows: the unit is almost certainly in continuity with the base of Micra Kameni, and the distance between the two points is too short to suppose that there was ever deep water between them. But what gives greater weight to the argument are Forbes' observations regarding the fossils and their state of preservation. The shells collected at Lophiscos form a thin layer within the pumice, and, in his opinion, this shows that at the time of emergence considerable time had elapsed since the last disturbance of the seafloor at this location. Forbes adds that the state of preservation of the bivalves, the fact that they are closed, and their well-preserved epidermis indicate they were killed suddenly, while the valves of the *Pectunculus* and *Arca* are found separated or at least wide open. There is no doubt, he says, that they were buried during deposition of the pumiceous ash in which they are now found.

Messrs. Reiss and Stübel remarked that the interval of 134 years between the eruptions of 1573 and 1707, which, in a geological context, is very short, corresponds rather well to the thin fossiliferous layer studied by Forbes and the small amount of material deposited on it, and they offered the opinion that a brief warming of the sea could abruptly have wiped out the mollusks whose shells are found in the deposit.

The following objections can be raised to these interpretations. Lophiscos was formed of pumice that had a thickness of at least 10 meters, so it owes it origin to an eruption that discharged torrents of pumiceous material. Neither the eruption of Micra Kameni nor that of Palaea Kameni, nor, for that matter, any other historical eruption within the bay of Santorini appears to have produced true pumice. No such material can be found in the rocks of any of the Kameni islands or in the two submarine cones between Thira and Micra Kameni. During my soundings of the submarine cone closest to Thira, I recovered fragments of lava, lapilli, dark-colored ash, and only a small amount of rounded pumice that probably came from debris eroded from the cliffs by waves; I found no coherent deposits of pumice.

The mass that made up White Island cannot, therefore, have come from ejecta of one of the eruptions that took place in the central part of the bay; it certainly cannot have been produced by the explosions that took place during the eruption of 1573, for it has no resemblance to the ejecta on Micra Kameni.

The hypothesis does not explain the occurrence of the fossils in a thin layer within the pumice. The mollusks lived on the surface of a pumice bed that formed the base of Lophiscos and were subsequently covered by a new deposit of pumice that now makes up the upper part of this mass. The lower and upper parts cannot, therefore, be products of the same eruption, and, since no volcanic activity occurred within the bay of Santorini between 1573 and 1707, if the basal section was formed at the time of the first of these eruptions, it is hard to see how the mollusks living on the surface of the first deposit could have been covered before 1707 by a new layer of pumice. Moreover, it is highly unlikely that a deposit several meters thick covering the fossils could be an avalanche from the cliffs of Thera or Therasia; the pumiceous part of the falling material would have drifted on the surface of the sea. How could a landslide have produced a mass of pumice 10 meters thick in the middle of a bay the size of that of Santorini? If such a phenomenon had occurred between 1573 and 1707, the inhabitants of Santorini would certainly have preserved some record of it. The idea that there was a brief warming of the seawater without a volcanic eruption during the same period of time seems no less implausible. So, whatever the cause of their death, the remains of mollusks and zoophytes within the pumiceous material of Lophiscos is inexplicable in terms of any hypothesis that attributes the pumice to the eruption of 1573.

Forbes has offered another possible explanation. He suggests that the floor of the bay at one time had a uniform depth suitable for the kind of life represented by the fossils and that the eruption of 1707 caused a depression of at least 100 fathoms which did not affect the islands around the bay or in its center but only part of the seafloor.

Seen in these terms, such an explanation is obviously unacceptable. The activity of 1707 was not violent enough to have produced a collapse of this magnitude. Besides, how could a depression of such importance take place without affecting any part of the emergent ground, and, more important, how could it have taken place without the population living nearby being aware of it?

The depression responsible for the superficial form and present depth of the bay of Santorini resulted from an extraordinary cataclysm that must be related to the huge outpouring of pumice covering the islands of Thera, Therasia, and Aspronisi. These explosive eruptions were exceptionally violent, but they were interrupted by several pauses, as one can see from the horizontal breaks in the pumice being quarried from the upper parts of the cliffs of Thera and Therasia. One can imagine that after the collapse that produced the bay, there could have been places where a shallow part of the bottom was covered by pumice and where mollusks could have developed only to be suddenly buried a short time later by the products of some renewed eruption. Thus, a layer of submarine pumice with a thin, intercalated bed of fossils could have been formed at a depth of about twenty fathoms. Then the lavas of the 1707 eruption could have pushed up a fragment of such a deposit to form White Island and, in turn, Lophiscos. What happened at the beginning of the 1866 eruption illustrates the mechanism that can lift such a part of the ground in this way.

In summary, I believe that the pumice of Lophiscos should be considered contemporaneous with that in the cliffs of Thera and Therasia and that after it was laid down during formation of the bay, the shoal became the habitat of marine life. The mass of pumice brought to the surface by the lavas of 1707 was completely hidden by the lavas of 1866.[5]

While I have not accepted Forbes' interpretation of events, I agree not only with the accuracy of his observations but also with the biological deductions he drew from them. It is important to note, however, that the levels he assigned for the various marine organisms living in the Grecian archipelago are based on a very limited number of species. One could also object that his conclusions regarding the occurrences of these marine species in the bay of Santorini could not be absolutely rigorous because of the modifications imposed by the temperature and chemical composition of the water near a volcanic vent. But these remarks cannot diminish the general significance of the conclusions that can be drawn from his research. The marine species he noted in the pumice of Lophiscos certainly lived at a shallow depth below sea level and not at 150 to 200 fathoms, like those found in almost all other parts of the bay.

The cove of Voulcano situated at the southern foot of Lophiscos was only a small inlet of the coast. Its north-south dimension was estimated to be about 150 meters. On the western side it was bounded by an abrupt bank formed by the edge of a lava flow a dozen meters or so above the water level. The cove ended at the northwest with a low beach of pumice and at the northeast at a kind of make-shift pier that formerly served Lophiscos. To the east, the shore was low and very irregular; about midway along its length a small promontory projected out toward the south-southwest, then turned toward the southeast.

The entrance to the cove was narrowed on the eastern side by several small islets. Between the westernmost islet and the opposite shore on the west, the width of the cove is about 90 meters. It was reduced to 50 meters near the small promontory, Beyond that, the end of the cove formed a wide cul-de-sac.

On the western side of the entrance, soundings indicated a depth of 13.5 meters, but the ground rose steeply toward the east. Across from the promontory, the maximum depth was still 16 meters, but farther in it diminished rapidly. The cul-de-sac was nothing but a channel, the depth of which did not exceed 4 meters on the south and was only a few centimeters on the north.

The low beach bordering the end of the cul-de-sac on the northwest scarcely rose above sea level, so when the wind blew from the southwest it was almost entirely under water. The surface of the beach was formed by grayish sand composed mainly of fragments of altered pumice. A pond of salt water, known locally as the muddy lake extended along the western base of Lophiscos. The beach that separates the pond from the end of the cove was about 20 meters long. The pond measured about 40 meters in length and up to 20 wide. As the map of Bory de Saint Vincent shows, its shape was triangular. The maximum depth was 4 meters. It is likely that it was formerly part of the cove and directly connected to the sea before it was cut off by the bar of sand. Boiling and emissions of gas continued in the part of the cove that was still connected to the sea. The water there was discolored and warmer than the sea.

The water in the cove of Voulcano had long been known by seamen to attack the copper sheeting of their boats; it eroded the metal and killed the organisms adhering to the hull. After a few days in the cove, all the material that reduced the speed of the boats could be wiped from the wet sides with a broom.

Lieutenant Leycester has left us a description of the cove of Voulcano (which he called the bay of gaseous emissions): "it is," he says, "wide enough to take a ship of the line. Close to the entry of this small bay, especially if strong winds make the sea choppy, one is struck by the reddish color of the water. The same color is seen on the surface of nearby blocks of black lava. Close examination reveals that bubbles are rising through the water and bursting at the surface. Not only are bubbles being released but occasional jets of boiling water come from a depth of 6 fathoms (10.8 m.) with enough force to rise an inch or two above the surface. What is even stranger, the water of these jets is not always red but sometimes green and at other times completely colorless. It is said that this water has a strong effect on copper. Some badly tarnished copper coins that we left in the water for a week became as shiny as if they had just come from the mint after wiping off a thin coat that could be removed by the touch of a finger. When the vent discharged violently, the jets of Voulcano Cove colored the seawater as far as the coast of Thera. The thermometer revealed no measurable difference between this water and that of the sea."

This last statement does not agree with that of Pègues, who noted there in 1836 a temperature notably hotter than that of the sea in the rest of the bay. During the month of March 1836, he found that the temperature of the water of Voulcano was around 20 degrees, while that of the sea in the middle of the bay was only 14.5 degrees and that of the air 16 degrees.

According to information I received at Thira, Pègues' observations were very precise, and the temperatures of the springs of Voulcano seemed to have risen even more in recent years. The maximum observed is said to be near the western shore of the cove. On that side one can stand in the water with warm feet while elsewhere in the cove ones feet will be cold even though the upper part of the body will be in warm water.

The waters of Voulcano Cove were known for their proven curative powers, particularly for illnesses related to lymphatism. People came to bathe there in the summer, and to avoid repeated trips between Thira and Nea Kameni, the main families of Santorini had built houses around the cove. In 1866 when the eruption began, about thirty of these houses had been constructed on the southeastern part of Nea Kameni, on the high ground around Voulcano Cove, near the foot of the cone of Nea Kameni, and beside the canal between Nea Kameni and Micra Kameni. The buildings were small, and most had only a single room. Like all the houses of Thera, they were built with irregular blocks of lava held together with mortar and pozzolana. The roofs were arched; the doors and windows were narrow and closed by light wooden shutters and doors. Several were surrounded by a small square courtyard with a wall of loose stone. In front, some had a shed that allowed the bathers to enter the water directly while sheltered from the wind and sun. Their orientations varied, though most faced the shore. They were in two groups, one on the eastern side of the cove and the other along the channel between Nea Kameni and Micra Kameni. Two houses that belonged to Mr. de Cigala were near the southeastern tip of the island separated some distance from the others.

In the middle of this group of houses were two chapels, one Orthodox and the other Catholic; they were located half way along a path that ran between the two groups of houses, about 7 to 10 meters above the shoreline, and was bordered on both sides by small walls. No houses had been built west of the cove, probably because the ground was higher and more broken, and the slope down to the shore was steeper. All that had been built on this southern side of the cove were three small sheds like those in front of the houses on the opposite bank*.

Very soon after the eruption of 1866, a pier was built on the southeastern coast of Nea Kameni on the channel between that island and Micra Kameni, and stone pillars were erected to tie up boats. Five of these posts were on the landing place between the houses of Dr. de Cigala and Nicolas Sfoscioti; two of them were completely covered by lavas of the last eruption. A sixth had been set up about 100 meters north of the house of Sfoscioti below the part of the shore that projects out east of the base of the cone of Nea Kameni. Three others were located in the first depression north of the previous ones. Another was on a point south of the eastern promontory of Nea Kameni, and the last still stands on the eastern side and about 150 meters from the end of the flow that forms the northern end of the island.

Four posts had also been set up on the southern coast of Micra Kameni on the other side of the channel. The southernmost one is now very

---

* The names of the owners of houses on Nea Kameni were: Pierre Rubin, Nicolas Barbarigo, owners of the storage places on Lophiscos; Auge Fumail, Constantine Carao, Caoustos, Michel Murat, Lazare Delenda, Antoine Rinaldi, Alexaki, Minetas, Pierre Rubin, Gaspard Delenda, Marquisini, owners of houses along the eastern shore of the cove; Petalas and de Cigala, whose houses were on the southeastern tip of the island. Gavallos, Sigalla, Foufonderos, Nicolas Basco, Cristophe Matha, Jacques Basco, Antoine Matha, Philopatridis, the Dimarchie of Phira, Nicolas Josika, Syrigo, Vitale, Christophe Matha, Nicolas Sfoscioti, owners of houses built along the channel of Micra Kameni. Between the house of Matha and that of Nicolas Sfoscioti there was a wide space occupied mainly by a walled enclosure for storing pozzolana.

close to the pass separating Micra Kameni from the lavas of Giorgio Kameni.

Finally, to complete this account of the posts, I should mention that there were a certain number of them around the harbor of Saint Georges. Two had been set up at the northern side of the entrance, five along the innermost shore, and two near the southern side of the entrance. Of the five inner ones, three were close to the small promontory where the chapel is located.*

Because the rocky bottom makes anchorage more difficult around Nea Kameni and in the channel separating that island from Micra Kameni than it is in the harbor, the posts are very useful for boats that tie up there. In certain seasons of the year, the waters around Nea Kameni are crowded with ships of all sizes; ships of moderate size, about a dozen in number, could find shelter in the harbor of Saint Georges, and one or two could anchor in the cove of Voulcano. But the most sought-after anchorage was in the channel between Nea Kameni and Micra Kameni. Its two entrances made it easy to get in and out, and it was deep enough to be accessible to even the largest ships.

The lavas of Nea Kameni are compositionally identical to those produced during the eruption of 1866; both are labradorite-bearing and have a variety of forms. Most are black, glassy, and either massive or scoriaceous. The dark color on the surfaces of clean fractures shows that they have resisted weathering for a century and a half. In a few very restricted areas they are red or yellow, owing to hydrothermal alteration by fumaroles that are now extinct. These colors can be seen all around the summit and inside the crater of Nea Kameni. The lavas that cover most of the surface of the island are so scoriaceous and blocky that it is difficult to get from one side of the island to another, but it is not impossible, for I have crossed it several times in various directions.

Rain water filters into the broken, rocky ground and quickly disappears, so there is no trace of gullying. Although it is covered with ash and lapilli, the surface of the cone is perfectly intact and lacks the erosional grooves one commonly finds on volcanic mountains.

The absence of alteration in the rocks of Nea Kameni and the rarity of ash make that island almost entirely sterile. One sees only a few tufts of

---

* Most of the posts set up around Nea Kameni are now hidden, so I have omitted several from this list.

meager grass on the flanks of the cone or a few bushes in the summit crater. Fig trees had been planted in the latter place, but the trees were few in number and grew with difficulty; they were killed by the gases of the last eruption.

GIORGIOS-KAMENI

INCLUDING THE LAVAS AND OTHER VOLCANIC MATERIAL OF GIORGIOS, APHROESSA, AND THE ISLES OF MAY

(Based on observations made in September and October 1875)

The discharge of fluid lava and the ejection of ash and scoria from 1866 to 1870 have profoundly modified the central part of the bay of Santorini. In places where the sea recently had a depth of more than 100 fathoms there is now a field of lavas covering about 50 hectares and rising about 80 meters above sea level. This enormous mass of superficially scoriaceous rock has a very irregular surface. In places it is crossed by irregular ridges; elsewhere it is grooved by deep crevices. Its northern part is surmounted by a cone reaching 126.5 meters above sea level near the widest part of the vent. The new lavas rose initially from two separate islets, Giorgios and Aphroessa, but today they are joined to the 1707 lavas and form only a small part of the total island. The main topographic effect of the last eruption was to enlarge the island of Nea Kameni from its size when formed in 1707. Its surface area has more than doubled, and it has a new cone 25 meters higher than the one formed in the last century.

To establish the present state of the volcano, I drew up, at a scale of 1 to 10,000, a plan of the island of Nea Kameni showing all the additions it has had in recent years (Plate XXIX). In addition, I have made, at a scale of 1 to 2400, a plan of the summit area of the new cone (Plate XXX). These two maps, which show the situation and all the irregularities in the form and depth of the craters, demonstrate the alignment of the main vents of the volcano and support the opinion of geologists who consider it a general rule that most new vents open along linear traces. This observation has all the more importance in this case, because until now the phenomenon has been noted only where the site of the eruption has been several kilometers from the sea. The new eruptive vent of Santorini was submarine in its initial stages, and although the eruption began under a layer of water, one can see that it broke out along a linear fissure, just as it would on dry land. As a result, the rule first proposed by Gemmellaro on the basis of his observations at Etna finds new support here at Santorini.

The lavas of Giorgios have completely covered the source of those of Aphroessa, so that the eruption of 1866 formed only a single cone in the northern part of the island near the edge of the new lava field. This cone occupies the site of the cove of Voulcano on the former southern shore of Nea Kameni. Its upper edge is elliptical measuring 190 by 120 meters. The largest dimension trends northeast joining it to the 1573 cone of Micra Kameni. This is also the orientation of the main craters at the summit of the cone. The largest of these craters, marked A on the map, is situated near the center of the upper truncation of the platform.* Its opening has an irregular shape, but both diameters are about 40 to 45 meters. It is divided into two almost equal parts by a small ridge oriented north-northeast. The highest part of the rim is on the south and the lowest on the east. The bottom of the crater is about 9 meters below the average elevation of the rim and 20 meters below the highest point on the cone. The walls are made up of fragments of lava thrown out from the crater. They are very abrupt, especially on west and north; the slopes range from 35 to 50 degrees. No traces of saline deposits or fumaroles are to be seen.

Crater B, the next in importance after the one just described, is situated at the north-northeastern end of the summit. It is elongated in the same direction. The largest diameter of the rim is 45 meters, the smallest about 30. The average elevation of the rim is about 7 meters below the highest point of the cone. It is breached on the north-northeast. The bottom of the crater, which is 10 meters deep, is littered with debris; the walls, which are made up of small fragments of ejecta, are the site of numerous fumaroles, particularly near the upper edge and on the side of the breach of the rim.

To the south of the junction of these two craters there is a large depression as deep as the two just mentioned. At first glance one might take it for another vent. The elongated cavity, marked C on the map, is composed of two parts, one much deeper than the other and extending east-west in the

---

* See Plate XXIX.

direction of the largest diameter. From its western end it seems to stand on the prolongation of the central crater described earlier. It is surrounded by walls with the maximum angle of repose for loose material; the walls extend around all but the northwestern side. It is not difficult to climb the slope. The largest diameter between the upper edges is about 50 meters, the smallest 30; the depth is 13 meters. The second part of this hole adjoins the eastern edge of the first. It differs in that its elongation is orientated toward an east-west direction. Its average depth is only 5 or 6 meters. It is elliptical with a breach on the southwestern side. The largest diameter is 30 meters and the smallest 20.

The upper part of the cone still has a line of less important depressions distributed in a chain around the northern and western crests. These depressions form a kind of moat about 4 to 5 meters deep and 20 meters wide.

The explosive eruptions during this activity were strongly influenced by the configuration of these vents. They were very unequal and intermittent. When their intensity was moderate, they were concentrated at points along a short part of the fissure marked by distinct vents, such as points A and B that I have just described or others, such as D, on the same alignment beyond the summit and at the crater that discharged lavas at Aphroessa. During periods of moderate activity, ejecta accumulated between the two orifices at A and B. During one of the explosions during this phase, the partition between the craters disappeared and the two craters merged into one. The top of the cone was then occupied almost entirely by a single large crater. A new period of relative calm followed, and the outlines of the vents were re-established. The smaller of the two became centered within the larger depression formed earlier. By this reasoning, the large crater marked D and the small depressions forming a semicircle around the northern and western rims would be the remains of a larger crater formed by strong explosions, and craters A and B would be the focus of more moderate activity. The form and position of crater D are not consistent with it having been a vent on the scale of the other two. The semicircular depressions around the north and west could also have another explanation. In the initial phases of the eruption, when the cone was still low, it discharged both gases and molten lava from the same fissure. The lavas had a temperature close to solidification and accumulated around the vent, forming a plug that impeded the release of volatiles. When the latter could escape only around the edges of the pile of lava, they formed a circular trough which remained in nearly the same place from the first months of 1867 until the end of the eruption. This trough was especially distinct in 1867, when it was in exactly the same position around the lava dome where the semicircular depression is seen today.

On the northeastern side, the slope descends at a constant angle of about 35 degrees from the top of the cone to sea level. A round boulder can roll unimpeded from the top to the base. The uniform mantle of lapilli that covers it is broken in only a few places by a few blocks that are slightly larger than average.

Toward the east and north the surface has the same appearance but the apparent base of the cone is somewhat higher. On the eastern side it abuts against thick flows of lava the lower members of which were discharged in 1866 and 1867 while the upper ones were erupted during the following years. The space between the lavas and cone has the form of a ravine sloping from an elevation of about 80 meters at the south down to sea level on the north. As a result, the line marking the base of the cone rises toward the south.

On the northern side, the cone of Giorgios abuts against that of Nea Kameni. In 1866 when it first began to grow, it was separated by a space of about 40 meters, but in March of 1867, the bases of the two cones first touched one another. Since then the growth of the new cone has caused it to encroach on its neighbor covering the southern flank of the latter to a height of 74 meters. The result is a kind of saddle with scattered blocks that have fallen there or rolled down the slope. A gently sloping path that goes up around the flank of the cone of 1707 provides easy access to this spot. It is the route by which the cone is usually climbed. The flank of this high place is less regular on this side. From top to bottom, it has long trains of large blocks that make the climb less difficult by providing stable foot holds. The slope of only 25 to 30 degrees is not as steep as that on the eastern side, and it becomes even more gentle toward the top where a stretch of ground extending east-west has an inclination of 5 to 16 degrees.

Toward the west, the line defining the base of the cone descends again to an elevation of about 10 meters. On that side the flanks of the cone are very abrupt; enormous blocks standing almost vertically look as though they are on the verge of tumbling down to the foot of the escarpment. The steepest part of the slope is surmounted by a small, nearly level space formed by nearly flat-lying lavas. As a result, the slope on this side is not as uniform as it is on the eastern side of the cone.

The form of the southern flank of the cone is quite different from that of the other three sides just described. It is there that the lavas broke out during the entire eruptions, and, in addition, the summit of this side of Giorgios is a continuation of the flat area formed by the accumulation of the high part of the lava flows. The peak is only about 30 meters higher, and toward the western end, the difference of elevation is even less. The southern flank of the peak is covered by lapilli. The slope is rather irregular except at the southeastern edge where the slope of the cone has a large breach that seems to have been the source of flows coming from well within the interior of the crater. Alternatively, it may correspond to a buried fissure running cross-wise with respect to the main fissure along which the eruption developed. At present, the ash and lapilli covering this depression make is impossible to be more definite about its basic nature. The upheaval of the ground that one sees there is consistent with either one of these possibilities, and the flows exposed a little lower cannot be related to it clearly enough to resolve the question.

The plateau has an average elevation of 85 meters. It extends from the cone toward the south, forming a crude triangle with the summit at the northern end and the base on the south defined by its east-west trending edge. The longest part of the plateau, corresponding to its base, measures about 900 meters; its smallest dimension, at the upper end of the triangle just mentioned, is almost 350 meters. Taking into account the irregularities of this area, its surface must total at least 18 hectares.

The plateau is covered almost entirely with ash, but it is grooved along the greater part of its length by three relatively narrow flows. These lavas are made up of large, black blocks, the dark color of which contrasts with the pale gray cinders around them. In any case, the slight tilt of the plateau from northwest to southeast seems to have controlled the courses of the three lava flows. The latter will be considered in greater detail below in order to determine what happened to their flow fronts as they moved toward the sea.

The most interesting part of the plateau is the northwestern edge. This area is occupied by a large, deep crater, the center of which is essentially on a straight line between the two main craters of Giorgios. I consider it as having opened, like the others, on the main eruptive fissure of the volcano, which is to the southwest and only a few meters from the base of the cone. The upper edge of this vent is 27 meters below the highest point on Giorgios. Its form is nearly circular, its average diameter 65 meters, and its depth 18. Its inner walls are very abrupt except on the northeast side. It is open to ground level, and its rim does not project onto the surrounding ground. It continues southward where it forms the highest part of the plateau, but only forty meters from the western edge of the crater, one encounters an escarpment of 40 to 50 meters that forms part of the eastern slope of the cone of Giorgios. The southern part of this steep, blocky cliff curves toward the southeast, so that one is tempted to consider it part of the western flank of a cone with the large crater just mentioned at its summit. The central mound at its top and its continuity with the plateau would explain the burial of the other outer walls of the cone.

After defining the southeastern edge of the plateau for about 100 meters, it turns sharply toward the west making a conspicuous curve. It maintains an east-west direction for about 100 meters then suddenly veers southward and follows that direction for a distance of 400 meters. Along most of its length the talus from the plateau forms a steep escarpment about 40 to 50 meters high. The point where the edge of the plateau forms a sharp angle has a very large, pillar-like block of lava.

Masses of lavas project out from the plateau at seven different places near its southern limit. Each of these is indicated by a number on my map and is also marked by special symbols indicating the nature of the fumaroles that were noted there.

The most notable and consistent feature of these centers is that they were places where much gas and steam were discharged from fumaroles that have since declined in strength. The upheaval of the ground seen there, the abundance of ash and lapilli in the surrounding area, and the alteration of the rocks by acidic fumes show that these were sites of explosive eruptions. This has been confirmed by several eye-witnesses. The Lazarists have told me that on several occasions during the eruption they could see violent explosions at different places within the area of the lava field and at considerable distances south of the top of the crater rims.

The most remarkable of these vents are those situated farthest toward the west; they are numbered I and II on Plate XXIX. At the southwestern tip of the plateau one sees large amounts of ash, scoria, and blocks of lava of all sizes. The talus slope extending from the southeast to the southwest forms a steep precipice 50 to 60 meters high. It marks the starting place of an important flow of lava that descends toward the west, first with a slope of about 25 degrees, then at 5 to 10 degrees to form a group of flows extending westward toward the southernmost tip of Palaea Kameni. The upper part of this 100-meter wide flow has a very distinctive dark color and two deep depressions that extend along its sides.

The second vent is situated northeast of the first and closer to the center of the platform. Its conical dome can easily be recognized from a great distance. It rises from the southern side on the edge of an almost vertical escarpment 50 meters high from the foot of which a deep gorge extends all the way to the shore. To the east, one sees emerging from its base a very narrow flow that is very narrow at its source but soon becomes wider toward the southeast; it continues in this direction as far as the end of the lava field. Along its course it sent off two strong branches that dropped toward the southern coast. These two secondary flows of dark black lava can easily be distinguished from the coast of Thera. On the other side they reunite midway to the coast with the flow that crosses the center of the plateau and become a single flow.

In the middle near the junction of these two lavas one of the vents rises from the eastern shore at the edge of the plateau; this one is marked IV on the map. Like the others, it is manifested by a mass of lavas that have been highly altered by acidic fumes. It rises 4 or 5 meters above the surrounding lavas. The absence of a layer of ash and lapilli in this area distinguishes this vent from all the others. It also differs in that it is not surmounted by a deep round depression open toward the sea. The depression near its southeastern base is a long, narrow, open channel 20 to 30 meters deep, sharply cutting through the middle of the lavas. This unusual feature may be associated with vent IV that has been surrounded by lavas that came from sources on higher parts of the plateau, and, as a result, if there was a wide round basin at its southeastern base it must have been filled in large part by the piling up of these lavas, and the channel just mentioned could be one of the last remaining signs of it.

The vents numbered II and III are remarkable for the wide extent of the deep depressions cut into their bottoms. The basin that can be seen from the southern side at the foot of vent III is divided into two unequal parts by a flow of lava situated at a level of 30 to 40 meters; it spreads out toward the sea. The lavas that cover the bottom of the depression had already been discharged in 1867, while those that go around their sides were discharged in the following years.

Vent VII takes up a broad area about 200 meters in diameter, surrounded by stepped talus slopes and connected to the sea through a narrow winding passage a short way to the northeast.

Finally, the closely spaced vents numbered V and VI are situated near the northwestern end of a small terrace covered with ash and lapilli and rising about 60 meters above a round depression that opens to the coast on its eastern side. The basin is unusual in that its bottom is below sea level. It forms a small, sheltered inlet surrounded by talus slopes 50 to 60 meters high that are at the maximum angle of repose of loose material. This harbor, which is about 60 meters wide and 110 meters long, is elongated in a N 55° E direction. It is used now as an anchorage by the ships that operate in the bay.

From all that has just been said, one can see that these vents are characterized not only by the explosive eruptions that came from them but also by their form and especially by the distribution of deep depressions that tend to be concentrated near the shoreline. The piles of ejecta produced by these eruptive centers have obstructed lava flows, deflecting them from their normal course and forcing them to flow in a lateral direction toward the sea. As a result, many of the flows coming from higher elevations abut against these volcanic cones, surrounding and piling up on the uphill sides while forming a barrier that has protected the downhill sides from being flooded with lava. Thus, while acting as an obstruction they have contributed on one side to the growth of the plateau and on the other they have produced deep, steep-walled depressions the outer sides of which are open to the sea. These relations do not necessarily imply that the vents are related to sources deep under the volcano. It is more reasonable to suppose that most of these bodies are situated on flows that have come a long way from their sources. Any obstacle that by chance stops the advance of a flow must cause a bulge in the surface and a depression on the down-slope side; in this way, the lavas pile up and exert their maximum force against the obstacle. If a lateral outlet is formed, they will continue to flow around one side or the other of the obstruction, sometimes both, as they move toward the base of the volcano. Whether the barrier remains effective or eventually fails, the lavas accumulated behind it become a reservoir of gases and water vapor. One can see, therefore, that strong explosions could be produced in these places and, more important, that the rocks would be altered by high temperature fumarolic emissions.

The distance between the point where a flow emerges from the ground and where it stops to form a secondary vent like those just described is, of course, quite varied. The most active secondary vents are those that appear to have had a direct connection with the main crater of the volcano. Vents I and II are especially instructive in this regard. They were the sites of strong explosions that threw out large blocks of lava, and their connections with the main craters are very evident. Between these two vents and crater D, the ground is riddled with cracks like plowed furrows running between the two places. Numerous fumaroles are aligned along these fractures. The same cracks, fumaroles, and broken ground can be seen on the north side of crater D, and they can be followed up the southern slope, then across the northern summit of the cone of Giorgios. Thus the continuity of this line of volcanic emanations can be seen from the northeastern edge of the large cone to the two secondary vents, I and II.*

The other vents, III, IV, V, VI, and VII, are much less important. They do not appear to have been the sources of lava flows. Their ejecta are much less abundant than those of vents I and II. The explanation I have proposed for the secondary vents on flows that have come far from their sources applies only to these particular vents.

The flows discharged during the course of the eruption have accumulated to great thicknesses. Compared with the lavas of Vesuvius or Etna, they are distinguished by their thickness and the short distances they have flowed. Like all lavas rich in silica, these have a weak fluidity; they reach the surface under conditions close to solidification. But it would be a mistake to believe that they are always thrown out as solid blocks. Observers that have followed the eruption from its beginning have generally seen only a growing mass of volcanic rock coming to the surface and piling up as incoherent blocks. One of these observers has called Giorgios a *homogeneous cumulus volcano*. But already at that time it was possible to see that at a few local spots the lavas still had a certain amount of fluidity and were spreading in specific directions. Using the point of a stick, one could make deep imprints into the blocks close to the main fissures where combustible gas was being released. As of now there is still no true flow and the lavas discharged in small volumes have cooled quickly on coming in contact with the sea to form two small, rounded features. But as soon as the emission of lava increases a

---

*Refer to the map of Nea Kameni and Giorgios, Plate XXIX, at the end of this volume.

little one can distinguish longitudinal tongues bordered by blocks and crossed by transverse furrows that are convex away from the source. These relations are clearly visible on the surface of the lavas erupted in the spring of 1866 from the center known as Aphroessa. Their source, which was close to vent I on the plateau, was later buried by lavas from Giorgios, but the flows that spread from it toward the northwest still remain exposed on the southern shore of the entrance to the harbor of Saint Georges where all the details I have mentioned can easily be seen. The longitudinal fissures along the margins, the curved transverse ridges, and the trails of blocks covering the sides are all clearly visible. The youngest flows from Giorgios have the most obvious characteristics of lavas that were very fluid when they were erupted. One can see this at a glance. For example, the two narrow flows that cross the plateau from northwest to southeast must have been remarkably fluid to be able to spread more than a kilometer to reach the sea instead of piling up close to their source. The average thickness of these flows is 5 to 6 meters and their width on the plateau about 30 to 40 meters. It should be noted that they spread on a very gentle slope along the middle part of their course. They built two ramparts of solid blocks at their margins, while the middle parts of the currents were still fluid. At that stage, the central part was fluid enough to solidify as a long strip that was first continuous but was then separated by withdrawing into large parallel sections normal to the direction of spreading. In places, the separation of the central part of the flow from the two bands that form its margins is unusually clear. In this respect, the lavas of Santorini have some unusual features. For example, in the middle of the lavas that spread northeast from vent VII on the plateau one sees a deep cut with two sharp vertical walls. The bottom is simply the interior of the flow and the sides are its margins. The deep depression in the center is due to draining of the still-fluid lava from the interior while the margins had become solidified.

The fluidity of the lavas of Santorini at the time of their emission is also illustrated by the curious drainage features one sees in them. For example, in the middle of the lavas that descended toward the sea below vent II on the plateau, one can see a vertical fissure that is scarcely a meter wide but 5 to 6 meters deep and 100 meters long. This crack, which is slightly curved in the direction of the flow, must have been due to a withdrawal following its solidification. Another example that is even more spectacular but more difficult to explain is seen near the southeastern point.

Further evidence of the initial fluidity of the lavas at the time of their eruption is found in certain blocks with crude scales similar to those on the convex surface of a piece of hot iron that has been violently bent while in a semi-plastic state. This is found on almost all the blocks, but the best example one can examine is a large mass of lava near the bottom of a ravine south of vent I on the plateau. This mass is part of the margin of a flow that is divided into thick sheets 2 or 3 meters thick and 20 to 30 meters long arranged like the pages of a partly open book.

Thus, examination of the lavas of the last eruption clearly reveals that they had a rather pronounced fluidity at the time they were discharged. The last lavas erupted formed the most characteristic flows, probably because at that time they were farther from the effects of the sea and cooled more slowly. Under these conditions they must have behaved like the lavas of volcanoes on land while still preserving the properties inherent in their high silica content and low degree of melting.

Owing to the presence of the cone built in 1707, the lavas of the last eruption could not spread toward the north. In the beginning, they could move toward the northwest from the eruptive vent located at Aphroessa, but from May of 1866 until the end of the eruption, the lavas spread only in the sector between the west and southeast that includes most of the southern slope. The ends of those flows that moved more toward the east were diverted farther toward the east and even toward the northeast as far as the base of the cone of Micra Kameni. Several flows discharged at different stages of the eruption came from the same source and followed the same course, so that one was superimposed almost exactly over the other. At times the new flows spread beyond the older ones, hiding them completely from view or leaving only small parts of their margins exposed. This is what happened in the southeastern part of the lava field. At other times they were stopped or diverted closer to their source leaving the extremities of the older flows exposed. This was particularly noticeable in the flows directed toward Micra Kameni. Examining the lavas that moved in this direction, one can easily recognize three separate overlapping tiers stepping back toward the south, one over the other. The lowest layer, which is about 20 meters thick, dates from 1867; the other two, which rise about 40 and 60 meters, were formed during the following years.

To understand the substantial thickness of the flows one need only note the chemical composition of the lavas and, at least during early stages of the eruption, the cooling effect of seawater. Other factors that appear to have been equally important must also be noted. The eruption temperatures of the lavas must have differed from one flow to another, but, unfortunately, we have no reliable way of demonstrating this. Their flow velocities varied within wide limits, and this had a strong influence on their ability to spread. For example, the lavas moving northwest from Aphroessa in April and May of 1866 appeared to be very fluid; this has been attributed to the rapidity with which they were discharged. On the other hand, a group of lavas that spread toward the south in 1867, south of vent II on the plateau, must owe their substantial thicknesses to the slow rate at which they were erupted.[6] During my stay on Santorini, I followed this extraordinary flow with great interest. Its terminus was more than 40 or 50 meters above sea level, while the rest of the flow rested on a base of 200 meters. The flow front formed an imposing cliff that slowly advanced at a rate of less than one meter a day. The fluidity of the interior of the flow at that time is shown by the deep grooves along its direction of motion, but its condition at the time it was discharged can only be guessed from the almost imperceptible velocity of its great mass. The occasional fall of blocks that formed its carapace was the only sign that it was mobile.

When one examines the outer limits of the lavas of the last eruption one is struck by the nature of the irregularities one sees on their surface. Flows formed at different times or even during the same stage of the eruption can be seen side by side, separated by narrow, longitudinal depressions that may be as deep as 50 to 60 meters and have blocky talus slopes at the angle of repose (about 35 degrees). The two most remarkable examples are those that came from vents I and II on the plateau and advanced as far as the southern coast.

The first, after having moved toward the southwest during the first half of its course, suddenly changed direction and turned southeast. Its total length is about 400 meters. Very wide at its source, it narrows greatly throughout most of the rest of its length. Its base is defined by the two blocky talus slopes that form its margins.

The second, which is even narrower, is more difficult to reach but more imposing; it is almost 500 meters long. It first moved toward the south for about 300 meters then turned to the southeast and eventually reached the sea.

The depressions between two flows are produced when the bases of two blocky talus slopes come together. We know from many observations at

Vesuvius and Etna that only a small obstacle is needed to arrest the advance of a flow or at least to modify considerably its direction. When a lava flow encounters a topographic high, an artificial barrier, such as a wall, or even a tree trunk, it is often seen to be deflected by the obstacle and diverted to a different course. The blocks it is carrying pile up against the obstacle and increase its solidity; they create a strong rampart of blocks and, in most cases, greatly augment the resistance of the original obstacle. A flow that is stopped in this way at its base sometimes manages to surmount the obstacle, but more often its forward motion is arrested, and it begins to send out lateral branches. At Vesuvius and Etna, it has often been possible to make use of this behavior to change the course of lavas that threaten to ruin dwellings and cultivated fields. When an obstacle of any kind, quite insignificant in appearance, is set up in front of the invading mass of lava, it can often deflect it. So what kind of an obstacle could a flow encounter that would be stronger than another flow against which it abuts, at either its side or extremity? When the sides of flows meet, the obstruction is so much stronger that the main force of the fluid mass in the interior is exerted in another direction. The slope of the ground causes the maximum pressure to be toward the front of the flow rather than toward the talus that encloses it on both sides. One should not be surprised, therefore, by the long ravines of the kind we have noted in these two examples.

The base of these flows is generally only slightly above sea level and at times may even be below it along much of its length. Most of the irregularities of the coastline correspond to flow fronts and submerged margins between them. Thus the deep embayment on the western shore of the lava field on the southern side of Giorgios owes the shape of its southeastern side to this effect. The point designated G on the map is almost in the middle of a ravine between two lava flows where the base of the flow first went below sea level. The flow that formed the marginal talus was largely overrun in 1867 and completely buried by the superposition of lavas erupted during the following years. The northern talus is made up of lavas erupted in March, April, and May of 1866; at that time Aphroessa was still a distinct center of activity, even though it had become connected to the older ground of Nea Kameni. The submerged part of the ravine is remarkable for its long, narrow shape. It is about 150 meters long but has an average width of only 10 meters and a depth of less than 4 meters.

Another partly submerged depression between lava flows comes into the innermost part of the first cove at the southeastern end of the harbor of Saint Georges and marks the boundary between Aphroessa and the older part of Nea Kameni. This depression, bordered by lower lavas than the former, was formed in March of 1866 by lava that poured out toward the northwest from Aphroessa and abutted laterally at their base against the former shore of Nea Kameni. In April of 1866, the lower part formed a long channel with its connection to the sea restricted only locally by fallen rocks. Since then, the rocky slopes have diminished the extent of this arm of water. Two small ponds, each about 10 meters long and 4 or 5 meters wide, are all that remain.

The narrow passageway between the lavas of the last eruption and the cone of Micra Kameni has a similar origin. In March of 1867 lavas began to escape from the southern flank of the cone of Giorgios. The easternmost flows went around the base toward the east and, heading northward, appeared to be on a course that would completely cut off the entrance to the channel between Nea Kameni and Micra Kameni. But as the lavas advanced in this direction, their motion was retarded and their thickness increased. Finally, when the space between the flow front and the base of Micra Kameni was reduced to about 10 meters, the lavas came to a complete stop. From that day on, the lavas flowed in different directions. Despite the debris falling down the slopes and reducing its depth, the passage can still be crossed in a small boat.

The tendency for lavas to form the sides of these deep depressions is not confined to the last eruption; it can be seen in other lavas in the bay of Santorini, most notably those erupted in 1707. The deep embayment that ends at the harbor of Saint Georges on the northeast, as well as another somewhat less pronounced corridor on the eastern side of Nea Kameni near the northern side of the point projecting farthest toward the east, are the main examples.

The topographic changes resulting from the recent eruptions also extend for considerable distances out from the shore. In order to assess these effects, I carried out a number of soundings in several areas where lavas had changed the form of the shoreline. In particular, I explored the channel between the two cones of Micra Kameni and Nea Kameni, the harbor of Saint Georges on the other side of the island, and the large channel between the new lavas and Palaea Kameni. The accompanying map shows the results.

Let us consider briefly some of the measurements. Before the eruption, the channel between the two cones of Nea and Micra Kameni opened directly toward the south with a depth, according to the English chart, of 12 directly toward the south with a depth, according to the English chart, of 12 to 13 meters near the middle. The slope from one end of the channel to the other was not great, but at both ends the bottom dropped sharply from a depth of 22 meters to 47 meters at the southern end and at the northern end from 22 meters in a very short distance to 58 meters.

The southern end of the channel is now barred by a thick lava flow that leaves only the narrow passage, already mentioned, between it and the base of the cone of Micra Kameni.

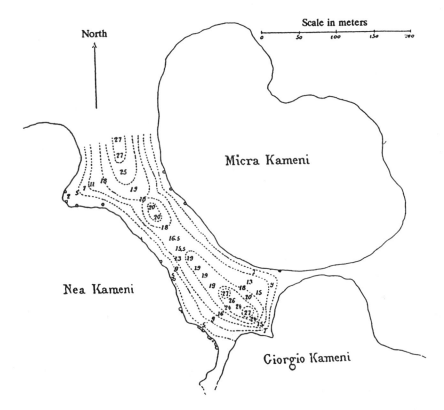

Soundings in meters of the channel between Nea Kameni and Micra Kameni

Subsidence of the older ground on the southeastern end of Nea Kameni is shown by the submersion of the pier and the houses that had been built near the shore. The less marked but still notable submersion of the posts used to tie up boats on the opposite bank of Micra Kameni shows that subsidence of the island, but the maximum displacement has been near the southeastern end of Nea Kameni. The pier, which was reached by three steps and was 50 centimeters above sea level is now completely covered with water. It is no longer horizontal but inclined toward the east at an angle of about 10 degrees and leans toward the south at about the same amount. The edge is 1.30 meters under water in front of the former guardian's house. The three posts set up for tying up boats dropped 30 to 50 centimeters. The nearby houses have been invaded by the sea. Around those in the southern part of the area shown by the map, the ground is covered by 2 meters of water, and since these houses had been constructed at least a meter above sea level, one must conclude that they have subsided at least 3 meters. The depression of that side is also shown by the formation of a small inlet extending to the base of the cone of Giorgios near the ruins of the Catholic church. This embayment is 90 meters long, 12 wide at the entrance, and 35 at the inner end. It occupies the site of the path between the Greek and Catholic chapels and did not exist before 1867. The water is not very deep; it does not exceed 2 meters at the entrance and a meter in the interior. On the other side of the channel, the posts on Micra Kameni have dropped much less. The masonry base of the southernmost one, near the entrance to the southeastern passage, is close to wave level. The others set up farther north have settled even less. The average height of the bases was 25 to 30 centimeters, so this is about the maximum amount of subsidence that could have occurred.[7]

At the present time, the deepest part of the channel is near its southern end. Soundings showed depths of 27 meters in two places in the small inlet running north-south. A little farther north in the middle of the channel there is another less prominent depression the bottom of which is at 20 meters. The minimum depth found there is 15 to 16 meters, so the slope of the ground becomes fairly even near the northern end, where a depth of 27 meters has been measured and then a little beyond it is much deeper.

The harbor of Saint Georges opens on the other side of Nea Kameni and, as noted earlier, is limited by the lavas of 1707 and some of the ones erupted from Aphroessa in 1866. The main effect of the new eruption was to extend the entrance in the form of an inlet toward the northeast. The tops of some of the posts installed near the shore are still visible, but most are entirely submerged and covered by 10 to 40 centimeters of water. All these posts were set up on a circular masonry base on dry land. Their bases were about 25 centimeters high, and the pillars about 50. So one can conclude that

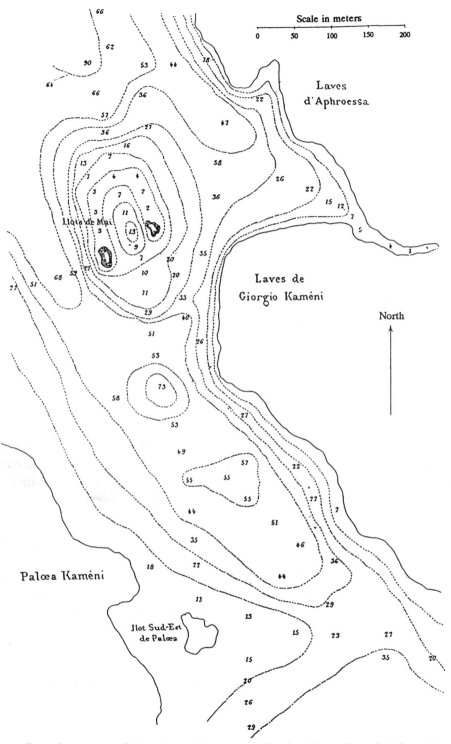

Soundings in meters of the channel between Giorgios Kameni and Palaea Kameni

the average subsidence of the ground near the harbor of Saint Georges was about a meter. It is difficult to provide more precise information on the posts, because they were set up at different elevations above water level, mounted on bases of different sizes, and had different heights. Some had pedestals composed of a series of circular steps.

The soundings indicate a notable depth of water but only in the vicinity of the entrance. In the old harbor, the bottom forms a broad basin with a depth of 13 meters. The bottom closer to the entrance is more elongated and reaches a depth of 40 meters. From the deepest point it rises rapidly toward the entrance and is only 13 meters deep in the middle of the channel where it opens to the sea.

The new eruption has brought great changes in the configuration of the channel between Palaea Kameni and the area covered by the new lava field. The greatest depth indicated by the recent soundings is no more than 73 meters, whereas it previously exceeded 182 meters. Before the eruption the bottom had the shape of a deep basin bounded on the north by a ridge that had an indentation on the north-northeast, and ended on the south with another underwater ridge. The latter was shallow on the Palaea Kameni side but dropped off sharply toward the east. Soundings in that area indicated a depth of 110 meters.

At present, the bottom of the channel still has the same general form but is considerably modified in detail. Depths have diminished everywhere but most notably on the southern side, where the underwater barrier is now much closer to the surface. It extends from a small islet near the southeastern point of Palaea Kameni at depths of 11 to 14 meters, drops to 25 meters, and rises again toward the new lavas. This saddle-like crest separates the two parts of the channel, so that one drops off toward the north and the other toward the south. The northern end is by far the largest. The newly formed islets known as the Isles of May rise in its middle. They originally formed two reefs running NNE-SSW. In 1867 they were still 60 meters long, although the sea had greatly eroded them. In October of 1875, the largest of the two, the one closest to Palaea Kameni, measured only 27 by 26 meters; it had since become almost circular.* The same was true of the other, which was 22 by 20 meters. They were 65 meters apart, and the largest was 108 meters from the new lavas.

* The map on the preceding page shows this islet as more elongated than it is now.

The soundings show two banks along the prolongation of the islands that are only three meters or so below the water and are separated by about 80 meters in a north-northeast direction. At their northern tip, they are linked by an underwater base where soundings show a depth of 3.5 meters. The submarine depression between them forms a channel, the bottom of which now has an average depth of about 11 meters. The central part of this depression is 13 meters below sea level. It extends between the two islets at a depth of 9 meters and ends a few meters farther south where soundings indicate a transverse rise at a depth of 7 meters. Thus, between the two islets and the banks that form their underwater extension there is a small, elongated, boat-shaped basin with raised sides that are partly submerged and ends that are open and entirely under water. All around the piles of rocks that make up the two islands and the bank on which they stand, the seafloor drops away abruptly. Less than 30 meters from all sides one finds depths of 18 to 27 meters. Between the eastern islet and the point closest to the shore line formed by new lavas, the greatest depth shown by soundings is 37 meters. Between the western islet and the closest part of Palaea Kameni, it is 67 meters.

South of the two islets, the bottom of the channel of Palaea Kameni has two depressions of unequal depth. The center of the northern one is 73 meters deep, whereas the second is only 55. These two small depressions are separated from one another by a submarine rise where a depth of 49 meters has been measured. From the southernmost of the two, the bottom rises gently with an average depth of 45 meters to the shallow bank forming the southern side of the channel.

Aside from the channel of Palaea Kameni, soundings at several places along the coast show a fairly steep slope down from the shoreline. The new lavas ending under water form slopes that are rarely less than 25 degrees and in some places are as much as 35 or 40 degrees.

At the time of my last trip to Santorini in the autumn of 1875, one could still see various emanations on and around the new lavas. The following is an account of their condition at that time.

I shall mention first the hot springs coming out on the older ground of Nea Kameni at the foot of Giorgios on the eastern side and near the pier that used to be opposite Micra Kameni. These springs had temperatures ranging between 45 and 60 degrees. At any given point the temperature changed very little from one day to the next, but they had small oscillations of 1 or 2 degrees that went on continually. The water had the same composition as that of the surrounding seawater; it differed only in the amount of iron bicarbonate it held in solution. Though clear when it first flowed out, the water took on an ocherous tint and precipitated hydrated iron oxides that colored the rocks at the water's edge. The hot water spread as a surface layer on the cooler water of the sea and formed long, reddish-yellow streaks that the wind drove as much as 2 or 3 kilometers downwind. This layer of warm ferruginous water was so thin that a swimmer in a vertical position had his legs in cool water while the water around his chest was warm. The thinness of the layer is also shown by the fact that each main branch split into smaller branches and exposed streaks and patches of the colder dark green water from underneath.

The springs near the harbor could be seen on both sides of the house of the guardian who formerly lived on the island. This was the northernmost of the group of buildings near the pier. Its southern side was next to a rectangular space about 10 meters long and 5 wide that was surrounded by a wall and used to store pozzolana. This enclosure is now occupied by a pond of water separated from the channel by a narrow sand bank rising about 10 to 15 centimeters above sea level. At certain times this small lagoon was completely enclosed; at other times, when the tide rose, it was connected to the channel and formed a stream of varying width across the sandbar, sometimes completely submerging it. During my last stay at Santorini, I was able to note that the state of the small body of water occupying the location of the former storage place changed with the conditions of the weather and sea, as shown in the accompanying table.

Except for minor variations in the extent of the sand bar, the conditions noted on the 13th of October 1875 did not change until the end of the month. The guardian's house could not be entered with dry feet during this time, owing to the continued elevation of sea level. Although the barometer rose to 766.2 mm on the 19th of October, the dominant winds came from the south and west and no doubt contributed to the high level of the water in the channel between Nea Kameni and Micra Kameni. It is noteworthy that on the 8th of October, when the water was not moving, the temperature of the lagoon fell to 25 degrees, in contrast to 47 degrees on the 27th of September, 55 degrees on the 2nd of October, 48 degrees on the 4th of October, and up to 61 degrees on the 13th of October and following days.

The warm springs feeding the lagoon appeared in spaces between the

Conditions of thermal springs near the pier of Nea Kameni between 27 Sept. and 13 Oct. 1875.

Date of observation	Condition of sea	Barometric Pressure	Temperature of air	Wind	Comments
27 Sept, 8 am	Low tide (about 30 to 35 cm below high tide. Measurement not precise	769.3	16°	N moderate	Stream connecting with sea, 65 cm wide and 12 cm deep; current 20 cm per second.
28 " "	---	766.9	18°	N weak	Stream 82 cm wide, 13 cm deep; flow 20 cm per sec.
1 Oct 8 am	---	761.0	23°	S weak	Stream in two branches, one 100 cm wide, 15 cm deep; other 30 cm wide, 30 cm deep, flows 25 cm per second.
2 " "	---	763.2	21°	N moderate	Stream in two branches, one 250 cm wide, 25 cm deep; other 150 cm wide, 8 cm deep, ave. flow 25 cm per second.
4 " "	---	766.4	21°	N moderate	Same conditions as 2 Oct.
6 " "	Water 35 cm below high tide	765.8	19°	N moderate	Stream 100 cm wide, 4 cm deep; flow 15 cm per sec.
7 " "	Water 32 cm below high tide	766.2	19.5°	N moderate	Stream 20 cm wide, 6 cm deep; flow 15 cm per sec.
8 " "	water 12 cm below high tide	766.6	19°	N weak	No flow of water
11 " "	water 20 cm below high tide	762.4	20.4°	E moderate	Stream 260 cm wide, 25 cm deep; flow 25 cm per sec.
12 " "	water 12 cm below high tide	759.4	25.6°	S moderate	Stream 300 cm wide, 30 cm deep; flow 25 cm per sec.
13 " "	water 5 cm below high tide	755.7	26.0°	SW moderate	Sand bar almost entirely covered.

rocks of the wall around the storage yard. At the beginning of October, they were seen coming out 4 centimeters above the surface of the water. On the 6th and 7th, the flow was only a dribble, and on the 8th it had completely stopped. On and after the 11th, the springs became very active again and were soon drowned in their own discharge.

The flow of water from these springs was accompanied by emissions of gases composed mainly of carbon dioxide. These gases came not only from cracks in the wall but from the hot water as well. They rose from almost the entire lagoon, increasing when the water was hot and declining when it cooled. On the 8th of October, the discharge stopped altogether, and not a single bubble broke the surface of the water in the former site of the enclosure.

On the 27th of September, inside the house of the guardian, there was a small, shallow puddle of dirty, smelly water. Close to the northern side of the house there was a pond of stagnant saltwater about 10 meters long and 4 wide. The water had a normal temperature and gave off no gas. Until the morning of the 11th of October, nothing unusual happened, but at that time, when I came in a boat to the front of the guardian's house, I was struck by a cloud of white vapor covering the stagnant water on the north side. I soon discovered that the temperature had risen to 44 degrees and that carbon dioxide was being given off. The same thing happened in the small puddle inside the house, and on the following days, this condition became more and more accentuated. The northern lagoon remained separated from the sea and produced no apparent flow of hot water, but the spring inside the house gave off a stream of hot water that came out of the house through a hole in the base of the wall.

I have noted already that depression of the ground had created an elongated inlet at the southwestern end of the channel between Micra Kameni and Nea Kameni. This cove did not exist in April of 1867 but was formed during the last years of the eruption. In 1875, it was the site of numerous hot springs and very productive gas vents. Like that in the storage yard, the spring water had the composition of sea water with dissolved iron bicarbonate. In a few places it came out without releasing carbon dioxide. A few meters behind the entrance to the cove near the shore of Nea Kameni, one could see several steady jets of hot water rising from below the surface of the sea but without releasing gas. These jets were particularly prominent at low tide, when they could rise as much as 6 centimeters above the surface. When the sea was high, however, they were still noticeable, and, though they did not rise as high, their force seemed to be greater.

The emissions of hot water and vapor were especially pronounced in the innermost part of the cove. In the irregularities of the southwestern shoreline, the quantity of hot water was at times so great that it flowed toward the sea like a river, and a little farther to the north carbon dioxide bubbled up through the water. At times there was so much gas that it was dangerous to remain on the shore.

On the 1st of October, the hot water flowing from the southwestern end of the cove formed a current 350 centimeters wide and 15 deep with a velocity of 25 centimeters per second and a temperature of 53 degrees. But on the 8th of October, there was only a meager stream 1.2 meters wide and 5 centimeters deep; the temperature had dropped to 44 degrees. Then on the 13th, the flow had returned to a width of 350 centimeters and a depth of 25 centimeters; the velocity was about 30 centimeters per second. At the point where most of the carbon dioxide was released, the water had a depth of 20 centimeters on the 1st of October and a temperature that varied between 50 and 53 degrees. The release of gas was least on the 8th of October.

From all this one can conclude that the proportion of hot water to gas is not constant but varies with the condition of the sea. It is a maximum when something causes the water level to rise. The tides have an influence but only a weak one, for I have not been able to find a variation during the course of a day. The main effect is certainly that of the wind, and it is complicated at Santorini by the peculiar conditions resulting from the shape of the bay and the islands in its center. Careful, long-term observations would be necessary to establish a relation between the wind direction and sea level. All I can say is that the strength of the emissions of gas and hot water seemed to me to be clearly related to the level of the sea in the area where they were discharged.

Is there a similar relationship to barometric pressure? I hesitate to say there was, because at the end of October the barometer remained for several days at a pressure of 764 millimeters with no decrease in the amount of gas or hot water.

One can conclude, therefore, that we have here an example of the penetration of seawater deep into the ground and its return to the surface. Salt water comes out again with an elevated temperature and an increased content of dissolved mineral components. The channels through which it descends and returns represent two equal branches of a siphon, the descending one being filled with seawater at normal temperature and the ascending one filled with the same water dilated by heat and bubbles of gas. Under these conditions, there could not have been equilibrium, and with the descending limb in communication with the inexhaustible reservoir of the sea, continued circulation would have to follow the opening of the other limb. A rise of the temperature of the walls of the ascending column, a decrease in atmospheric pressure allowing expansion of the gas in the rising liquid, and an elevation of sea level that increases the height of the two columns are also factors that favor circulation of such a system. If these forces are strong, the flow will be rapid; if they are weak, it will be slower and may cease altogether. It is easy to see how this theory can account for the observations cited earlier.

A release of gas that is weaker than that just described was found on the other side of the area of activity in the passageway that forms the end of the bay opposite the Isles of May. The temperature of the water where the gas escaped differed from place to place by as much as 27 to 35 degrees. At the same time, it gave off water charged with iron bicarbonate, for all around it the rocks along the shore were covered with an ocherous deposit of hydrated iron oxide. An even weaker emission of gas was seen in the inner southeastern part of the harbor of Saint Georges where a low place separates the lavas of Aphroessa from those of Nea Kameni. The gas was carbon dioxide mixed with about 2 parts in 100 of nitrogen. The rocks along the shore were colored yellow by ferruginous deposits. The temperature of the sea at this place was 39 degrees. The gas mixture contained a trace of hydrogen sulfide. Paper impregnated with lead acetate was quickly turned black, either by exposure to the gas or by the water through which the bubbles rose. The two ponds of water that were the remains of the channel that, in 1866, separated Aphroessa from Nea Kameni were filled with stagnant water that gave off no gas.

Finally, a mixture of gases, composed mainly of carbon dioxide, was given off in the small harbor of Saint Nicolas on Palaea Kameni. Gas was already being emitted here before the 1866 eruption.

All these gases consisted mainly of carbon dioxide with various proportions of nitrogen, oxygen, and occasional traces of hydrogen sulfide. Their temperatures were rarely above 60 degrees.

The gases given off by the volcanic cone and lavas that came from it had somewhat higher temperatures. It is more appropriate to refer to these emissions as fumaroles because of the rather dense fumes they gave off.

Some of these fumaroles were distinguished by temperatures corresponding to incandescence*. These were found only on the cone of Giorgio-Vouno, and at night they could be recognized from a distance by the light they gave off. All the rocks in the middle of the noisy gas vents glowed in the dark, and when they were moved with an iron rod they resembled a pile of coals in a fire. A very small fumarole of this kind was seen near the southern edge of the central crater only 5 meters from the crest of its talus slope. Two others of the same kind but much larger were located in the wide opening on the outer southeastern slope of the cone about 30 meters from the southern crest. The largest had a surface area of about 18 square meters, the other scarcely four. A piece of wood that I tossed into one immediately ignited. The gas is colorless and has no effect on reactive test-paper. Nothing was condensed in the sampling apparatus, so the gases must be completely dry. An analysis carried out in the field indicated a composition very different from that of the atmosphere. These are dry fumaroles of the type found on Vesuvius by Mr. Ch. Sainte-Claire Deville in 1855. They deposit no solid material on coming in contact with the air, and the rocks between which they come out are black and appear to be unaltered.

Fumaroles with temperatures between 100 and 300 degrees were much more common than the hotter ones. They were rich in water vapor, so they produced thick, white clouds of condensed steam when they came in contact with cool air. They were composed of various proportions of sulfur dioxide, hydrochloric acid, and carbon dioxide. The hottest ones had the most hydrochloric acid, but without exception all had large amounts of carbon dioxide. They immediately turned litmus paper red. Rocks they came in contact with were altered and whitened on the surface or coated with iron chloride and oxide and with sulfates, mainly of calcium, impregnated with free sulfuric acid. Much of the calcium sulfate was almost pure, and parts of the summit of the cone were covered with bright-white deposits resembling a layer of fine crystals of fresh snow.

The third group of fumaroles was weakly acidic and had temperatures between 90 and 98 degrees, never in excess of 100 degrees. Carbon dioxide

---

* The number and output of fumaroles have varied widely during the course of the eruption. The description that follows is that of conditions after the end of explosive activity. The observations were made in September and October of 1875.

with hydrogen sulfide and water vapor are their principal constituents. The fumes from these vents are white and rather dense, and the deposits they formed consist of crystals of sulfur, often associated with calcium sulfate and small amounts of iron chloride and oxide. Fumaroles of this type are very numerous. It is notable that one finds among their products only insignificant traces of salts of ammonia.

None of the fumaroles that are still active has any trace of flames of the kind seen in the beginning when they were produced at all the eruptive centers by burning hydrogen and methane. During one of the nights I spent on the volcano in 1875, however, I was able to see a pale, flickering glow from a spot on the northern slope of the cone where there had been many acidic fumaroles. I have no doubt that this weakly luminous flame was from burning sulfur or hydrogen sulfide. At the same locality I found the remains of globules of molten sulfur, indicating that the temperature had reached at least 113 degrees.

BANCO

The summits of some of the cones produced by submarine eruptions never reached the surface of the sea, while others that are just barely emergent tend to be quickly eroded by waves. While making soundings in a volcanic area one frequently finds conical features a short distance off shore and just below sea level. The bay of Santorini has two examples of these. Between the Kamenis and the large island of Thera there are two underwater cones that owe their origins to these eruptions. The one closest to the new lavas, called Banco, is a slightly truncated cone, the top of which is at a depth of 11 meters. The flat part of the summit has a surface area of about two hectares. Until now, this has been the only spot in the bay where naval ships have been stationed when visiting Santorini; it is also the only place where large commercial ships can anchor. It is dangerous during storms, for it is exposed to northern winds and to the full force of unobstructed waves coming from farther out in that direction and passing through the northern entrance of the bay.

The figure of 9 meters that we have indicated as the depth of the summit of Banco is taken from the chart prepared in 1870 by Captain Germouny, commander of the Austrian gunboat *Reka*. The English chart published in 1848 by Captain Graves shows an almost uniform depth of only seven to eight meters for the entire summit. Thus it appears that the top of Banco has dropped about two meters in twenty-five years, but if measurements carried out at other times are considered, this conclusion is much less attractive.

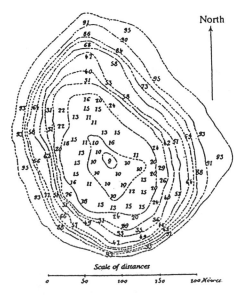

Soundings in meters of Banco made in 1870 by Captain Germouny.

In 1829, when the geologist with the Morée expedition, Mr. Virlet d'Aoust, made soundings of Banco, he found a rocky bottom covered with marine plants at a depth of 8.2 meters (4.5 fathoms). In 1830, when carrying out the same measurements, he noted that the top of Banco was only 7.3 meters (4 fathoms) below the surface, and he concluded that the ground had been uplifted by a meter. At his request, Admiral Lalande carried out new soundings of Banco in 1835 and found depths of no more than 4.5 meters (2.5 fathoms). So in five years there seems to have been 2.8 meters of uplift. This observation seemed to support so well the conclusions that Mr. Virlet had drawn previously that he felt justified in predicting that Banco would become emergent in 1840. But, far from coming out of the water, the top of the cone seems to have dropped again to a depth of 7.3 meters (4 fathoms), as indicated by Captain Graves' work, and, finally, in 1870, the depth had again increased, as shown by the measurements of Captain Germouny. According to those results, there would have been an uplift of Banco between 1829 and

1835 and a depression from that time until 1870.

Such ground movements are certainly not impossible, but their validity seems to me far from proven. The kind of evidence I would like to see is that sought by Lieutenant Leycester, who observed Banco closely in 1848 and said of Mr. Virlet's conclusions, "I do not attach the least importance to the differences of half a fathom that Mr. Virlet found between his two visits; a small increase or decrease of sea level would explain this difference. Considering that eighteen years passed between the second measurements and the time when Lieutenant Mansell made his careful bathymetric survey without any difference of depth being observed, one is justified in dismissing, at least for the present, the notion that part of Santorini is on the verge of a new eruption. I cannot explain how it is that Admiral Lalande found only two fathoms, and I would be happy to see his official report. Several times when the weather was calm I have been able to see the bottom in this region." This last remark certainly indicates that if there were a place where the bottom was at two fathoms, it could not have escaped the notice of an observer.

Another factor makes the measurements cited by Admiral Lalande even more unreasonable. If any part of Banco had been as shallow as two fathoms, it would have been very dangerous for ships, and sailors familiar with Santorini would certainly have made a note of the bottom being so shallow.

Finally, it would be quite extraordinary if a local uplift had been followed by a depression of at least two and a half fathoms on a totally inactive cone when nothing comparable is seen anywhere else in that area.

In summary then, if one disregards the measurements reported by Admiral Lalande and accepts the differences noted by Mr. Virlet as due to the difficulty of repeating accurate measurements at exactly the same place on a very uneven surface, one must conclude that it is very likely that the depth of the submerged surface of Banco has been about eight meters for the entire period between 1829 and 1848 and that since that time the depth has increased about a meter as a result of erosion by the sea and anchors digging into the surface. There may also have been a general subsidence of this central part of the bay during the eruption of 1866.*

---

* Mr. Gorceix, in a letter addressed to Mr. Ch. Sainte-Claire Deville and published in the Comptes Rendus of the Academy of Sciences, session of 7 February 1870, stated, in speaking of Banco, that the ground at that time had been uplifted and that the anchorage in that area had become more extensive. No accurate measurements were cited in support of this observation which conflicts with the evidence provided by soundings carried out during the same period by the officers of the *Reka*.

SUBMARINE CONE NEAR THE HARBOR OF THIRA

This cone is shown on the English chart as a hill, the highest point of which reaches a depth of 36 meters.[8] The small number of measurements shown on the chart are not enough to define its true form, and the soundings I carried out in this area had the purpose of obtaining additional information. They show that it is an irregular cone standing on a base with the same form. The top is more or less circular with a diameter of about 200 meters. The outer slopes are very steep, dropping away at angles of 30 to 35 degrees.

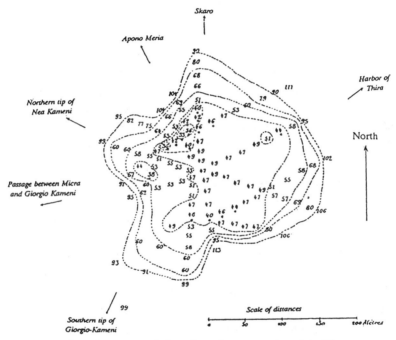

Soundings in meters of the submarine cone closest to Thera

The top is almost flat and is at a nearly uniform depth of 43 meters. It is most likely a crater, for the contour of the flat part is formed by a belt of rocks that seem to be the remains of a raised circular rim. The highest point and most continuous part of the edge is on the northern side. Soundings indicated that on this side a strip of rocks about 110 meters long and 40 wide has high points rising 12 to 14 meters above the center of the plateau. On the south, one sees another high place about 40 meters long, ten meters or so in width, and averaging 5 to 6 meters in height above the central area of the summit. To the east and west, the survey revealed only a few isolated high points. The shallowest depth found was 33 meters. The surface of the plateau is covered with a mixture of sand and volcanic lapilli, and the sides are rocky. Each dredge brought up fragments of a small delicate type of coral that grows on the entire surface. The cone stands on an earlier platform at a depth of 90 to 110 meters beyond which the slope continues to various distances before reaching a general depth of 180 to 220 meters.

In summary, the cone has a total height of 140 meters. It rises gently from its base up to two-thirds its elevation then more steeply in its upper third. The crater, the rim of which is still preserved at the summit, must originally have been a good bit deeper and more eccentric. It has been leveled by waves that eroded loose material from the rim. Owing to its broader, flatter surface and the protection from northeasterly winds provided by the nearby cliffs of Thera, the area of the upper part of the cone seems to have been provided more favorable conditions from anchoring ships than the other small cone closer to the Kamenis. These features, together with the high points of the rim, the rocky bottom, and an adequate depth of the water that provide good anchorage at times of storms, seem to have made it a favorite place for stationing ships. The center of the summit is 1350 meters from the passage between Micra Kameni and the lavas of Giorgios and about the same distance from the pier at Thira. From this point one can see the peak of Profitis Ilias in the direction of Cape Alonaki; the saddle between Megalo and Kokkino is seen above the small rock forming Cape Tourlos; the northern cape of Nea Kameni opposite a point on the coast of Therasia lies 6 degrees south of a straight line running south from the southern peak of the island. The highest point on the cone is at the intersection of two lines, one running from Cape Alonaki and following the east side of the channel of Apano Meria and the other connecting the middle of the base of Nea Kameni with the foot of the slope at Thira.

SUMMARY

The distribution and forms of the various volcanic cones within the bay of Santorini have several interesting features.

First, one is struck by their distribution along a common axis. They are arranged in a strip extending from the ENE to WSW.[9] This alignment is, of course, far from straight; it has a substantial width, but it is well expressed in the relief of both the emergent topography and the bathymetry of the bottom. It forms a ridge that runs between the cliff at Thira and the island of Aspronisi dividing the bay into two unequal parts. I have no doubt that this line follows the preferred direction of fractures in the upper levels of the crust, and I firmly believe that all the eruptions in the middle of the bay opened along a series of parallel, closely spaced fissures. In a word, I regard the essentially rectilinear fissures as conduits for each of the historic eruptions that have occurred in the bay along the same orientation and not as part of a set of fissures radiating from the center. The modern volcano of Santorini no longer has a central vent, and none of the cones seen there today resembles those of Vesuvius or Etna, from which the eruptive fissures spread outward.

Second, there is a notable difference in the form of the two submarine cones between Micra Kameni and Thira on the one hand and that of the four subaerial cones that make up the Kamenis. The two submarine bodies have a nearly regular conical form without the bulges and projections that are normally produced by lava flows. Each of the Kameni islands, however, has a lava field extending from the cone and running down to sea level. One must conclude, therefore, that submarine lavas probably solidify quickly on contact with seawater and pile up close to their source, while those erupted subaerially retain their fluidity longer and can therefore flow farther from the point of emission.

Finally, one might wonder why the eruptions that formed the three original Kamenis - Palaea Kameni, Micra Kameni, and Nea Kameni - all produced lavas that flowed from south to north. It is interesting that on each of the three islands one sees the surface of the lavas inclined in the same direction and the flows always following the same course.

We should also note that among the lavas of 1866, those that were emitted from the eruptive vent known as Aphroessa also conform to the same pattern noted in the others. They all flowed primarily northward. But lavas produced by the other vent that was active during this episode moved toward the south, because the position of their vent at the foot of the older cone of Nea Kameni made it impossible for them to do otherwise.

Thus, of the five subaerial flows we find four that moved toward the north and only one that took the opposite direction (owing to obvious topographic conditions). The explanation for this offered by Messrs. Reiss and Stübel was that the direction of the flows had been determined by the inclination of the surface on which they spread, but this is purely hypothetical, for although the northern part of the bay is deeper than the southern part, there is nothing to indicate that at any particular place the ground has a uniform slope from south to north. Moreover, we know that the eruption of Nea Kameni in 1707 broke out at a point where the water was quite deep and that the same was true of the vent that produced Aphroessa in 1866. In both cases, it is certain that the vents opened on a high place on the bottom and that the lavas could have flowed southward just as well as toward the north. If one assumes that prior to the eruptions there was a small submarine ridge running east-northeast, it is understandable that the flows would have moved perpendicular to that direction, but it is more difficult to explain why they tended to flow northward. For example, one can find nothing in the submarine topography to explain why nearly all the lavas of Aphroessa moved in that direction. In that case, they followed the western shore of Nea Kameni and spread along the flank instead of following the direction where the slope was steeper. (Their high viscosity explains this apparent violation of hydrostatic principles.) There must therefore have been some other factor, possibly resulting from the particular conditions and the way the primary or secondary fissures opened, which tended to send all these flows in the same direction.

CHAPTER FIVE

CHEMICAL STUDY OF THE PRODUCTS OF THE RECENT ERUPTION OF SANTORINI

LAVAS AND OTHER CRYSTALLINE SILICATES

The common lava of this eruption normally takes the form of black obsidian spotted with small crystals of clear feldspar and, more rarely, dark minerals, mainly pyroxene. Thus, to the naked eye or even under a hand lens, this lava gives the impression of a very dark, homogeneous glass containing scattered crystals, but when seen under a microscope, even with a magnification of no more than two or three hundred power, the material that seemed amorphous is seen to be a tangle of small, colorless microlites with elongated shapes and inclined crystal faces. These are associated with myriads of opaque iron oxides and set in a matrix with small amounts of yellowish gray amorphous material. The small, colorless prisms revealed by the microscope differ from the crystals of feldspar visible under the hand lens; in places, they penetrate the latter, and a few are entirely enclosed. They rarely exceed 0.1 mm in length and 0.03 mm in width, whereas almost all those of the other type are considerably larger, with lengths of about a millimeter and widths of half a millimeter or so. The first are normally separate crystals; the larger ones tend to be in clusters. It seems likely, therefore, that we are dealing with two distinct minerals, probably two different feldspars.

It would be of interest to determine the compositions of these minerals, as well as that of the pyroxene, which the microscope shows to be present in greater amounts than one would expect from examining the lava with the naked eye or even with the aid of a hand lens, but the rock is so fine-grained and the crystals so intermeshed that every attempt to separate them has failed. This is why none of the many scientists who have studied the lavas of Santorini have even attempted to do this. The only analyses we have had are of bulk rocks, and these give only the composition of the mixture of all the components combined. I have sought to carry this investigation further by two new methods, one for extracting the iron-bearing components of the rock and another for separating minerals that have little or no iron. The first makes use of concentrated hydrofluoric acid, while the second is based on the strong electrolitic attraction of certain components.[1] In order to use either of these methods conveniently, one must first reduce the rock to particles of uniform size. Sieves of appropriate sizes are used to eliminate grains that are too large and also those that are only fine dust. Because most over-size grains are composites of more than one component and are attracted by an electrical charge but only weakly affected by hydrofluoric acid, they cannot be effectively separated by either technique. The fine dust cannot be used, because it is so hygroscopic that it tends to agglutinate, and even when one takes the precaution of sintering it slightly, the grains take on an electrical charge and adhere to one another. It is also difficult to use hydrofluoric acid with fine powder, for the effect of the acid on iron-bearing components is reduced, and, in addition, the separated mineral grains are so small that they are easily lost in the washing that is required to get them completely clean.

When the rock has been crushed, the grains that are best suited for separating the minerals are those that have a small size but are free of dust. The absolute size of the crushed material depends, of course, on the size of minerals contained in the sample.

To treat the rock with hydrofluoric acid, it is very important that the particles be of uniform size, because the rapidity with which the acid attacks them depends not only on the nature of the minerals but on the grain size as well. It is also very advantageous to work with a rather small amount of material in order that the temperature can be raised to the boiling point where the reaction proceeds more rapidly. Under these conditions, the minerals are evenly attacked on all sides of each grain. The surfaces of the crystals remain smooth and when they break down, they split cleanly along their cleavage directions, so that the faces remain sharp. In this way the variations in the effects of acid on different minerals are accentuated to the maximum degree.

When the lavas of Santorini are treated in this way, the electrostatic method is most effective when the grains of the crushed rock have a diameter of about 0.2 mm. To treat them with hydrofluoric acid, a grain size of about 0.1 mm is best.

When treating the powdered sample with hydrofluoric acid, if the temperature of the mixture does not rise spontaneously to 100 degrees, it is helpful to increase the temperature by means of an external heat source.

After a procedure of this kind was carried out on the lava from the last eruption of Santorini, a residue of pyroxene, olivine, and iron oxides remained. A weakly magnetized bar removes the iron oxides. The small proportion of light yellow olivine crystals are readily distinguished from the dark green pyroxene; they are easily sorted by hand. By this procedure, 100 grams of the Santorini lava yielded on average 230 milligrams of pyroxene, 80 milligrams of iron oxides, and 10 milligrams of olivine. When the separated grains are examined under the microscope, a certain number of them are seen to be broken and corroded, but most are perfect crystals with clean surfaces and sharp edges.

For the second operation, one uses a large electromagnet connected to six large Bunsen elements.[2] By successive passes over the crushed lava, the magnet removes the iron oxides, pyroxene, olivine, interstitial material, and microscopic prisms of feldspar that have inclusions of iron oxides. What remains is a perfectly white crystalline powder that, under a hand lens, is seen to be pure feldspar. A quick inspection under a low-power microscope is needed to eliminate impurities, such as fragments of feldspar with too many inclusions of pyroxene, feldspar in microscopic prisms, and, above all, vesicular glass with or without gas bubbles.

In the mineral separations I have carried out on the recent lavas of Giorgios, Aphroessa, and the Isles of May, I processed 2 kilograms of crushed lava. This amount is needed to obtain 2 grams of feldspar that is perfectly free of inclusions and suitable for chemical analysis.

From 30 grams of this crushed lava one can obtain about one gram of feldspar, but most of the grains have glassy inclusions and are not suitable for an exact analysis.

When the grains of feldspar obtained in this way are placed in hot, moderately concentrated hydrochloric acid, they remain completely intact if the action of the acid is not too prolonged, and their analysis, which is given in a later section, shows that they have the composition of labradorite. But in other cases, particularly when this treatment is applied to feldspar obtained from lavas of Aphroessa, some of the grains are attacked. They swell and quickly become milky white. The proportion of grains that are altered in this way reaches more than two percent of the total amount of treated feldspar. A chemical examination of the dissolved material and of the alumino-silicate framework that remains shows that the composition of the grains that are attacked is anorthite.

To complete the separation of crystalline components, one must isolate the microscopic laths of feldspar. To do this, I take fragments that, to the naked eye, appear to be completely glassy and after reducing them to grains about a quarter of a millimeter in diameter, I verify with a strong lens that they are entirely free of visible feldspar. I then reduce them to a fine powder and, using an electromagnet, separate the iron oxides, pyroxene, olivine, and part of the amorphous material. The residue is a light gray powder composed of microscopic feldspar crystals and small amounts of glass. These last two components cannot be totally separated, but in working with samples of some of the scoriaceous rocks in which the crystals are grouped in tight sheaves in a sparse matrix, what remains from electromagnetic separation is seen in polarized light to be almost entirely crystalline. Thus the analyses of these separates are fairly representative of the microscopic feldspars.

I have not been able to separate the groundmass matrix but by following the procedure just outlined for samples of dense rocks in which the crystals are clustered together in a more abundant matrix and by separating and analyzing this mixture of crystals and amorphous material, one can compare its composition with that obtained for purer separates of feldspar and infer the probable composition of the matrix material. The results of this comparison show that its composition resembles that of albite but is richer in silica. It is probably this excess silica that impeded its crystallization.[3]

The two operations, for which I have just outlined the principles and immediate results, can be applied in various ways. I have found the following procedures most useful.

To treat a powdered rock with hydrofluoric acid by the method laid out above, one uses a platinum crucible with a capacity of about 150 grams and places in it 60 grams of concentrated fuming hydrofluoric acid. To this liquid one adds about 30 grams of the powder and mixes the two with a wooden-handled platinum spatula. One cannot pour the powder into the acid all at once, for the heat generated would be so great that the boiling acid would overflow the container. Nevertheless, the operation should be carried out as quickly as possible. As soon as the boiling has stopped, the reaction should be arrested by immersing the crucible in a pan of water that fills the container and quickly causes the liquid to overflow. Most of the gelatinous material formed by the attack of the acid on the powdered sample becomes suspended and flows out with the liquid. One can facilitate this separation by continuing to agitate the contents with a platinum spatula. Soon, nothing will remain but the unaffected part of the powder and a small amount of gelatinous material.

If the action is interupted too soon and, for that reason, is judged to be insufficient, one should pour the water from the crucible and recover the unaffected residue for a second treatment with concentrated hydrofluoric acid. In this way there is no risk of adding more acid than necessary. Finally, when the effect is thought to be adequate, the action of the acid is stopped by flooding the vessel with water as just described. It is difficult to eliminate the last bits of gelatinous material mixed with the remaining crystals. Several methods for doing this can be employed in the following order.

1. Rub the remaining material with a finger under water. This loosens the small amount of gelatinous material still adhering to the crystals.

2. Decant the water from the crucible and dry the residue; this will cause it to become a white powder. Then fill the crucible with water again and repeat the process of loosening it with the finger.

3. Pour out all but about two cubic centimeters of the water and gently agitate the crucible. This will cause the crystals to separate from the small amount of remaining gelatinous material, so they can be isolated by wiping the latter away with the tip of the finger.

If this procedure is followed, most of the crystals that are separated will be unbroken and perfectly clean.

With a little practice, one learns how to stop the operation before it is complete and in this way dissolve only the amorphous material without affecting the feldspars.

Pyroxenes, amphibole, olivine, biotite, and iron oxides are so resistant to hydrofluoric acid that they are much easier to separate. Quartz is also much more resistant than feldspar. The pyroxenes and amphiboles that are least affected by acid are those containing the largest proportions of iron.

After the powdered rock has been treated with acid, it contains both fluorides and fluorosilicates, but the former are much more abundant. If one evaporates a bit of this liquid on a piece of glass covered with hard Canada balsam and examines the product under the microscope, one will see cubic crystals that are not potassium fluorosilicate but calcium fluoride. In addition, hexagonal prisms of sodium fluorosilicate will also be noted along with quite a few simple or twinned monoclinic prisms of calcium fluorosilicate. If one allows the liquid to stand a while before taking a small sample for microscopic examination, it will normally be covered by a thin layer composed chiefly of calcium fluorosilicate. At the same time, most of the crystals in the liquid will have become increasingly pure fluorides.

Separation of iron-bearing minerals by means of the electromagnet is effectively carried out as follows. The electromagnet is suspended from the ceiling by means of a flexible line so that the lower edge is a few millimeters above the surface of a table at which the operator is seated. A hand-held switch is connected to one of the lead wires connecting the electromagnet to the batteries. The crushed sample is placed on a thin sheet of cardboard beneath the electromagnet. As soon as the power is turned on, part of the powdered material will become attached to the electromagnet. The latter is then moved by hand until it is over a sheet of paper placed on one side. Using the switch, the power is turned off and the grains adhering to the electromagnet are released and allowed to fall. The operation is repeated several times until all the sample has been exposed to the attraction of the electromagnet.

This procedure could be made more efficient by employing a motor-driven rotating device that at the same time operates the electrical switch. If the battery were replaced by a stronger power source, one could increase the force of the magnetic field; in this way, an electromagnet with a strength equivalent to that of fifty large Bunsen cells would be capable of quickly picking up the biotite in crushed granites, but owing to the limitations imposed by the size of the coil of the transformer, which becomes quite warm, there is danger of melting the rubber insulation around the wires. Moreover, one rarely needs to resort to such a powerful field.

Even though its color is not as dark as that of augite, olivine tends to be more readily altered. In some rocks, however, the relations of the pyroxenes and olivines are reversed. I have noted in particular that in lherzolites it is easier to remove the green diallage (sometimes referred to as diopsidic pyroxene) than the enstatite and olivine in the same sample.

By using the power of only two large Bunsen cells one can easily pick up iron oxides, amorphous material, and some of the other iron-bearing minerals in a crushed Santorini lava. What is left is a mixture of feldspar, pyroxene, and olivine. These three minerals are easily separated. The best way is to spread them over a plate of rough glass overlying a sheet of white paper. Under a large, low-power magnifying glass, the grains of pyroxene and olivine are recognizable by their color. They can be removed with a small wooden stick with a sharp, slightly moist point. As one or more grains are picked up in this way, they are placed on a watch glass partly filled with water and placed conveniently nearby. The grains drop from the wooden stick and fall to the bottom when they come in contact with the water. One watch glass is used for the grains of pyroxene and another to collect the olivine. Separations carried out in this way become a relatively simple operation.

Minerals of the lavas of Giorgios

Crystalline pyroxenic material separated by HF.   Density = 3.477

	Composition in percent	Oxygen				
SiO_2	50.12	26.0		Ratio of oxygen in SiO_2 to that in CaO, MgO, and FeO		
Al_2O_3	2.12					
Fe_2O_3	1.60					
FeO	23.59	5.4			Calculated	Observed
CaO	10.49	2.9	12.6		2:1	2:0.97
MgO	11.05	4.3				
Na_2O	0.67					
	99.64					
Loss	0.36					

The alumina, ferric iron, soda, and a small part of the silica and ferrous iron probably come from inclusions of feldspar and iron oxides in the analyzed crystals. Inclusions of this kind can be seen under the microscope.

When doing this work, I considered the pyroxenic material, an analysis of which is given immediately above, as a single type of crystalline species, augite. Since then, further study has convinced me that there are two separate pyroxenes in the common lavas of Santorini, a monoclinic augite and another that must be orthorhombic hypersthene, because it contains more iron.

When the lava of Giorgios is crushed to a convenient grain size and treated with concentrated hydrofluoric acid, and the action of the acid is allowed to continue, one finds a crystalline residue composed almost entirely of hypersthene; augite, being more readily dissolved by the acid, disappears from the assemblage. By using greater caution, one can remove both minerals from the rock and then separate them from each other by mechanical sorting.

Hypersthene forms brown prisms with sharp faces and good terminations with low pyramidal forms. These prisms average 0.4 mm in length and 0.1 mm in width. Under crossed nicols, they have an extinction parallel to their long sides and are strongly dichroic. They are dark brown when their long edges are perpendicular to the polarizer (with the analyzer removed); their color is greenish when these sides are parallel to this principal section. Their composition is essentially that given above, for in the mixture that was analyzed, there were very few crystals of augite.[4] When these coarse crystals are isolated by use of hydrofluoric acid, one can mount grains on the point of a needle and, turning them between the crossed nicols, observe them in all orientations. When they are rotated about an axis parallel to the direction of elongation, one sees that they are strongly dichroic and have extinction parallel to their edges. The augite is green and very weakly dichroic, especially when it is oriented with a face elongated in the zone ph_1. Concentrated hydrofluoric acid attacks it less rapidly than feldspar but more so than hypersthene. Its surface becomes pitted with grooves parallel to the longitudinal faces of the crystals.

The augite has a specific gravity of 3.372; its composition is as follows:

	Composition in percent	Oxygen		Ratios of oxygens of silica, alumina, and monoxide cations	
				Calculated	Observed
SiO_2	50.2	26.9			
Al_2O_3	3.3				
CaO	26.4	7.5		2:1	2:1.07
MgO	12.2	4.9	14.1		
FeO	7.8	1.7			
Sesquioxide	0.4				
	100.3				

The two pyroxenes are more easily distinguished as separated minerals than they are in thin section.

There is nothing unusual about the inclusions in these two minerals. They are mainly inclusions of colorless, pale green, or brown glass with or without bubbles of gas and commonly with one or two grains of iron oxides. One also finds in them long, colorless microlites, some of which must be apatite, while others seem to be oligoclase.

I have not seen fluid inclusions in any minerals of the lavas of the last eruptions of Santorini. It is remarkable that the glassy inclusions in the pyroxenes of these lavas are homogeneous, clear, and almost colorless, even when the glassy inclusions of the feldspars that accompany them are brown and filled with tiny globules.

The augite is commonly twinned in the normal fashion, that is with h_1 as the twin plane. The cleavages parallel to the faces of the prism are well developed, as are the faces h_1 and g_1.

The hypersthene has smooth faces suggesting that it has little cleavage.

Labradorite crystals that are visible under a hand lens have been separated with the electromagnet. They have a density of 2.702 and the following composition.

Labradorite crystals visible under a hand lens and separated by electromagnet. Density = 2.702

	Composition in percent	Oxygens		Ratio of oxygens with silica, alumina, and monoxide cations	
SiO_2	55.12	28.63			
Al_2O_3	29.92	13.97			
Fe_2O_3	0.35			Calculated	Observed
CaO	9.45	2.65		1 : 3 : 6	0.92 : 3 : 6.14
MgO	0.79	0.31	4.27		
Na_2O	5.08	1.30			
K_2O	0.08	0.01			
	100.79				
Excess	0.79				

The small excess of silica is due to inclusions of albite and groundmass material that were not entirely removed from the analyzed labradorite.

Most phenocrysts of labradorite and anorthite in the recent lavas of Santorini are euhedral, but broken crystals are not uncommon. The fragments are elongated and tend to be in clusters. The g_1 face seems to be well developed, but it is not strong enough to be apparent in most sections; in phenocrysts seen in thin section, the three faces p, h_1, and g_1 are about equally developed. The extinction directions observed under crossed nicols correspond to those given for labradorite by des Cloizeaux and Michel-Lévy.

The cleavages tend to be weakly developed, but in some instances, particularly in single crystals, they are remarkably clear and exactly like those of sanidine. I have no doubt that this is why the phenocrysts in the lavas of Santorini have been mistaken for sanidine, even by eminent petrographers.

Carlsbad twins are very common. Albite twins are also rather common, in some cases as coarse twins, in others as thin and multiple twins. Baveno twins are rare. They are seen mainly where only two crystals are twinned. Pericline twins are commonly associated with albite twins.

Glassy inclusions are very abundant and in almost all cases contain bubbles of gas. A few have more than one of these bubbles. Some inclusions are irregular, but most fill cavities with shapes controlled by the crystallographic elements of their host, and in a sense, they can be thought of as negative crystals. In sections normal to the plane of symmetry, it is not unusual to find beautiful inclusions with the forms of hexagons and dodecahedrons in which one can recognize the h_1, m, t, g_1, g_2, and $_2g$ faces.

Inclusions tend to be aligned parallel to the faces of the crystals and distributed in three directions parallel to the planes p/m, p/g_1, and p/t.

They are normally brown, and although the color may be even, more commonly it is patchy. Around the outer edges there may by a mantle that is colorless in the middle and with almost all the brownish material concentrated in countless small globules. In many cases this border covers so much of the walls of the bubble that it makes the latter almost opaque. Many of the inclusions, especially those in anorthites, are distributed in quadrants that are alternately brown and colorless. Gas bubbles are also seen in labradorite and anorthite; a few are of great size and have the forms of the crystals in which they are enclosed.

The crystalline inclusions consist of iron oxides, pyroxene, apatite, and albite or orthoclase.

Colorless prismatic microlites with very small amounts of matrix. (Rock from Giorgios where the microlites are so abundant that amorphous material cannot be distinguish, even with high magnification.)

Density = 2.556	Composition in percent	Oxygens	Ratio of oxygens in SiO_2 and Al_2O_3 to monoxide cations	
SiO_2	67.07	34.84		
Al_2O_3	18.61	8.69		
Fe_2O_3	4.91		Calculated	Observed
CaO	1.02	0.28	1 : 3 : 12	0.93 : 3 : 12.02
MgO	1.73	0.64		
Na_2O	5.62	1.44 } 2.58		
K_2O	1.33	0.22		
TiO_2	0.51			
	100.80			
Excess	0.80			

Colorless prismatic microlites with a large proportion of matrix (Rock from Cape Tino of Therasia, very rich in amorphous material)

Density = 2.555	Composition in percent	Oxygens	Ratio of oxygens in SiO_2 and Al_2O_3 to monoxide cations
SiO_2	68.30	35.49	
Al_2O_3	17.69	8.26	
Fe_2O_3	4.20		0.95 : 3 : 12.88
CaO	1.33	0.37	
MgO	1.66	0.64	
Na_2O	5.47	1.47 } 2.69	
K_2O	1.33	0.21	
	99.98		

It is impossible to obtain a perfectly clean separate of feldspar microlites from an aphyric rock, either by the electromagnet or by use of hydrofluoric acid. Whatever one does, a certain amount of amorphous material always adheres to the crystals, but the amount is so small that its effect on the analysis is insignificant. The microlites separated in this way yield essentially the same results in the two samples above. Thus, there is little doubt that the microlites of the last lavas of Santorini have the composition of albite.

We consider next whether these chemical results agree with the optical measurements.

Under the microscope, these microlites are seen to be very elongated, with one dimension much greater than the other two. It is known that, as a rule, when crystals grow in elongated forms, the direction of elongation is the same for all crystals produced under the same conditions. This is precisely the case for the microlites in the lavas of Santorini; their uniform appearance indicates that they belong to the same stage of crystallization during which they all took on the same form. They are undoubtedly one or more different species of feldspar. This is shown not only by the chemical analyses but also by their optical properties. Under crossed nicols, they have the conventional aspects of feldspars. Many are twinned parallel to their long axes, commonly in binary twins of the kind so typical of feldspars. The question to be resolved is whether they belong to one or more species of feldspar and, if so, what are the main types.

One can state at the outset that most are triclinic, for when viewed under high magnification and crossed nicols, a certain number of them have multiple twins of the kind that results from albite twinning of several lamellae of triclinic feldspars, and those that are untwinned tend to have oblique terminations that are equally characteristic of plagioclase.

We can assume, therefore, that, pending further evidence to the contrary, all the microlites are triclinic. If so, we note that almost all the twinned and untwinned microlites have small extinction angles nearly parallel to their long axes. For the great majority, this angle is less than 3 degrees, but in a few it is as much as 18 degrees.

Since all the microlites are elongated in the same crystallographic direction, they must all show faces in the same zone when examined under the microscope. If we consider the characteristic properties of triclinic feldspars, a single zone, pg_1, will have extinction angles within the range of those just noted. This is shown by the work of Mr. des Cloizeaux and by the optical studies of Mr. Michel-Lévy. These scientists have shown that albite and oligoclase are the only triclinic feldspars for which the extinction angle is between 0 and 19 degrees in the zone pg_1.[5] This eliminates labradorite and anorthite as possible compositions, and it only remains to determine whether the feldspar microlites have the optical properties of both oligoclase and albite or whether only one of these two minerals is present.

First of all, they cannot all be oligoclase, for the extinction angles of the feldspars in the zone pg_1 do not exceed 3 degrees, so if one considers all the microlites with extinction angles of less than 3 degrees to be oligoclase (and these are by far the most numerous), there still remain some with larger extinction angles that would have to be albite. Certainly, if the chemical analysis had not established the albitic composition of the microlites, the explanation that would best fit the optical properties would be that they are a mixture of oligoclase and albite with the former being more abundant.

To reconcile the optical and chemical observations, one must choose between the two following hypotheses.

1. One can assume that the microlites in question are flattened parallel to g_1 in such a way that they are visible only when they are oriented along p or the faces of the zone pg_1 close to p. On these faces their extinction angle is at its minimum and they can easily be confused with oligoclase while the properties are actually those of albite. Many microlites present the face g_1 or those of the nearby zone pg_1, but unless g_1 has its normal thickness, they will not be visible under polarized light.

Observations show that in fact a certain number of microlites, although of considerable dimension, remain dark in all positions of the rotating stage. In some positions, they appear to be much less numerous in polarized light than in natural light, whatever the orientation of the flow structures may be with respect to the directions of the polarizer and nicols. If one assumes that the microlites are albite and that they are flattened on g_1, one can explain in this way the optical observations as well as the analytical results, but it is unlikely that oligoclase is altogether absent, and I am strongly inclined to believe that oligoclase is commonly present in notable amounts among the microlites.

This is rendered especially reasonable by the observation that the extinction angles of both members of binary twins are generally smaller than that indicated by the curves defined for albite by Michel Lévy. One notes, in fact, that in binary twins, the extinction angle of one of the individuals is very close to 0 degrees, whereas the other in some cases may be as large as 5°. According to the accepted optical properties of crystals, the extinction angles of the two individuals are smaller than the reported values for albite and strongly indicate that a certain number of the microlites are oligoclase.

2. Another possible explanation that could account for the small extinction angles assumes that many of the microlites are monoclinic feldspars. The extinction angles on various faces in the zone pg_1 range from 0 to 5 degrees for this feldspar. But this explanation seems unlikely if one con-

siders the rarity of rectangular terminations of the prismatic microlites and if one recalls that the amount of potassium in the chemical analysis of the microlites is small.

In summary, I shall continue to identify as albite the principal microlites of the common lavas of the last eruption of Giorgios, as well as those of Aphroessa and the Isles of May.

I have also used the two techniques to separate other components of the common lavas of Santorini. One of these is the gray lava that is found as rounded masses up to a cubic meter in size within the obsidian and is composed of crystals, up to 2 millimeters in size, of pyroxene, sphene, olivine, and iron oxides, in some cases with very little interstitial material and in others with considerable amounts of amorphous groundmass. Even in these rocks in which the crystals are relatively large, the separation procedures outlined above have an advantage, for they are quicker than simple hand picking and yield a purer product.

Another material for which I have also used these methods is the drusy masses of anorthite, pyroxene, sphene, olivine, and iron oxides that are rather common in the Santorini lavas. The crystals of these assemblages are loosely joined and large enough to use goniometric measurements to determine their forms. Despite the apparent separation of the crystals in their original form, the techniques described above offer the same advantages as they do for the preceding material.

Analysis of crystals of the gray lava of Santorini

Anorthite reduced to fragments a quarter of a millimeter in size and purified with an electromagnet. Density = 2.782.

	Composition in percent	Oxygens	Ratio of oxygens of silica and alumina to those in monoxide cations	
			Calculated	Observed
SiO_2	45.93	23.09		
Al_2O_3	36.60	17.00	1 : 3 : 4	0.9 : 3 : 4.2
Fe_2O_3	0.88			
CaO	16.09	4.05 } 5.1		
MgO	1.29	0.06		
	100.79			
Excess	0.79			

The excess silica and alumina come from a small amount of amorphous material and a few microscopic crystals of more silica-rich feldspar included in the anorthite.

Pyroxene separated by hydrofluoric acid from the crushed lava. Density = 3.358.

	Composition in percent	Oxygens	Ratio of oxygens of silica to those in monoxide cations	
			Calculated	Observed
SiO_2	51.15	26.57		
Al_2O_3	3.65			
Fe_2O_3	0.95			
FeO	5.68	1.29 ⎫	2 : 1	2 : 1
CaO	25.44	7.14 ⎬ 13.27		
MgO	13.52	4.84 ⎭		
	100.39			
Excess	0.39			

The alumina and ferric iron probably come from inclusions of foreign material that is visible under the microscope.

Olivine separated by hydrofluoric acid from the crushed lava and separated from pyroxene by hand picking. Density = 3.603.

	Composition in percent	Oxygens	Ratio of oxygens of silica to those in monoxide cations	
			Calculated	Observed
SiO_2	38.17	19.83		
Al_2O_3	1.66			
FeO	21.82	4.97 ⎫	1 : 1	1.02 : 1
CaO	2.90	0.81 ⎬ 19.43		
MgO	35.27	13.65 ⎭		
	99.82			
Loss	0.18			

Analyses of crystals in druses in the lava of Santorini

Crystals of anorthite after crushing and purification with an electromagnet. Density = 2.756.

	Composition in percent	Oxygens	Ratio of oxygens of silica and alumina to those in monoxide cations	
			Calculated	Observed
SiO_2	44.25	22.98		
Al_2O_3	37.00	17.26		
Fe_2O_3	0.43			
CaO	18.98	5.32 ⎫	1 : 3 : 4	0.94 : 3 : 3.99
MgO	0.07	0.02 ⎬ 5.41		
Na_2O	0.28	0.07 ⎭		
	101.01			

Very dark green crystals of pyroxene after purification with hydrofluoric acid. Density = 3.358.

	Composition in percent	Oxygens	Ratio of oxygens of silica to those in monoxide cations	
SiO_2	48.87	25.36		
Al_2O_3	7.21			
Fe_2O_3	1.28		Calculated	Observed
FeO	8.20	1.87	2 : 1	2.11 : 1
CaO	27.77	7.80 12.03		
MgO	6.09	2.36		
Na_2O	0.07			
	99.49			
Loss	0.51			

The notable quantity of alumina in this pyroxene indicates the presence of inclusions of anorthite. This inference is supported by the fact that before treating it with hydrofluoric acid, an analysis of the same material yielded an even greater amount of alumina, 13.54 percent. Hydrofluoric acid dissolved all the feldspar inclusions exposed on the faces of the crystals of pyroxene but not the ones entirely enclosed in the interior. Neverthess, most of the alumina must be considered as part of the true composition of the mineral.

Olivine purified with hydrofluoric acid. Density = 3.136

	Composition in percent	Oxygens	Ratio of oxygens of silica to those in monoxide cations	
SiO_2	39.41	20.51		
Al_2O_3	2.96			
FeO	13.86	3.14	Calculated	Observed
CaO	3.97	1.11 19.96	1 : 1	1.03 : 1
MgO	40.60	15.71		
	100.80			
Excess	0.80			

These analyses show that several species of feldspar, hypersthene, augite, and olivine are found in the lavas of Santorini. The common lava of Santorini contains albite and oligoclase microlites along with phenocrysts of labradorite and anorthite. Moreover, it contains hypersthene with a remarkably large amount of ferrous iron, as well as calcium-rich pyroxene, apatite, titaniferous iron oxides, and an amorphous material that resembles albite in composition but is richer in silica.[6]

The anorthite-bearing lava and the material in drusy cavities contain very calcium-rich pyroxenes that also have large amounts of magnesium. The two olivines differ mainly in the proportions of ferrous iron. That of the common lavas is probably even more different, but I have not collected a sufficient amount to analyze it.

The Santorini lavas also contain another variety of pyroxene that differs from the other two in its crystalline form, color, and chemical composition. It is found in elongated geodes of various sizes up to 2 or 3 decimeters in length and 7 or 8 centimeters in diameter. It forms small, light-green crystals no more than a millimeter in length with the form of fassaite but a chemical composition closer to augite.

These crystals are remarkable for their relatively low silica and magnesium contents and for the notable amount of alumina that is an essential part of their composition. They are associated with very small amounts of anorthite. They have a density of 3.257 and the following composition.

	Composition in percent	Oxygens		Ratio of oxygens of silica to those in monoxide cations	
SiO_2	43.61	22.65			
Al_2O_3	14.70	6.87			
FeO	13.71	3.06		Calculated	Observed
CaO	22.84	6.61	12.00	2 : 1	1.9 : 1
MgO	5.59	2.23			
Na_2O	0.32	0.09			
	100.77				

Seen under the microscope, these crystals appear to be much purer and freer of inclusions than all the other pyroxenes in the lavas of Santorini. In order to determine whether the alumina in them is the result of contamination with feldspar, they were treated with hydrofluoric acid and reanalyzed. This treatment had little effect on their composition, as the following results show.[7]

	Composition in percent	Oxygens		Ratio of oxygens of silica to those in monoxide cations	
SiO_2	43.56	22.6			
Al_2O_3	12.40	5.8			
FeO	17.33	2.4		Calculated	Observed
CaO	22.65	6.4	10.6	2 : 1	2.1 : 1
MgO	3.89	1.8			
	99.83				

Summary table of the compositions of pyroxenes de Santorini

	Albitic lava with labradorite phenocrysts	Lava with anorthite phenocrysts	Geodes with anorthite and black pyroxene	Geodes with green pyroxene and rare anorthite
SiO_2	50.2	51.1	48.9	43.6
Al_2O_3	3.3	3.6	7.2	14.7
Fe_2O_3	0.4	1.0	1.3	0.0
FeO	7.2	5.7	8.2	13.7
CaO	26.4	25.4	27.8	22.8
MgO	12.2	13.5	6.1	5.6
Na_2O	0.0	0.0	0.1	0.3
Density	3.372	3.358	5.364	3.257
Color	mottled dark green	dark green almost black	dark green almost black	light green

Compositions of olivines

	Olivine in anorthite lava	Olivine in anorthite lava
SiO_2	38.17	39.41
Al_2O_3	1.66	2.96
FeO	21.82	13.86
CaO	2.90	3.97
MgO	<u>35.27</u>	<u>40.60</u>
	99.82	100.80
Density	3.603	3.136

Composition of hypersthene in albitic lava with labradorite phenocrysts

SiO_2	50.1
Al_2O_3	2.1
Fe_2O_3	1.6
FeO	23.6
CaO	10.5
MgO	11.0
Density	3.477
Color	brownish

Compositions of anorthites

	Anorthite in dense blocks	Anorthite in drusy cavities
SiO_2	45.93	44.25
Al_2O_3	36.60	37.00
Fe_2O_3	0.88	0.43
CaO	16.09	18.98
Na_2O	-	0.28
MgO	1.29	0.07
	100.79	101.01
Density	2.782	2.756

In summary, one sees that there are three kinds of triclinic feldspars in the recent lavas of Santorini: albite, labradorite, and anorthite. Oligoclase may also be present, for many microlites have extinction angles close to zero. Monoclinic feldspar may very well be present as well, both as phenocrysts and microlites, but it has not been identified with certainty either by chemical analysis or by optical means.

In addition, these lavas contain titaniferous iron oxides and occasional apatite and sphene. The latter mineral is most common in the anorthite-bearing lavas and druses.

Chemical reactions indicate the presence of apatite in greater amounts than one would suspect from a microscopic examination. But in many cases apatite is easily recognized under the microscope, because of its typical elongated, prismatic form, transverse cleavage, lack of color, and longitudinal extinction. It is rare to find hexagonal, isotropic sections, but they are seen from time to time. Apatite is ordinarily devoid of inclusions, but I have occasionally seen tiny violet-colored grains aligned parallel to the sides of the crystals. In one specimen, a crystal of apatite of relatively large width (0.4 mm) has the form of a sheath the interior of which is filled with amorphous material.

The lava of the last eruption normally has a microlithic, fluidal structure. Less commonly it is simply glassy with diverse crystalline forms, such as laths and trichites, but I have never seen a rock of true magmatic origin with a granular or microfelsitic texture.

Nodules of wollastonite, fassaite, melanitic garnet from the lavas of Santorini

These nodules are of two kinds, one hollow and the other solid. The first are rough and no larger than one's fist. Their outer part forms a complete crust that is broken by a number of openings. The interior is filled with a mesh of very small crystals among which one can distinguish the following.

1. Prisms of wollastonite up to 4 mm in length but rarely wider than 0.3 mm or thicker than 0.2 mm. Under the microscope, these inclusions are colorless and transparent. The best-developed face is almost always the base p; the elongation is parallel to the intersection p/h_1, and the angle of the intersections p/g_1 and p/h_1 is commonly modified by the inclined intersection p/e_1.* Many are free of inclusions of foreign material, but it is not rare to see small gas cavities and inclusions of glassy solids. These cavities and inclusions are rounded or polygonal with diameters of at most 0.01 to 0.02 mm.

2. Crystals of green pyroxene with the form of fassaite are imbedded mainly along the edges of the elongated prisms of wollastonite.

3. A few yellowish-green, somewhat turbid globules may have well defined polygonal outlines. They have no effect on polarized light and dissolve readily in acids. They tend to be embedded in the p faces of the wollastonite.

4. A light yellow, transparent mineral is well crystallized in square plates, some of which are truncated at angles of 135 degrees by square edges. These crystals are either separate or distributed in groups of many individuals. They are riddled with irregular inclusions of solid translucent material, normally with a dark color. They are soluble in acid and show no birefringence in polarized light. They are not abundant, but are found mainly where the wollastonite crystals are attached. Their diameter is at most 0.6 mm. They are calc-silicates containing sodium and a little chlorine.[8]

5. Melanite garnets in dodecahedral rhombohedrons are very well developed. Found in only a few specimens, they normally rest on the projections of sheaves of wollastonite prisms or are lodged in the interstices. Their diameter is about half a millimeter. They are fusible into a dense, black glass that is slowly dissolved by hydrofluoric acid. Under the microscope they are a clear homogeneous greenish-brown.

Nodules of the second kind are welded to the lava that envelops and penetrates them. They are light yellowish green with spots of white and veins

---

* Hessenberg has been able to study much larger crystals and measure several modifying faces.

of gray. They resemble fragments of limestone but do not give off carbon dioxide on contact with hydrochloric acid. They consist mainly of a mixture of all the components noted in the porous nodules except garnet, which is not seen. The crystals of wollastonite and the turbid, yellowish-green globules dominate the assemblage. They are not in any fixed proportions.

Some of the white spots are due to small clusters of wollastonite crystals; others are irregularly arranged grains of milky white quartz, the largest of which have a diameter of not more than 1 millimeter. Their reaction to a blowpipe or hydrofluoric acid and their appearance under the microscope leave no doubt as to their nature. They are strongly birefringent, and they have many gas bubbles and solid inclusions of a vitreous appearance but none that are fluid.

The gray veins that cut the nodules are formed by the surrounding lava. This lava is strongly modified, as is the coating in direct contact with the surface. Crystals of feldspar, pyroxene, and iron oxides, which are less numerous there than in the rest of the lava, are spotted with transparent amorphous material that has no birefringence. Their faces are corroded, especially near the corners. The amorphous material is filled with small gas bubbles and solid inclusions of various colors; some of the latter have bubbles of gas of questionable mobility. These inclusions are for the most part isotropic.

	Very clear wollastonite	Cracked wollastonite with inclusions	Wollastonite associated with fassaite		Wollastonite associated with melanite	Average of five analyses	Corresponding oxygens	
SiO_2	46.2	45.5	43.9	45.7	43.5	45.0	24.0	
CaO	41.8	43.0	41.3	42.2	42.3	42.1	11.8	12.4
MgO	1.5	0.8	2.0	1.9	2.0	1.6	0.6	
Al_2O_3	7.1	7.2	9.5	8.6	8.1	8.1	3.7	4.5
Fe_2O_3	2.9	2.8	2.5	2.5	3.3	2.8	0.8	
	99.5	99.2	99.2	100.9	99.3	99.6		
Density	2.910	2.906	2.915	2.913	2.920	2.913		

Average ratio of the proportions of oxygens Si : R^{2+} : R^{3+} = 3.9 : 2 : 0.7

The average loss on calcination is 0.8 percent. It is due mainly to release of sodium chloride that seems to impregnate the surfaces of the crystals.

	Melanite garnet		Fassaite pyroxene		Greenish yellow material (mixture of turbid globules and crystals of wollastonite). Average of three very similar analyses	Amorphous material from thin grey veins in the nodules
		oxygens		oxygens		
SiO_2	35.6	19.0	46.8	24.9	35.6	66.8
Al_2O_3	12.2	5.7 ⎫ 9.0	10.1	4.6	15.8	16.5
Fe_2O_3	16.8	3.3 ⎭	10.4 FeO	2.1 ⎫	4.4	2.9
CaO	33.3	9.3 ⎫ 9.8	24.9	7.1 ⎬ 11.9	41.1	3.9
MgO	1.2	0.5 ⎭	6.8	2.7 ⎭	1.8	0.9
Na_2O	0.0		0.0		0.3	7.4
K_2O	0.0		0.0		0.0	1.5
	99.1		99.0		99.0	99.9
Density	3.330		3.253		2.850	2.550

Ratio of oxygens	Ratio of oxygens	The black glass of
Si : R^{3+} : R^{2+} = 2.1 : 1 : 1.1	Si : R^{2+} = 2 : 0.96	the groundmass
Calculated = 2 : 1 : 1	Calculated = 2 : 1	lost 2 percent

The following conclusions can be drawn from these analyses.

1. The wollastonite and the fassaitic pyroxene associated with it are both very rich in alumina, even though these minerals are well crystallized and very pure. The alumina cannot come from inclusions of other minerals accidentally enclosed in them. It is actually an essential chemical component. It is noteworthy that the chemical formula is analogous to that of pyroxene and of wollastonite (M SiO_3), if one considers it as a combination of two oxides of alumina, Al^{1+} and Al^{2+}. The absence of isomorphism of these separate minerals does not necessarily mean they could not be linked.*

2. The greenish-yellow globules and light yellow isotropic crystals that accompany wollastonite in the nodules are even more basic than the wollastonite.

3. The garnets have very little manganese. They contain no alumina in excess of that required by the normal formula. In this respect, they are a striking contrast to the associated wollastonite and fassaite.

4. The amorphous material in close contact with the wollastonite nodules differs very little in composition from the normal lava of the same eruption despite the modification of its physical properties. The only difference is that it has 3.9 instead of 1.3 percent lime, and iron is slightly more oxidized.

5. The abundance of lime in all the silicates and the presence of quartz

---

* This interpretation of aluminous pyroxene was formulated long ago by Rammelsberg.

in the nodules seem to indicate that they owe their origin to the blocks of siliceous limestone caught up and melted by the lava.[9]

Oligoclase-bearing nodules in lavas of the last eruption of Santorini

The oligoclase-bearing nodules brought up in lavas of the recent eruption have rounded shapes with sizes ranging from less than a cubic centimeter to several cubic decimeters. These nodules are strongly attached to the surrounding lava which seems to have significantly altered their contact. To the naked eye, they are grayish-brown, finely scoriaceous, and almost homogeneous. Their very pronounced crystallinity is seen only under the microscope. They contain great numbers of feldspar crystals together with lesser amounts of what appear to be pyroxenes and iron oxides. All these crystals are set in a brownish-yellow, glassy material.

The crystals of feldspar form elongated prisms modified by terminal facets. Some are separate, others in clusters. Most of the latter form twins according to the albite law, but a few are joined on one of the inclined terminal planes of the prisms. The length of the prisms rarely exceeds 0.5 mm, and only a few are less than 0.1 mm. Their width ranges between 0.3 and 0.5 mm. In short, they have a moderate size range that is rarely exceeded in either direction. Under crossed nicols, they have strong birefringence and inclined extinction in all orientations, so they must be triclinic. They commonly have the characteristic bands of the sixth crystal system of the feldspars.[10] Crystals twinned on g_1 and oriented so the section is in the zone perpendicular to g_1 have extinction angles as large as 26 degrees from the trace of the twin plane. These crystals are quite resistant to boiling nitric acid. Many have inclusions of the surrounding glassy material and, more rarely grains of pyroxene. The glassy inclusions have no regular form but tend to conform to the symmetry of the crystal. Most contain a small gas bubble. Among the glassy inclusions without bubbles, a few with diameters of less than 0.003 mm and lengths of 0.01 mm are remarkably elongated. They are distributed in rectilinear strips parallel to the edge p/g_1 and p/m. These same feldspars have numerous gas-filled cavities without a glassy enclosure.

The pyroxenes are small yellow or light green crystals. Most are prisms modified by terminal faces. Their dimensions are almost the same as those of the feldspars, and, like the latter, they have a limited range of sizes. Under the microscope, they have strong birefringence. At least 90 percent have parallel extinction. The small number that do not have these properties have an extinction angle of about 35 degrees from their long axis. These latter crystals are light green and less pure. They have more small inclusions of iron oxides. They also have less perfect crystalline forms and most often have granular shapes. If grains separated by hydrofluoric acid are examined under the microscope in orientations where they are not in extinction, they are seen to have brilliant colors irregularly distributed over their surface owing to their irregular thickness. They are not dichroic. In short, they are true pyroxenes more or less flattened parallel to the g_1 face and with poor crystal forms.

Other crystals with a green or yellow color and extinction parallel to their long axis are weakly dichroic. In polarized light without crossed nicols they are green when the extinction direction closest to the long axis is oriented parallel to the polarizer and yellow when it is perpendicular. Almost all the crystals have good forms, but most have various kinds of inclusions, mainly of iron oxides or glassy material with or without gas bubbles. These crystals are very resistant to concentrated hydrofluoric acid. One sees no gradation between them and the crystals that are definitely pyroxenes. They have none of the striations or cleavages that are so distinctive of pyroxenes and amphiboles. The weak dichroism and form of the terminal faces of these crystals rule out any possibility that they are amphiboles. If they are pyroxenes, it is strange that in thin sections they are seen to have lamellae parallel to the zone ph_1 but not in other sections. This is quite unusual. The only remaining possibility is that they are rhombohedral. This conclusion is consistent not only with their optical properties but also with the small amount of calcium they contain. (See the analytical results).* In addition to this evidence, which I mentioned in a note to the Academy of Sciences, I will add that I have since been able to observe this mineral under crossed nicols on all the faces of the zone g_1h_1 by turning the crystals parallel to the edge g_1/h_1, and from this I have noted that the mineral has all the properties of hypersthene. I have also found hypersthene accompanying augite in the anorthite-bearing nodules.

Well-crystallized iron oxides are also very abundant.

The amorphous material is pure and clear even though it has a strong brown color. Apart from the crystals just described, it contains many sheaves of prismatic crystals that are very elongated, transparent, and colorless. These crystals, which attain lengths of 0.5 mm but are never more than 0.01 mm thick, are so thin that few show any birefringence. The thickest ones are white under crossed nicols and have parallel extinction.[11]

---

* Mr. des Cloizeaux, to whom I submitted these crystals, did not hesitate to offer the opinion that they are hypersthene.

	Composition of feldspar	oxygens	Orthorhombic pyroxene-like crystals	oxygens		Bulk composition of the nodules
SiO_2	59.7	31.5	48.6	25.9		58.4
Fe_2O_3	0.4		21.3	4.6		8.1
Al_2O_3	23.2	10.7	6.0			20.7
CaO	7.9	2.1	3.2	0.9	13.5	6.2
MgO	1.0	0.1	20.0	8.0		2.7
Na_2O	6.6	1.7	trace			3.7
K_2O	0.8	0.1	0.0			0.5
	99.6		99.1			100.3
Density	2.629		3.472			2.687
Oxygen	$Si : R^{2+} : R^{2+} = 8.83 : 3 : 1.12$		$Si : R^{2+} = 2 : 1.05$			

Saline deposits on lavas of the last eruption of Santorini

Substantial saline deposits, many of them bright-white, were found in 1867 in the central areas of the most recent lavas, particularly those of Aphroessa. The main concentrations were near a more or less linear crack several meters long and about 10 centimeters wide, and about 35 meters above sea level. According to local inhabitants, these deposits were first noticed in May 1866 when the adjacent lavas were still very hot. The aridity of the climate and the narrow shape of the crack explain the preservation of the mixed salts, several of which have the superficial appearance and compact nature of molten material.

Only a small part of these deposits can be dissolved in cold water. These soluble components are mainly sodium chloride and lesser amounts of sodium sulfate. Of seven samples analyzed quantitatively, the soluble fraction of six contained a small amount of magnesium carbonate associated, in some cases, with magnesium bicarbonate; the sample without magnesium bicarbonate had sodium carbonate instead. Magnesium chloride was found in three samples.

The insoluble fraction is composed of magnesium carbonate (with or without aluminum sulfate), traces of hydrated iron oxide, and calcium sulfate. No iodine, bromine, or salts of potassium were detected in one-milliliter samples of solution concentrated by evaporation from 120 grams of dissolved material.

It is unlikely that the potassium salts have disappeared from these mixtures as a result of atmospheric humidity, for in that case magnesium chloride would also be absent. Hence, these deposits cannot be due to simple evaporation of seawater. They must have come from the same deep-rooted vents that discharged the lavas, or, if not, the incandescent lava must have affected their chemical composition.

It is noteworthy, however, that these deposits are richer in magnesium compounds than those normally found in fumaroles on volcanoes. The presence of magnesium chloride seems to rule out high-temperature conditions.

This combination of seemingly contradictory evidence can be explained as a reaction of seawater with molten lava according to the principles laid out by Gay-Lussac. The unusually sodic character of the Santorini lavas accounts for the scarcity of potassium in the volatile products, the potassium being fixed by the hot lava with which the water vapor is in contact, while the sodium and magnesium escape this effect entirely, owing to their greater initial proportions. As for the magnesium chloride, it is true that it could not be volatilized in nature in the presence of water vapor, but magnesium hydroxide entrained by volcanic gases is easily replenished after being deposited on the surface of the lava, as either magnesium bicarbonate or magnesium chloride, for it is exposed to the action of carbonic and hydrochloric acids given off at the same source where the thermal conditions favor precipitation of mixtures of salts.

When seawater infiltrates the depths of a volcanic system, contact with the magma causes the proportions of components that are volatilized or entrained in the gases to be modified when they return to the atmosphere. These successive effects could account for some of the observed relations.

These conclusions are based on the analytical results compiled in the following table.

Mixtures of salts in fine-grained porous aggregates of crystals.

	No. 1	No. 2	No. 3	No. 4	No. 5	No. 6 Compact, milky white	No. 7 Compact, translucent
Soluble Fraction							
Magnesium bicarbonate	2.2	1.4	1.4	1.1	0.4	0.3	0.0
Magnesium chloride	2.1	0.0	0.0	0.0	2.8	4.5	0.0
Sodium sulfate	5.5	1.2	0.6	0.1	1.9	1.7	1.6
Sodium chloride	73.8	74.2	95.4	88.7	86.6	81.4	95.0
Sodium carbonate						0.0	0.7
Insoluble Fraction							
Magnesium carbonate	15.5	21.5	2.1	8.7	8.3	12.1	2.7
Alumina, iron oxides, and possibly sulfuric acid	0.4	0.8	0.2	0.7	tr	tr	0.0
Calcium sulfate	tr	tr	tr	tr	tr	-	-
	100.0	100.0	100.0	100.0	100.0	100.0	100.0

The physical and chemical properties of the salts listed in this table allow no other qualitative interpretation of the analytical results.

Study of the gases of the last eruption of Santorini

Studies of volcanic gases involve two separate operations, first collecting the samples and then analyzing them.[12] It is almost impossible to carry out complete and precise studies of this kind in the field, so one normally seeks to collect the gases in containers that can be hermetically sealed and taken to the laboratory for analysis. It is useful, however, to carry out semi-quantitative analyses in the field to determine the nature of the gases and the size of samples needed for more precise analyses.

Moreover, all volcanic gases are not of equal interest. For example, a sample of carbon dioxide mixed with a little nitrogen is obviously of less interest than one containing combustible components, such as hydrogen and hydrocarbons; so a rough analysis in the field can show which of the many vents one normally finds around an active volcano are worth sampling for a more complete analysis. I shall deal first with these semi-quantitative analyses and then turn to methods of collection and laboratory analyses.

Field analyses are carried out by means of tubes 15 centimeters long graduated in units of 0.5 cc and with a capacity of 8 cc. The tube can be filled with gas under two types of conditions, one wet and the other dry. In the first case, the gas is collected by first filling the tube with the same water through which the gas is escaping and attaching a small funnel that is placed over the point where the gas is being released. The bubbles of gas collected by the funnel rise into the tube, gradually filling it. When the operation is completed, the tube is plugged with the thumb and moved to a small pan filled with the same gas-saturated water. The funnel is then removed and the analysis begins. The water of the basin must be close to the ambient temperature. The tube filled with gas is thrust in and out of this water several times until it comes to the same temperature. The tube is then raised until the level of the water is the same inside and out. The level read on the graduations gives the volume of gas analyzed. The gas is then exposed to different absorbing reagents. These reagents are introduced into the graduated tube by means of another smaller tube closed at one end, 2 cm long, and 3 to 4 mm in diameter. After inserting the reagent into the small tube, the rest of the volume is filled with water from the basin. Then, holding the opening upward, it is introduced into the graduated tube. The mouth of the latter is then sealed with the thumb, and the tube is lifted from the basin and shaken vigorously. The reagent in the small tube comes out and is exposed to the gas, which it then absorbs. The graduated tube is returned to the basin with the small tube at its lower end, and the thumb is gently removed so the small tube can slide out and be removed. The graduated tube is then raised and lowered several times to equilibrate its temperature and placed at the proper level to measure the volume of gas. The difference between this value and the first indicates the amount of gas absorbed by the reagent.

The following reagents are used in succession to complete the analysis.

1. Granular lead acetate with a few drops of acetic acid is used to absorb hydrogen sulfide.

2. Small pieces of potassium hydroxide are used to absorb carbon dioxide.

3. Potassium pyrogallate is used to absorb oxygen. This reagent is prepared by mixing pyrogallic acid and potassium in a small tube. Potassium must be present in excess.

After these three operations, one is normally content to determine whether the remaining gas contains combustible components. This is done in a place that is as dark as possible. Holding the graduated tube with its opening upward, the thumb is removed and a match placed near the opening. The residual gas will burn if it contains significant amounts of combustible components.

In the case just described, we have assumed that the mixture of gases is discharged through a small depth of water and can be collected by kneeling over the source. If the discharge comes up through a much larger body of water and the bubbles of gas appear at long intervals in different places, one must use a funnel of much larger size (as will be explained in a later section devoted to these instruments). Gas collected in this way is transferred to the graduated tube using the method normally employed in chemical laboratories, and the subsequent operation is the same as the one already outlined.[13]

A different method must be used when the gas is released under dry conditions, as, for example, when fumarolic gases are collected for a rough field analysis. The procedure currently used by geologists for such conditions is very ineffective and should be completely abandoned. I describe it here only for the record. It consists of filling with water a graduated glass tube and emptying it over a source of gas. The result is a gas mixed with a large amount of air and altered by the spilled water. If the tube is held by hand and capped with the finger before pouring out the water, one risks being badly burned if the fumarole is hot.

I recommend the following procedure, which has the advantage of avoiding these inconveniences.

The graduated tube, having been filled with air, is fitted with a plug having two short pieces of glass tubing the end of one of which nearly reaches the closed end, while the other extends only slightly beyond the plug. On the outside the ends of these tubes project only 2 or 3 centimeters. Each is attached by means of rubber tubing 50 centimeters long to a straight glass tube of normal diameter. One of these is very short (about 10 cm); the other, which serves to draw in the gas, is much longer (about 120 cm). The free end of the latter is placed in the orifice giving off gas, or if the temperature at this point is hot enough to melt the glass, one can connect another tube of por-

celain to lead from the fumarole to the glass tube. In this case it is a good idea to cool the glass tube with wet paper to prevent it from affecting the rubber connections.

The mixture of gases from the fumarole comes into the graduated tube from the collecting tube and goes out through the other displacing the original air with the volcanic gas. To facilitate the operation, it is almost always well to obstruct as much as possible the orifice of the fumarole around the glass or porcelain collecting tube. In most instances it will also be useful to apply a slight suction either with an ordinary aspirator or, more conveniently, with the mouth. In the latter case, a slight suction is applied at intervals while taking care to have the mouth filled with water in order to avoid any harmful effects of the gas.

After a certain time, usually a half hour to an hour, the operation is complete. The rubber tube is clamped near the junction to the aspirating tube. The gas stops flowing through the graduated tube, which is then cooled. When the cooling is complete, the rubber section of the collecting tube is clamped and both rubber tubes are cut as close as possible to the outer sides of the clamps.

Having done this, the sealed graduated tube is moved to a basin of water to remove the plug along with the glass tubing that goes through it.

If one is careful to cool the graduated tube before closing the second tube, the sudden absorption of gas in contact with water is due almost entirely to absorption of sulfurous and hydrochloric acid in the mixture. This has been confirmed by careful experiments that will be described in a later section.

The rest of the analysis is carried out in the same manner as when gas is collected from an underwater source.

The water container used for analyses should be rather large, at least in its upper part, in order to make handling the apparatus easier. It should also be deep enough to immerse the entire graduated tube, and it should have the smallest capacity possible, so that the least amount of water is required to fill or clean it. On high volcanoes like Etna or Tenerife, or even moderately high cones, such as Vesuvius or Stromboli, there is little water near the fumaroles, and it is necessary to have it brought up from below; it is therefore important to use as little as possible.

These three requirements are met by a container of tinned sheet metal the lower part of which is a cylinder 12 centimeters high and 4 in diameter surmounted by another cylinder 4 centimeters high and 10 in diameter. It is easy to carry out the manipulations in this container. The total height of 16 centimeters is greater than that of the graduated collecting tube, so less than half a liter of water is needed to fill it. It has a supporting ring for the base, which also serves to hold a supply of rubber tubing.

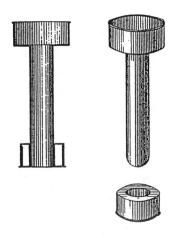

On the right, a basin for analyses in the field; the support is shown separately. On the left, a section through the basin and its support.

The tests just described give only approximate indications of the composition of the mixtures of gases. To obtain samples for precise laboratory analyses, one must use a special collecting method.

The simplest, though not necessarily better than other ones that are sometimes used, requires only a funnel and some flasks with ground-glass connectors. If the mixture of gases is rising through a layer of water, it is collected by means of a funnel placed under an inverted flask. The latter is first filled with the same water through which the gas is rising. When the flask is filled with gas, it is closed under water with a stopper that is well coated with grease. It is then turned upright and wiped dry, and, as a precaution, the stopper is tied with string and sealed around its upper edge with wax or instrument grease.

The same type of stoppered flask can be used to collect mixtures of gases discharged under dry conditions. The neck is fitted with a stopper of cork or rubber with two holes and pieces of glass tubing. The rest of the arrangement is the same as that outlined above for filling graduated tubes for analyses in the field. When one sees that the volcanic gases have displaced the air in the flask, the two-holed stopper is replaced with one of ground glass coated with grease. In all other respects, one follows the procedure described above.

A better apparatus than the flask just described is one equipped with glass tubes that can be closed with a flame after being filled with gas.

Mr. Bunsen, in his well-known memoir on his scientific work in Iceland, used tubes of two kinds, one to collect gas rising through water and the other for gases discharged under dry conditions.

The first has a capacity of 15 to 20 cubic centimeters and a length of 15 centimeters. It is closed at one end and drawn out at the other to provide a narrow neck equal in length to the main part of the tube and with a narrower section about 2 or 3 millimeters in diameter near the middle. The walls of the narrow part have almost the same thickness as the rest of the tube.

The entire tube is filled with the same water the gas is rising through and inverted over the source from which bubbles are rising. A funnel fitted to the neck of the tube facilitates collecting the bubbles of gas, which can be made to pass through the neck and into the main section by shaking the apparatus slightly. When the tube and the narrow constriction are both filled with gas, the latter is heated until it melts and seals the gas until it is reopened in the laboratory.

The tubes that Mr. Bunsen uses to collect gases released to the atmosphere are slightly larger than those just described.

They are open and pointed at both ends. A small funnel fitted to one end is placed over the source and insulated as well as possible with clay soil packed around the sides. The neck of the funnel is connected by a tube of unvulcanized rubber to one of the pointed ends of the collecting tubes while the other end of this tube is connected to a second one of the same type. In this way, several tubes can be connected in series with the end of the last one placed under a small amount of water held in a flask. The gases are allowed to flow through this entire system. After the gases have flowed through the tubes long enough to displace all the air, both ends of each tubes are sealed with the flame of a lamp.

The main disadvantage of Mr. Bunsen's tubes lies in the difficulty of closing them. To melt the glass he uses a small oil lamp, the flame of which is enhanced by means of a blowpipe. Even with this arrangement, however, the points of the tubes used by Mr. Bunsen are so thick that melting them in the open air is very difficult. In volcanic regions the air is almost always quite gusty, so to melt the glass one must shelter the flame from the wind by nestling it under some sort of cover. The tubes that are closed at one end are especially difficult to use. The water flows through them very slowly and filling them with gas is a very tedious task.

The collecting tubes preferred by Mr. Ch. Sainte-Claire Deville do not have these inconveniences. They are evacuated ahead of time by means of a pneumatic device and the pointed ends are sealed with a flame. Their capacity is about 100 cubic centimeters. The procedure for filling the tubes with gas differs depending on whether the gas is released into the air or under water. In the first case, one places a funnel with a slender mouth over the source and seals it as well as possible with soil. When it is thought that the air in the funnel has been flushed out by the gas, the narrow end of a collecting tube is attached to the stem of the funnel with a piece of rubber tubing. The point of the tube is then broken by pressing it obliquely inside the rubber, and the gas immediately flows into the tube and fills it. The narrow part where the glass is very thin is then closed with the flame of a lamp.

A bell jar equiped with a valve and filled with water is used to collect gas rising through water. It is placed over the source with the valve closed, so that the gas rises into it, gradually displacing the water. The closed pointed end of an evacuated tube is attached above the valve with a rubber connector then broken in the same way as before. The valve is opened, gas flows in, and the tube is sealed with the flame of a lamp.

I have often used this method, and despite its advantages for collecting the gas and sealing the glass containers, I think it should be abandoned because of the serious problems associated with it. In the first place, it is difficult to obtain a vacuum that does not still have troublesome amounts of atmospheric gases. When one breaks the point of the tube to allow the gas to enter, it comes in so quickly that air almost always leaks in at the same time through the rubber connectors or, if working in the open air, through openings around the base of the bell jar.

For my own work, I have adopted tubes that are open at both ends and have a narrow section with thin walls at each end similar to those used by Mr. Bunsen to collect gases released to the free air. I make the diameter of the narrow part small and the walls very thin. I also find it convenient to use a bell jar, as Mr. Deville does, instead of funnels, which are difficult to pack in a traveling case and are very easily broken.

When collecting gas rising through water, a bell jar is placed over the source, and the valve is opened to allow water to rise to the top. The valve is then closed again. A tube is filled in the same way and connected to the outlet of the bell jar. The free end of this tube has a section of rubber tubing that is placed under 3 or 4 centimeters of water held in a small container. The valve of the bell jar is then opened; the gas rises to the top and flows into the collecting tube, slowly displacing the water. When the tube is filled, the valve is closed and the end of the rubber tube under water is plugged with a small glass rod. The rubber tube attached to the valve is detached while clamping it tightly between the finger, and its end is placed under water and plugged in the same way as the other end. Finally, the narrow sections at the ends are heated with a lamp and sealed. If one wishes to collect several samples of the same gas, it is a simple matter to connect more tubes in a series with short pieces of rubber tubing.

A bell jar is not normally required when working in open air. One of the ends of a glass or porcelain tube can be inserted directly into the orifice where the gas is discharged and the other end connected to a series of tubes open at both ends, as already described. The collecting tubes are linked together with rubber tubing. If the gas is flowing under pressure, the free end of the last tube has a piece of tubing that is placed under water and, if the pressure is not sufficiently greater than that of the atmosphere or if the gas is escaping freely through cracks in the ground, it is necessary to aspirate the gas. The air originally in the tubes is driven out and gradually replaced by the volcanic gas and by the liquid produced by condensation of water vapor.

When the air is completely expelled, each end of the rubber tubing is clamped in two places and the rubber between them is cut. The sample tubes are thus closed off from one another, and as an added precaution the ends are plugged with glass rods so the clamps can be removed. Finally, the pointed ends are sealed with a flame.

When transporting the tubes, it is a good idea to protect the thin ends with a glass cover about 6 millimeters in diameter and held on with mastic.

To attach or detach the covers, it is sufficient to warm them slightly with the same lamp that is used to seal their ends. Each tube is then wrapped in tissue

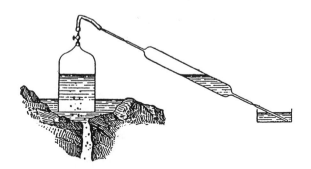

A bell jar with a valve is used to collect gas that rises through water.

A series of tubes is used to collect gas that is discharging to the open air.

paper, labeled, and placed in a cardboard case marked according to the label on the tube. When traveling, a case is used to hold the tubes in their cartons. It is a good idea to place the tubes in a metal box large enough to hold about ten sampling tubes, which is about the number needed in one day of work in the field. The box has a strap so it can be carried over the shoulder.

The figures above illustrate the operations just described.

To seal the tube by heating the narrow sections of their stems I use an alcohol lamp with a narrow wick. Since the alcohol flame is scarcely visible in daylight and moves with the slightest breeze, I have had to enclose it in a shield that permits the gas tube to be introduced easily into the flame. This apparatus is a cylinder of sheet metal with, in its lower third, two diametrically placed holes through which the tube can be inserted. Near the upper and lower ends it has a number of smaller openings to allow air to circulate and support the combustion. It has metal covers that fit over each end and form

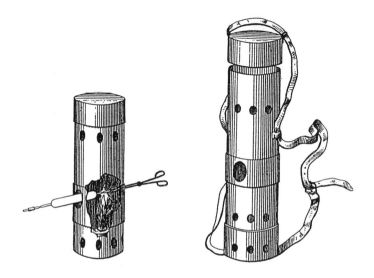

On the left, the lamp housing is shown as it is used to fuse the ends of gas tubes. A part of the wall is cut away to show the interior. On the right, the three parts of the shield are shown separated from one another.

the two bases. The lower cover has holes corresponding to those in the lower part of the cylinder, so that they can be conveniently matched to each other. The upper cover has no holes and must be raised slightly when in use, so that it does not close the upper openings of the cylinder.

When using this apparatus, it is essential to regulate the circulation of air to support the flame by superimposing more or less exactly the lower air holes. It is also necessary to place the tube and holder in a suitable position above the alcohol lamp before closing the cover. If the wind is not too strong, the operation can be carried out without having the cover in place.

When working near a building or a place that is well sheltered from the wind and collecting gas rising through water, it is easy to fill a bell jar with gas and then place the cover of the cylinder beneath it. While the valve on the bell jar is closed, the metal cover is filled with water. One can then carry the sample to a sheltered place to work as one would in the laboratory.

A bell jar with a valve and the cover of a heater serving as a water basin.

In addition to the apparatus just described, two other small items of equipment are needed to study volcanic gases. A few large-mouthed containers with ground-glass stoppers are required for the reagents: caustic soda, pyrogallic acid, lead acetate, quick lime, silver nitrate, barium nitrate, and acetic acid. Another container is needed for reactive paper, graduated tubes for field analyses, small tubes to introduce reagents into the larger tubes, forceps for holding the potassium, tongs with rounded ends, sealing compound for joints, small thermometers, a pair of scissors to cut rubber tubing, a package of small gummed labels, a bar magnet, pencils of various colors, a pocket knife, a stick of Chinese ink, a brush, a few pads of colored paper, and so forth.

I had Mr. Wiesnegg construct a box of small size to carry all these items except the tubes for gas samples. The walls of the box are formed by the lamp shield described earlier. Inside it are placed:

1. the vessel for field analyses
2. its support
3. a bell jar and a metal cover lined with flannel to protect the glass
4. an alcohol lamp
5. the necessary reagents
6. various instruments

The illustration on the following page shows a cut-away view of how the compartments are arranged.

The only instance in which I have found this equipment inadequate was when I was collecting small amounts of gas rising through deep seawater during a season when the sea was always rough. In that case I had to employ a metal funnel with a diameter of 60 centimeters and a narrow stem. One end of a long rubber tube was fitted to the stem while the other end was connected to the neck of the valve on the bell jar. Everything was first filled with water, then the bell jar was placed in the a water pail that the boatman carried. The funnel had a lead ring near its rim and was supported by three cords that were joined near the mouth.

The best thing to do when one is anxious to collect all the acid components in a mixture of gases is to pass the mixture through a sodic solution. The best apparatus to use for this is a simple test tube with a two-holed stopper of unvulcanized rubber. One can put several of these in series. To

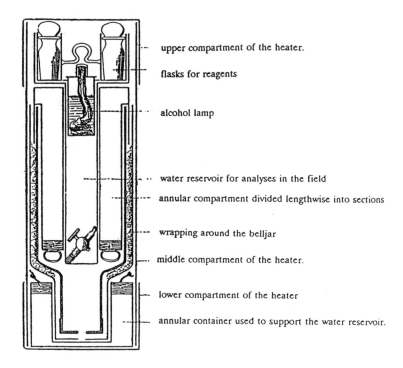

- upper compartment of the heater.
- flasks for reagents
- alcohol lamp
- water reservoir for analyses in the field
- annular compartment divided lengthwise into sections
- wrapping around the belljar
- middle compartment of the heater.
- lower compartment of the heater
- annular container used to support the water reservoir.

prevent heating of the liquid in the tubes it is a good idea to place them in a jar of cold water that can be changed from time to time.

To carry out analyses of the gas in the laboratory I have used a Doyère apparatus that I highly approve of. It gives results with an accuracy of 1 in 500, which is quite adequate for my types of analyses.

The volume of gas used in each analysis was 30 cubic centimeters. Hydrogen sulfide is absorbed in a very small amount of concentrated solution of copper sulfate, carbon dioxide in caustic soda, and oxygen in sodium pyrogallate. After that, the gas is ignited in air, oxygen, or in a reservoir to which these have been added.

After measuring the decrease of volume after ignition, the carbon dioxide produced by the reaction is absorbed with caustic soda and the excess oxygen with potassium pyrogallate.

The results of these tests have always been perfectly consistent; the proportion of oxygen used for ignition has always been what one would expect from the proportion of carbon dioxide that was formed and the decrease of volume assuming the combustible gases in the mixture were entirely hydrogen and methane.

Each analysis has been repeated with almost identical results. I have also experimented with the Bunsen apparatus and that of Regnault. The slightly greater precision they have does not make up for the longer time they require. But by modifying the chamber of the Regnault apparatus to allow the tip of a Doyère pipette to be inserted into the controller, I found that this apparatus could be made much more useful without sacrificing its accuracy. In this way one can use the Doyère pipettes for absorption while still using the Regnault apparatus to measure volumes, but one can also carry out the absorptions in the apparatus itself using the Regnault method and thereby obtain the maximum accuracy.

I have used the Doyère pipettes in the form he and Ettling originally described. With a few precautions and a little practice, the use of these instruments presents no problems. The following pages give my results.

Study of the chemical composition of gases emitted by Santorini between 8 March and 26 May 1866

The eruption of Santorini has been remarkable for the unusual quantities and compositions of gas accompanying the lavas. The location of the eruptive center in the middle of the sea prevents air from reaching the interior of the volcano. As a result, the combustible gases that normally are burned and consumed by the oxygen that mixes with them at high temperatures are not notably altered, and their complete combustion has taken place only when they came in contact with the air at high temperatures. When that happens, jets of flames have on occasion burned at the surface of the new mound of lava, while at other times the gas bubbled up through the seawater and ignited on contact with the blocks of incandescent lava. Never before has such an abundance of combustible gas been observed and until now, apart from a bluish glow from the combustion of sulfur or hydrogen sulfide, no true flames have been reported as coming from an active crater. Two geologists, first Mr. Pilla, then Mr. Abich, and later Mr. Verdet, a distinguished physicist, whose recent death has been a great loss to science, have noted flames at night at the summit of Vesuvius and suggested that they came from burning hydrogen or hydrocarbons, but to convince everyone that this is true, samples of these gases would have to be collected and analyzed.* The exceptional circumstances of the Santorini eruption have enabled me to carry out such tests. At certain places around the new eruptive vents when the gas rising from greater depths often comes in contact with the air, the temperature is too low for the gases to ignite, and I was able to collect them unaltered.

Most of these gases have been collected at the surface of the water, either around the newly formed islands or from cracks where the older ground of Nea Kameni has been opened in the space between the two main centers of activity, Giorgios and Aphroessa. Some, however, come from discharges into the open air on the flanks of the cone of Giorgios or in the neighboring older ground of Nea Kameni and are always mixed with a considerable proportion of air.

These gases have been collected, some during March 1866 when very abundant flames were seen throughout the active area, others two months later when they had almost completely disappeared.

The following table gives the results of their analyses.

Gases collected at the surface of the water on 17 March 1866

	No. 1	No. 2	No. 3	No. 4
Sulfuric acid	trace	trace	trace	trace
Carbon dioxide	37.04	37.24	36.42	35.60
Hydrogen	27.10	28.12	29.43	30.09
Hydrocarbons	0.43	0.47	0.86	0.81
Oxygen	0.41	0.51	0.32	1.46
Nitrogen	35.02	33.66	32.97	32.04
	100.00	100.00	100.00	100.00

1. In a fissure in the old ground at the southern end of Nea Kameni between Giorgios and Aphroessa at the surface of water very rich in sulfur and with a temperature of 73 degrees.
2. In a fissure in the old ground of Nea Kameni between Giorgios and Aphroessa at the surface of water very rich in sulfur and with a temperature of 75 degrees.
3. In the northernmost fissure in the old ground of Nea Kameni between Giorgios and Aphroessa at the surface of water very rich in sulfur and with a temperature of 78 degrees.
4. In the channel between Aphroessa and the southwestern tip of Nea Kameni at the surface of water made milky by decomposition of sulfuric acid. Temperature = 61 degrees.

* It is possible that these observers witnessed combustion of hydrogen sulfide. On several occasions, Mr. Elie de Beaumont, Ch. Sainte-Claire Deville, and I have noted that in the crater of Vulcano in the Eolian Sea that gas burns with a bluish flame and produces sulfurous acid.

Gases collected at the surface of the sea

	No. 5	No. 6	No. 7
Sulfuric acid	trace	trace	0.00
Carbon dioxide	0.07	1.49	78.44
Hydrogen	1.62	0.00	0.00
Hydrocarbons	0.71	0.42	0.64
Oxygen	21.56	18.45	3.37
Nitrogen	76.04	79.64	17.55
	100.00	100.00	100.00

5. 13 March 1866 near the northern shore of Aphroessa; milky water due to decomposition of sulfuric acid; temperature very variable; about 60 degrees at the surface and not more than 20 degrees at a shallow depth.
6. 10 March 1866 near the island of Reka; milky water due to decomposition of sulfuric acid; temperature very variable from 50 degrees at the surface to not more than 20 degrees at a depth of a few decimeters.
7. 13 March 1866 near the shore of Palaea Kameni in the small harbor of Saint Nicolas; water not saturated with sulfur; temperature 19 degrees; this release of gas preceded the present eruption.

Gases collected on Nea Kameni

	No. 8	No. 9	
Sulfuric acid	trace	1.64	
Carbon dioxide	50.41	17.28	
Hydrogen	16.12	0.49	The hydrocarbons make up most of this mixture, which contains only traces of pure hydrogen.
Hydrocarbons	2.95		
Oxygen	0.20	14.12	
Nitrogen	30.32	66.47	
	100.00	100.00	

8. 25 March 1866 in the northernmost fissure of Nea Kameni; water very rich in sulfur; temperature 69 degrees.
9. 25 March 1866 in the open air on the western flank of Giorgios at a place covered by a thick layer of sulfur; temperature 160 degrees.

Gases collected at the surface of the water

	No. 10	No. 11	No. 12	No. 13
Sulfuric acid	traces	traces	traces	traces
Carbon dioxide	90.78	95.37	86.76	84.85
Oxygen	0.88	0.49	2.01	2.31
Nitrogen	8.34	4.14	11.23	12.84
	100.00	100.00	100.00	100.00

10. 4 May at the bottom of a fissure on Nea Kameni (probably where sample no. 3 was collected); water rich in sulfur; temperature 65 degrees.
11. 4 May at the bottom of a fissure on Nea Kameni (possibly the middle one); sulfur-rich water; temperature 56 degrees.
12. Pond of water remaining from the channel between Aphroessa and Nea Kameni; sulfur-rich water; temperature about 50 degrees. Gas collected on 4 May.
13. Same pond as no. 12; same characteristics; gas collected 12 May.

Gases collected 12 May at the foot of Giorgios near the base of the cone of Nea Kameni

	No. 14	No. 15
Sulfuric acid*	0.42	0.90
Carbon dioxide	5.88	12.24
Oxygen	18.99	16.41
Nitrogen	74.71	70.45
	100.00	100.00

14. Small sulfatara surrounded by crystals of octahedral sulfur at the level of the orifice; temperature 87 degrees.
15. Small sulfatara surrounded by deposits of sulfur that was partly fused and partly crystalline; temperature 122 degrees.

* In gases no. 9, 14, and 15, sulfuric acid was determined at the site and the numbers for it are less precise than those given for other components.

The last five analyses have only doubtful traces of hydrogen and hydrocarbons.

At the time of their emission, the gases of no. 9, 14, and 15 were accompanied by a considerable proportion of water vapor. The condensed water was strongly acidic and gave a thick, white precipitate with silver nitrate after adding nitric acid, and a weaker one with barium chloride. Before the addition of nitric acid, it darkened paper soaked with lead acetate.

Four cubic centimeters of condensed water from the location of gas sample no. 9 contained, after oxidation with nitric acid:

> Hydrochloric acid.....9 mg
> Sulfuric acid...............2 mg

Four cubic centimeters of condensed water from the location of gas sample no. 14 contained, after oxidation with nitric acid:

> Hydrochloric acid.....3 mg
> Sulfuric acid...............1 mg

Conclusions:

1. The results given above clearly show the important role played by hydrogen in the eruption of Santorini, since the mixtures of gases coming from points closest to the eruptive centers have about 30 percent of this gas.[14]

2. The numbers also demonstrate the remarkable relation between the hydrogen and hydrocarbons that were discharged with it. Previous studies of gases from secondary vents of Vesuvius and Etna had already led me to think

that hydrogen belonged to a higher level of volcanic activity than that which is dominated by carbon gases. This relationship between the relative proportions of the two gases, for which Mr. Chevreul has given a rational explanation, is so evident in the gas samples collected at Santorini that it is now beyond doubt. In fact, in the central parts of the eruption at the bottoms of fissures between the two main vents, the proportion of hydrogen rose, on the 17th of March, to 29.43 percent. Closer to one of the vents (Aphroessa) and along the same line, we found on the same day a gas with 30.09 percent. Then as one goes farther from the main eruptive fissure, the proportion of hydrogen diminishes with increasing distance, while the relative proportions of hydrocarbons and carbon dioxide increase. Hydrogen is completely absent from the samples collected near Reka and in the small harbor of Palaea Kameni, while hydrocarbons and especially carbonic acid are seen in notable amounts.

At a given point, the eruption seemed to become weaker with time, and we noticed changes of the same order in the gases given off.

On the 17th of March 1866 in the northern-most fissure of Nea Kameni, when the water was at 78 degrees, the gas contained 29.43 percent hydrogen and 0.85 percent hydrocarbons.

At the same place on the 25th of March, the temperature was no more than 69° and the gases contained 16.12% hydrogen and 2.95% hydrocarbons.

On the 4th of May in a fissure I believe to be the same as the preceding one but modified by ground movements, the temperature was 65 degrees and the gas contained no combustible components but had a large amount of carbon dioxide.

The information obtained on gases from the bottoms of fissures on Nea Kameni in the middle part of the eruption are of all the more value because the gases were discharged through a minor amount of water with only indirect communication with the sea. Consequently, the compositions of the gases and their temperatures could have undergone only slight modification by the water they went through.

3. Despite the imperfections of the procedure for determining sulfuric acid, analyses no. 9, 14, and 15 show that the proportions of this gas decline with falling temperature.

4. Although the sulfuric acid and sulfurous gases may not have been very abundant during the entire course of the eruption, we see from the results shown above that hydrochloric acid was always relatively more important than acids of sulfur.

Gases Collected in 1867

In the spring of 1867, incandescent flows poured into the sea south of Giorgios with much noise and sharp hissing. At these places the copious discharges of gas could be clearly seen to move each day as the flow front advanced. Samples of these gases, collected in tubes sealed with the flame of a lamp, have been analyzed in the laboratory by means of the apparatus designed by Doyère. All had notable amounts of free hydrogen and smaller amounts of methane, as well as quantities of oxygen and nitrogen in more or less atmospheric proportions.

The first of these gases (no. 1) was collected at the toe of the flow moving toward Cape Akrotiri; the source was sampled only once, on the 5th of March 1867. The second (no. 2) comes from the front of the flow headed toward Balos. The table below presents the compositions of samples collected at this point on three different occasions, the 3rd, 5th, and 7th of March 1867. The third (no. 3) comes from the end of the flow moving toward Athinios; it was taken on the 7th of March 1867.

	Gas No.1	Gas No. 2			Gas No. 3
		3 March	5 March	7 March	
Carbon dioxide	0.00	0.19	0.25	0.57	0.22
Oxygen	24.94	20.09	20.41	18.65	21.11
Nitrogen	72.12	64.34	64.36	65.51	21.90
Hydrogen	1.94	14.98	14.70	14.96	56.70
Methane	1.00	0.40	0.28	0.31	0.07
	100.00	100.00	100.00	100.00	100.00

When exposed to a flame, sample no. 3 ignited with a strong explosion; no. 2 also burned but less violently. (The combustibility on contact with air of the residual gas after removing carbon dioxide and oxygen was apparent in the field.) All these gases were collected in tubes evacuated to 2 millimeters; small amounts of oxygen and nitrogen indicated by the analyses must therefore come from air still in the tubes, but this small source of error in no way alters any conclusions regarding the gases collected in this way.

All the gases were discharged from well-defined sources through seawater very near the ends of the flows. The places where they emerged moved with the advance of the toe of the flow, and their development can only be explained by supposing that they were enclosed in the molten lava and were suddenly released by the rapid cooling and cracking of the lava on contact with the seawater.

In 1867, weaker gas emissions were still located around the entire periphery of the eruptive zone, but they seemed to be formed by air that was caught up in the lavas and modified more or less by its passage through sea water. The compositions of three samples of these gases are given below. The first two (nos. 4 and 5) were collected on the 5th and 7th of March 1867 at points where the water was milky white due to decomposed hydrogen sulfide.

	Gas no. 4	Gas no. 5	Gas no. 6
Carbon dioxide	0.00	0.00	0.16
Oxygen	20.62	20.58	12.65
Nitrogen	79.38	79.32	87.19
	100.00	100.00	100.00

In 1867, I collected another sample of gas that was quite different from all others in its composition and the location of its vent. Its source was near the harbor of Saint Georges at the end of the former channel between Nea Kameni and Aphroessa at a point where gas samples had already been collected the previous year, the first time in March 1866 when the adjacent lavas were still incandescent, and a second time in May 1866, when they were almost completely cooled. The compositions of three samples taken in 1867 are given in the following table.

	No. 7 3 Mar 1867	No. 8 5 Mar 1867	No. 9 7 Mar 1867
Carbon dioxide	61.29	60.63	56.63
Oxygen	0.50	0.73	1.84
Nitrogen	37.99	38.26	41.41
Hydrogen	0.11	0.17	0.00
Methane	0.11	0.21	0.12
	100.00	100.00	100.00

At the time the first sample was taken, the gas escaping in the small harbor of Saint Nicolas of Palaea Kameni had the following composition.

Carbon dioxide	79.27
Oxygen	2.21
Nitrogen	18.30
Methane	0.25
	100.00

Finally, I have determined the composition of a gas obtained on the 5th of May 1867 by distillation of a quantity of seawater taken from the front of a lava flow heading toward Balos. A liter of this water yielded 38 cc of a gas that had the following composition and was quite different from the gas given off naturally about 10 meters away.

Carbon dioxide	83.58
Oxygen	3.79
Nitrogen	0.25
	100.00

The analytical results given above point to the following conclusions.

1. They confirm the laws of compositional variation for volcanic gases first recognized by Ch. Sainte-Claire Deville.

2. They show that Santorini lavas that were still partly molten must have carried combustible gases for distances of several hundred meters from the place where they emerged from the underlying vent.

3. Hydrogen and methane are present, and the amount of hydrogen seems to increase with the temperature at which the gas comes from the lava.

4. The composition of sample no. 3 clearly shows that hydrogen and oxygen coexist in these mixtures without reacting, probably because of the high temperature of the lava. Thus it is reasonable that the water vapor that escapes in such quantities from all the active craters and recent lava flows is in a dissociated state in the magma in the interior of the volcano.[15]

Gases collected in September 1875

These are generally mixtures composed mainly of carbon dioxide and various proportions of nitrogen, oxygen and occasional traces of hydrogen sulfide. Their temperatures at the point of emission rarely exceed 60 degrees.

Most of the places where gas was given off in 1866 and 1867 were covered by the 1875 lavas of Giorgios. The ponds of water in the former channel between the lavas of Nea Kameni and Aphroessa southeast of the harbor of Saint Georges no longer gave off even a single bubble of gas.

Gas was still being discharged in the harbor of Saint Nicolas on Palaea Kameni. Because of the curious compositional variations of this gas in the course of the eruption, it is interesting to examine the results of analyses made at various times, as shown in the following table.

	CO_2	C_2H_4	O	N
Gas collected in March 1866	78.44	0.64	2.37	17.55
" " " May 1866	76.06	0.00	12.39	11.55
" " " May 1867	79.27	0.25	2.21	18.30
" " " October 1875	70.29	0.00	2.18	27.53

It is evident from these results that the proportions of gas varied with time. Methane appeared and disappeared twice. In May 1866, the mixture contained an extraordinary amount of oxygen. Reiss and Stübel, who also noted this, attributed it to decomposition of carbon dioxide coming from marine plants, but plants are so rare in this area that this explanation cannot be correct. Oxygen was released here under eruptive conditions that were the same as those where I collected gases in 1867 at the southern tip of Giorgios. It is only in the samples from Saint Nicolas, where oxygen is not accompanied by hydrogen, that there is any reason to suppose that it does not come from dissociation of water, as it does in the other cases.

ASH OF THE 1866 ERUPTION

Ash fell on Santorini on several occasions during the last eruption; some parts were so fine that they were impalpable dust; others had a range of grain sizes. At the town of Thira, ash varied widely in grain size, even though the distance from the vent was constant. The ash fall of 14 November 1866, for example, was a fine powder composed of particles with an average diameter of 0.01 mm, but that of 26 November had grains of 0.1 mm. All this ash consisted of irregular fragments with freshly fractured surfaces. It has all the normal components of the most recent lava, but the most abundant part is amorphous material with granules of iron oxides and microlites of albite. One also finds crystals of feldspar and pyroxene, either widely separated or attached in clusters to the amorphous material. In polarized light, the feldspar has the multiple bands and extinction characteristics of labradorite. Most commonly, it is elongated parallel to g_1, and yet it has an extinction angle typical of labradorite; the angle is 27 degrees on the face pg_1. One must conclude, therefore, that these are simple crystals. If the crystals are twinned, the twinning is not evident; the crystals do not go to complete extinction in any orientation but have only four positions where the illumination has a marked minimum, as though one of the two sets of bands making up the twin dominated the optical properties of what appears to be a single crystal.[16]

Some crystals elongated in the g_1 direction go to extinction parallel to the face pg_1 and are either sanidine or, more likely, oligoclase, as indicated by the test of Professor Boricky (based on examination of crystals treated with fluosilicic acid) and that of Professor Szabo (based on the color they give to flames).[17]

By treating the ash with concentrated hydrofluoric acid, one can extract the crystals of pyroxene and iron oxides and obtain beautiful little prisms of hypersthene that are strongly dichroic and have an extinction direction parallel to their long axes.

There is no trace of mixed pumiceous material. Among the samples of Santorini ash described by Mr. Vogelsang, the glassy material he mentions seems to me to be accidental. The ash examined by this petrologist was probably collected from the surface of a concrete terrace, where it was mixed with pumiceous dust.

In summary, the ash of the last eruption of Santorini has exactly the same texture and composition as a lava that was pulverized at the time of its eruption. There is nothing in its nature that resembles the fine pumice formed by violent exsolution of gas bubbles in a completely molten lava.

In this respect, the present ash differs enormously from the older ash seen in various levels of the cliff of Thera and Therasia. It was formed by pulverization of lava that was already almost entirely crystalline, whereas the older ash was formed from magma that was still almost entirely liquid.

The grain size and proportions of mineral components of a single pyroclastic unit vary from place to place, according to its distance from the volcano. Fragments with the smallest size and density have been transported farther than the others.[18]

CHAPTER SIX

DESCRIPTIONS OF THE OLDER PARTS OF THE SANTORINI ARCHIPELAGO

Thera, Therasia, Aspronisi

The group of older islands collectively referred to as Santorini comprises two main islands, Thera and Therasia, and a small islet, Aspronisi.* The first of these is by far the most important; its name is often used for the entire group. It encloses the bay on three sides, the north, east, and south. To the west are found Therasia and Aspronisi, as well as the passages that link the bay to the open sea.

The largest diameter of the bay is 11 kilometers north-south, the smallest about 7 kilometers east-west. Thus the total surface area, leaving aside the Kamenis, is around 80 square kilometers.

Three passages communicate with the sea. The first, between Thera and Therasia, opens to the northwest; its width where it connects with the bay is 1730 meters. Bounded by the two islands, it is about 1700 meters long and widens slightly toward the sea, so that its surface area is about 3 square kilometers.

The second opening is between Aspronisi and the southern coast of Therasia; its width at the entry to the bay between Aspronisi and Cape Tripiti at the southern tip of Therasia is some 2540 meters. It widens considerably toward the interior owing to the northwesterly trend of the southern coast of Therasia, so that its width reaches 4500 meters between the western end of Aspronisi and the cape at the southwestern end of Therasia. Its total surface area is approximately 5 square kilometers.

The third passage is situated between Aspronisi and the southwestern end of Therasia. It is 2250 meters wide where it opens into the bay between the eastern end of Aspronisi and the cape of Thera. It widens toward the interior to 2880 meters between the western end of Aspronisi and Cape Akrotiri on Thera. Its surface area is about 2.5 square kilometers.

---

* The small islands known as Christiania should also be included with Santorini, but having never visited then, I shall not discuss them here. A brief description can be found in the work of Messrs. Reiss and Stübel.

The island of Aspronisi is so small that in reality the latter two channels just described can be thought of as a single opening forming the principal entrance to the bay.

The cliffs that make up almost all the inner side of the islands around the bay of Santorini are remarkable for their great height and for the rocks exposed in them. A person coming into the bay for the first time is immediately struck by the strange character of these sharp escarpments made up of dark black lava, layers of reddish scoria, and a cap of bright-white pumice on which houses and a variety of other buildings have been constructed.

The inner side of the island of Thera has the shape of a horseshoe concave toward the west, but its general form is modified by many local irregularities. The northeastern angle of the bay is an indentation, known as the gulf of Mousaki, where the coast changes direction abruptly. The northern shore between the point of Apano Meria[1] and the apex of this angle is about 4 kilometers long. Although the form of the shore is fairly continuous, one can distinguish two parts separated by a promontory known as Cape Perivola.

A small, open bay extending west of this promontory has a pebbly beach at the foot of the cliff. The bay takes its name of Armeni from a building that stands on its shore. East of the cape, the cliff becomes less abrupt and higher while maintaining this form to the innermost part of the gulf of Mousaki. In this interval one can see at the base of the escarpment a small promontory called the Cape of Seven Children (Heptapedio). It is a small, rocky salient, half-eroded by the sea. On its eastern side is a wave-cut indentation and a small beach, a few square meters in area, on which a chapel has been built.

The eight kilometer-long eastern shore of the bay is sharply divided into two parts by the promontory of Skaros. The part north of this point forms the gulf of Peristeria. Along this northern part, the cliffs plunge directly into the sea, so that a small boat can enter all the inlets and caves that have been carved into the coast by waves. Small, narrow beaches have been formed by fallen debris, particularly in the section below the highest parts of the eastern cliffs.

Near the southeastern angle of the gulf of Peristeria, there still remained at the end of the nineteenth century, a mass of prominent rocks with an almost flat surface where the wealthy people of Santorini, then living at the crest of Skaros, descended in the summer in search of shade and refreshing breezes. These rocks were known then as Archicondivari. Shaken by earthquakes and battered by waves, they have since disappeared into the shallow waters at their base. The cliffs are now almost inaccessible on this side.

The second part of the eastern shore of the bay extends from the promontory of Skaro to the inlet northeast of the small harbor of Athinios. Along this section the cliff shelters the landing place for the island's main town, Thira. A narrow, uneven pier borders the shore for about 100 meters. The houses and warehouses built along it are constantly threatened by falling rocks.

Farther south, the point of Alonaki projects out from the coastline. In this part of the shoreline, the coast is abrupt and the cliffs very high, but in a few limited places one can find a gravelly beach at the foot of the escarpments, particularly in the section below the heights of Merovigli and south of Cape Alonaki.

The coast then takes a southwesterly trend and continues in that direction for about 4 kilometers. The cliffs becomes less steep and lower. The zigzag trail from Athinios is much slower than that connecting Thira with its harbor. It takes almost half an hour to climb, while the other takes scarcely a quarter of an hour. The slopes at certain places around Plaka and Therma are gentle enough to support a little cultivation.

Finally, the coast continues in an east-west direction for about five and a half kilometers. The cliffs have their lowest elevation at the point where they make this bend, then they become quite high a little to the west of Balos before descending again and finally rising at Cape Akrotiri, the farthest point on the southwestern side of the bay. The slope toward the bay along this section of the coast is highly varied; in certain places it is quite sharp, in others, especially near its end, it is very gentle. A belt of fallen debris extends along the foot of the most abrupt parts. An irregularity of the shoreline produces the small harbor of Balos. The small island of Aspronisi is elongated east-west and contributes only in an insignificant way to the enclosure of the bay on the western side.

The coast of Therasia extending between the two capes of Tinó and Tripiti bounds the bay on the west for a distance of about 5 kilometers. It is bordered by very high, steep cliffs. The pronounced promontory of Simandiri divides it into two parts and marks the northern end of the gulf of Manolas, named after a village situated on the crest of the island. A few long, gravelly beaches are found at the foot of the rocks below the village of Kera and especially at the innermost part of the gulf of Manolas. At the latter place, a zigzag foot path leads to the upper parts of Therasia.

The island ends on the south with very steep cliffs that can be reached on foot at only a few places. These escarpments are very high at the southeastern corner but decline to sea level toward the western end. This part of the coast has many irregularities that are constantly being modified by the action of waves.

The passage opening to the northwest is bounded by a fairly unbroken coast on the side toward Thera. The coastline on the other side, however, has a cove that cuts deeply into Therasia. The small island of Ayios Nikólaos stands on the eastern side of the entrance of the channel into the bay. Nearby, a small inlet on the coast shelters the landing place of Apano Meria. This town is situated on the crest of the adjacent cliff. It is reached by a zigzag path like those of Thira and Athinios. The cliffs of Thera are quite high at the angle near Apano Meria and descend gradually toward the north to a height only slightly above sea level, but the entire slope is steep. The shore of Therasia on the other side is steep only at the two ends of the channel. In the cove, it is formed by a low beach that rises gently toward the interior of the island.

One can think of the two islands of Thera and Therasia as consisting of high ridges, very steep on the side toward the bay and everywhere else sloping gently outward. But the island of Thera is also divided into two parts by a rocky, narrow ridge that crosses it from northwest to southeast starting near Athinios and extending to the southeastern point dominated by the ruins of the town of Mesa Vouno. This transverse chain begins on the northwestern side of the cove at an elevation of 568 meters. The elevation of the hill of Mesa Vouno is 375 meters above sea level.[2]

Thera has other notable peaks. In its southwestern part, the village of Akrotiri stands on the slope of a mass of volcanic tuff, the highest point of which has an elevation of 210 meters. The ridge crest at 180 meters has a gentle slope extending down to the bay where a steep cliff forms the west side of the harbor of Balos. On the opposite side, it is continued as far as the coast by three hills that are separated from their common point of origin by deep ravines and descend toward the coast in a series of tiers. Two of these hills end in abrupt cliffs; the westernmost one is cut almost vertically at its southern end, forming an escarpment 85 meters high and is crowned by a conical peak, Mount Arkhangelos, with a summit at 166 meters.

A plateau between the ridge of Akrotiri and the cape of the same name has an average elevation of only about 80 meters, but the ground rises again at the end of the island. The top of the cliff at Cape Mavros is at 115 meters and that of Cape Akrotiri 127 meters above sea level.

In all other parts of the island, the highest points are not far from the coast on the side toward the bay. The greatest elevations are at Merovigli, which is 360 meters high, the small peak of Mikro Profitis Ilias, which reaches nearly 320 meters, and at Megalo Vouno, and Kokkino Vouno the summits of which are 324 and 288 meters respectively.

The belt formed by the islands surrounding the bay has the overall form of a large ellipse, the large diameter of which measures 17.5 kilometers from the channel of Apano Meria in the northwest to Cape Exomiti on Thera in the south-southeast. The smaller diameter is 13 kilometers. The crossing point of these two axes is essentially in the center of the bay very close to the two cones of Nea Kameni (the cone of 1707 and Giorgios).

The total surface area of this encircling belt is about 170 square kilometers. The areas of its individual parts are:

Water of the gulf	76.3 km²
Channels	10.5 "
The Kamenis	3.0 "
Thera	72.0 "
Therasia	9.0 "
Aspronisi	0.12 "

The area occupied by water is about the same as that of land, 86.8 km² for the bay and channels and 84.13 for the emergent parts.

The bay has a circumference of about 34 kilometers divided as follows:

Inner coast of Thera	21.5 km
Channel between Thera and Aspronisi	2.2 "
Inner coast of Aspronisi	0.1 "
Channel between Aspronisi and Therasia	2.5 "
Inner coast of Therasia	6.0 "
Channel of Apano Meria	1.7 "

From this we see that the bay is surrounded by islands for eighty percent of its circumference.

The island group of Santorini has an external circumference of 53 kilometers, which breaks down as follows:

External coast of Thera	37.6 km
Channel between Cape Akrotiri and Aspronisi	2.9 "
External coast of Aspronisi	0.1 "
Channel between Aspronisi and Cape Kiminon on Therasia	4.5 "
External coast of Therasia	5.5 "
Channel of Apano Meria	2.4 "

The bay of Santorini is much deeper than the channels connecting it with the open sea. The underwater relief can best be illustrated by considering what would result from different amounts of uplift of the bottom.

An uplift of 25 meters (13.7 fathoms) would be sufficient to connect Thera and Therasia by way of Aspronisi. The bay would be completely closed on the west and southwest, and the only connection between the bay and open sea would be the channel of Apano Meria. The bay would be a large gulf with a single opening and two inlets on the west and southwest at the present sites of the channels on either side of Aspronisi. Micra Kameni and Nea Kameni would be connected, and the present channel between the two islands would be dry along its entire length. The same would be true of the port of Saint Georges, where a small pond of water would be completely separated from the sea. The Isles of May between Palaea Kameni and Giorgio Kameni would be attached to the latter by a reef and would rise to their highest point as a small, rocky mass elongated north-south and cut by a trough. A little farther to the south, Palaea Kameni would be connected to the southwestern point formed by the 1866 lavas from Giorgio-Kameni with a rocky bar just below the surface forming the southern side of a small gulf between the two islands. A narrow opening between Palaea Kameni and the promontory formed by the Isles of May would be the entrance to this gulf. Thus, all the islands in the center of the bay would be joined, and there would be only a single mass with several peaks and deeply indented on the north by part of what is now the channel of Palaea Kameni. East of the Kamenis, the summit of Banco would form a new island, while the summit of the other cone presently beneath deeper water would still be under 8 meters of water.

The shores of Thera and Therasia would change little on their bay sides but would be considerably extended on their outer sides. The island of Thera would extend farther to the south beyond the present cape of Exomiti.

An uplift of 100 meters (55 fathoms) would expose the submarine cone opposite the landing place of Thira and complete the junction of the Kamenis by making the Palaea Kameni channel disappear completely. Banco would still be a separate island. The emergent parts surrounding the bay would be greatly increased, particularly toward the south. Thera would be lengthened in that direction by about 7 kilometers, and its surface area would be almost doubled. Toward the west, the bay would be closed by a continuous strip of land, 2 to 4 kilometers wide. To the north, the eastern shore of the channel of Apano Meria would be further extended to a length of 5 kilometers. The interior form of the gulf would remain more or less the same as after an uplift of 25 meters; the greatest changes would be in the shape of the outer coast.

With an uplift of 100 meters, Santorini would take on the form of a large, elliptical island cut by a broad bay with three islands of unequal size in its center. The large island produced by uniting Thera, Therasia, Aspronisi, and the beaches around them would have the following dimensions.

Maximum north-south dimension	26.5 km
Average " " "	21.5 "
Maximum width	19.0 "
Average "	18.0 "

The surface area would be about 350 square kilometers. Thus the area of the emergent parts would be four times greater than is today.

The central bay would have about the same dimensions as the present one but would be augmented by the extension formed by the channel between Therasia and Aspronisi. Its surface area would be 81 square kilometers. Its importance becomes apparent if one compares the tabulation above with the corresponding numbers below for the present area enclosed by the outer contours of the islands of Thera, Therasia, and Aspronisi.

Maximum north-south dimension	16.1 km
Average " " "	14.4 "
Maximum width	14.0 "
Average "	12.0 "

One can judge the extent of the ground that would be exposed by an uplift of 100 meters by scanning the following table showing the distances between the two inner and outer shores of the islands bordering the bay of

Santorini, first for the present condition and then for a hypothetical uplift of 100 meters.

Directions in which measurements are taken	Present Condition	After 100 m. of uplift
Northern point	2.5 km	5.0 km
Northeastern corner	1.5 "	3.0 "
Eastern coast	5.5 "	7.0 "
Southeastern coast	5.5 "	8.5 "
Southern point	1.2 "	11.0 "
Southwestern coast	1.5 "	2.5 "
Western coast facing the channel between Therasia and Aspronisi		2.0 "
Western and northwestern coast	2.5 "	5.5 "

From this it can be seen that the increase would be mainly toward the south.

After an uplift of 100 meters, the combined Kamenis would compose a round central island with a jagged shoreline and a surface area of 4.5 square kilometers. Its highest point, the cone formed by the last eruption, would rise 236 meters above sea level.

Banco and the other cone now under water would form two separate round islands, the first 98 meters high and 0.7 square kilometers in area and the second 76.5 meters high and 0.35 square kilometers in area.

An uplift of 140 meters would result in the bay being completely closed. With an uplift of 200 meters (110 fathoms), the outer coast of Therasia would retreat another 2 kilometers. The entrance to the channel of Apano Meria would be barred by a sandy ledge 3 kilometers across, the lowest point of which would be 70 meters high. There would be less important changes to the outer coast on the eastern and southern sides of Santorini.

If the lake produced at the present site of the bay maintained the same level as the surrounding sea, the emergence of Santorini would produce marked changes. The central lake would be divided into two basins by the Kamenis and a strip of ground connecting them with the nearest coast of Thera. The cones would form a peninsula, and the two basins would be connected by a narrow channel a few meters deep between Aspronisi and the western coast of Palaea Kameni. This isthmus would be made even narrower by a small island. The southern basin would have an irregular shape with its largest dimension east-west. Its area would be 15 square kilometers and its depth about 82 meters. The northern basin would correspond to the part of the bay between the Kamenis, the inner cliffs of Therasia, and the northern and northeastern parts of Thera. It would form a long northwestern arm where the Apano Meria channel is now and, to the southwest, another indentation where the present channel passes between Aspronisi and the coast of Therasia. The total surface area of the basin would be 39 square kilometers made up as follows:

Central part of the basin	23 km
Northeastern prolongation	7 "
Southwestern prolongation	9 "

The area of the two basins together would be 54 square kilometers. The depth of its central part would reach 185 meters and that of the two prolongations about 110 meters.

Supposing that the water level remains the same inside and outside the island, an elevation of 300 meters (165 fathoms) would eliminate the southern basin and divide the northern one into two. A northwest-elongated lake at the present site of Apano Meria would have an area of 19.5 square kilometers and a maximum depth of 87 meters. To the southwest a strip of land connecting Palaea Kameni with Cape Tripiti of Therasia would follow a body of water 3.5 square kilometers in area and tens of meters deep.

An uplift of 390 meters would be necessary to put the entire amphitheater of Santorini above water.

Having seen the effects of such elevations, let us examine the effects of subsidence. Lowering everything by 100 meters would cause Aspronisi, Micra Kameni, and Palaea Kameni to disappear, leaving only the summit of Giorgios above water. The peak of Palaea Kameni would be a reef just below the surface. The southern side of the former cone of Nea Kameni would form a narrow bar only a meter or two above the water. Giorgios would be a small island of about 5 hectares with a maximum height of 27 meters. The island of Thera would be substantially reduced to a long, north-trending ridge with a bulge toward the east near Merovigli and an indentation near Thira. Only the chain of Mt. Profitis Ilias would remain on its transverse part. The outer shore would diverge under Pirgos to follow the chain around a bay open to the northeast. The southern side of the rocks of Mesa Vouno and of Mt. Profitis Ilias would form steep cliffs. The shore would extend from there westward and join the bay a short distance south of Plaka at what would be the southern end of Thera. Of the plain at the southern foot of Mt. Profitis Ilias, only the highest point at Platanymos (Mt. Gavrillos) would remain as a small island, and all the southwestern part of Thera would be islets, the most important of which would be the village of Akrotiri, the peak that overlooks it, and the surrounding region as far as the southern shore of Thera. The summit of Mt. Arkhangelo would stand at the southwestern end with a surface area of 212 hectares. And finally, the two capes of Mavro and Akrotiri presently at the western tip of Thera would be two small, low islands.

Most of Therasia would also be gone; there would be only a strip with the same inner coast as the present one but a very irregular shore between Cape Tinó at the northeast and the middle of the southern coast.

A lowering of 200 meters would make Thera smaller but would not change its general form. It would have two main parts connected above Athinios: a ridge extending along the bay and a rocky peak crowned by Mt. Profitis Ilias. Of all the region southwest of Thera, only a very small islet would remain at the present site of Loumaravi. The Kamenis would be under deep water, and Therasia would be reduced to two small islands, one the peak of Simandiri and the other a ridge between the southern edge of the village of Manolas and the southern coast.

A drop of 300 meters would leave no trace of Therasia or the area around Akrotiri. The part of Thera formed by Mt. Profitis Ilias would have about the same form as before but would be divided into three parts by water cutting it between Pirgos and Mt. Profitis Ilias and between Mesa Vouno and the eastern flank of the same mountain. To the north, the peak of Megalo Vouno would be an island of 2 hectares. To the east, the twin peaks of Mt. Mikro Profitis Ilias would form two islands connected by a reef close to sea level. A short distance to the south, Merovigli would form another elongated island with an area of 90 hectares.

Lowering the region 400 meters would submerge Mesa Vouno and Pirgos. One would see an island formed by Mt. Profitis Ilias with an area of 50 hectares, and, with a subsidence of 500 meters, this island would have only 23 hectares.

The foregoing considerations are based mainly on soundings and the English map of Captain Graves, but owing to the highly varied bathymetry around the Kamenis after the latest eruption, I have had to use other sources, notably the soundings carried out in 1870 by Captain Germouny, commander of the Austrian ship *Reka*. Banco, the harbor of Saint Georges on Nea Kameni, and the channels of Micra Kameni and Palaea Kameni have been surveyed by this distinguished officer. In 1875, I took up this work and completed it in so far as I could. I also carried out soundings in the strip between the harbor of Thira and Aspronisi, then between Aspronisi and Cape Akrotiri. My results were identical to those shown on the English map; they confirmed its accuracy in showing that the eruption did not change the bathymetry outside the immediate area of the Kamenis.

To measure elevations in 1866 and 1867, I used a Fortin barometer and in 1875 a Gay-Lussac barometer. I also completed several triangulation measurements of elevation. Finally, I made a great number of photographs and dark-chamber drawings to record profiles.

It was not possible to make barometric readings simultaneously at a fixed station and at the points of interest in order to compensate for the abrupt meteorological changes that affect Santorini. This makes all the corrections questionable. To see this one need only glance at the range of elevations indicated for the same point by the different naval officers and scientists who have been at Santorini. For example, the heights given for the highest point on Palaea Kameni by Messrs. Schmidt, von Seebach, Reiss, and Stübel range from 92.6 to 101.9 meters. The difference of 9.3 meters is about a tenth of the measured height. I found that even with all possible precautions, one often gets different values for the same point, though the relative error is seldom as great as in the example just given. This is why I attach little importance to isolated barometric measurements and consider them only approximations. I have tried to repeat observations at certain established points, hoping that the random errors would by averaged out. The precision of my barometric measurements is illustrated by the averages I obtained in 1866, 1867, and 1875 for the terrace in front of the door of the Lazarist monastery at Thira.

```
March and May, 1866, average of 15 measurements......258.2 m
February, March, 1867,    "      " 12    "     .............. 257.9 m
September, October, 1875, "      " 16    "     .............. 257.7 m
Maximum difference for values of 1866 ........................ 10.5 m
    "         "        "    "    " 1867 ........................  7.5 m
    "         "        "    "    " 1875 ........................  6.3 m
```

The following table is a list of elevations I have been able to establish by repeated measurements:

Localities	Elevation	Number of measurements	Max. difference of measurements
Terrace of Lazarists at Thira	257.93	43	10.5 m
Highest point on Merovigli	359.6	12	7.0 m
Northern peak of Mt. Mikro Profitis Ilias	300.1	6	6.4 m
Chapel of Mt. Mikro Profitis Ilias	300.1	6	10.3 m
Summit of Megalo Vouno	323.8	8	13.7 m
Summit of Kokkino Vouno	288.1	8	14.0 m
Top of cliff east of Athinios	305.7	10	8.2 m
Door sill of monastery of Profitis Ilias	568.5	4	12.1 m
School of Martinos	346.0	4	15.0 m
Top of Mesa Vouno	378.2	4	13.0 m
Grotto of Mesa Vouno	325.0	4	12.7 m
Saddle between Mesa Vouno and Profitis Ilias	274.6	4	13.2 m
Lowest point on cliff east of Akrotiri	62.3	15	5.4 m
Summit of dome west of the village of Akrotiri (Loumaravi)	210.2	8	7.2 m
Summit of Mt. Arkhangelos	165.6	8	4.3 m
Cape Akrotiri	126.8	8	6.8 m
Top of Palaea Kameni (1866, 67, & 75)	98.4	6	4.9 m
Highest point of Micra Kameni (1866)	70.9	8	3.8 m
" " " " (1867)	69.8	6	3.7 m
Bottom of crater, Micra Kameni (1866)	30.1	4	2.8 m
Highest point of Nea Kameni (1866)	105.0	6	2.6 m
" " " " (1867)	103.4	8	4.2 m
" " " " (1875)*	101.2		
Highest point of Giorgios (1875)	126.5	20	9.0 m

Having taken this broad overview of the archipelago of Santorini, we now turn to a detailed description of each of the main parts of this group of islands and set out their individual features.

The study of the older part of the archipelago is greatly facilitated by the form of the bay, which, by sharply truncating the rocks along their long axis and greatest heights, reveals them in complete, well-exposed sections. For this reason, we begin the study of each area with a consideration of the parts of the interior cliffs and follow this with a discussion of the information provided, usually in less detail, by a traverse across the islands and the outer cliffs.

* By leveling on the northern flank of Giorgios from a point of equal elevation.

THERA

This island is divided into two very different geological parts. One, consisting of metamorphic rocks, is represented by the chain of Mt. Profitis Ilias extending from the cliffs at the harbor of Athinios to the point of Mesa Vouno. The small ridge of Platanymos (Mt. Gavrillos) south of the village of Emporion and several calcareous rocks rising through the tuff on the other side of the chain are part of the same group of rocks.

The rest of the island consists of rocks containing volcanic material, but while in the north and east these are products of subaerial volcanism, such as lava flows, scoria, cinders, lapilli, and pumice, the southwestern area has a considerable amount of fossiliferous tuff laid down by submarine eruptions. The great geological diversity found in different parts of Thera, together with complexity of the constituents mixed together in each group of rocks, forces us to make the following additional subdivisions:

1. the northern part of the island, extending from the point of Apano Meria to the base of the gulf of Musaki
2. the eastern region, extending from the gulf of Musaki to the metamorphic unit of Athinios
3. the area of metamorphic rocks
4. the southern area, starting at the western flank of the island and extending to the small harbor of Balos and the village of Akrotiri
5. the southwestern region, extending from the harbor of Balos to the western tip of the island

The Northern Part of Thera

This part of the island is elongated east-west. Its longest dimension, measured from Cape Kolombos to the channel of Apano Meria, is 5 kilometers. Its maximum width, between Cape Mavropetra and the cliff of Marmarini (Cape Perivola), is 2.4 kilometers. One can distinguish a lower section in the northern part of this area from an upper one that follows the shore of the bay and connects with the other on irregular, ravine-cut slopes. The upland areas continue to rise eastward from the point south of Apano Meria as far as the summit of Megalo Vouno, the highest point of the region. The elevations of the principal points on the ridge crest are as follows:

Extreme tip of the cliff of Apano Meria	141.8 m
Cliff of Cape Perivola	170.9 m
Chapel of Agios Elias	228.0 m
Chapel of Stavros	305.9 m
Peak of Megalo Vouno	323.8 m
Peak of Kokkino Vouno	288.1 m
Saddle between Megalo Vouno and Kokkino Vouno	269.0 m

This zone is best developed toward the east, where it is widest and the slopes are steeper. The inclination of the ground surface along the cliff bordering the channel of Apano Meria is only 4 or 5 degrees. It becomes at least 15 degrees on the northern flank of Megalo Vouno and 20 degrees on the northern and northeastern slopes of Kokkino Vouno.

The low pediment at the foot of the escarpment has an average elevation of about 15 meters and is only gently inclined. The break of slope at the base of the rocky section is as prominent here as it is farther east. It has a maximum width on the west and narrows between Kokkino Vouno and the sea. It is enclosed on the east by a ridge that begins at the shore near the southern slope of Kokkino Vouno and, inclined outward, abuts against Cape Kolombos. The area of the rocky section is about 400 hectares and that of the lower pediment 600, so the ratio of these two surfaces is about 2 to 3.

The whole region is covered with the same layer of white pumice found on the surface throughout Santorini. This prehistoric deposit is missing in only a few places, where it has been eroded away by the runoff of rain. For example, bits of this tuff can be found even at the top of Megalo Vouno. Most of the summit of Kokkino Vouno is also covered, and in the depression between these two peaks the tuff has accumulated to a thickness of several meters, so that the residents have been able to cut vaulted caves into it. The pumice layer tends to be thick and continuous on the ridge crests of the western part of this region, while it is thin and discontinuous on those of the eastern part. Near the western end below the town of Apano Meria it reaches a thickness of 60 meters. Numerous blocks of lava and a few nodules of aragonite can be found in this pumice.[3]

Pumice eroded from the upper elevations has been deposited below where the major part of the soil is made up of water-worn fragments. At several places on the northern coast, the pediment is bounded by small cliffs formed by the mouths of ravines cut into this material, but at the same time, one finds there rounded pebbles and cobbles, worn pieces of limestone and gypsum, as well as blocks of lava with barnacles adhering to their surface.

This material deposited by the sea is now found 10 to 12 meters above the level of the water. There is little doubt, therefore, that the ground has been uplifted. This is in accord with several other observations that Mr. de Verneuil and I have made on the eastern coast of Thera, where evidence of uplift can be seen along the entire northern and eastern part of the outer shore of the island. We have estimated that this uplift was as much as 12 meters! It seems to have been greatest around Cape Kolombos.

The eastern shore of the channel of Apano Meria has two small bays known as Pirgouna and Amoudhi. At the innermost part of the second of these is the wharf for Apano Meria, from which the road rises to the town. A great many of the buildings of Apano Meria are situated in amphitheaters dug into the pumiceous tuff and facing the bay.

From Cape Mavropetra to the level of the chapel of Saint-Joanni, the cliff is composed exclusively of layers of pumice. Farther toward the south the pumice is underlain by darker beds with a dip that is less than it is at a lower level. Despite erosion, the pumice is at least 40 meters thick in the cliff overlooking the northeastern part of the bay of Amoudhi. It is underlain by the following sequence:

1. a bed 20 meters thick composed of lapilli and brownish scoria with bits of lava
2. a layer of lava 2 to 3 meters thick
3. a substantial deposit of bright red scoria about 40 meters thick
4. eight flows of black lava separated by red scoria and having a total thickness of about 40 meters at the southwestern end of the gulf of Amoudhi

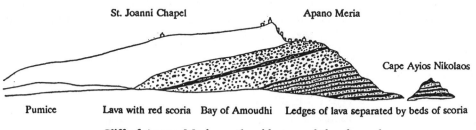

Cliff of Apano Meria on the side toward the channel

Cape Ayios Nikolaos is composed of debris from four of these units. One of the flows filled a depression in the original surface and accumulated there in an irregularly shaped mass. The lava making up these units is rich in glass. The dominant feldspar phenocrysts are triclinic and insoluble after being in boiling nitric acid for an hour. Judging from the resemblance of the rock to other lavas of Santorini, one would surely conclude that this feldspar is labradorite. Pyroxene is rather common, and iron oxides are also present, but olivine is completely lacking. The feldspar microlites are albite, and there is no pyroxene in the groundmass. Glass is very abundant.

These lavas are also visible on the inner coast next to the bay, where Cape Ayios Nikolaos turns toward the east. The upper ledges jut out a bit beyond the lower ones to the east. This group of lavas must come from a source near the inner side of the bay. They attest to the uniform tilt of these layers toward the interior. It should also be noted that the red scoria of this same vent forms one of the main basal units of the cliff.

The mineralogical composition of this scoria is the same as that of the lavas in Cape Ayios Nikolaos, but its texture and color show that it was ejected explosively by a sizable eruption of fluid, gas-rich magma. The eruption was subaerial, but water and acidic gases played an important role. The concentration of this thick unit in a limited area near the cape at the entrance of the bay indicates the area of the vent from which it came.

Below Apano Meria on the inner side of the bay, the lower layers of lava not far from the cape give way at the same level to a scoriaceous conglomerate overlain by a layer of thinly bedded ash. The red conglomerate is well developed. The lava flow resting on it is composed of several sheets that rise as they continue toward the east and seem to belong mostly to the same eruption that laid down material on a gentle slope on the flanks of the main structure of this region.*

The layer of red scoria resting on these lavas in the cliff bordering the channel of Apano Meria is missing in the part of the escarpment facing the bay beyond Cape Ayios Nikolaos, or it is reduced to a few very thin remnants lying on or between the layers of lava. The upper pumice is well developed in this area. A vertical section in the cliff at the eastern edge of the town of Apano Meria is shown, from the top down, in the table below.

Upper pumice	50 m
Layers of lava	10 m
Red scoria	50 m
Finely bedded ash	25 m
Scoriaceous conglomerate	20 m

Nearby, one can see in the cliff a dark black dike (Dike No. 1), about a meter thick, with nearly vertical, slightly undulating sides trending N5°W. This dike is exposed up to the top of the lavas underlying the upper pumice

---

* See plate XXXI at the end of this volume.

and probably fed some of the flows. It is composed of labradorite basalt.

Quite close to the dike on the east, the layers of lava, partly concealed by talus, can be seen at sea level. The scoriaceous conglomerate of the lower part of the cliff section has disappeared. A little farther on, the red scoria thins and gives out against a series of ash layers and lavas dipping westward. The latter appear to be from vents near the eastern coast of Megalo Vouno.

The cliff about 100 meters east of the hamlet of Armeni before the limit of the red scoria shows the following section.

Upper pumice	40 m
Bed of lapilli	10 m
Layers of lava	15 m
Red scoria	30 m
Fine-bedded grey ash	20 m
Two sheets of lava separated by a thin bed of scoria	15 m
Talus	30 m

In summary, the group of lavas and red scoria making up what we refer to as the eruptive unit of Apano Meria rests discordantly on basal ash and lavas discharged from vents farther to the east and is in turn covered by a layer of lavas that came down from sources in almost the same exposures. Finally, all of this is surmounted by the pumice layer that is seen at the surface of the entire island and must have been discharged by the eruption that produced the bay.

A short distance from the point where the red scoria of Apano Meria disappears, the coast turns toward the interior of the bay and forms Cape Perivola. On the eastern side of the cape, ash is the dominant unit in the cliff section. It has various colors, has thin concordant bedding, and dips toward the west. Some of these beds have a very light shade and seem to be made up of pumiceous material. The lavas that can be traced from Armeni in the lower half of the cliff are hidden by ash and appear to come from deeper in the interior of the island. The section at the point where the talus ends is as follows.

Upper pumice	40 m
Lavas with scoriaceous interbeds	30 m
Finely bedded gray ash	40 m
Light-gray ash and pumiceous debris	10 m
Ash and brown lapilli	20 m
Thick layers of lava	20 m

The eastern slope of Cape Perivola is cut by a 3-meter-thick dike (dike no. 2) that one can trace up the cliff to the height of the uppermost unit. This dike trends N10°W. It is a coarse-grained black rock with streaks of fault gouge and a scoriaceous surface. It is exposed along a short interval where it rises through the loose material under which it is partly buried. The rock is rich in olivine and has phenocrysts of anorthite.

From Cape Perivola to Cape Heptapedio, the layered rocks of the cliff rise steadily toward the east. The upper pumice becomes thinner and thinner, while the lava immediately underlying it becomes thicker. Below, ash still makes up an important part of the section. Several more or less continuous lava flows are separated by layers of lapilli and thin beds of scoria and scoriaceous conglomerate. While the ash and upper group of lavas dip westward, the lower lavas, lapilli layers, and scoria beds dip toward the south and southwest. These products came from different vents. The upper lavas, which descend to the shore of the channel of Apano Meria, start at a point on the ridge between the chapel of Ayios Ilias and that of Stavros. They must have come from one or more of the dikes seen in this part of the island and almost certainly from the dike that is designated below as number 16. The underlying ash forms a layer of varied thickness which, becomes coarser and turns into lapilli as one follows it toward the summit of Megalo Vouno. This ash bed is relatively thin below the chapel of Ayios Ilias and as far as the end of the lavas it rests on, but it reaches its greatest thickness at the level of the peak of Megalo Vouno, where it has a thickness of 90 meters. It was no doubt produced by a late eruption from a parasitic cone, of which Megalo Vouno is an imposing remnant. The ledges of lavas that can be seen in the two lower tiers of the cliff between Cape Perivola and Cape Heptapedio must have been erupted from the large number of dikes that can be seen near the latter cape.

I will refer to the part of the cliff just described as Foinikia after a village situated in this region on the other side of the crest.

The prominent heights of Megalo Vouno are distinguished by greater proportions of scoria in the interval between dikes and lava flows and by the distribution of ash and lapilli that cover them. These rocks not only crown the summit of the mountain but are seen on the buttress slopes facing the bay and descending toward the south. The highest point of Megalo Vouno is not exactly at the top of the cliff; it is another 100 meters away, and one can follow the road from Thira to Apano Meria on the southern side of this crest.

On the dome-shaped summit, a mantle of cinders and lapilli drops away on all sides from the peak. One can see the limits of this feature on both the western and eastern faces of the cliff, and on the western side one can distinguish very clearly the superposition of the Megalo Vouno beds over those of Foinikia. In general, the slope of the lower beds of Megalo Vouno is steeper than that of the upper part of the same series, especially those that are inclined from the coast of the bay.

The dikes numbered 3 through 16 seem to be related to the rocks making up the Foinikia section. Most are very closely spaced, but an interval of about 60 meters separates dike 16 from dike 17. Another group of closely spaced dikes, numbered 17 to 40, can be seen below Megalo Vouno, where they are concentrated below the western side of that peak. Also on this side are the products of the many successive eruptions that produced Megalo Vouno. One can see from the distribution of dikes and from the exposures on the upper surface that the position of the vent must have shifted many times.

The dikes associated with Kokkino Vouno also form a separate group extending 350 meters from those of Megalo Vouno; they are numbered 40 through 52. Several of these dikes cut lavas and scoria of the Megalo Vouno system. From this, one would conclude that Kokkino Vouno was formed later than Megalo Vouno, but the banks of lava and scoria beds cut by the first of these features are also cut by the dikes reaching the peak of the second. The series of eruptions at these places produced overlapping and interfingering material, and if, as seems likely, Megalo Vouno began to form before Kokkino Vouno, it is certain that the later eruptions from these two cones were essentially contemporaneous.

The mineralogical aspects of these 52 dikes are described in detail in the following chapter.

In the middle of the upper part of the cliff of Megalo Vouno, one can see a voluminous body of rock with the plug-like features of a feeder vent for lavas. There, more than anywhere else, the continuity of the extensive series of lavas is quite apparent. Each ledge is composed of several flows resulting from multiple discharges. It seems that most lavas making up this thick group were erupted with a very limited fluidity.

The same irregularity and limited extent of the lavas can be seen on the northern slope of Megalo Vouno. Several flows seen in the lower slopes on that side of the mountain or next to the rocky ridge on which the chapel of Stavros is situated are remarkable for their short length and the steep slopes that mark their flow fronts.

The cliff shows only transverse sections through the flows that spread directly toward the south and the center of the bay. One cannot, therefore, determine their original extent with any degree of certainty. It is likely, however, that they, too, were short and irregular, for the central part of the bay must always have been the main locus of the vents of the Santorini complex. Since lavas normally flow radially outward from the main eruptive centers, it is difficult to imagine that flows coming from a parasitic cone, such as Megalo Vouno, would be diverted toward the south and that they would find a more favorable route in that direction. Moreover, the substantial southward inclination of the beds exposed in the cliff face of Megalo Vouno shows that the cone was equally steep on this side, especially in the later periods of its development.

Water and acidic gases appear to have played an important role in the eruptions of Megalo Vouno and Kokkino Vouno. This is shown by three different observations: first, the abundant proportion of ash and lapilli among the eruptive products; second, the bright-red color of some of the lavas and associated scoria; and, third, the abundance of tridymite in several of the dikes.

The amount of ash and lapilli ejected from a volcano is a function of the amounts of gas and steam that are the main agents responsible for pyroclastic eruptions.

The reddish color of scoria and lava is due to oxidation of these products. It is not atmospheric oxygen that is responsible, at least directly, for this effect. Incandescent lava, when it is ejected pyroclastically, even in the finest fragments, has only a very superficial interaction with the atmosphere. The iron oxides that this produces are confined to a thin coating of the fragments, and one finds only a slight alteration of the more or less dark grey interior of the fragments. The red color of certain scoriaceous, fragmental lavas can be attributed more reasonably to oxidation due to the dissociation of water picked up at depth along with other erupted material. We know that water breaks down at high temperatures into its two elements, hydrogen and oxygen, and the fact that this dissociation can take place within the eruption column is demonstrated by the composition of mixtures of gases collected in 1867 close to active lava flows. The dissociation of water that takes place in the magmatic reservoir can lead to various chemical reactions, including oxidation, when the oxygen produced in this way is at a sufficiently high temperature in the presence of elements of the rock that are susceptible to alteration.[4]

But the clearest evidence of oxidation, which one invariably sees in all eruptions, is the effect of acidic gases, particularly hydrochloric acid, on the minerals of the rock. Acids contained in the semi-molten magma attack all the ferriferous minerals and produce compounds of iron that, on contact with water vapor, produce bright-red hydrated iron oxides, the color of which varies according to the degree of hydration and the physical condition under which they form.

As for tridymite, its presence in vesicles or veinlets cutting the rocks indicates that it results from the action of water in the magma that drew silica from the rock and deposited it in these places. It is especially abundant in vesicles that are only partly filled and clearly owe their development to expansion of gas and water vapor. It is worth noting that samples of lava having the greatest amount of tridymite also have other signs of alteration. The ferromagnesian minerals, such as pyroxene and olivine, are discolored, the latter normally being the most seriously attacked. Magnetite has disappeared, and the rock is shot through with irregular grains of limonite. In these ways, all the different effects of water on the magma are manifested in the same rock.

The Eastern Part of Thera

We can divide this region into three subdivisions: the first is confined to Mt. Mikro Profitis Ilias and the area around Kolombos and Tokaiou, the second extends from the southern edge of Mt. Mikro Profitis Ilias to Thira, and the third starts at Thira and extends as far as the point known as Athinios.*

The first sub-region corresponds to the inner side of the gulf of Mousaki. Its hilly part is much more developed than the lower section. The twin peaks of Mikro Profitis Ilias are the culminating points. The northern one has an elevation of 319.6 meters, and the southern one is only a meter lower. The lowest elevation in the section between Kokkino Vouno and Mikro Profitis Ilias is only 190 meters above sea level. The spurs buttressing the ridge on the inner side of the island are very well developed and reach the sea at the point of Kolombos and at that of Tokaiou at the foot of Mikro Profitis Ilias.

* See plate XXXII at the end of this volume.

The surface area of the hilly section is about 140 hectares and that of the lower part only 36 hectares. It is the narrowest part of Thera.

The lower slopes of Kolombos end east of a cape that has been notably cut back in a short period of time. The mineral spring indicated on the English map of 1848 can no longer be seen; it is under the water. In very calm weather it can be detected about 25 meters off-shore by the elevated temperature and lower salinity of the seawater at that point. This spring was not situated at the very tip of the cape in 1848, but in a small cove on its southern flank, so that one can estimate the width of ground carried away by wave erosion in that period as 50 meters.

The ridge linking Kokkino Vouno with Mikro Profitis Ilias consists mainly of ash and lapilli, principally in its most northerly part. Toward the south near the top of Mikro Profitis Ilias, it has ledges of lava dipping toward the east at angles of less than 20 degrees.

The cliff at the northeastern corner of the bay has a series of layers consisting of gray ash, black and red lapilli that rise toward the south, and a few thin, interlayered lavas dipping in the same direction. The five dikes numbered 53, 54, 55, 56, and 57 come up through the middle of this section.

Several prominent faults are concentrated in a limited part of the escarpments making up this part of the cliff. On the southern side of the main fault, dike number 55 can be seen several meters above the water level cutting through the middle of beds of reddish scoria and lava. This part of the cliff is one of the most disturbed. The faults cutting it testify to the displacement that has taken place between the northern area and the eastern part of Thera. Between the two large masses of Megalo Vouno and Mikro Profitis Ilias one finds an even greater structural complexity than that mentioned in my earlier description of the other side of Megalo Vouno below the ridge between the chapels of Ayios Ilias and Stavros. This complexity must be due to the superposition of eruptive products coming from several different sources, the earlier ones from Megalo Vouno, others from Kokkino Vouno, Mikro Profitis Ilias, vents near the center of the present bay, and even from small local eruptions.

On the eastern side of this zone, one can see the same beds of ash and lapilli along with the corresponding red scoria. These products are especially well exposed in the ravine extending along the lower northern flank of Kolombos. One can distinguish three units of red scoria, each from a different eruptive center. Two other groups that are even more sharply defined are seen in the slopes closer to Mikro Profitis Ilias. The substantial pile of ash forming the lowermost flank of Kolombos is difficult to explain unless one assumes that the ash was ejected from one or more vents in the immediate vicinity of these deposits. In short, it can be seen that a series of small eruptions centered at Megalo Vouno, Kokkino Vouno, and Mikro Profitis Ilias filled the space between these large parasitic cones.

The base of Mikro Profitis Ilias is formed by a large lava dome. The light reddish-gray rock has phenocrysts of anorthite that are readily decomposed by boiling acid. Dike number 58 can be seen on the northern edge of this dome, and the interior is cut by dikes 59, 60, 61, 62, and 63. On the southern side are three more dikes, the last of which, number 66, is remarkable because the head of its outcrop swells out in the shape of a mushroom just below a scoriaceous lava.

Some of the rocks that form the dikes in this region contain labradorite, others anorthite. Tridymite tends to be abundant in the former, but it is not altogether absent from the latter, for under the microscope one commonly finds in a single thin section of anorthite-bearing, olivine-rich lava both tridymite and olivine within the same field of view. The compositions of these dike rocks will be discussed in greater detail in the chapter that follows.

A series of sub-parallel lava flows and thin scoriaceous interbeds is draped over the lava dome forming the base of Mikro Profitis Ilias. One of these flow units, exposed about two-thirds of the way up the cliff, is composed of large numbers of irregular blocks with a total thickness of almost 15 meters.

In the upper part of the cliff where the road along the crest leads from Thira to Apano Meria one finds under the northern peak a black volcanic conglomerate composed of isolated, fist-size blocks overlying red scoria, a bed of gray ash, and then light-gray platy lavas dipping toward the north, east, and even southeast at angles of 25 to 30 degrees. These lavas contain labradorite. The massive lavas are well exposed at the northern side of Mikro Profitis Ilias; scoria and detached blocks make up the major part of the southern peak, although steeply dipping lavas can also be seen on its back side.

From what has been said here, one can see that, just as at Megalo Vouno, the earliest eruptions from Mikro Profitis Ilias seem to have produced mainly anorthite-bearing rocks, whereas most of the later lavas contain labradorite.

All around the summit of Mikro Profitis Ilias one finds bits of pumice wherever there is the slightest crevice or irregularity in the surface of the lava. It is certain, therefore, that the pumice once covered this hill; it is missing only from the highest points, where it must have been eroded away by the action of rain or where the slope of the surface was too steep for it to be retained.

The lavas of the lower dome at the base of the cliffs of Mikro Profitis Ilias not far from their southern end must have came from the three dikes just mentioned. They were later covered by ash. The deposits of ash around the lower part of the dome are inclined toward the south and, in a few places, toward the bay, just as the underlying lavas are. A short distance to the east, they rise toward the point where the three dikes emerge and rest discordantly on the massive lavas, scoria, and lapilli that were intruded by the dikes .

One sees from this that Mikro Profitis Ilias is the product of multiple eruptions from various different vents. The main source was close to the center of the complex, and the first products were the lavas of the lower dome and the thin beds of ash that rest directly on them and, in a few places, dip toward the bay. The orientation of the scoria beds and lavas forming the top of Mikro Profitis Ilias show that the main vent was situated somewhere in the interior of the present cliff. The vent itself must have been close to the present peaks. One is led to this conclusion by the distribution of red scoria beneath the upper lavas and by the very restricted underlying layer of ash and lapilli.

The central part of eastern Thera is more extensive than the part just discussed. To delineate it, one must imagine two straight lines cutting across the island, one from the southern end of Mikro Profitis Ilias and the other from the harbor of Thira.

The highest point of the cliffs in this area is at the peak of Merovigli (359.6 meters) facing the abrupt peninsula of Skaro. The upper part of this point terminates in the thin remnants of a layer of lava resting on a double layer of scoria of similar extent. The island's most important center of population was formerly around these fragmentary exposures at an elevation of about 280 meters. During the Venetian occupation, this place was a fortress where the inhabitants of Santorini took refuge from the attacks of Muslims; later it served as a shelter from the pirates that once infested the archipelago. With Greece's declaration of independence and the more active role played by European navies in this region, better security was assured throughout the Cyclades, and there was no longer any reason to remain on this rock, which was becoming increasingly dangerous, owing to the landslides constantly cutting back the cliff. Very strong earthquakes at the beginning of this century caused great damage to the group of houses built on Skaro. Large parts of the rock fell into the sea, carrying with them the buildings that had been constructed on them. The houses and public monuments left standing had undergone so much damage from the earthquakes that their repair would have meant total reconstruction. By common accord, the people decided to abandon the place, and the population moved to the present site of Thira, which soon thereafter became the most important town on the island.

The elevation of the summit of Skaro is 303 meters; that of the saddle between Skaro and Merovigli is 253 meters.

This area is very narrow on the side of Mikro Profitis Ilias but widens greatly toward the south. The distinction between the lower part and the steeper section becomes less accentuated in this region and, despite the height of Merovigli, which is higher than Mikro Profitis Ilias, the settlement there occupies a relatively larger area than the one around Mikro Profitis Ilias. The lower part has an area of about 600 hectares and the mountainous part 400.

This region of Thera is the one that has the greatest accumulation of lavas. The middle part of the cliff has a series of thin flows that appear to be continuous but in reality are made up of narrow flows the regularity of which is only apparent. Some of the lavas, however, can be followed as far as 200 meters showing that the sections through them in the cliff face must coincide more or less with their long axes. All the flows that are cut by transverse sections have a much smaller width. These lavas are 18 to 20 in number, of which some are less than a meter thick, and only a very few reach thicknesses of 4 or 5 meters. They are separated by thin beds of lapilli and scoria. Thicker flows are found in the upper levels as well as near the base of the cliff. At water level, one can also see sizable bodies of lava, some of which are accompanied by irregular deposits of scoria and appear to come from near-by vents. Other beds of scoria are seen higher in the cliff. In one place they are immediately overlain by the pumiceous tuff.

The pumice deposit, which is all but absent on Mikro Profitis Ilias, becomes progressively thicker in the direction of Merovigli, where it reaches a thickness of about 30 meters. The pumice, as well as the upper group of lavas and scoria beds, are missing from the top of Skaro.

The lavas of the region just described are very similar to those of the modern eruptions; they are dark with small crystals of feldspar scattered through a black matrix that appears glassy and crystal-free under a hand lens but under the microscope is seen to be rich in feldspar, pyroxene, and iron oxides. The feldspar phenocrysts are labradorite, and one finds tridymite in many samples.

The scoria is composed of the same material. It differs from the lava that it is associated with in having a greater porosity and a reddish color due to oxidation of iron.

Most of the lavas are massive with an irregular fracture. Some have suggestions of prismatic jointing. On the southern slope of Skaro close to the water level and below the village of Thira Stephani, a ledge of very thick lava is formed by a light-colored rock that readily splits into plates a few centimeters thick. It is quarried for flagstones, some of which are plates measuring a square meter in size.

At least a few of the rocks in the lower part of the cliff come from sources not far from their present exposures, but the lavas overlying them come from a more distant source. The latter are the ends of flows that came from vents some distance to the west of the present bay side of the island. These flows must have spread in a direction that was more or less from west to east.

A unit in the upper part of the cliff seems to have had a quite different origin and mode of emplacement. This lava is in direct contact with the pumiceous tuff and above the bed of red scoria already mentioned. Very little of it is exposed in the cliff face northeast of Merovigli, but it is quite prominent in the cliff south of that point. It only begins to appear in the cliff facing Skaro, disappears for a short distance under Merovigli, but reappears on the southern coast, first as a thin layer, then becoming thicker until just below Thira it reaches its maximum thickness of 40 meters. It flowed from north to south and is cut lengthwise by the cliff. It reveals all the irregularities of its internal structure and surface. It can be seen to twist and turn, inflate, rise against obstacles or pass over its own blocky debris that accumulated at its margins. It is rare to see such an instructive example of a longitudinal section through a lava flow. Its source was somewhere under the present site of the bay of Peristeria, probably near the point of Merovigli.

This lava is almost identical to the ones erupted in 1866; it would be impossible to distinguish one sample of these rocks from the other, even though their ages are very different.*

The underlying bed of scoria, which is very thick at the point of Merovigli, thins toward the south and finally disappears near the path leading up to Thira.

The section of lavas just described rests at its northeastern end on the massif of Mikro Profitis Ilias; the superposition is apparent in the lowermost units and becomes less clear in the upper layers, which seem to be concordant with those forming the upper part of Mikro Profitis Ilias. In any case, it is clear that this large mass already formed an elevated volcanic feature before the eruptions that discharged the lavas in the middle of the eastern part of Thera.

A similar situation is seen at the southern end of the same area. The ledges of lava making up the cliffs of Merovigli give out toward the south below Thira Stephani, where they abut against the thick layers of loose material that were already in place at the time of their eruption. They terminate at their northern end in a talus with a lower slope of about 40 degrees but only 5 to 10 degrees in its wider, upper part. The effect of this is that the front of the lavas abutting against this loose debris extends farther toward the south at higher levels. The lower lavas extend only to Thira Stephani, whereas the upper ones can be followed as far as the village of Thira.

As a result, beyond that point the section in the cliff where the path goes up to Thira is complex. In its middle and lower parts it is made up of beds that are part of the fragmental series extending along the southern half of Thera, while the upper part is made up of ledges of lava and a few thin layers of scoria belonging to the group that reaches its maximum thickness near the point of Merovigli. All this section has a thin mantle of pumiceous tuff.

All the naturalists who have studied Santorini have published descriptions of the section near the path up to Thira, and most of them have described in minute detail the units that compose it. I, too, must mention its salient features, but in doing so I shall add a few details that have a bearing on the origins of the rocks found there.

* See plate XXXIII at the end of this volume.

At the base of the cliff there is a gray, more or less decomposed, tuffaceous mixture of ash and scoria. The dominant feldspar in the blocks is labradorite; pyroxene and iron oxides are equally common; olivine is rare and badly altered. The labradorite and pyroxene are visible with a hand lens, but the other crystals can be recognized only under the microscope. The same is especially true of the feldspar microlites in the very altered amorphous matrix. This tuff is very coherent. Storage places and dwellings have been dug into it. It is about 50 meters thick.

The north-dipping, 10-meter-thick lava just above has a very remarkable lithologic character. Its top and bottom surfaces are parallel and even. They have none of the irregularities and cavities that give most lavas their scoriaceous appearance. The rock itself is compact, black, glassy, and spotted with small crystals of clear feldspar. It has many inclusions of other lavas that in places give it the appearance of a breccia. Under the microscope one sees that it is composed essentially of brown, amorphous material in which are scattered crystals of triclinic feldspar, pyroxene, and iron oxides. The glassy groundmass has dark and light areas; here and there in the brown part there are beautiful clusters and moss-like aggregates of crystallites. Feldspar microlites are equally common. All of these components are distributed in tortuous bands and the glassy material, even when it contains nothing but small globules, has a very pronounced fluidal structure. This lava can best be described as a tachylite that picked up fragments of older lavas and at the same time had a marked tendency to precipitate crystals of the minerals normally found in volcanic rocks. Among the phenocrysts there are a remarkable number that seem to have been caught in the process of forming. Among these are the iron oxides around which one sees a clear, colorless zone as if the mineral had drawn all the ferruginous components from the magma surrounding it. Others, however, appear to have undergone corrosion by the melt in which they were contained. In particular, one sees numerous crystals of triclinic feldspar with resorbed edges. Their lamellae have rounded ends and are unevenly resorbed. Elsewhere, this lava has essentially the same silica content as the crystalline lavas of the Kamenis, so the difference in texture has nothing to do with chemical composition but rather the physical conditions that governed the formation of the lava. The marked fluidity it had when erupted is demonstrated by the form of the layer, by the presence of numerous inclusions, and, above all, by the fluidal texture and large amounts of glass. This fluidity can be explained only by the temperature of the lava, which must have been greater than that of most Santorini magmas when they were erupted. This rock resembles in all respects the submarine lavas which, like this one, are quite massive and glassy, and if one also considers the tuff on which it rests, one could easily assign it a subaqueous origin, but it must have been subaerial, for otherwise it would be impossible to account for the foreign inclusions that are present in such large numbers. The margins of a lava discharged under seawater would be solidified so quickly that they could pick up very few fragments that were already solid.[5]

A breccia, about 60-meters thick, directly overlies this flow. Most of the blocks in it are angular. They are enclosed in a reddish, tuffaceous matrix composed of small particles of the same volcanic detritus but in a more advanced state of alteration. The largest blocks are concentrated near the base of the breccia, and they become less abundant upward and are completely absent from the uppermost levels where one finds, instead, thin layers of ash and lapilli with a cross-stratification similar to that of coastal sands. It seems likely that the upper parts of the breccia were laid down close to sea level.

Other observations lend support to this hypothesis. The ash beds contain pisolites, the formation of which requires deposition in agitated water and corresponds well with a coastal setting.[6] Moreover, in the middle of these beds of fine-grained, light-colored ash one sees very thin zones with a darker color. These small layers consist of compact lithic grains separated from the more pulverized part. Thus, within the middle of this fragmental unit there was a density sorting of the pyroclastic material according to the laws governing gravity separation, and this could only have occurred in water. Finally, I must add that there is evidence of a shore line at this elevation elsewhere on the island of Thera.

If one accepts the explanation I have just proposed and assumes, at the same time, that the underlying lava flow was discharged under subaerial conditions, one must conclude that the ground level had been depressed during formation of the breccia and that later uplift exposed these beds where they are seen today.

The part of the cliff near the path up to Thira is also of interest because of a well-developed fault that is clearly visible in the recessed area north of the path.* This fault has cut all the units exposed in the cliff with

---

* See plate XXXVIII at the end of this volume.

the southern side displaced downward with respect to the northern side. The offset is greater for the lower part than for the layers closer to the crest. The difference in elevation between the two sides of the fracture cutting the lower lava is about 12 meters, but the two sides of the large upper flow are almost at the same level, and one would be tempted to believe that the fracture is older and did not cut the upper levels of the section. This is not the case, however. The fault is represented by a crack that is still open and cuts the entire rocky ridge bordering the bay of Santorini. Underground openings are found whenever work is done on the terraces in the Lazarist monastery and the two adjacent properties to the east and west. At ground level in the garden of the Lazarists, the width of the crack is 60 centimeters, and it is so deep that one can hear stones dropped into it falling for several seconds. This experiment was tried by the present Father Superior of the Santorini Lazarists, and the result seems to indicate that the crack extends at least to sea level. At the time of my visits to Santorini the top of this crack was closed. The opening had been covered with masonry, but traces of it could be seen very clearly along the wall of the garden and in the cellar of the convent of the Sisters of Charity. The same crack has been found on the eastern slope of the island, notably in the church of Kondo-Chirio. Whenever there is an earthquake at Santorini, part of it reopens. So there can be no doubt about its geological importance.

The eastern slope of Thera is almost totally covered by pumice, but it is cut by shallow ravines that widen and become shallower down slope. The upper ends of these ravines are below the ridge crest. Only rarely do they expose the units below the pumiceous tuff, but at Vourvoulos the upper lavas are exposed in horizontal cuts.

The subzone south of the eastern part of Thera extending from the harbor of Thira to that of Athinios is bounded on the south by the chain of hills that includes Mt. Profitis Ilias. Its surface area is about 250 hectares, of which the rocky part accounts for roughly a third. It is the most prosperous part of the island, with the most fertile vineyards and the most numerous villages.*

The ravines are similar to those in the previous zone. They do not reach the upper edge of the cliff but begin a short distance to the east. They cut deeply into the pumice and, in places where the slope steepens abruptly, they even expose the underlying beds. Groups of houses have been built in

* See plate XXXIV at the end of this volume.

the deepest parts of these ravines, because they provide shelter from storms, and one can easily dig cellars into the pumice. This latter factor is especially important in a place where wood is scarce. In addition, the cisterns needed to collect water can be easily constructed where rain water flows naturally. These seem to have been the reasons why the villages have grown up in such strange places.

The cliff below this zone is formed essentially of a succession of stratified pyroclastic layers. The units differ from one another in color, volume, crystallinity, porosity, degree of alteration, and mode of stratification. Viewed in the larger sense, the cliff has four major parts that overlie one another in the following upward succession. The lowest zone is composed of reddish-brown tuff with many large blocks that give it the overall character of a volcanic breccia. This unit is very irregular; in places it has interlayered lavas or ash beds of contrasting color. The lava flows are thin and have wavy surfaces with scoriaceous crusts. They belong to moderately large flows discharged from vents located to the northwest of the cliffs in the area now occupied by the bay. Of these lavas, the one that the Lazarist building overlooks merits special mention. It consists of a very light-colored, very fissile rock that is quarried for flagstone. The interior of the slabs is dark blue, whereas the surface is reddish-yellow. Its plates have a rough coating made up of crystals of monoclinic and triclinic feldspar, pyroxene,[7] hypersthene, iron oxides, and, less commonly, sphene and tridymite. The microscope reveals that these same minerals make up the interior of the rock, and it also shows myriads of microlites of feldspar, hypersthene, and augite, most of which are only two or three hundredths of a millimeter in length and, at most, 0.005 mm long. The feldspar microlites have inclined extinction at angles of not more than 10 degrees to their long axis. These are probably crystals of albite or oligoclase similar to those in the lavas of the last eruption from the Kamenis. The glassy groundmass has few spherulites, but it has a marked tendency to separate into crude hexagonal grains. This is no doubt due to the great number of very thin hexagonal crystals of tridymite making up the amorphous matrix and binding the mineral grains together.

Most other lavas in this section of the cliff closely resemble the younger lavas of Santorini, but a porous gray unit seen at the base of Cape Alonaki is distinguished from the rest by its special characteristics. The small crystals of feldspar, barely visible to the naked eye, are anorthite, not labradorite. This rock has the other unusual features of anorthite-bearing lavas that were pointed out in the earlier discussion of dikes in the northern part of the cliff of Thera. It is rich in rounded crystals of olivine about 0.05 mm in diameter and bordered by a yellow zone of alteration. Great amounts of pyroxene can be seen, not only with a hand lens but as microlites. Large crystals of iron oxides are rare, but oxides are abundant as small grains in the groundmass. This assemblage is characteristic of all the anorthite-bearing basalts of Santorini. This particular lava is the southernmost anorthite-bearing unit exposed in the eastern cliff of Santorini. It should be noted that despite the calcium-rich composition of the feldspar, the rock commonly contains tridymite in small, irregular cavities.

Most of the lower conglomerate rests on a varicolored layer of ash. It is overlain by a light yellow tuff composed of pumice fragments. This tuff forms a continuous bed that can be followed from the Lazarist monastery of Thira almost as far as the southwestern end of the island. Its great extent and regularity, together with the blocks it contains, show that it was thrown out by a strong eruption; it is not the product of a satellite cone. It certainly came from a very large central vent near the center of the present bay. Its conspicuous blocks are riddled with large cavities. Their exterior surface is gray, but on freshly broken surfaces the partitions and walls of the vesicles are a dark gray. It is composed of a finely crystalline pumice. Under the microscope one finds a light brown amorphous glass in which are scattered innumerable tiny microlites of albite (0.01 by 0.002 mm). These are accompanied by crystals of a triclinic feldspar that is not attacked by acid and by scattered crystals of augite and hypersthene averaging 0.1 by 0.03 mm in length. Glassy vesicles are fairly common, and the matrix is crowded with gas bubbles. It also has a great many small, dark brown nodules, about 0.05 mm in diameter, that are seen to be isotropic in polarized light and must owe their color to local concentrations of dark components. With sufficient magnification one can see that some have broken down into masses of small globules, a few of which are crossed by the microlites that abound in the surrounding matrix.

The crystallinity of this pumice, the brown color of its glass, and above all the concentration of this dark material in nodular patches are also characteristic of the upper pumice covering the islands of Thera, Therasia, and Aspronisi. In order to distinguish it from the latter and emphasize its position in the stratigraphic sequence, it will be referred to from here on as the lower pumice.[8]

Just above the lower pumice is a series of thin beds of banded ash distributed in small horizontal layers that are marked by distinctive colors and grain sizes. Their regularity, considerable lateral extent, and location in the section suggest that they too come from the same crater that produced the lower pumice. The thin bedding shows that they were produced by many separate explosions.

Finally, we come to the upper pumice.

This group of beds takes on a regular appearance only in the section beyond Cape Alonaki. Between that point and the path up to Thira, small eruptions centered a short distance from the present position of the cliffs produced a very irregular distribution of the ejecta that can be seen along the escarpments. To appreciate the complicated nature of this section one need only compare the two sections shown below.

CAPE LAZARET

Northern slope of the promontory of Lazaret

Southern slope of the promontory of Lazaret

What strikes one's attention in the sections on the two flanks of Cape Lazaret[9] is the absence in the southern face of the upper lava exposed on the opposite side and then the greatly reduced thickness of the breccias in the upper part of the cliff and their replacement on the south side by ash beds.

The two sections of the cape between the Lazaret and Alonaki show how much thicker the lower pumice has become at this place. Rather similar

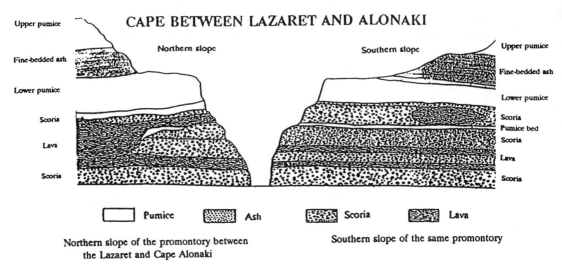

Northern slope of the promontory between the Lazaret and Cape Alonaki

Southern slope of the same promontory

in their upper parts, they differ notably at their base, mainly in the lava flows they contain.

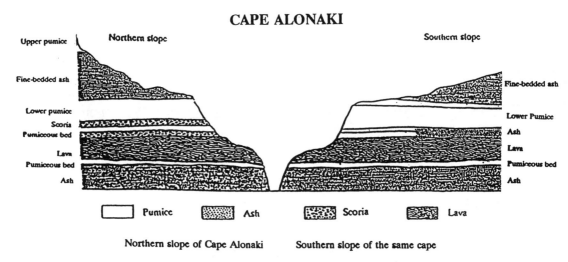

Northern slope of Cape Alonaki

Southern slope of the same cape

The northern and southern faces of Cape Alonaki show great similarities; the brown lower breccia in the two previous sections is replaced there by beds of ash.

In the reentrant angle of the shore line where it turns from north-south to northeast-southwest, a short distance northeast of Athinios, a voluminous mass of red scoria in the middle and upper parts of the cliff indicates that there was a small separate eruptive vent nearby that ejected this material. This unit lies within the banded ash beds between the lower and upper pumiceous tuffs. At the site of the leper asylum, two similar small bodies occupy the same level in the cliff.

The blocks making up these local deposits are filled with gas bubbles and have a bright red color. Under the microscope they are seen to have a vesicular texture in which the silicate fraction is limited to thin walls separating large holes. All this glassy material is brownish yellow and contains many scattered feldspar microlites with parallel or slightly inclined extinction (up to 8 or 10 degrees). In addition to simple crystals and twins, there is pyroxene and a triclinic feldspar that is not affected by acids.

The Metamorphic Zone of Thera

This zone, which divides the island of Thera into two unequal parts, is composed of a main chain trending northwest-southeast and a small branch off the central part that forms the hills of Platanymos (Mt. Gavrillos).[10] Its surface area is about a thousand hectares. All of it is rocky, and only the western part is extensively cultivated.

Apart from a few local cultivated fields, the zone is too rocky and broken to have workable organic soil. For example, the highest parts of the chain have vineyards on their western slope thanks to the banks of pumice that rise that far. Elsewhere, around the eastern base of Mesa Vouno one finds a few cultivated fields, particularly in the ravine between Mesa Vouno and Mt. Profitis Ilias. Bands of altered schist interlayered with limestones support a meager cultivation.

The elevations of the main points of the region are as follows.

Threshold of the monastery of Mt. Profitis Ilias.	568.5 m
School of Martinos (west slope of Mt. Profitis Ilias).	346.0 m
Highest point of Pirgos	375.0 m
Saddle between Pirgos and Mt. Profitis Ilias	293.0 m
Highest point on the cliff east of Athinios	366.0 m
Highest point of Mesa Vouno	378.2 m
Grotto of Mesa Vouno (southern side of the hill)	325.0 m
Saddle between Mesa Vouno and Mt. Profitis Ilias.	274.6 m
Emporion	80 m
Summit of Platanymos (Mt. Gavrillos).	136 m

The rocks of this region are limestones, quartzites, and mica and talc schists. They are exposed in many places, but at one time they were no doubt completely covered by pumice, for fragments of pumice can be found on all the hill crests and even at the entrance to the monastery of Mt. Profitis Ilias built on the higher point in the region. The abundance and wide distribution of the pumice below the escarpments of metamorphic rocks show that great amounts have come down from these places and collected at the base of the slopes. Today most of the western side of Mt. Profitis Ilias is still covered by a mantle of pumice that increases in thickness toward the base of the mountain. The village of Pirgos is also built on a high place covered by pumice and caves have been excavated into it on the northern side of the hill. So it is certain that the upper pumice fell on all the highest points of the island and that its local absence today can be attributed to the steepness of the slopes, which were stripped clean of it by erosion. The slope of the road going up to Mt. Profitis Ilias on the side toward Pirgos is 17 degrees where the pumice can be found today, but 22 degrees where it has been removed by erosion.[11]

Examination of the cliff section has already enabled us to see that before the eruption of the upper pumiceous tuff there were repeated eruptions of ash, lapilli, and more or less scoriaceous material, some of which resembles the pumice in appearance and, in some instances, even in composition. So one must wonder whether this material was also deposited on the highest parts of the island or, in other words, whether the explosions that produced it had intensities comparable to that responsible for the upper pumice. I have not been able to find much useful information bearing on this question, because the ash deposits and lower pumice are visible on the surface at high parts of Thera only where the upper pumice has been removed by erosion, and in these same places the underlying beds would only rarely have been preserved. I was fortunate, however, to find a place where they have survived. Quite close to the top of Mt. Profitis Ilias on the southeastern side the limestone beds stand almost vertical, and there are narrow cracks where these layers have been spread apart. One of these was open at the time of eruption of the lower pumice and the ash layers that rest on it. These ejecta filled the crack and lime precipitated from the water seeping through them cemented the particles together. The fine-grained breccia formed in this way contains fragments of all the varieties of lava found in the ash layers beneath the upper pumice as well as fragments from the lower pumice layer. None of the debris I observed resembled the upper pumice; so the latter must have been erupted after the crack had been completely filled by the products of earlier eruptions. The position of this crevice close to the crest of the ridge and on the side away from the bay indicates that the explosions that ejected this material were very powerful and comparable to those responsible for the upper pumice.

The carbonates in the metamorphic zone of Thera normally form thick layers, but they also occur as thin lenses, separate grains, and veinlets within the schists. The schists have a foliated, greenish appearance and a smooth feel. Under the microscope they are seen to consist of highly birefringent chloritic minerals with a laminated texture. They have a strong, wavy fabric, and the quartz, white mica, and feldspathic carbonates have the usual optical properties in polarized light. A few veinlets of calcite cut the schistose bands. The quartz has many fluid inclusions with gas bubbles that move spontaneously. The liquid of the inclusions is not very rich in volatiles, for the bubbles do not disappear when heated to a temperature of 100 degrees. They probably consist of water with dissolved salts of the kind normally found in such inclusions. Small opaque grains, probably iron oxides or sulfides, are scattered throughout the schistose bands but are visible only under the microscope.

A dike containing galena and cupriferous pyrite has recently been discovered in the zone of metamorphic rocks close to the sea and a few meters above the warm mineral spring of Thermia. At the time of my last visit to Santorini in the autumn of 1875, a financial company had just obtained authorization to exploit the deposit. At that time, the work had scarcely started.[12]

The summit of Mt. Profitis Ilias, the hill of Mesa Vouno, and the small ridge of Platanymos (Mt. Gavrillos) are formed mainly of crystalline carbonates. Schists are well developed between Mt. Profitis Ilias and Mesa Vouno, and their greater susceptibility to erosion is probably responsible for the depression between the two hills. A two-meter thick bed of mica schist that is exposed there has thin layers separated by beds of flattened carbonate pebbles. The large dimensions of the pebbles are oriented parallel to the plane of schistosity.

The tilt of the layered rocks is quite steep throughout Mt. Profitis Ilias. On the southern side in particular, marble layers stand at angles of 50 to 70 degrees. The dip is normally toward the E20°S. On the opposite side, the slopes are so gentle that in antiquity the northern flank of the mountain could be used as a burial field. The marble is cut into narrow ledges on which one can still see the cavities of tombs. Pumice is essentially absent from surfaces of the marble that are inclined more steeply than 30 degrees.

At mid-elevation on the southeast side, two grottos that penetrate about fifteen meters into the schist and marble seem to be products of fracturing of the schist, which was not as readily deformed as the marble. These grottos, which are 3 to 4 meters wide at their entrances, quickly narrow and become lower, so that a short distance from the entrance one cannot move without bending over. Water saturated with calcium bicarbonate filters into them, and in one it has formed small pools surrounded by white travertine.

Because the slopes of Mt. Profitis Ilias are less steep on the northwest, that is the side where the road leads up to the monastery.

To go from Mt. Profitis Ilias to Mesa Vouno one can follow a path along the crest of the ridge and pass by the ancient burial places cut into the rock. Today, Mesa Vouno is a barren rock with only a single dwelling that is rarely occupied.

The summit is formed by a rather broad plateau covered with ruins. Trunks of columns, fragments of statues, and bits of pottery are everywhere. The cellars of houses and public buildings are perfectly recognizable. The inscriptions found in this place and the architectural style of the monuments show that the town flourished mainly during the period of Roman occupation.[13] A thin layer of pumice that covers the ground on which the town was built has still not disappeared completely. The marble is exposed only near the peak, mainly on the south and west. On the eastern and northeastern sides, the pumice comes halfway down the slope. Below that, the slope becomes steeper and the mantle of pumice disappears. The upper part of the western flank is an almost vertical face of marble with two bands of schist in the middle.

The lower levels have a pediment deposit of pumice. In the twisted part between Mt. Profitis Ilias and Mesa Vouno a conglomerate layer ten meters thick is formed entirely of carbonate rocks and schist. It underlies the pumice and is therefore older than the great eruption that poured out the upper pumice.

TABLE OF PLATES

Although some of the original plates are not reproduced here, the original numbering and captions are retained.

Pl. I. Topographic map of Santorini based on the map of Captain Graves with modifications resulting from the eruption of 1866.

Pl. II. View of the bay and islands of Santorini from the monastery of Mt. Profitis Ilias, the highest point on the island of Thera. From a camera obscura drawing made in 1875 by L. de Cessac.

Pl. III. The Kamenis seen from the lazaret of Thira after a photograph taken by Constantinos in March 1866. In the middle, a puff of smoke rises from the cone of Giorgios. In the right foreground, Micra Kameni with Nea Kameni behind. At the left, Palaea Kameni.

Pl. IV. The southern end of Giorgios. In the foreground, the older ground of Nea Kameni. Small ponds where warm water was welling up. In the middle ground on the left, part of the former ground invaded by the sea as a result of subsidence. A house at this place is almost entirely underwater. In the background, recent lavas from Giorgios. The base of the pile is partly obscured by the warm vapor coming from sea water in contact with the lava. From a photograph taken by Constantinos at the beginning of March 1866.

Pl. V. Lavas of Giorgios and the eastern dock of Nea Kameni from a photograph taken by the author in March 1866 from the summit of the cone of Micra Kameni. In the foreground is the channel between Nea Kameni and Micra Kameni and houses near the pier. At the extreme right is the house of Mr. Sfoscoti. Toward the left, the surface of the pier is not yet covered by the sea; posts for tying up boats are still visible. Behind the houses at the foot of the new lavas is an Orthodox church on the left and a Catholic church on the right. Flames rise from the highest point on the Giorgios lavas. To the right, the lower slopes of the old cone of Nea Kameni. In the background, the cliffs of Thera between the harbor of Balos and Cape Akrotiri.

Pl. VI. The southern end of Giorgios from a photograph taken by the author in April 1866 from a point behind the houses at the pier of Nea Kameni. At the left, a cove and ponds of water caused by sinking of the ground. The half-submerged houses shown in Plate V have completely disappeared under the water. To the right, the Orthodox church, the roof of which has been broken by a volcanic bomb, is in immediate contact with the new lavas from Giorgios. In the left background, the cliffs of Akrotiri are obscured by steam coming from the warm sea water.

Pl. VII. The northeastern extremity of Giorgios from a photograph taken by the author in April 1866 from a point behind the houses at the pier of Nea Kameni. In the foreground are ponds produced by sinking of the ground. The Catholic church is half demolished by falling bombs and the lava that overruns it. In the right background is the southern flank of the older cone of Nea Kameni.

Pl. VIII. The summit of Giorgios seen from its western flank in April 1866. From a photograph taken by the author. The lower two-thirds of the escarpment is made up of blocks covered with saline deposits. At the junction with the upper third one sees the ditch-like crevice where ejecta were being thrown out explosively.

Pl. IX. The southeastern part of the pier of Nea Kameni from a photograph taken by the author at the end of March 1866 from a point behind the pier. In the foreground the older surface of Nea Kameni is crossed by a road from the pier to the Orthodox and Catholic churches. The ground is littered with volcanic bombs. This part of the pier is entirely submerged. The house on the left is half flooded, mainly on its southern side. The sea also penetrates the other house, but the water is shallower. In the distance are the cliffs of Thera between the routes up to Thira and Athinios.

Pl. X. Pier of Nea Kameni and the cone of Micra Kameni from a photograph taken by the author on the same day as the preceding one. In the foreground are the older ground of Nea Kameni and ruined houses along the pier. These are the same houses shown in plate V, but they are seen from the opposite side and about a month later. In the middle ground is the cone of Micra Kameni and the southern rim of the crater. In the distance are the cliffs of Thera and the zigzag path up to the town of Thira.

Pl. XI. The 1866 cone of Giorgios and the 1707 cone of Nea Kameni seen from the west. From a photograph by Paul Desgranges from Palaea Kameni in December 1866. In the foreground, lavas of Aphroessa are still steaming. On the right, the cone of Giorgios is covered with ash and a trail of blocks indicates the path of the lava. The summit of Giorgios is covered with fumes. On the left is the cone of 1707. In the background are the cliffs of Thera between Merovigli and Thira, with the houses of Thira at the top.

Pl. XII. The channel between Micra Kameni and the lavas of Giorgios from a photograph taken by the author in March 1867 from the eastern part of the former pier of Nea Kameni. On the left is the cone of Micra Kameni; on the right are the lavas of Giorgios. In the background are the cliffs of Thera between Cape Alonaki and Athinios.

Pl. XIII. The eastern pier of Nea Kameni and the flow from the southern flank of Giorgios from a photograph taken by the author in March 1867 from the western foot of Micra Kameni. The houses are the ones shown in Plate V. The subsidence of the ground had continued so that the houses are lower than in Plate V. The house on the far left has been invaded by the sea as shown in Plate XII, but it was still dry in the photograph taken in 1866. The pier is entirely submerged.

Pl. XIV. A view in February 1867 from the same point and in the same direction as in Plate X taken in 1866. From a photograph taken by the author. In the foreground are seen the surface of Nea Kameni and ruined houses along the former pier. On the right, the subsidence has led to flooding behind the dwellings. In the middle distance on the left is the cone of Micra Kameni; on the right is the flow from Giorgios. In the distance one sees a small part of the cliffs of Thera.

Pl. XV. The cone of Giorgios looking toward its eastern slopes and the former pier of Nea Kameni, from a photograph taken by the author in February 1867 from the foot of the cone of Micra Kameni toward the northwest of that cone. In the foreground is the channel between Micra and Nea Kameni. Near the shore on the right is the house of Sfoscoti that is shown in Pl. V. This house is partly demolished by volcanic bombs. In the middle ground on the right is the southern flank of the cone of 1707, in the middle the cone of 1866 (Giorgios). This illustration is interesting because it shows these hills covered with ash, while in March 1866 (see Pl. IV, V, VI, VII, and VIII) Giorgios was composed almost entirely of angular blocks piled up along their path of descent.

Pl. XVI. View of the eastern slope of Giorgios from a photograph taken by the author in February 1867 from the tip of the promontory that forms the northern border of the cove of Saint Nicolas of

Palaea Kameni. In the foreground, the Isles of May. In the middle ground is the coast formed by the lavas of Aphroessa with the cone of Giorgios in the right rear and the cone of 1707 on the left. In the background between the two cones can be seen a small part of the cliffs of Thera.

Pl. XVII. Views of the cone and lavas of Giorgios from the top of Palaea Kameni. From photographs taken by the author, the first during the eruption at the beginning of February 1867 and the second in September 1875 after the eruption had ended. The dark areas are products of the eruptions of 1707 and 1866. The line in the background marks the cliffs of Thera between the point of Apano Meria and the harbor of Athinios. Comparison of the two figures shows the increase in height of the cone of Giorgios between 1866 and September 1870 and the great mass of lava that spread during this time. The figures also show the changes of the Isles of May as a result of the action of waves.

Pl. XVIII. *Fig. 1.* View of the cone and lavas of Giorgios taken in September 1875 from the top of the cliffs of Palaea Kameni about 100 meters north of the summit. This figure shows very well the form of the high plateau formed by the cone and the central part of the lavas of Giorgios. *Fig. 2.* View of the central part of the bay of Santorini from the harbor of Balos in September 1875. *Fig. 3.* View of the central part of the bay of Santorini from Manolas (on the island of Therasia) in 1867.

Pl. XIX. View taken by the author in September 1875 from the top of the cliff above the cove of Saint Nicolas on Palaea Kameni. The ship in the foreground is between the two promontories of lava that form the cove of Saint Nicolas. In the middle ground is the channel between Palaea Kameni and the lavas of Giorgios; one can see the two Isles of May. Farther back, the highest point is formed by the cone of Giorgios with the cone of 1707 on the left. In the background are the cliffs of Thera.

Pl. XX. The northeastern slope of the cone of Giorgios from a photograph taken in September 1875 from the foot of the lavas of Giorgios at the southern end of the channel separating Nea and Micra Kameni. In the foreground is the cove produced by subsidence of the older ground surface of Nea Kameni. At the foot of the cone are the ruins of the Catholic church shown in Pl. V and VII as it was in 1866. On the right is the saddle between the two cones of 1866 and 1707.

Pl. XXI. The eastern slope of the cone of Giorgios in September 1875. From a photograph taken at the entrance of the channel separating the cone of Micra Kameni from the lavas of Giorgios. Near the shore are the ruins of houses by the former pier of Nea Kameni. To the left of the houses is the cove shown in the preceding plate. In the middle is the cone of Giorgios with trails of blocks coming from its upper part. On the left is the lava from Giorgios.

Pl. XXII. View in September 1875 from the same point as Pl. X and XIV taken in 1866 and 1867. In the right and middle foreground is the cove formed by subsidence of the older ground surface of Nea Kameni. (See Pl. XX and XXI.) On the left is part of the still emergent surface of Nea Kameni. In front is a boat entering one of the house near the former pier. In the middle ground on the left is the cone of Nea Kameni; on the right are the lavas of Giorgios. A small boat can still pass between the base of the cone and the lavas. In the background are the cliffs of Thera.

Pl. XXIII. A volcanic bomb on the plateau of Giorgios in September 1875. See page 89 of the text.

Pl. XXIV. Map of the progressive growth of Nea Kameni as a result of the rise of the Isles of May and lavas spreading from Aphroessa and Giorgios during the last eruption. Note especially the changes of form of the harbor of Saint Nicolas and the channel between Micra and Nea Kameni. The western shore of this channel is indented by invasion of the sea into the area of subsidence.

Pl. XXV and XXVI. Successive views of the Kamenis from the east from the beginning until the end of the eruption of 1866. From photographs taken in 1866 and 1867 from the terrace of the Lazarist monastery at Thira and from a camera obscura drawing made at the same place in 1875 by L. de Cessac. In the foreground, Micra Kameni is shown in dark gray. In the middle ground on the left the cone and lavas of Giorgios are shown in a reddish tint and on the right are the cone and lavas of 1707 (Nea Kameni). Palaea Kameni can be seen behind Giorgios. Farther in the background on the left a promontory is formed by the southwestern end of Thera (the district of Akrotiri). The small islet of Aspronisi can be seen in the middle behind Palaea Kameni. On the right is the southern end of Therasia (Cape Tripiti). In the far distance between Cape Akrotiri and Therasia, one can see the island of Ascania (Christiania), one of the distant islands of the Santorini group.

Pl. XXVII and XXVIII. Successive views of the Kamenis from the southeast from the beginning until the end of the eruption of 1866. From photographs taken by the author in 1866 and 1867 from the top of the cliff of Athinios and from a camera obscura drawing made at the same point in 1875 by L. de Cessac. The lavas of Giorgios are in the middle foreground. In the figure showing the Kamenis before the eruption, the lavas have not yet appeared and one can see in the right foreground the cone of Micra Kameni. To the right are the cone and lavas of 1707 (Nea Kameni). Slightly farther away and far to the left is Palaea Kameni. The island of Therasia is in the middle background.

Pl. XXIX. Map of Micra Kameni and Nea Kameni on a scale of 1 to 10,000 showing the distribution of lavas erupted from the latter during the eruption of 1866. Based on geodetic and hydrographic measurements by the author in September and October 1875. See pages 132 and following.

Pl. XXX. Map of the summit of Giorgios at a scale of 1 to 2400 based on geodetic measurements by the author in September and October 1875. See the text on pages 157 and following. The numbers on the map are elevations.

Pl. XXXI. Profile of the inner cliffs of the island of Thera. *Fig. 1* shows the section from Apano Meria to Cape Perivola. *Fig. 2* extends from Cape Perivola to the Gulf of Musaki. Based on photographs taken by the author and camera obscura drawings by L. de Cessac. See page 247 and following.

Pl. XXXII. Continuation of the profile of the inner cliffs of Thera. *Fig. 1*, from the Gulf of Musaki to the cove of Arkicondivari, and *fig. 2*, from the cove of Arkicondivari to Cape Tourlos. The numbers shown on this and the preceding plate indicate the order of dikes and lavas starting from Apano Meria. See page 255 and following.

Pl. XXXIII. Continuation of the profile of the inner cliffs of Thera. *Fig. 1*, from Cape Tourlos to Thira and *fig. 2*, from Thira to Cape Alonaki. See page 259 and following.

Pl. XXXIV. Continuation of the inner cliffs of Thera. *Fig. 1*, from Cape Alonaki to the harbor of Athinios; *fig. 2*, from the harbor of Athinios to Therma; *fig. 3*, from Therma to the lowest part of the cliff in the vicinity of Balos. See page 268 and following.

Pl. XXXV. Continuation of the inner cliffs of Thera from the vicinity of Balos to Cape Akrotiri. See page 276 and following. See Chapter VII for discussions of the rocks of the Akrotiri region.

Pl. XXXVI. Profile of the cliffs of Therasia. The scale is the same as for the cliffs of Thera. The dike of Kera has inadvertently been omitted from this plate. See page 286 and following.

Pl. XXXVII. View of the southern exterior cliffs of Thera. From a camera obscura drawing by L. de Cessac made from one of the islets facing the rock of the Obelisk. See pages 280, 283, and following. See also Chapter VIII for a discussion of the rocks of this region.

Pl. XXXVIII. View of the harbor of Thira from a photograph taken by the author in September 1875 from the end of Cape Lazaret. Below is the pier of Thira. At the top of the cliff is the town of Thira. The figure shows the zigzag road leading to it. On the left of the houses near the pier, a large landslide occupies the position of a fault cutting the cliff. (See page 263.) To the left in the middle ground is a promontory ending at Cape Tourlos and surmounted by a large, barren rock around which can be seen the ruins of the town of Skaro. In the distance above Cape Tourlos, one can see a small part of the cliffs of Apano Meria.

Pl. XXXIX. Jars and fragments of pottery from the prehistoric structures of Santorini. *Figs. 1* and *5* show pottery from the excavations carried out by Messrs. Gorceix and Mamet. *Figs. 2, 3,* and *6* are of pottery collected by the author in the ravine east of Akrotiri. See Chapter III.

Pl. XL. Prehistoric pottery from Santorini. The jars shown in *figs. 1* and *3* were collected by the author. The jar in *fig. 2* comes from the excavations of Messrs. Gorceix and Mamet.

Pl. XLI. Prehistoric pottery from Santorini. The fragment in *fig. 1* was collected by the author on Therasia; the jars in *figs. 2* and *3* come from the excavations of Messrs. Gorceix and Mamet; the jars in *figs. 5* and *6* are from the structures found in the ravine of Akrotiri; the jar in *fig. 5* was collected by the author; the one in *fig. 6* was given to Mr. L. de Cessac by Mr. Chigi of Thira.

Pl. XLII. Prehistoric pottery from Santorini. The jars in *fig. 1, 5, 6,* and *7* come from excavations carried out by Messrs. Gorceix and Mamet. The pottery in *fig. 2, 3,* and *4* was collected by the author in the ravine of Akrotiri. The jar in *fig. 4* is a common type in the deposits of both Therasia and Thera.

Pl. XLIII. Thin sections of prehistoric pottery of Santorini. Magnification 80 diameters. Drawing by Jacquemin. See page 125. Magnification 80 diameters.
Fig. 1. Slide is seen in natural light. Brown argillaceous material makes up most of the pottery. Reddish-brown fragments of altered mica schist. Several colorless fragments of marble with cleavage traces of calcite. Small, colorless pieces of quartz and feldspar in the main material of the pottery.
Fig. 2. Slide is seen in natural light. Spherulite composed of two concentric zones and enclosed in the main brown material of the pottery. Crystals of labradorite and pyroxene are at the center of the spherulite. Radiating fibers are composed of fine labradorite in a matrix of exceedingly fine sectors of brown amorphous material. In the material surrounding the spherulite are a few crystals of labradorite and augite. (See text, p. 366.)
Fig. 3. Slide is seen in polarized light between crossed nicols. Centered on the spherulite is a large, poorly defined black cross that varies in size and shape as the slide is rotated and the orientation of the sheaves of labradorite changes. The lateral dimensions of the cross are determined by the extreme extinction angles of the labradorite in the zone pg_1, owing to their elongation parallel to the pg_1 face. (See text, p. 366).
Fig. 4. Slide is seen in natural light. A fragment of trass enclosed in the brown ceramic material of prehistoric pottery from Thera. Near the center of the fragment is a crystal of quartz with a rounded hexagonal cross section. Surrounding it are microlites of feldspar, most of which are swollen and deformed. The surrounding brown matrix contains small pieces of quartz and feldspar. (See p. 362.)
Fig. 5. Slide is seen in natural light. The brown matrix contains a crystal of green augite cut nearly

parallel to its lateral faces. Black crystals of iron oxide are partly or totally enclosed in the augite, as well as inclusions of round, gray amorphous material. The brownish matrix contains bits of feldspar and quartz.

Fig. 6. Slide is seen in natural light. Pumice contained in pre-historic pottery from Santorini. Fluidal structure is defined by stretched-out microlites surrounding small patches of glass. The surrounding material contains pieces of quartz and feldspar. (See text, p. 363.).

Pl. XLIV. Thin sections of prehistoric pottery of Santorini. Magnification 80 diameters. Drawing by Jacquemin. See text on page 125. Magnification 80 diameters.

Fig. 1. Slide is seen in polarized light between crossed nicols. Fragments of labradorite show characteristic albite twinning. On the left, a rectangular crystal of sanidine with Carlsbad twinning. Below is a large piece of feldspar with very marked growth zones. Fragments of hornblende and augite. The matrix is in extinction.

Fig. 2. Slide is seen in natural light. A sample of trass impregnated with opal and cemented to a crystal of labradorite. Small globules are still preserved in the less altered parts of the trass. The labradorite has inclusions of grayish glass. The brown groundmass contains fragments of quartz and feldspar. (See the text on page 364.)

Fig. 3. Slide is seen in natural light. Pottery with an inclusion of pumice. The fluidal pumice contains tear-shaped clots of glass charged with crystallites. Bits of feldspar are seen throughout the groundmass. (See the text on page 363.)

Fig. 4. Slide is seen in natural light. Large crystals of labradorite with glassy inclusions that contain several bubbles of gas. A crystal of brown hornblende is cut almost perpendicular to its lateral faces and shows the typical cleavage of this mineral. Black iron oxides and bits of feldspar and quartz are common in the groundmass. (See the text on page 344.)

Fig. 5. Slide is seen in polarized light between crossed nicols. The sample is the same as the one shown in natural light in fig. 4. The labradorite has the characteristic albite twins of this mineral.

Fig. 6. Slide is seen in natural light. At the upper left is a rounded piece of marble with the cleavage of calcite. In the middle of the section, a piece of obsidian has microlites of oligoclase. Below it is a large crystal of labradorite. The brown matrix contains bits of feldspar and quartz.

Pl. XLV. Organic material seen in the thin sections of the various pieces of pottery from Santorini. Magnification 80 diameters. Drawing by Jacquemin. See the text on page 125.

Pl. XLVI. Thin sections of the common lava of the eruption of 1866. Reproduced from photomicrographs by the author using natural light. Magnification 80 diameters. See Chapter V.

Pl. XLVII. Thin sections of the common lava of the eruption of 1866. Reproduced from photomicrographs by the author. Magnification 80 diameters. See Chapter V.

Pl. XLVIII. Photomicrographs of thin sections. Magnification 80 diameters. Drawing by Jacquemin.

Fig. 1. A block of wollastonite included in lavas erupted in 1866. The slide is seen in natural light. The white, fibrous crystals are wollastonite, the green crystals fassaite, and the brown ones melanite garnet. The straw-yellow crystals are a cubic species, possibly hauyne. (See the text on page 207.)

Fig. 2. Thin section of a lava from one of the islets off shore from the Obelisk. The slide is seen in polarized light. Reddish-green crystals of augite, colorless prismatic crystals of untwinned labradorite. A globule of opal with a black cross. (See the text on page 370.)

Fig. 3. A druse of wollastonite extracted from a lava erupted in 1866. The slide is seen in natural light. The loosely joined crystals are the same species as in fig. 1. The large brown crystal is melanite. (See text on page 207.)

Fig. 4. Thin section of the selvage of dike no. 14. The slide is seen in natural light. Greenish yellow crystals of augite. Colorless crystals of labradorite. Amorphous material with a cracked, bulbous appearance contains sparsely scattered microlites of albite and oligoclase. The bulbous form is due to spreading of the colorless and ferruginous components of the amorphous material of the matrix. The fretted appearance is from contraction and small cracks that were filled by colorless glassy secretions when the lava was not yet entirely solidified. (See the text on page 304.)

Fig. 5. Crystals from a wollastonite druse selected to show the forms of the wollastonite and fassaite pyroxene. The slide is seen in natural light. The other species are as indicated for fig. 1.

Fig. 6. Thin section of an anorthite-bearing rock from dike no. 4 in the northern cliff of Thera. The slide is seen in natural light. Green augite shows the usual cleavage traces. Two large, colorless crystals of olivine with a rough appearance. The crystal on the left is rounded; that on the right is crudely prismatic. Each has is a rounded, gas-filled cavity. Large, colorless crystals of labradorite with two sets of cleavage at angles close to 90 degrees. Elongated, colorless microlites of labradorite. Greenish microlites of augite. Grains of iron oxide. In the space between the crystals, the amorphous material contains innumerable scattered globulites. (See the text on page 310.)

Pl. XLIX. Microscopic sections of rocks and volcanic material seen in polarized light between crossed nicols. Magnification the same as in plate XLVIII. Drawing by Jacquemin. The principal directions of the polarizers are parallel to the edges of the page.

Fig. 1. Slide is seen in polarized light between crossed nicols. Anorthite-bearing lava. The same thin section is shown in natural light in pl. XLVIII, *fig. 6*. Olivine and pyroxene are brightly colored. The anorthite has the triclinic bands characteristic of albite twinning.

Fig. 2. Slide is seen in polarized light between crossed nicols. Isolated crystals of a druse of wollastonite, fassaite, and melanite garnet. The same thin section is shown in natural light in pl. XlVIII, *fig. 5*. The garnets and yellow crystals are in extinction; the wollastonite and fassaite have brilliant colors.

Fig. 3. Slide is seen in polarized light between crossed nicols. The same thin section is shown in natural light in pl. XLVIII, *fig. 4*.

Fig. 4. Slide is seen in polarized light between crossed nicols. A druse of wollastonite. The same thin section is shown in natural light in pl. XLVIII, *fig. 3*.

Fig. 5. Slide is seen in polarized light between crossed nicols. Globules of opal with black crosses. Crystals of augite have multicolored tints. Labradorite in small white and grayish crystals. The same thin section is shown in natural light in pl. XLVIII, *fig. 2*.

Fig. 6. Slide is seen in polarized light between crossed nicols. A compact wollastonite-bearing rock. The same thin section is shown in natural light in pl. XLVIII, *fig. 1*.

Pl L. Thin section of a siliceous concretion collected on the southern slope of Loumaravi. The specimen is seen in polarized light between crossed nicols. The directions of polarization are the same as in the preceding plate. Magnification 120 diameters. Drawing by Jacquemin. Chalcedonic spherulites with a black cross. In the parts of the spherulites with large individual crystals in the middle of the figure one can distinguish the form and the zones of growth of the crystals. Near the left and right edges of the figure are lamellae of quartz. Toward the lower right is a spherulite with concentric growth zones.

Pl. LI. Fragments produced by pulverizing the perlite of Balos. The specimen is seen in polarized light between crossed nicols. The directions of polarization are the same as in the preceding plate. Magnification 120 diameters. Drawing by Jacquemin. See the text on page 353. Perlitic globules with black crosses. Crystals of hornblende and augite with their characteristic cleavages. Some large crystals of untwinned feldspar show the usual cleavage traces of this family of minerals.

Pl. LII. Thin section of a sample of the perlite of Balos in the southwestern part of Thera. The specimen is seen in natural light. Magnification 120 diameters. Drawing by Jacquemin. Perlitic globules have scattered inclusions of very fine crystals some of which are arranged in lines. Large, colorless crystals of labradorite in multiple twins have glassy inclusions and indications of cleavage that cut them at angles of about 90 degrees. Brown amphibole, green augite, and elongated prisms of apatite with transverse cleavages. See plate LI and the text on page 353.

Pl. LIII. Thin section of pumiceous breccia from the Arkhangelos in the southwestern part of Thera. Specimen is seen in natural light. Magnification 120 diameters. Drawing by Jacquemin. Large colorless crystals of labradorite with signs of cleavage at angles of about 90 degrees. Large crystals of brown amphibole some of which are cut almost perpendicular to the base of the prisms; the cleavage traces seen in these orientations are at angles close to 120 degrees. Green augite and black iron oxides. Fluidal bands of amorphous material makes up the groundmass. Very fine crystallites are arranged radially around the rounded parts of the chains of fluidal inclusions. See page 356.

Pl. LIV. Thin section of a siliceous andesite from Loumaravi as seen in natural light. Magnification 120 diameters. Drawing by Jacquemin. Large, colorless crystals of labradorite with cleavage traces at angles of about 90 degrees. One of the crystals at the lower right of the figure, is corroded and separated into two parts that are partly resorbed. Microlites of oligoclase are seen, particularly in the left part of the view. Greenish crystals of hornblende, some cut along their length and others crosswise (the first with cleavage along their length, the second in two directions making an angle of 124 degrees). A spherulite is seen on the right. A vein of hyalitic opal crosses vertically through the middle. Near the bottom are druses lined with various crystals, including stilbite and possibly tridymite. See the text on page 356.

Pl. LV. Thin sections of various rocks of Santorini. Magnification 80 diameters. Drawing by Jacquemin.

Pl. LVI. Various rocks and pottery. Magnification 80 diameters. Drawing by Jacquemin.
Fig. 1. Slide is seen in polarized light between crossed nicols. The sample is an ejected block from the ash at the base of the cliffs of Therasia. Labradorite has the characteristic albite twinning. Black iron oxides. Augite with bright tints of various colors. (See the text on page 376.)
Fig. 2. Slide is seen in polarized light between crossed nicols. The sample is an ejected block from the ash at the base of the beds that border the southern shore of Thera between Cape Mavro and the Obelisk. Labradorite has the characteristic transverse albite twinning in the upper part of the slide. Sanidine has many fine cleavage traces inclined from top down and from left to right. Quartz with a pale yellow or grayish color has no cleavage and only a few scattered inclusions of glass. Diallage, seen at the lower right, has inclusions aligned in two directions. Augite with bright colors and irregular cleavages. Sanidine near the middle of the slide is in extinction. (See the text on page 375.)
Fig. 3. Slide is seen in polarized light between crossed nicols. The sample is an ejected block collected from the ash at the base of the beds east of Balos. To the upper left and lower right are large crystals of oligoclase twinned according to the albite law. In the interstices is micropegmatite grading into microgranulite. Above, a crystal of diallage cut parallel to its longitudinal faces. Below, other crystals of diallage are cut transversely and show the inclusions characteristic of this mineral. (See the text on page 373.)
Fig. 4. A section of pottery collected from the excavations in the ravine at Akrotiri. The slide is seen in natural light. A spherulite is imbedded in two crystals of feldspar (probably sanidine). The radial crystallites of the spherulite are made up of small prisms of labradorite. (See the text on page 366 and plate XLIII, *figs.* 2 and 3.)

Fig. 5. Slide is seen in polarized light between crossed nicols. The sample is a piece of pottery collected from the excavations in the ravine of Akrotiri. A piece of common lava with microlites of albite and oligoclase and grains of iron oxide is identical to the lavas of the 1866 eruption. In the surrounding groundmass are bits of feldspar and quartz. (See the text on page 310 and plate XLVI.) *Fig. 6.* Slide is seen in natural light. The sample is from pottery collected in the ravine at Akrotiri. A piece of perlite from Cape Akrotiri is enclosed in the matrix. (See the text on page 368 and plate LV, *fig. 4*.)

Pl. LVII. Thin section of dike no. 7 in the northern cliff of Thera. Magnification 120 diameters. Drawing by Jacquemin. *Fig. 1.* View in natural light. The brecciated rock is made up of microscopic fragments of various lavas welded together. These fragments differ from one another in their relative proportions of microlites and phenocrysts, in the varied sizes of the microlites, the proportions of glassy matrix, and in the nature of the matrix, which in some fragments is colorless and glassy and in others charged with small globules. See the text on page 302. *Fig. 2.* The same thin section seen in polarized light.

Pl. LVIII. Thin section of a lava from the islet facing the Obelisk. Magnification 120 diameters. Drawing by Jacquemin. *Fig. 1* Specimen in natural light. Phenocrysts and microlites of labradorite. Greenish augite. On the left, a clot of chlorite surrounded by small crystals of augite and crossed by a crystal of labradorite. Iron oxides form large crystals and small, very thin granules. Innumerable crystallites are scattered throughout the amorphous matrix. In a few places it is stained yellow by limonite. In the phenocrysts of labradorite are inclusions of augite and glassy material. Bubbles of gas can be seen in the inclusions of the labradorite crystal in the lower part of the figure. *Fig. 2* The same specimen seen under crossed nicols. Most of the phenocrysts of labradorite are twinned according to the albite law; others are simple crystals with conspicuous growth zones.

Pl. LIX. Thin section of a reddish lava exposed between the harbor of Balos and Cape Akrotiri. The sample is exceptionally rich in olivine. Magnification 120 diameters. Drawing by Jacquemin. *Fig. 1.* View in natural light. Microlites of labradorite and anorthite; the latter are especially numerous. Phenocrysts of olivine with rectangular twins and shapes elongated in the direction of the face pg_1. Extinction is parallel to this face, which is sharply terminated by the g_2 faces. Microlites of augite. Granules of iron oxide altered to limonite. The amorphous matrix is speckled with these granules. See page 372. *Fig. 2.* The same slide in polarized light between crossed nicols.

Pl. LX. Crystals separated from the lavas of Santorini by means of hydrofluoric acid. Natural light. Magnification 120 diameters. Drawing by Jacquemin. *Fig. 1.* Crystals of hypersthene and iron oxide separated from a lava with phenocrysts of labradorite. *Fig. 2.* Crystals of olivine, augite, and iron oxide taken from a lava with phenocrysts of anorthite.

Pl. LXI. Geologic map of the Santorini group of islands.

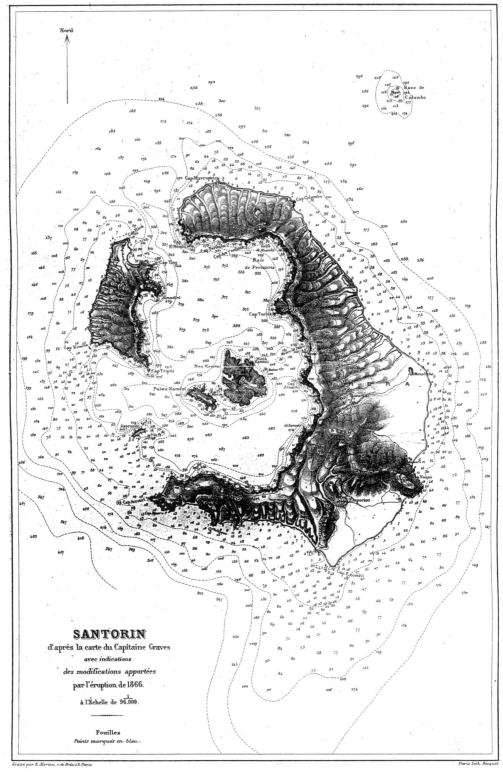

Plate I. Santorini, after the map of Captain Graves, with modifications showing changes resulting from the eruption of 1866. Scale of distances: 1/96,000

Plate III. The Kamenis seen from Thira

Plate IV. The southern end of Giorgios, March 1866

Plate V. Lavas of Giorgios and the eastern dock of Nea Kameni, March 1866

Plate VI. The southern end of Giorgios, April 1866

Plate VII. The northeastern extremity of Giorgios, April 1866

Plate VIII. The summit of Giorgios seen from its western flank, April 1866

Plate IX. The southeastern part of the pier of Nea Kameni, March 1866

Plate X. Pier of Nea Kameni and the cone of Micra Kameni. These are the same houses shown in Plate V about a month later.

Plate XI. The 1866 cone of Giorgios and the 1707 cone of Nea Kameni, December 1866

Plate XII. The channel between Micra Kameni and the lavas of Giorgios, March 1867

Plate XIII. The eastern pier of Nea Kameni and the flow from Giorgios, March 1867

Plate XIV. A view in February 1867 from the same point as Plate X

Plate XV. The eastern slope of Giorgios, February 1867

Plate XVI. The Isles of May, February 1867

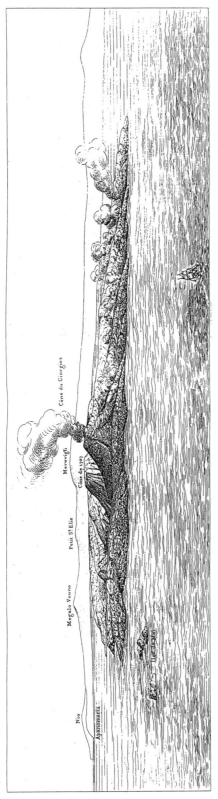

Plate XVII. *Fig. 1.* View of the cone and lavas of Giorgios from the top of Palaea Kameni in February 1867

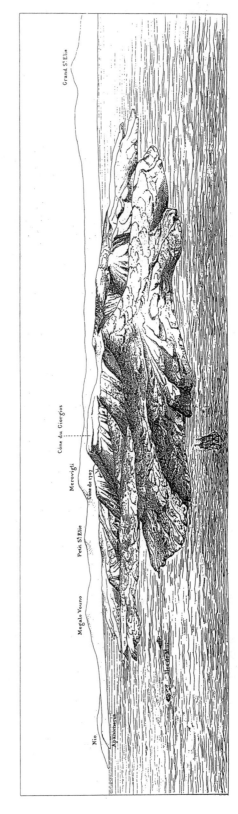

Plate XVII. *Fig. 2.* View of the cone and lavas of Giorgios from the top of Palaea Kameni in September 1875

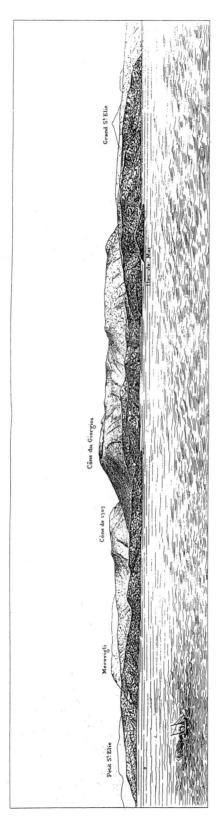

Plate XVIII. *Fig. 1.* View of the lavas of Giorgios taken from the top of the cliffs of Palaea Kameni in September 1875

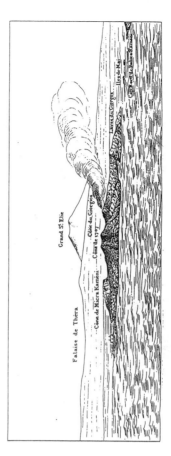

Plate XVIII. *Fig. 3.* View of the Bay of Santorini taken from Manolas in February 1867

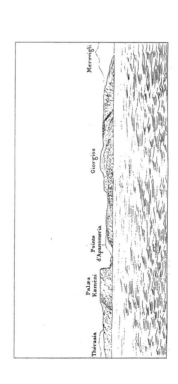

Plate XVIII. *Fig. 2.* View of the Bay of Santorini taken from Balos in September 1875

Plate XIX. The western coast of Giorgios taken in September 1875 from above the cove of Saint Nicolas of Palaea Kameni

Plate XX. The northeastern slope of Giorgios, September 1875

Plate XXI. The eastern slope of the cone of Giorgios, September 1875

Plate XXII. View in September 1875 from the same point as Plates X and XI

Plate XXIII. A volcanic bomb on the plateau of Giorgios, September 1875

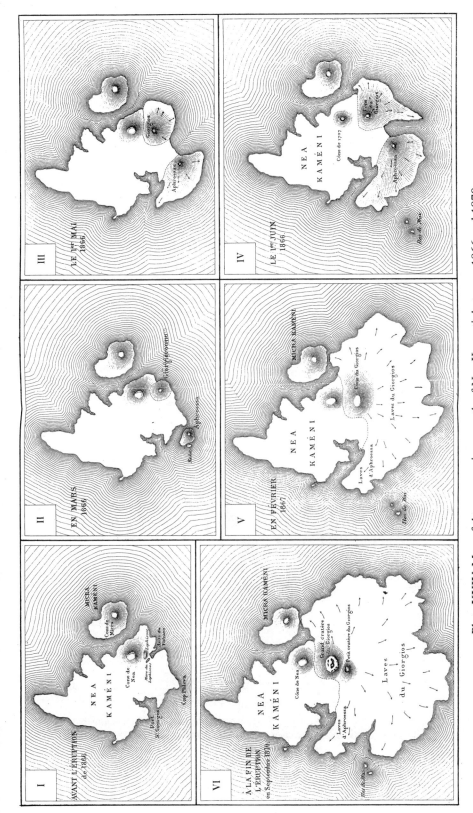

Plate XXIV. Map of the progressive growth of Nea Kameni, between 1866 and 1870

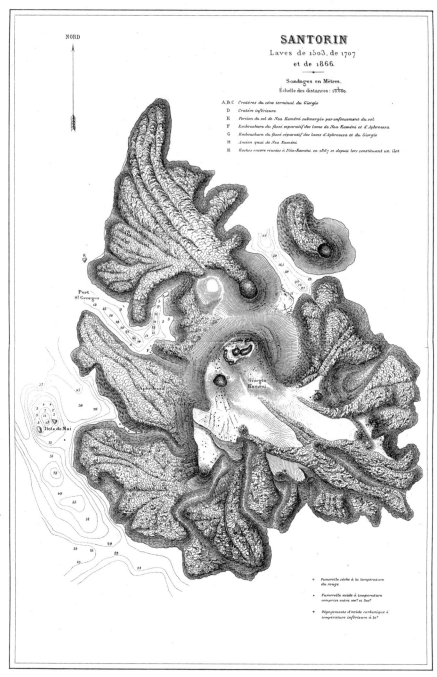

Plate XXIX. Map of Santorini, Lavas of 1503, 1707 and 1866
(A,B,C) Craters of the summit cone of Giorgio; *(D)* Lower crater; *(E)* Part of the submerged surface of Nea Kameni; *(F)* Opening of the crevice separating the lavas of Nea Kameni and Aphroesa; *(G)* Opening of the crevice separating the lavas of Aphroesa and Giorgio; *(H)* Former pier of Nea Kameni; *(K)* Rocks still part of Nea Kameni in 1867 but since then forming an islet. Symbols: ○ Dry fumarole at red temperatures; ● Acidic fumarole at temperatures between 100° and 300°; + Emissions of carbon dioxide at temperatures below 50°. Soundings in meters. Scale of distances: 1/10,000

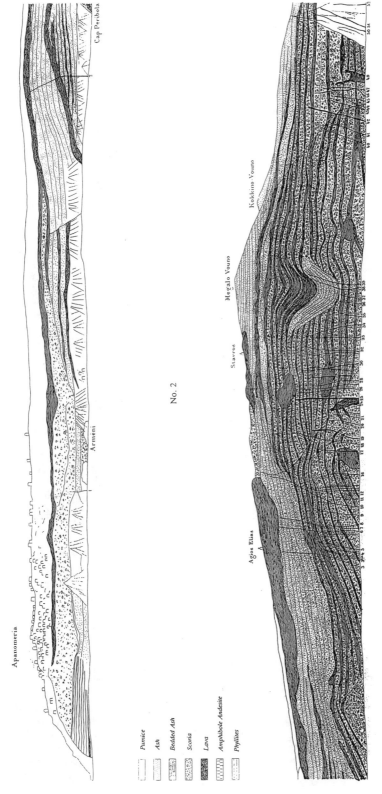

Plate XXXI. Profile of the inner cliffs of the island of Thera: *(top)* from Apano Meria to Cape Perivola; *(bottom)* from Cape Perivola to the Gulf of Musaki

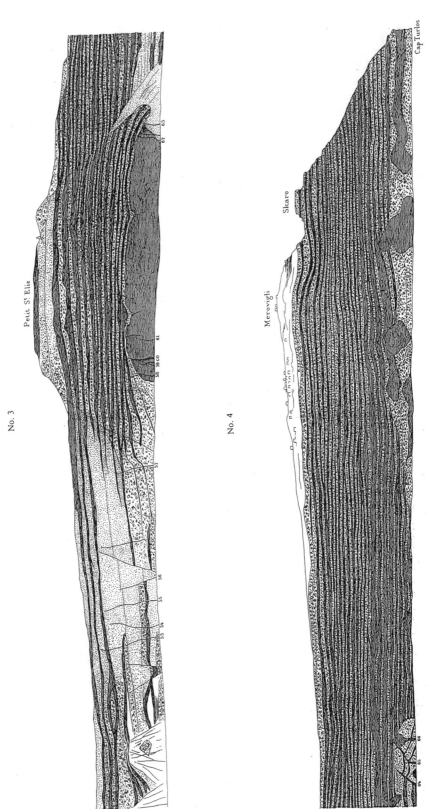

Plate XXXII. Continuation of the profile of the inner cliffs of Thera: *(top)* from the Gulf of Musaki to the cove of Arkicondivari; *(bottom)* from the cove of Arkicondivari to Cape Tourlos.

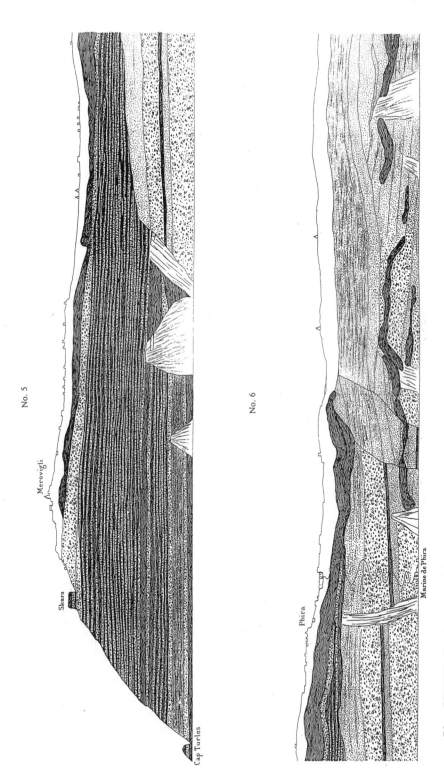

Plate XXXIII. Continuation of the inner cliffs of Thera: *(top)* from Cape Tourlos to Thira; *(bottom)* from Thira to Cape Alonaki

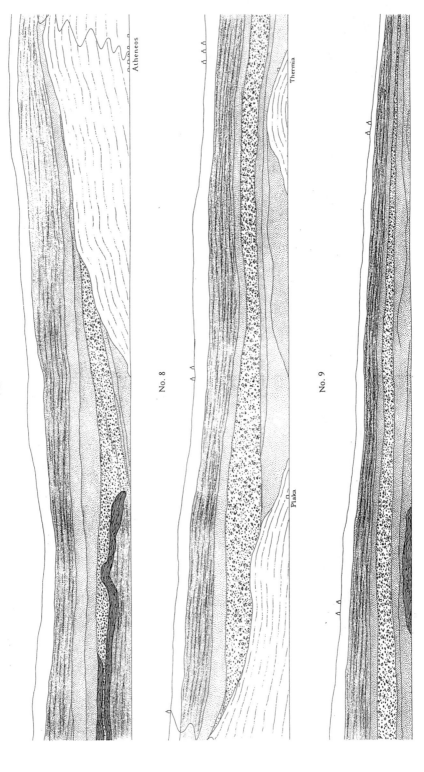

Plate XXXIV. Continuation of the profile of the inner cliffs of Thera: *(top)* from Cape Alonaki to the harbor of Athinios; *(middle)* from the harbor of Athinios to Therma; *(bottom)* from Therma to the lowest part of the cliff in vicinity of Balos

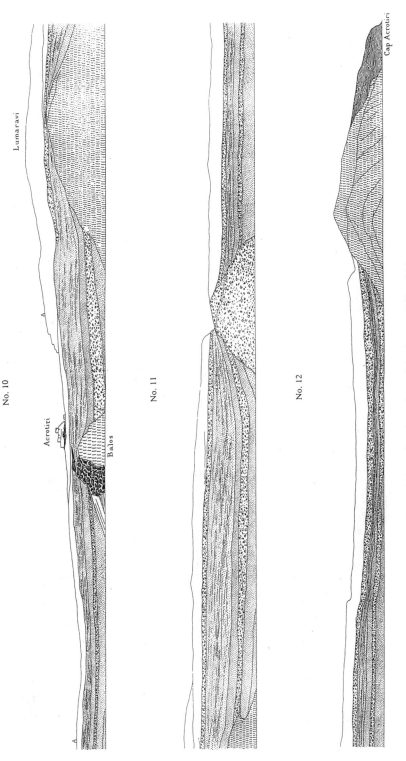

Plate XXXV. Continuation of the inner cliffs of Thera

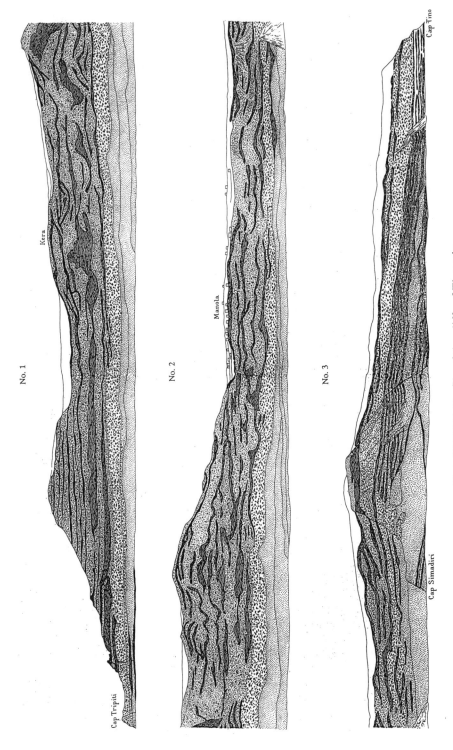

Plate XXXVI. Profile of the cliffs of Therasia

Plate XXXVIII. View of the harbor of Thira

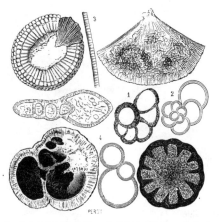

Fig. 1. — 1. *Rotalia lepida*. — 2. *Rotalia senaria*. — 3. *Cocconema*. — 4. *Globigerina*.

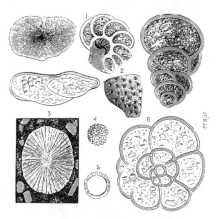

Fig. 2. — 1. *Rotalia*. — 2. Fragment de test de *Globigerina*. — 3. Plaque de grès avec *Coscinodiscus*. — 4. *Lithasteriscus* (?). — 5. *Gallionella varians*. — 6. *Planulina globularis*.

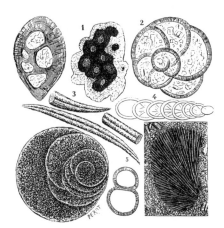

Fig. 3. — 1. (Indéterminé) — 2. *Rotalia*. — 3. *Spongolithis acicularis*. — 4. *Vaginulina*. — 5. *Gallionella distans*.

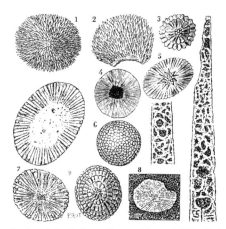

Fig. 4. — 1, 2 et 5. *Frustrella concentrica*. — 3. *Lithasteriscus tuberculosus*. — 4. *Gallionella varians*. — 6, 7 et 8. *Coscinodiscus minor*; 9; *radiolatus*.

Plate XLV. Organic material in various pieces of pottery from Santorini

Plate II. View of Santorini from the monastery of Mt. Profitis Ilias

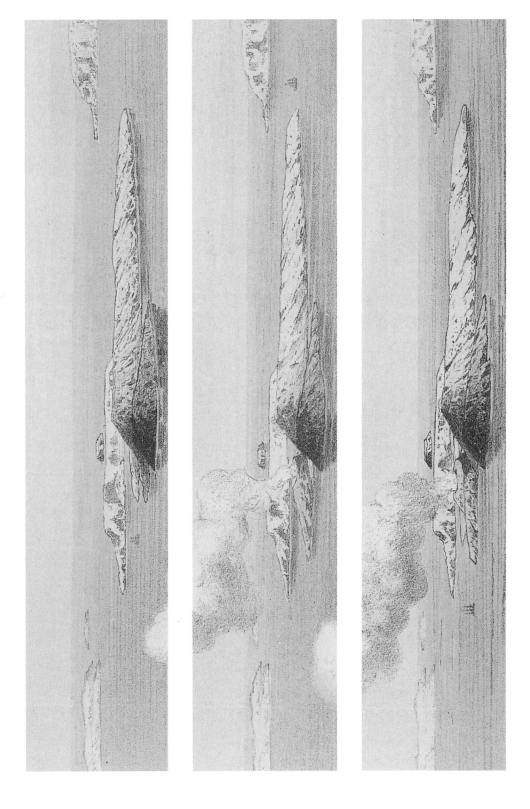

Plates XXV and XXVI. Successive views of the Kamenis from Thira from 1866 to 1870: (*top to bottom*) before the eruption of 1866, March 1866, May 1866, February 1867, after the eruption

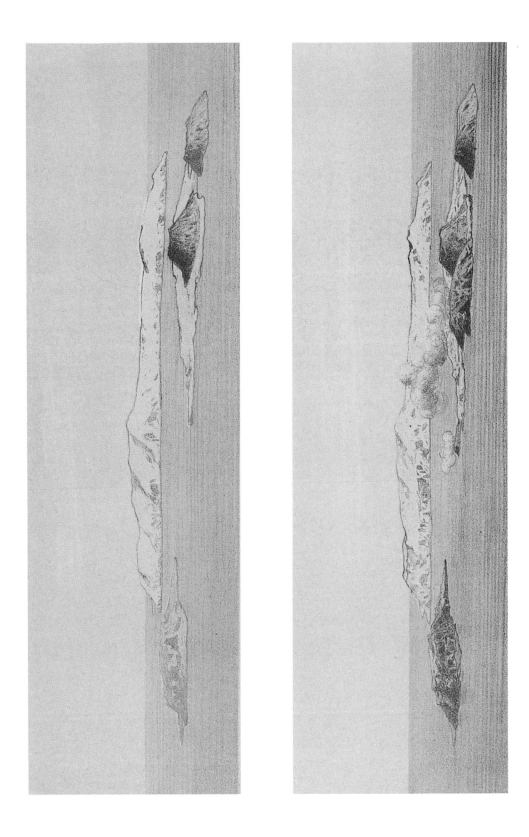

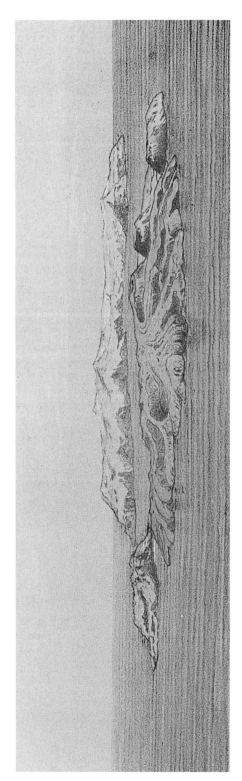

Plates XXVII and XXVIII. Successive views of the Kamenis from Athenios from 1866 to 1870: (*top to bottom*) before the eruption, May 1866, February 1867, after the eruption.

Plate XXXVII. View of the southern exterior cliffs of Thera

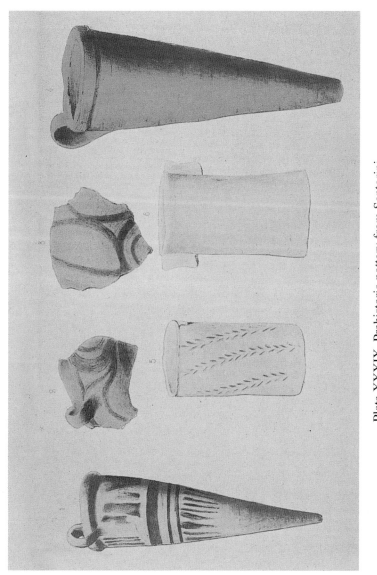

Plate XXXIX. Prehistoric pottery from Santorini

Plate XL. Prehistoric pottery from Santorini

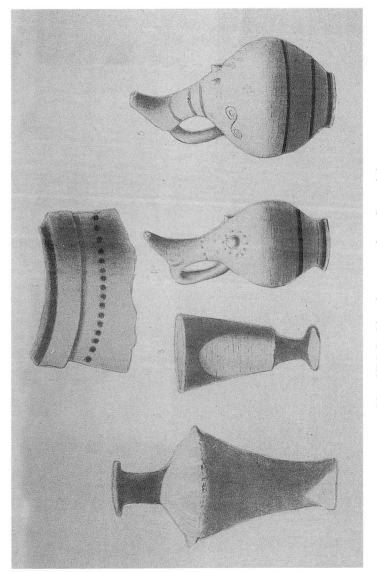

Plate XLI. Prehistoric pottery from Santorini

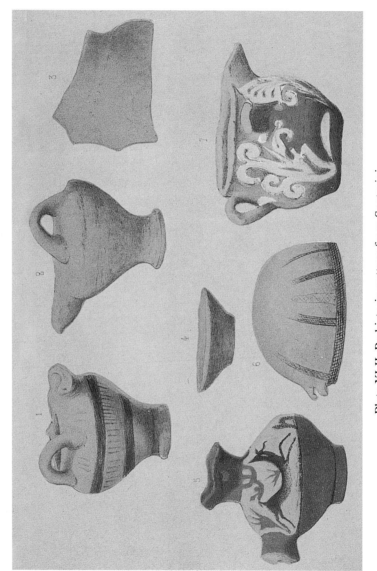

Plate XLII. Prehistoric pottery from Santorini

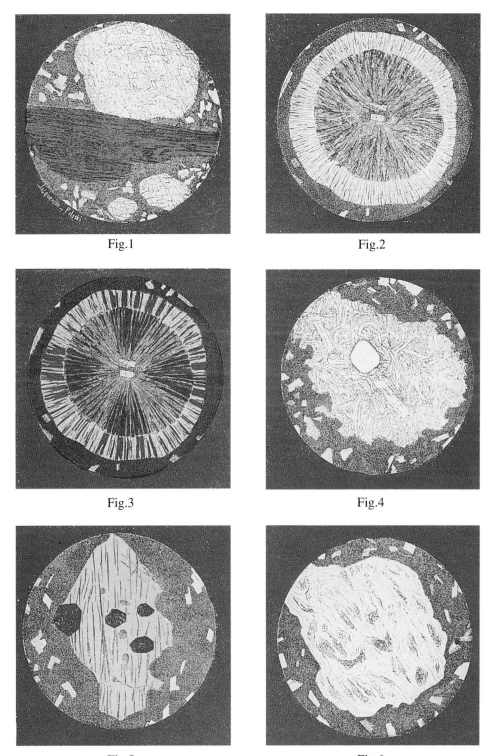

Fig.1 Fig.2
Fig.3 Fig.4
Fig.5 Fig.6

Plate XLIII. Thin sections of prehistoric pottery of Santorini

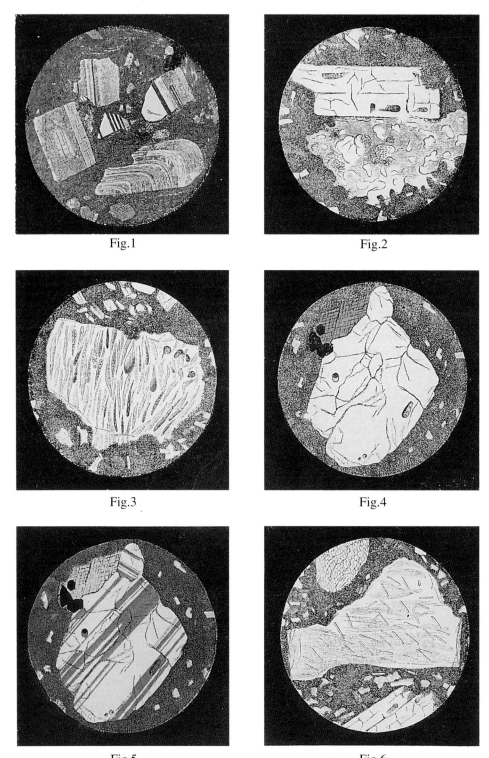

Plate XLIV. Thin sections of prehistoric pottery of Santorini

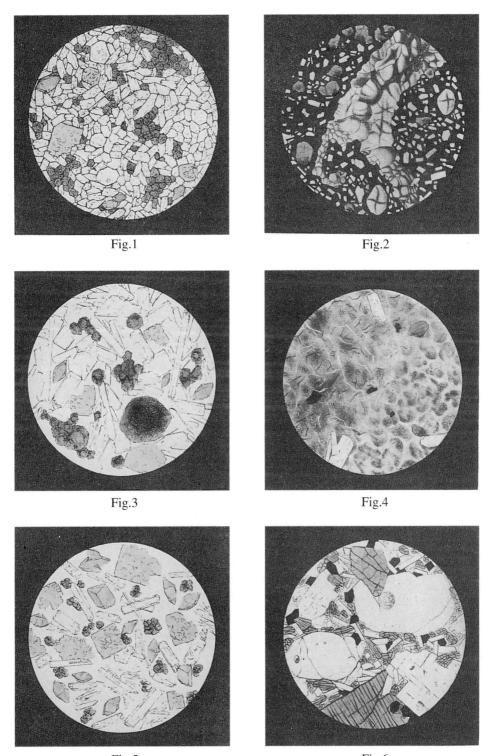

Fig.1

Fig.2

Fig.3

Fig.4

Fig.5

Fig.6

Plate XLVIII. Various photomicrographs of thin sections

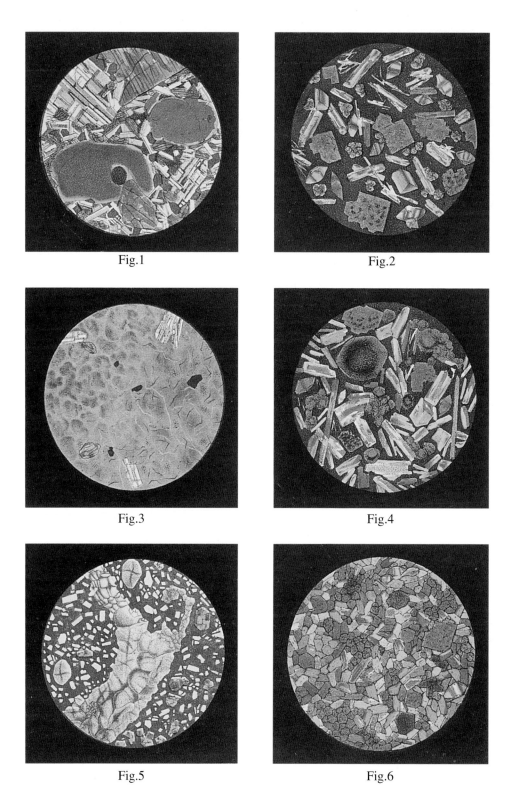

Plate XLIX. Microscopic sections of rocks and volcanic material

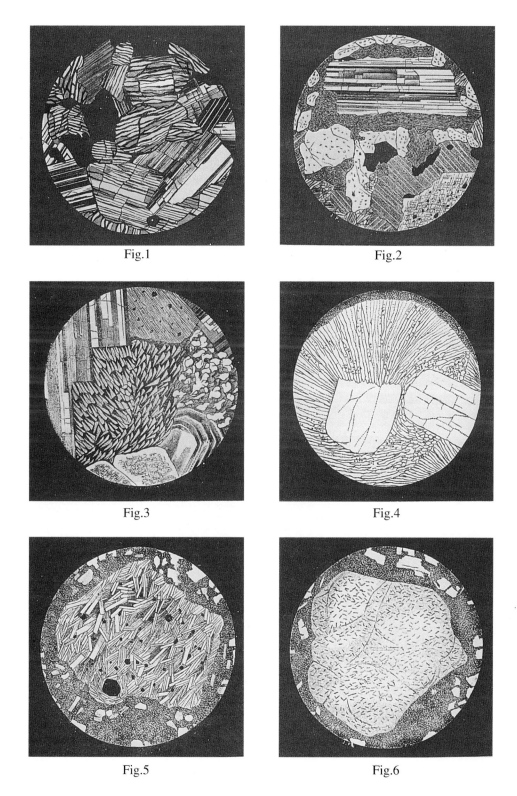

Plate LVI. Thin sections of various rocks of Santorini

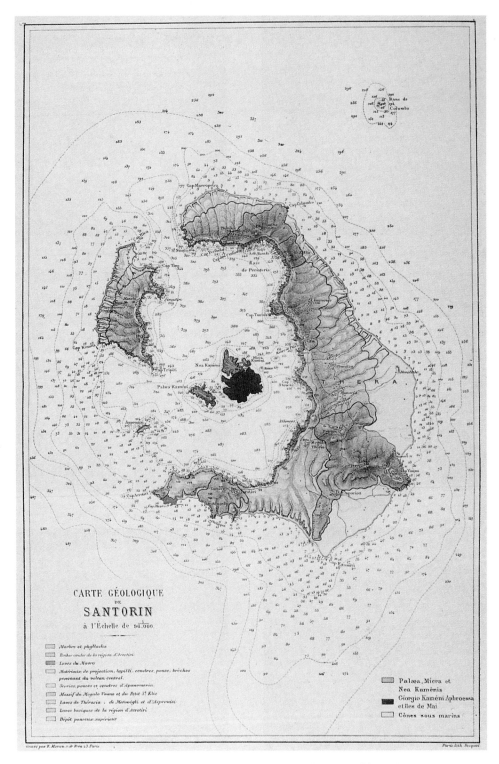

Plate LXI. Geologic map of the Santorini group of islands

The marble monolith of Saint John, a rock 30 meters high, rises in the middle of the plain northeast of the chain of Mt. Profitis Ilias not far from the point of Kamari (Monolithos). The same type of rock forms the lower eastern slopes of Pirgos and is quarried at Exo Gonia. Thus, marble is the dominant rock on the northern slope of the metamorphic ridge, but schists are not entirely absent. One sees the schist, for example, in the ravines that striate the slopes between Mt. Profitis Ilias and Pirgos and between Pirgos and Exo Gonia, but it tends to be badly altered. In places it is even transformed into a yellowish clay that is so decomposed that one can imagine it being used for pottery.

Along the length of the cliff bounding this part of the island, the metamorphic rocks are exposed in two unequal bodies, both of which are buried under volcanic debris. The first and by far the largest of these is in the cliffs at Cape Athinios; it extends almost to the top of the cliff. The second, that of Cape Therma, is not nearly so high. At Cape Athinios, the volcanic beds are banked against both flanks of the metamorphic rocks. The deposits of ash and conglomerates that form the lower unit in that part of the escarpment pinch out upward. The lower pumiceous tuff layer becomes thinner but crosses over the top of the schists and carbonates, and the overlying beds of ash and lapilli do the same; then the upper pumice covers everything. The volcanic rocks that blanket the Athinios body are not horizontal but must have been tilted and faulted by uplift of the older rocks, for if they had been that way originally one would see on each of the flanks of the metamorphic unit a suite of inclined beds crossed here and there by transverse fractures but still preserving their parallelism and normal thickness. In fact, however, one sees that the lower volcanic layers are only interrupted, while the upper ones wrap around the metamorphic body without losing their continuity but diminishing in thickness where they are draped over projections of the schist and carbonates. Moreover, the beds terminate against the contacts at sharp angles, and their inclination is correspondingly steeper than in their normal stratigraphic position at a higher structural level.

Thus, the eruptions that discharged the ash, lapilli, scoria, and pumice that make up the volcanic part of this part of the cliff took place after the metamorphic mass had been uplifted and had taken on more or less its present form.[14] The unusual features of the upper volcanic deposits are those that normally affect any deposit of loose pyroclastic material that falls on pre-existing cone-shaped surfaces. The sloping ash beds are stretched out by an amount that depends on the amount of ejecta, and each layer has a minimum thickness where it is closest to the top of the underlying conical feature.

The mineral spring of Plaka is located close to sea level at the southwestern end of the metamorphic complex of Athinios. Its water has a temperature of 33 degrees; it is alkaline and very rich in chlorine but has little sulfur. It is a mixture of seawater that infiltrates through all sorts of sand and gravel and substantially modifies the temperature and composition of the spring water.

The Athinios massif can be followed along the coast for about 1700 meters. Its highest known elevation is 223 meters.

The small body at Cape Therma is 900 meters farther to the southwest. It is exposed for 360 meters along the coast and the highest elevation it reaches is no more than 55 meters. The metamorphic rocks have all the same characteristics as those of Cape Athinios, but carbonate rocks are proportionately less important, and the schists are more altered.[15] As we have noted above, they are cut by a mineralized vein. Close to this vein at an elevation of a few meters above sea level, a sulfurous alkaline spring has a temperature of 35 degrees. The foliation of the schists at this locality dips toward the southeast.[16]

The Southeastern Volcanic Region

This region is divided into two parts by the small carbonate ridge of Platanymos (Mt. Gavrillos); it extends from the chain of Mt. Profitis Ilias to the village of Akrotiri and the lower slopes that start a little before the village on the east and extend southward to the sea. It has an area of about 1500 hectares. The part east of Platanymos (Mt. Gavrillos) forms a low plain that is totally covered by pumiceous tuff. The only notable thing one sees there is the trace of an ancient shore line 30 to 40 meters above present sea level along the marble escarpments of Mt. Profitis Ilias. It is formed by a bed of sand and cemented gravel that can be followed at the same level for a distance of 150 meters.

The area west of Mt. Profitis Ilias and Platanymos (Mt. Gavrillos) has a more irregular topography. The surface slopes gently toward the southwest and south. Two steep-sided ravines have been cut into the upper pumice beginning near Pirgos and extending as far as the village of Emporion where they come out on the plain just mentioned. At several places they reveal beds of ash and lapilli beneath the pumice. A dozen or so ravines in the area near the southern coast are not as deep or steep-sided, so it is easier to walk along them.

The shore of the bay ends in an abrupt cliff, but the escarpment has little height. The lowest point, which is about 600 meters from the village of Akrotiri, has an elevation of 62.3 meters.

About 360 meters southwest of Cape Therma the cliff has a separate body of lava that forms a thick ledge at sea level. Beyond this enclave of lava and the metamorphic rocks of Athinios and Therma the cliff has only continuous beds of loose material that are extensions of those already noted on the other side of Athinios. The lower unit forms the most disturbed part of the section; its stratification is very irregular. The thickness and structure of its beds vary considerably from place to place, with local discordancies and small interbeds of scoria within the ash. But many of these irregularities are more apparent than real. In many instances they result from the coast intersecting the section along a very broken line with many indentations and projections, so that the section cuts the beds at different distances from the center of the bay and where the layers have different thicknesses.

It is worth noting that ash and small lapilli are proportionately more important in this zone than the coarser fragmental debris. One can conclude from this that this part of the island is farther from the main eruptive center than is, for example, the area around Alonaki, for wherever fragments of differing size are thrown out of a crater, the largest fragments go only a short distance while the lapilli are carried farther, and ash reaches even greater distances.

The yellowish lower pumice visible in the middle part of the cliff, the thin-bedded ash that rests on it, and the upper pumice have quite an uneven appearance and indicate that the eruptions that discharged them had a greater intensity.

It is mainly in this part of Thera that the upper pumiceous tuff is quarried for pozzolana. Large barges constructed especially for this purpose tie up at the foot of the escarpments and receive the material that is caused to slide down the slope. From there it is transported to ships anchored in the bay and destined for Alexandria, Port Saïd, and Trieste. The pumice is excavated in open quarries that are chosen for their lack of blocks of lava. The verticality of the cliff, its substantial height, and the deep water just offshore are also favorable factors.

The lava at the water's edge between Therma and Akrotiri is part of a flow composed of irregular blocks. These are compact, reddish lavas with rough surfaces. Under the hand lens they are seen to contain great numbers of feldspar crystals the longest of which reach 0.5 mm and are firmly embedded in the glassy groundmass. Without resorting to a magnetic separator, one could never hope to separate enough for a chemical analysis. The microscope shows that the feldspar is triclinic and the analysis shows its composition to be that of labradorite. The microscope also reveals microphenocrysts of augite, microlites of feldspar with parallel extinction, and iron oxides but no olivine. In short, this lava appears to be identical to certain of the recently erupted varieties.

The Akrotiri Region

This part of Thera begins 200 meters east of the village of Akrotiri and extends to the extreme western end of the island. Three subzones can be distinguished. The first, which we refer to as the zone of Akrotiri proper, includes the village of the same name, the dome of Loumaravi that overlooks it on the west, the escarpment of Balos on the north, and, toward the south, the hill of Arkangelos and its lower slope descending to the sea south of the village. The second corresponds to the western point of the island and the two promontories above Capes Akrotiri and Mavros, and the third includes the intermediate area between the two other subzones.

The rocky Akrotiri subzone has an area of about 250 hectares. The summit of Loumaravi has an elevation of 210.2 meters, while that of Mt. Arkangelos is 165.6 meters. The very steep cliff on the interior coast next to the bay has a height of about 95 meters; immediately west of the small harbor of Balos and a little farther west and above a large mass of trass[17] and altered trachyte it has a height of 162 meters. It is there that it reaches its maximum elevation. Other points in the same region have the follow elevations.

Akrotiri church house	96 m
Entrance to Fortress in Akrotiri	108 m
Mill to the west of the village	160 m
Chapel of St. Michael west of Loumaravi	142 m

The southern slope of this part of the island is grooved by large, deep ravines with a form that must be due entirely to denudation. It is likely that the surface waters have only widened and deepened the interval between the volcanic centers or various lava flows of different forms and mineralogical composition. This is especially true of the three deep valleys that descend from the village of Akrotiri and the dome of Loumaravi, while the ravine that goes around the west side of Mt. Arkangelos appears to have been cut entirely by water.

The corresponding part of the cliff on the side toward the bay has one of the most interesting sections. From the lowest point on the coast to the harbor of Balos, the deposits of ash, lapilli, scoriaceous conglomerates, and pumice discussed earlier continue through this section with only minor local variations. I will draw attention only to the appearance of black scoria and ash of the same color between the upper pumice and the underlying beds. But near the harbor of Balos the regularities of the layers suddenly disappear. The units forming the cliff rise from their original horizontal layering, and their inclination becomes even steeper than that of layers at lower levels. An ash bed composed of large numbers of fine layers forms the basal unit of the same system that we have already examined along the southeastern part of the cliff and dips at an angle of 27 degrees toward the east. It rests on a small, granular deposit of trachyte-like material with layers tilted in the opposite direction at an angle of 8 degrees. This deposit, which is only a few meters wide, is cut off a short distance to the west along with the units described earlier by a sizable mass of dark brown, blocky scoria that rises almost to the height of the cliff and accompanies the masses of lava at sea level on the western side of the harbor. The cabins that serve as shelters and storehouses for the fishermen of Akrotiri have been constructed in the shelter of these rocks.

The scoriaceous body is about 60 meters wide. It is bounded on the west by another mass of almost equal size composed of a whitish trachyte-like rock similar to that of the smaller body mentioned earlier but with more regular stratification. It has only a few divisions that are limited to its interior. The main mass is formed of material of only moderate strength that is strongly altered and has no visible stratification.

At sea level near the foot of the tuffaceous unit just mentioned there is a layer of grayish-white rock crossed by brown veins that are probably due to infiltrations of siliceous solutions. It seems at first that this layer extends toward the south below the cliff, but actually it is only a fragment of the lower, stratified rocks on each side. The body does not extend from the side toward the bay; ledges and dark-colored breccia are banked against it on this side. It is likely, therefore, that the eruption of the tuffaceous rock took place at the same place or at least very close to the present body.

West of the greenish-white body of Balos the beds of ash, lapilli, breccia, and pumice return to their normal appearance and stratification, but the ash layers in the lower part of the cliff are replaced by an unusual local variety of dark-colored breccia that seems to be related to the underlying dark scoria described earlier. This relation, together with the situation of the lavas of Balos near sea level, indicates that the products of this eruption were laid down before the loose, stratified material that makes up most of the southern part of Thera.

The units along this part of the coast rise gently toward the west; the ash beds rise under the lower breccia and, finally, about half a kilometer from Balos all the beds abut against a tuffaceous body that is much larger than the previous one, even though it has the same physical characteristics. This large mass is exposed for 800 meters along the coast; it has a visible height of 150 meters. Its various parts have very different appearances owing to differences in their original structure and to the more or less advanced alteration that it has undergone. The colors and textures are especially varied. They have every shade from pure white to yellow-green and gray, and in places they seem homogeneous, while in others they resemble breccia or sandy clay. One cannot fail to be impressed by its unique character.

The only question to be decided is whether this mass originated as a compact body of trachyte or as a jumble of incoherent blocks. The second hypothesis seems much more reasonable to me, for it is difficult to see how such a sizeable body could be completely transformed into argillaceous material. Such a change can be better explained if one assumes that there was an accumulation of moderate-sized blocks through the interior of which steam, gas, and reactive fluids were able to circulate.

The body of dark scoria at Balos was evidently emplaced in a fissure cutting the same tuffaceous white material that comes up through the trachytic tuff and separates it into two unequal parts.

Could the large body of a similar nature situated farther to the west be connected in some way with the other two? The answer to this question should certainly be affirmative if one considers that these bodies extend into all parts of this subzone and constitute almost all of the terrain. They are nothing but the salient points jutting out of a single mass that has been cut along its edge by the cliff.

The subzone around Akrotiri has a variety of amphibole-bearing andesites, most of which are highly silicified and either massive or vesicular. They differ widely in their crystallinity as do the white and yellow tuff that are the pyroclastic products of the same eruptions that produced the andesitic lavas.

Volcanic material of this kind greatly dominates the section in all of this part of Thera, but mafic rocks are not altogether absent. One finds lavas with labradorite and augite phenocrysts and microlites of albite closely resembling the modern lavas in mineral composition. Some even contain anorthite.

These mafic rocks are exposed below the village of Akrotiri and both the interior and exterior coasts. The flows are aligned along a single fissure that extends from the harbor at Balos and continues as far as the point of Mavrorachidi.

One of these mafic lavas that is exposed in the upper part of the cliff of Balos has an outward appearance unlike that of any other volcanic rocks of Santorini. It is dark black and scoriaceous with thin beds and a crazed, ropy surface. It is the sort of lava one would expect to see on a volcano that produces basaltic lavas of low viscosity. Some parts of the surface of this rock have a dusty red coating that owes its color to the effects of oxidation. Judging from the concentration of this dusty material on the projecting parts of the flow, I am led to think that it was caused by ash falling on the still-viscous surface. The lava is rough to the touch where it has this reddish deposit, but elsewhere it is a bright, shiny black. It contains very few crystals apart from a few glassy grains of feldspar that are visible under a hand lens.

The labradorite-bearing lava below the village of Akrotiri, particularly on the southern side of the Kastron (the ancient fortress) forms a small body of massive, black rocks that look like basalt. It is probably part of a flow that was erupted from a nearby source that was later buried beneath the pumice. The main eruptive vent that discharged this lava appears to have been close to a body of red scoria about 200 meters from the west side of the village of Akrotiri on the eastern flank of Loumaravi.

This eruption was probably related, not only to the ropey flow at the top of the cliff at Balos but also to the thick mass of brown scoria that ends toward the south at Cape Mavrorachidi and may possibly be the swollen toe of a flow that came from the vicinity of Akrotiri. But the mantle of amphibolitic andesite that covers the southern slope of Loumaravi seems to be unbroken between the village of Akrotiri and Cape Mavrorachidi; so I prefer to consider the body at the cape as a separate product of the same eruption that discharged the other mafic lavas just mentioned. If so, a long fissure may have broken the surface irregularly and discharged lavas at several places where the openings were most favorable.

Mt. Arkangelos, like the dome of Loumaravi, is composed of amphibole andesites, in either massive rocks or pumiceous tuff. The top is formed by a brown, andesitic breccia divided into small vertical prisms about 10 centimeters in diameter and 50 centimeters long. At first it looks like a thin ledge of trachyte with columnar structure, but on closer examination one sees that the prisms give a false impression; the vertical separations are due to grooving.

On the highest point of the hill the pumice is like that of the layer covering the rest of the island; it contains a large block, about a cubic meter in size, with phenocrysts of labradorite and augite and identical to the modern lavas.

The lower slopes around the base of Mt. Arkangelos turn toward the southeast and end at a remarkable high cape. The middle part of the cape is composed of rocks resembling trachyte and with an upper surface that seems to be composed of loose blocks, while the base is a compact rock with closely spaced prismatic joints. Where breccias composed of this same rock are exposed on the eastern side, they enclose a layer of pumiceous tuff, and on the west it is overlain by an enormous mass of white tuff that rises like a high wall running east-west. When the western side is seen from a boat its profile looks like a gigantic vertical column. This is why the local people call it The Obelisk. The breccia east of the main body of trachyte dips from there toward the east. The pumiceous tuff continues toward the west under layers of yellow ash and beds of fine scoria that are mainly from distant sources and cover the upper pumice.*

Offshore from the cape of the Obelisk several islets composed of scoriaceous, reddish-gray blocks form ledges dipping toward the coast. Where the same rocks are seen close to sea level at the base of the Obelisk, they have the same southerly dip.

It is in the region properly called Akrotiri that one finds fossiliferous outcrops that help establish the geological age of the Santorini pumiceous tuff. These fossil beds are three in number. The first is the most extensive; it can be seen on the western flank of Loumaravi on a surface at an elevation of be-

---

* See plate XXXVII at the end of this volume.

tween 156 and 213 meters. Most of the fossils there are broken fragments distributed in small beds of limited extent.

The second locality is in a belt around Mt. Arkangelos. The fossils at this locality are fairly well preserved, especially on the northern side of the hill.

The third is within the high part of a large mass of tuff visible in the cliff west of Balos. The fossil beds are inaccessible, but one can find plenty of debris at the foot of the cliff on a small terrace slightly above the water level. Most of the fossils embedded in the pumiceous tuff are in a good state of preservation and easy to extract. Sea urchins are common; they tend to be more broken and deformed than the shells of mollusks.

The following fossils, collected from the beds at Balos, have been identified by Mr. Munier Chalmas:

Cacharias sulcidens. Agass.
Ditrupa sp.
Turritella subangulata. Brocchi
Scalaria pseudoscalaris. Brocchi
Lucina borealis. Linné.
Thracia convexa. Wood.
Pecten opercularis.
" jacobœus (common in the archipelago)
" polymorphous
Ostrœa lamellosa (common)
" cochlear. Poli.
" sp.
Terebratula ampulla. Broc. (common)
Brissopsis lyrifera. Ag.
Schizaster canaliculatus (common)
Cidaris melitensis. Agas. (common but mostly broken)
Echinus sp.   "   "   "   "
Psammechinus sp.   "   "   "   "
Cidaris sp.   "   "   "   "

This fauna belongs to the late Pliocene and is identical to the one identified by Palerme.*

---

* When I visited Santorini, I did not intend to publish a list of fossils; this is why I am far from reporting all the species that I might have collected, particularly in 1867 when the cliff had a large landslide. The species that I found at the Arkangelos and Loumaravi are the same as those listed by the German scientists.

Before I found these fossils, Messrs. Reiss, von Fritsch, and Stübel had discovered the beds at Loumaravi and Mt. Arkangelos and had collected there the following species, which were identified by Mr. Mayer of Zurich.[18]

Schizaster minor. Mayer (common)
Terebratula vitrae. Linn.
" septata. Phil.
" euthyra. Phil.
" caput serpentis. L.
Ostrœa hippopus. Lamk.
Anomia patelliformis. E.
Pecten similis. Lamk. (common)
" septemradiatus. Müll. (common)
" pseudamusium. Chemn.
" varius. Penn.
Avicula sp.
Arca barbara. L. (fragment)
" pectunculiformis. Scac.
Nucula sulcata. Bronn.
Leda nitida. Brocchi.
Cardium edule. L. (fragment)
" roseum Lamk.
Lucina astensis Bronn.
" spinifera. Montf.
Venus, cyth. or circ. sp.
Venus
Venus Gallina. L. (fragment)
Cardita.
Corbula.
Dentalium tetragonum. Brocc.
" dani. Hark.
Turbo sanguineus. L.
Rissoa (imprint)
Assiminea littorea. Delle Ch.
Vermetus glomeratus. L.
Mytitus galloprovincialis Lamk.
Cluster of foraminifera: ovulina, glandulina, nonionina, rotalia, spiriloculina, triloculina, and quinqueloculina
Bryozoan: ceriopora, eschara.

Mr. von Fritsch found the remains of fucoïdes in the large tuffaceous body west of Balos.

The fossils in this second list belong to the same fauna as those of the preceding one, but they are represented by other species. I believe, however, that the difference is not great enough to allow one to judge from this evidence alone. Among the damaged and fragmentary specimens I found at

Balos and left in place, there are a number of species that are also in the Loumaravi deposits and even more of those in the one at Mt. Arkangelos. I also noted that several of these fossils are identical to those of Milo.

The small subzone of Capes Mavro and Akrotiri includes the southwestern point of Thera beginning at the ravine that cuts the cliff east of Cape Akrotiri and ending on the eastern flank of the small headland at Cape Mavro. Its surface area is about 30 hectares. It has two high points that correspond to eruptions of different ages and types of rock. The elevation of the top of Cape Akrotiri is 126.8 meters, that of Cape Mavro 114.2 meters. The depression between these two points has an elevation of 90 meters.*

The rocks of Cape Akrotiri are identical to those of Loumaravi and Mt. Arkangelos. They are white tuff, gray andesites, and liparites with pale tints of rose, yellow, and green.

The major part of the cape is formed by tuff. On the inner side of the bay they can be seen in a series of superimposed units. At the point of the cape they dip toward the west at an angle of 20 degrees and are mixed with rounded blocks, but as one follows these cliffs farther east, they become finer grained and their dip becomes shallower until, farther to the east where the beds are covered by younger deposits, the dip is in the opposite direction. Where seen at the edge of the bay in the cliff facing the islet of Aspronisi they are more rounded and the beds on each side dip toward the center of the mass. The same cape is composed of a rock that looks like trachyte and is exposed at the highest point of Cape Akrotiri, where it turns toward the northwest in the form of a flow ending at the point of the cape. On cooling, this rock formed crudely prismatic contraction joints. On the southwestern face of the cape the andesitic lava of Cape Akrotiri is still seen on the flank of the body of pumiceous tuff close to sea level. This rocky mass may be a thick, irregular dike oriented about N20°W, which could be the course it followed to reach its vent near the ridge crest. It could also be a simple ramification of the large flow just mentioned. This rock is too deeply buried in the mass of tuff for this problem to be resolved with certainty.

At the surface of the cliff bordering the west side of Cape Akrotiri, erosion by the sea has triggered occasional landslides. The lower units have been carried away and broken up, and the upper ones have slumped. Large masses detached from the surrounding rocks have dropped to lower levels, so that the layers in debris and in the facing cliff are at different levels.

* See plate XXXV at the end of this volume.

The promontory of Cape Mavro is formed by an agglomeration of blocks of differing sizes, mostly rounded and embedded in a cement of the same composition. Most of the blocks are fist-size, but a few reach diameters of a meter or so, while others are much smaller. Almost everywhere, a certain amount of the weakest part of the matrix has been removed from the space between the blocks, so that the latter project out from the surface of the outcrop. In addition, the sea has worn away and undermined the base of the promontory, leaving the upper parts like vaults arching out from the cliff. Passing close to the shore in a boat one cannot avoid being fearful, for on looking up one sees a cornice 40 or 50 meters overhead projecting out for several meters. This overhanging ledge is all the more intimidating because of the obvious inhomogeneity of the rocks. Black rocks hanging from the lower surface like pendants from a ceiling seem on the verge of falling. The cement holding them in place resists weathering better than that near the base, evidently because the waves do not reach that height and the siliceous cement is more resistant.

The promontory of Mavro is behind and to the south of a thin dike. It seems to correspond to a fissure through which the lavas forming the cape rose to the surface. The dike is a meter thick and oriented N25°N. It is composed of a black central mass divided into small, transverse prisms about 20 centimeters in diameter. It has two dense but altered selvages about 25 centimeters thick. It rises like a perfectly regular wall in the middle of the conglomerate and where the waves have worn away the rocks it stands isolated and exposed on both sides.

Between Cape Mavro and Cape Akrotiri alternating layers of ash, pumice, and conglomerates turn up on both sides of the points, particularly on the coast next to the promontory of Akrotiri. The way the beds are turned up and terminate sharply in this area indicates that the massif of Akrotiri was formed earlier than the fragmental rocks that border it. The same relations are less evident along the coast at Mavro. Although there is actually something analogous to what is seen at Cape Akrotiri, the effect is too poorly developed to be attributed to distortion of the older beds by the eruption responsible for the large amount of lava seen here. It is also possible that the stratified upper units correspond to the eruption at Cape Mavro; this would explain their upwarping on both sides, while the lower beds are certainly formed by the same units as those at Akrotiri.

The subzone of Thera between the dome of Akrotiri and the small zone at the cape of the same name is uniformly covered by a blanket of pumiceous tuff. Its irregular surface has an area of about 460 hectares and an average elevation of about 80 meters. Up to the point where the coast facing the bay changes direction from east-west to northeast-southwest the ground slopes rather evenly toward the south, but beyond that point a ridge extends to the promontory of Cape Akrotiri, and the ground slopes away on both sides. The slope toward the interior of the gulf is pronounced enough near Cape Akrotiri to form a kind of valley ending at the sea and bordered on the southwest by the andesitic lava flows mentioned earlier. The slope is gentle toward the south. Ravines that cut the surface are generally straight and not very deep or numerous. A single wide, deeply incised ravine borders the western side of Mt. Arkangelos. On the southern side as far as the mouth of the ravine it reveals a series of pumiceous tuffs, ash, and pumiceous lapilli that differ in color and grain size. But the sequence of the main units is the same as that in the cliff east of Balos. The upper pumice covers the highest parts; the lower part is a zone of thin-bedded ash resting on the lower pumice and, below that are beds of ash and various volcanic breccias. A lens of black scoria between the upper pumice and the fine-bedded ash can be followed as far as the vicinity of Cape Akrotiri. We should also note that the partings of the fine ash layers become less marked toward the west. On that side the bed takes on a uniform grayish tint and varies notably in thickness. It is well developed between the large tuffaceous mass of Balos and the first cape toward the west facing the reef shown at a depth of four fathoms on the English map. It then thins considerably and disappears entirely not far from Cape Akrotiri. The same is true of the lower pumiceous layer on which it rests. The latter becomes thinner and loses its distinctive white color. Varicolored ash and breccias that occupy the lower part of the cliff dip slightly toward the west between the large tuffaceous body of Balos and the precipitous cliff. They turn up sharply there and bank against a mass of reddish brown scoria.

There was evidently an important eruptive vent at this place, and although it was subaerial, steam and acidic gases were quite active agents. The abundance of scoria making up this body and its more or less surficial red color similar to that commonly seen in modern subaerial volcanoes support this interpretation. The center of the body has no regular form, but the same is not true of the flanks; on the western side especially, the lapilli are stratified and form beds that terminate abruptly upward.

The surface of this region is covered by the same pumiceous tuff, but the underlying beds are exposed, at least in part, at the southern shore between Cape Mavro and the Obelisk. The low cliff along this coast is bordered by a small beach of rounded pumice and various kinds of water worn volcanic pebbles. At the base of the escarpment, two small conical hills are composed of a gray, trachyte-like rock with prismatic jointing.* A few meters offshore, two small islets of the same rock are certainly related to the same units.

These rocks are identical to those offshore from the Obelisk; both have patchy incrustations of quartz, calcite, and zeolites. Superficially, they look rather different from the lava of Mavro, but a microscopic examination and chemical analysis show that, where they are not too badly altered, they are essentially identical. They are overlain by beds of ash and pumiceous scoria that can be seen to the east and west. One of the beds, composed mainly of granular ash, is remarkable for its great number of rounded, fist-size inclusions. These are mostly gray with spots of green and have a friable granitic texture. They are obviously fragments of older rocks belonging to the basement series under Santorini. They were picked up and erupted in the same explosions that produced the ash in which they are found. They can be collected very easily, not only at this place but in an ash bed about half way up the path on the cliff at Balos on the opposite coast of Thera. Similar inclusions can be collected from ash beds in the lower part of the eastern cliff of Therasia.

THERASIA

The island of Therasia has the shape of a parallelogram with its northern and eastern sides indented by large, deep embayments. The average north-south dimension is about 5 kilometers and the width about 2 kilometers. The total surface area is approximately 9.5 square kilometers. The surface slopes westward from high cliffs on the side toward the bay at an average angle of about 6 degrees. The eastern coast is cut by the gulf of Manola

---

* See plate XXXIV at the end of this volume.

where a straight, pebbly beach extends along the base of the cliffs. At the inner side of this large inlet a zigzag path ascends to the village of Manola at the crest. Everywhere else the eastern cliff is inaccessible except for a narrow place at the foot of the escarpment below the village of Kera. Every day, the people of the village climb up and down this dangerous cliff, often carrying heavy burdens. The rim has two culminating points, one to the north and the other south of the village of Manola. The northern one has an elevation of 292 meters, the other 275.4 meters. The lowest part of the cliff at the inner side of Therasia near the village of Manola has an elevation of 160 meters.

The northeastern coast has two reentrants. The one extending between Cape Simandiri and Cape Tino is bordered by high escarpments, while the other, situated between Cape Tino and Cape Riva, can be reached along its middle part by a beach from which the ground rises gently toward the interior of the island.

The western coast has no prominent inlets but is bordered by small low cliffs cut by the outlets of ravines.

On the southwest the crest of the abrupt, broken shoreline rises progressively from Cape Kiminon to the point overlooking Cape Tripiti. The surface of Therasia, like that of Thera, is blanketed with pumice except where the topography is too irregular to retain the unstable deposits. There are also places where the pumice has been removed by surface waters. This is seen mainly where ravines coming down from the high points of the island expose the underlying layers of lava. The two large ravines sheltering the villages of Potamos and Agrilia are also cut partly into the pumice. Most of the dwellings of these two villages are little more than caves that have been cut into this layer.

Along the southwestern cliff, the lava flows can be followed for considerable distances and can be seen descending from east to west at an angle of no more than 5 degrees. These flows have quite varied forms with swells of substantial thicknesses. In some places, these seem to be due to the lava filling depressions in the underlying surface; in others the lava, instead of spreading evenly, seems to have been so viscous that it piled up in bulges 30 meters thick. Some of these thick bodies can be seen at sea level leading one to believe they had local sources, but in most places the flows can be traced continuously to the east and west, so it is likely that they are simply thick tongues of lava flows coming from the central volcano. Beds of reddish scoria and brown ash are interlayered between the denser units. Among the varieties of lava seen on this coast, some of the banded rose-colored rocks form long, continuous layers, but most consist of rounded, platy blocks. The lava flows are like this as far as point Kiminon, where they form the crest of a reef just offshore.

The cliff on the eastern side of the island is composed almost entirely of lavas of very unequal thickness that seem at first glance to be horizontally continuous, but as with the lavas in the cliffs of Thera, their continuity is only apparent. Each layer is actually composed of narrow interfingering or superimposed bands that produce a false stratification. The face of the cliff shows cross-sections of narrow flows, some thin and others more substantial, which are superimposed on each other in the spaces between older flows. Nevertheless, in a part of the cliff near Cape Tripiti the lavas are exposed in longer sections, probably because in this place they are not cut exactly crosswise. The lavas coming from the northeast must have been cut obliquely by the cliff face.

A particularly noteworthy feature of the lavas of Therasia is the frequency of rounded blocks ending in bulges with a concentric structure. Several of these can be seen in the cliff face or where they have been detached and rolled to the foot of the escarpment. Some seem to come from the toes of flows; others appear to be debris from the margins of flows and have the form of two bulges joined by a flat section.

In some places there is nothing between the lavas in the cliff but a bed of scoriaceous detritus a few centimeters thick; in others one sees thick, irregular deposits of scoria or dark lapilli that result mainly from disaggregation of the upper surface of the flows and from displacement of the rough blocks on its surface. It is possible that fragmental ejecta from the volcano contributed some of this material, but, if so, such an origin must have been of quite subordinate importance.

The lower part of the cliff exposes two layers that could have been produced by powerful explosions. The higher of the two is a bed of scoria and red lapilli that can be traced along the cliff from Cape Tripiti to the middle of the shore bordering the bay between Cape Tino and Cape Riva.[19] At the southern tip of the island, this conglomerate rests on a few small flows of lava close to sea level and makes up almost all of the point of Cape Tripiti. It is seen somewhat higher in the face of the cliff and ends along the coast below two high points of the island. At this place it is about 30 meters thick and rests on lavas at an average elevation of 60 meters. One can follow it all around the gulf of Manola. In the innermost part of this gulf it loses elevation but soon rises again rapidly, and at the point south of Cape Simandiri it develops into an enormous mass that rises as high as 180 meters in the upper part of the cliff. In the face of Cape Simandiri, the conglomerate rests on a great mass of locally derived lavas, so that one can no longer make out the form of a large band that is situated at an upper level and in large part masked by talus. Beyond the cape, the conglomerate takes on its normal character and returns to a lower level, but its thickness and, especially its importance relative to the other strata making up the island continues to increase. At Cape Tino, the height of the cliff is about 130 meters and that of the conglomerate 60 meters. At the same time, the conglomerate separates into distinct zones. In its upper and lower parts there are two layers of dark brown ash and lapilli. The upper one is separated from the conglomerate by a thin layer of lava. Then on the other side of the cape facing the gulf of Millo, the conglomerate, still very thick, dips westward along with the lapilli beds that accompany it and disappears below the water in the southwestern corner of the gulf. The upper bed of lapilli, which is much thicker here, then disappears a little farther on. The red conglomerate reaches its maximum thickness between Cape Tino and Cape Simandiri. The rock is in all respects identical to that which forms the prominent mass of scoria on the opposite shore of Thera.

Another unit formed by volcanic ejecta is found at the base of the eastern cliff of Therasia. It is a bed of light yellow ash overlain by thin lavas that separate it from the red conglomerate just described. This layer does not appear at Cape Tripiti; it probably exists there but is below sea level. It only appears a kilometer and a half toward the north below the village of Kera near the point corresponding to the southernmost of the twin peaks of Therasia. It is slightly undulating and rises somewhat in front of the northern peak while remaining at a very slight height above sea level. Lavas can be seen below it in this area. The layer of yellow ash, which in this part of the cliff has an average thickness of tens of meters, contains local zones of lava, lapilli, and conglomerate. In certain places it takes on a banded appearance and separates into yellow and black beds. To the south at the entrance to the gulf of Manola it rests on lavas at an elevation of about 15 meters. Along the inner part of the gulf, it descends until it no longer comes above the surface of the sea. In this part of the cliff, it has beds of scoria in its upper part as well as at its base. On the northern side of the gulf it rises again to an elevation of about 30 meters near the southern point of Cape Simandiri. At the same time its normal yellow color changes to bright-red indicating that it has been locally modified after deposition, probably as a result of reheating. Beyond the cape it returns to its normal character and remains at an elevation of 30 meters almost as far as Cape Simandiri, where it suddenly drops below the water level again. It is in this region that the bed is thickest and has such a quantity of scoriaceous blocks that there can be no doubt that the vent from which it came was situated in this part of the bay of Santorini. Near the middle of Cape Simandiri it is separated from the red conglomerate by a series of lavas coming from a nearby dike that will be described later. The emplacement of this dike and eruption of lavas that came from it took place after deposition of the beds of yellow ash and before that of the red conglomerate.

The cliff of Therasia exposes four dikes, two at Cape Tripiti, a third below the village of Kera, and a fourth on the eastern side of Cape Simandiri about 100 meters from its southern point.

The first dike, which is at the very end of the cape, cuts a lava at sea level and a small bit of red conglomerate that rests on it. It bifurcates. The main branch measures a meter across, while the other, which is only 30 centimeters thick, veers a meter or so toward the north. The space between the two branches is filled by the red conglomerate. The dike is black and badly weathered by the sea and atmosphere.

The second dike is about a dozen meters from the northern coast. Like the first, it cuts the lower lava and red conglomerate. It is 4 meters thick and has two glassy black selvages about 15 centimeters thick. The rock making up its interior is black but weathers yellow on the surface.

The third dike is in the innermost part of a small cove close to the place where the path to the village of Kera starts up the rocky slope. It is 8 meters thick and rises vertically to a height of 38 meters through the lava, ash bed, and red conglomerate. Its top is in a thick mass of yellowish rock and seems to be the source of nearby lavas.

The last dike at Cape Simandiri rises to a height of about 60 meters, but I believe it continues farther into the cliff and produced two interlayered lavas between the yellow ash and red conglomerate. It is less than a meter thick and has two very distinct selvages that in most places are separated from the main part of the dike by a gap that has resulted in the central part becoming detached and leaving an open space between to two intact selvages. The rock is black and massive.

All four of the dikes on Therasia trend toward the northeast - southwest.

All the lavas of Therasia have phenocrysts of labradorite and microlites of albite-oligoclase. The only exception I have found is the dike at Kera and a few lavas in the lower parts of the cliff. These uncommon rocks contain anorthite. The dike at Kera is one of the rare products of Therasia in which one finds olivine. Iron oxides are completely lacking as phenocrysts but form abundant octahedra, 0.005 millimeters in diameter, in the groundmass. Tridymite is much less common than it is on Thera. I have found it in notable amounts only in the gray platy lava forming the flows in the upper part of the inner cliff of the gulf of Manola.

Elsewhere on Therasia one finds a variety of labradorite-bearing lavas similar to those of Thera. Some, such as the ones at Cape Tino, have a proportionately large amount of dense, light-colored glassy groundmass. In others, such as the lava in the inner part of the gulf of Manola, crystals make up almost the entire rock. Some of these rocks are rich in pyroxene; others have very little. The varieties with many vesicles have a dark color, whereas the denser ones tend to have a lighter color.

No submarine deposits of the kind seen on Thera are found on Therasia. Carbonate blocks in the upper pumice are accidental inclusions that come from the basin of brackish water that will be explained in a later section.

ASPRONISI

This island is 680 meters long and 250 meters wide; its area is about 2.5 hectares, and its maximum elevation 71 meters. The coast line is almost vertical, and the flat surface of the interior is slightly inclined toward the southeast. The rocks are mainly ash beds, scoria, and lava overlain by the upper pumice. The lower units rise toward both the west and east but especially in the latter direction with the result that in the middle of the island they form a depression, the axis of which is oriented north-northeast. This form is less pronounced in the beds at higher elevation and cannot be seen in the configuration of the upper pumice layer.

The lower units, which consist of brown and dark gray ash similar to that on the opposite coast of Akrotiri, have well-developed traces of cross-stratification. On the southern coast of the island Mr. Reiss collected from this same deposit the remains of unidentifiable branches stripped of their leaves. One can also find there a layer very rich in pisolites similar to those in the ash on the path going up to Balos. Above this are layers of lapilli and scoria, and blocky, contorted lavas like those above the route up the cliff at Balos. Beneath the pumiceous tuff a layer of lava that spread as a very fluid flow has some of the features typical of trachyte. A few samples consist entirely of brown glass that is almost opaque, even in thin flakes. This glass is crowded with round cavities, about a millimeter in diameter, filled with transparent glassy material similar to the surrounding material and with a brown spherical nodule at its center. Rare microlites are detectable only in polarized light. In other samples, however, feldspar, very fine-grained iron oxides, and pyroxene have grown to visible sizes and in a few specimens are found as phenocrysts. The feldspar phenocrysts, which are not affected by acid, have the properties of labradorite, and the microlites have parallel extinction. In short, the rock has the mineralogical composition typical of the labradorite-bearing lavas of Santorini.

The pumice bed that covers Aspronisi is about 25 meters thick.

The friable nature of the ash beds, along with imprints of plants that are found in these beds, demonstrates their subaerial origin. The concave form of the basal unit also supports this conclusion, for loose material such as this would have been quickly leveled by waves if it had been deposited in water. Finally, the presence of rounded detrital blocks in the scoriaceous conglomerate shows that there was communication with the former mountainous part of the island.

THE UPPER PUMICE LAYER

The layer of pumice that covers Thera, Therasia, and Aspronisi is one of the most important geological elements of the archipelago. It reaches a maximum thickness of 30 meters in the southwestern cliffs of Thera and, on the other side, in the cliff of Akrotiri near Balos. It is composed mainly of fragments with a wide range of sizes. Most are less than fist-size, and many seem to have been ejected as fine dust. Almost all the interstices between fragments are free of foreign material, and there is little cement to contribute to its coherence; instead it is held together by compaction under the weight of overlying layers. It is a dry deposit modified by rain water that has filtered down through it over the centuries. It has marks of regular stratification where fine-grained beds alternate with others containing large blocks. This is seen particularly well in the large escarpments on the southwestern coast of Therasia. The lines between these beds are essentially horizontal, but they are not distinct enough to suggest that the layers came from eruptions separated by any notable lapse of time. Several are marked by a pale yellowish tint without any change of grain size across the boundary. It is probable, therefore, that the pumice beds correspond simply to successive pulses of a single eruptive event that, like most modern ones, may have lasted several years and had periods of relative calm alternating with stronger activity.[20]

The upper pumice layer has a great many inclusions of black labradorite basalt. These blocks are found not only in the middle of the layer but even where the pumice has been stripped away by water. On Megalo Vouno, Kokkino Vouno, Mt. Mikro Profitis Ilias, the two peaks of Therasia, Loumaravi, Mt. Arkangelos, and even at the entrance to the monastery on the peak of Mt. Profitis Ilias one finds these blocks, many of which have large dimensions. One on Loumaravi has a volume of several cubic meters. While some are porous and scoriaceous, most are a compact, glassy black lava like that of the modern eruptions. They represent all possible varieties of the labradorite-bearing lavas of Santorini. A certain number of these blocks that are still in their original positions in pumice that shows no sign of reworking are worn and rounded as if they had been transported by swift currents or rolled by waves. Neither of these two explanations can account for these unusual features.[21]

Among the foreign material found in the upper pumice, blocks of labradorite basalt are by far the most common, but they are always subordinate to the pumiceous material in which they are embedded. The pumice is quarried on a large scale at Santorini; ships of substantial tonnage are continually transporting it to Egypt, Constantinople, Pireas, Trieste, and even

Marseille. The pumice is mined in open quarries, where it is loosened with picks, pushed from the heights of the cliffs, and allowed to slide down into barges at the base of the escarpment. The barges transport it to ships anchored offshore or in a cove of the Kameni islands. On Thera, pumice is taken only from near Balos at the highest part of the cliff where very large quarries are now being worked. The same method is used at two places on Therasia - along the southwestern coast and at the northeastern end near Cape Tino - but only the workings in the southwestern region have been profitable; those near Cape Tino seem to be almost abandoned. Various conditions have determined where the quarries have been developed. The most favorable sites are at the edge of a cliff where the pumice is very thick. In addition, the pumice should be as free as possible of blocks of lava. And, finally, an important factor is the ease with which the pumice slides down to the shore. The preferred parts of the escarpment are those that have a minimum height and a vertical drop-off below the shoreline that permits boats to come close to the shore.

The upper pumice bed of Santorini contains not only labradorite basalt but also fragments, normally rounded, of all the rocks that make up the islands of Thera and Therasia. Fragments of anorthite-bearing lavas have a range of dimensions, but few are larger than fist-size. These rocks are very crystalline and identical to those found as inclusions in the lavas of the most recent eruption. They are rich in pyroxene, olivine, and iron oxides. Fragments of marble are very rare, and the same is true of schist. Few are found outside the area of Messaria and Mesa Vouno, and they are much less common than they are in ash layers in the lower part of the cliff. The rarity of metamorphic rocks in the pumice demonstrates that the eruption came up through a section composed mainly of older volcanic rocks. The subsurface system of Mt. Profitis Ilias either did not extend as far as the vent of the volcano or more likely, had already been carried away by earlier eruptions.

Fragments of diabase collected from the pumice at Messaria show that the phyllites of Santorini were probably cut by dikes of this rock. Also found are crystalline rocks similar to those that are common in the lower ash at Balos and on the southern slope of Thera; their physical properties and mineralogical compositions have been described in detail.

Inclusions consisting of crystalline aggregates of anorthite, fassaite, augite, wollastonite, and garnet are identical to those in the lavas of 1865.

Finally, one finds unusual blocks of dark gray, porous carbonates that look pisolitic and have cavities filled with aragonite. They have a concentric structure. Von Fritsch, who was the first to report them, has noted that they contain two types of fossils: *Bythinia ulvoe* (Penn. sp. turbo) and *Cerithium conicum* (Blainv.), *mamillatum* (Risso). They are found mainly on Therasia and around Apano Meria and Foinikia on Thera, but a few have also been collected at Merovigli, Vourvoulo, Thira, and Messaria, all in the southern part of Thera. Some of the nodules that I found near Apano Meria have a volume of several cubic decimeters and seem to be formed of a large block of aragonite enveloped in a thin fossiliferous layer of gray carbonate. These samples evidently come from a deposit that was largely destroyed by the explosion that produced the pumice. The two fossils are modern forms, and *Cerithium ulvoe* is a brackish-water species. These carbonate rocks cannot be related to the marine deposits of Akrotiri, and nothing analogous has been found in the cliffs around the bay. The most reasonable explanation is that the carbonates were formed in some sort of depression in the ancestral mountain, and, because the debris is found mainly on the coast of Apano Meria and in the southern part of Therasia, one must conclude that the carbonate deposit they came from was formed by sedimentation on a tidal pond on this coast under conditions of restricted communication with the sea.[22]

The pumice of Santorini, when seen by the naked eye, or even under a hand lens, seems to lack crystals. It is silky, fibrous, and very vesicular. In thin section it is seen to have clear crystals of monoclinic and triclinic feldspar, augite, iron oxides, and hypersthene. These crystals are so scarce that they are seen only in random samples. To study them, one must treat the pumice with hydrofluoric acid; in this way very pure crystals can be obtained.

The monoclinic feldspar seems to be more abundant than the triclinic one, for if one examines under crossed nicols feldspars extracted from the rock by hydrofluoric acid and gives special attention to lamellae that correspond to the face g_1, one sees that the extinction direction of almost all grains is essentially parallel to the pg_1 direction. From that one can conclude that most of the feldspar grains are probably monoclinic, but it is possible that a certain number are oligoclase, for the extinction direction on the g_1 face of that mineral is also parallel to pg_1.

The color of the pyroxene is pear-green and noticeably pleochroic, but it has the shape, extinction angle, twinning, and cleavage of augite.

Small brown prisms of hypersthene measure 5 to 7 millimeters in length and 1 to 3 in width. The best-developed lateral faces are g_1 and h_1, then m, and in some samples g_3, g_2, and h_2. The terminations are formed by a great number of facets belonging to the usual forms.

On all the lateral faces of the prism, dichroism is very pronounced and extinction is parallel to the edges.

The glass of the Santorini pumice has the following composition:

Silica	71.0
Titania	0.5
Alumina	16.8
Fe_2O_3	0.8
Lime	0.8
Magnesia	0.7
Soda	7.4
Potash	2.0
	100.0

If the Santorini pumice is crushed and treated with concentrated hydrofluoric acid and the action of the acid is interrupted at the appropriate time, one obtains a mixture of all the mineral constituents of the rock. 100 grams of pumice yields an average of 3 grams of crystalline residue.

The crystals obtained in this way are perfectly intact. Their edges are straight, their faces are clean, and there is no trace of alteration. Among the minerals found in great numbers are small prisms of brown, transparent hypersthene.

Turning one of these crystals in crossed polarized light so that the axis of rotation corresponds precisely to the longitudinal edge, one can see that on each face the extinction is perfectly parallel to the axis of elongation. In this way one can verify in plain light that all the faces are equally dichroic.

This mineral is therefore truly orthorhombic and not simply augite viewed on one of its faces in the zone ph^1, as one might think if the mineral were seen only in thin section.

The composition of this hypersthene is as follows:

Silica	49.8
Alumina	2.3
Fe_2O_3	0.8
FeO	25.0
Lime	10.8
Magnesia	11.2
Soda	0.5
	100.4
Specific gravity	3.485

This hypersthene is found in the pumice with a true pyroxene[21] that has a green color and little dichroism. It is also accompanied by iron oxides and feldspars.

The grains of feldspar extracted with hydrofluoric acid are a mixture of labradorite and a feldspar that is richer in silica. When seen on a g_1 face, the latter has extinction parallel to the edge pg_1. This feldspar contains very little potash and is probably oligoclase.

The mixture of feldspars had the following composition:

Silica	57.9	30.9	
Alumina	26.7	12.3	
Fe_2O_3	0.6		
Lime	8.6	2.4	
Magnesia	0.9	0.4	4.1
Soda	4.8	1.2	
Potash	0.7	0.1	
	100.2		

Specific gravity 2.612   Ratio of quantities of oxygen:
Si : Al : R = 7.5 : 3 : 1

CHAPTER SEVEN

PETROGRAPHIC STUDY OF THE DIKES IN THE NORTHERN PART OF THERA

The restricted dimensions and regular form of the Santorini archipelago indicate that it represents the remains of a single island, and the proximity and continuity of lavas of various compositions found there make it unlikely that the island was the product of several different underground sources. It was a single volcano that has remained active from the beginning of the Quaternary period down to the present day. At the same time, however, the intensity of eruptions, the duration of periods of repose, and the locations of the vents have varied from one century to the next. The chemical and mineralogical composition of the erupted lavas during each period of activity had distinct characteristics.[1] Some of the more conspicuous differences have been pointed out in the preceding chapter. The eruptive material seen in the cliffs enclosing the bay differs in important ways from one end of the island of Thera to the other. Toward the north, basic and intermediate compositions are most abundant, whereas to the south, felsic lavas and siliceous tuffs are the dominant components of every section.

In this chapter, I am concerned with a group of rocks in the north and northeast that have the greatest affinity to the products of the last eruption. These rocks are so wide-spread and complex that they merit special consideration. They are exposed in 66 dikes that are well exposed in the cliffs. Most of these dikes are vertical or only slightly inclined. They attain thicknesses of up to 10 meters and may be as thin as 20 centimeters, but most are between one and two meters thick. They rise to various heights. Some stop at a relatively low level where they seem to have been the feeders of the lavas in the lower part of the cliffs; others can be followed to heights of 200 to 300 meters near the top of the escarpment bordering the bay and can be traced into the lava flows that erupted at that level. Where they are in groups, it is possible to determine their relative ages. Their trend is between northwest and northeast, with the great majority striking either N30°W or N30°E.

The purpose of my studies has been to determine the orientation, sizes, and mutual relations of the 66 dikes and their relationship to their wall rocks. The results of these observations are given, both in the following detailed descriptions and in plates XXXI through XXXVI at the end of this volume. The latter illustrate the stratigraphic sections in the northern and northeastern cliffs between Apano Meria and the foot of the escarpment of Merovigli.

The mineralogical study of the rocks in these dikes was carried out mainly by means of the microscope. A thin section of each was examined very thoroughly at several magnifications in plain light and under crossed polarizers. These detailed studies were started in 1866, and since that time I have continued to pursue them, even though other work has occasionally intervened.

I have also used the microscope to great advantage in examining the residue left after each sample had been pulverized and treated with an electromagnet. When crushed, all of the dike rocks of Thera can be separated into their components according to the differing iron contents of the minerals. In the case of the lavas of Santorini, the amorphous material, even though it contains substantial amounts of feldspar microlites, is strongly attracted to the electromagnet. The magnetic properties of this material are probably due mainly to the numerous submicroscopic grains of iron oxides in the glass, but it must also be a result of the small isotropic globules of brown and violet glass that tend to be abundant in the clear glassy matrix that makes up the amorphous groundmass. Even where microscopic grains of iron oxide seem to be relatively rare, the magnetic attraction is quite pronounced when these globules (the diameters of which only rarely exceed .003 millimeters) are strongly colored and sufficiently abundant. In treating 100 grams of material, for example, one quickly extracts from the rock the iron oxides and almost all the amorphous material as well as the crystalline microlites it contains. All that is left are a few grams of grayish residue made up of crystals that are less strongly attracted to the magnet. When this latter fraction is treated again with the electromagnet, certain parts of the mixture can be moved while other crystals are completely inert.

In this way, augite, olivine, and hypersthene can be separated from the feldspar. If the rock contains hornblende or magnesian mica, these minerals go along with the other ferromagnesian minerals; if it contains calcium carbonate, tridymite, or zeolites, these remain with the feldspar. It is worth noting that olivine is entrained with pyroxene, even when it is much less iron-rich and more weakly colored. It might seem that one could use this procedure to effect a complete, quantitative separation of the mineral components of the rock. Unfortunately, this is not the case. The procedure is not that effective; even when the rock contains only amorphous material, iron oxides, pyroxene, olivine, and feldspar, the separation cannot be carried out perfectly. Among the small fragments produced by pulverizing the rock, many contain two or more minerals. It turns out, for example, that a fragment composed partly of amorphous material and partly of feldspar may be separated along with the first components picked up by the electromagnet, or it may be left with the nonmagnetic residue, depending on the proportions of iron-rich and iron-poor material in the individual fragment. One must realize, therefore, that a magnetic separation does not yield representative fractions of all components; it gives only an approximate idea of the upper or lower limits of the proportions of the minerals, and results vary from one rock to another.

On the other hand, when combined with the use of the microscope, it provides the best means of:

1. carrying out *qualitative* mineral analyses of a rock and
2. preparing purified material for chemical analyses of different minerals that can be extracted in this way.

The technique is a great help in qualitative mineralogical analyses, because it often enables one to detect the presence of rare accidental minerals. Suppose, for example, a sample of rock the size of one's fist contains only a single crystal of quartz a tenth of a millimeter in diameter. Such a crystal is almost certain to escape notice if the rock is examined under a hand lens, even if the sample is broken into numerous pieces. The same would also be true if, instead of being broken, the rock were cut to make several thin sections. It is quite unlikely that a random slice would intersect the crystal. If, however, the sample is pulverized and the resulting powder treated by the magnetic procedure, it is almost certain that one or more fragments of the crystal would be found in the final non-magnetic residue, which at most would amount to a few grams. When this residue is examined microscopically under crossed nicols the mineral would be immediately apparent.

In the particular case we are concerned with here, use of the electromagnet has enabled me to recognize the presence of olivine in rocks that contain only traces of this mineral. Thus, I have extracted a few grains that are perfectly characteristic of this mineral from a lava of Mikro Profitis Ilias, when a dozen thin sections of this rock provided no clue whatever of its presence. Several other dikes on Thera have yielded similar results. Had I not used this procedure, I would have reported incorrectly that this mineral was not present in any of the rocks that I had studied. The same could be said of tridymite.

Examination of feldspars separated by means of the electromagnet has made it possible to demonstrate quite easily that almost all the crystals obtained in this way are labradorite. The crystals tend to be flattened parallel to their plane of symmetry, so that, under the microscope, they are seen to be aligned parallel to their g_1 faces. Feldspars viewed on this face have extinction angles and other optical characteristics that vary with their composition, enabling one to determine the species to which the feldspar must be assigned.

The magnetic separator is a very useful tool for preparing the various mineral components of a rock for study. Chemical analyses would be quite impossible without this procedure. Use of the electromagnet has made it possible to extract from any of the Santorini lavas in less than half an hour as many feldspar crystals as one could separate in a month or more of tedious hand picking. I hasten to add that the procedure must be followed by a microscopic examination of the separated crystals in order to eliminate all the grains that have large numbers of foreign inclusions. This examination and the sorting that accompanies it are the most delicate and pains-taking steps in preparing material for chemical analysis, but both are indispensable, for an analysis of impure material has no value.* Failure to do this carefully has resulted in mineralogy being burdened with many analyses that are not internally consistent and cannot be reconciled with a proper chemical balance. I have made every effort to avoid being accused of this.

The rocks of the dikes of northern and northeastern Thera fall into two groups, each of which has its own petrographic texture and mineral composition.

All the samples have an amorphous groundmass containing myriads of colorless, prismatic crystals up to several hundredths of a millimeter in length and a few thousandths in width and thickness. The matrix also contains greater or lesser amounts of granular iron oxides, the sizes of which rarely reach a hundredth of a millimeter. Various larger crystals distributed through this groundmass normally have dimensions in excess of 0.01 mm. The distribution of the microlites, as well as that of the phenocrysts demonstrates the original fluidity of the rocks. The microlites tend to be aligned in swarms parallel to certain directions in a more or less wave-like fashion that was

---

* Use of a solution of diiodide of mercury in potassium iodide, according to the method of Mr. Thoulet, greatly simplifies the last step of this procedure. It is especially useful when the feldspar is mixed with only a small amount of magnetic material. There is even some advantage in using this solution directly when the magnesian minerals contain little iron, as they normally do in older eruptive rocks.

determined by the direction in which the molten material was flowing. Thus the matrix was still fluid when the crystals, both large and small, had formed. The phenocrysts (i.e., crystals larger than 0.01 mm) were carried along in the flowing lava, for one commonly observes that they are broken and their fragments, though still side by side, are separated by the glassy material or by other crystals that could only have intervened between them through the motion of the viscous material in which they are enclosed. In addition, the microlites are commonly seen to swerve about the edges of larger crystals, showing that they took part passively in the general motion of the matrix that was carrying everything along. In many instances the large crystals of feldspar continued to grow after they had been broken into fragments. They have a zonal structure in which the overgrowths follow not only crystal faces that are still intact but are concentric about the fracture surfaces as well.

As the rock cooled and solidified, the crystals embedded in it still preserved their distribution at the time the fluid motion came to an end, and one can clearly see under the microscope that all the dikes of Thera contained a molten matrix crowded with great numbers of crystals that had already acquired their individual identities.

Some of the samples of dikes on Thera show quite clearly that the rocks were formed from molten material that carried solids in suspension. In several of these dikes one sees extraneous fragments derived from other lavas. These are of various sizes, but some reach quite notable dimensions, up to fist-size or even larger. They seem to be quite random, and one gets the impression that they are accidental debris torn from the walls of the dike, but the microscope reveals that at times the phenomenon was more pervasive than it at first appears. For example, in several thin sections cut from a specimen taken from dike no. 47* one sees a great many small inclusions of distinctive lavas that come from rocks very different from that forming the main part of the dike. The diameters of these inclusions rarely exceed 0.01 mm; they have irregular shapes, and their sharp edges have been rounded. The essential host material is a very fine-grained rock approaching rhyolite in its chemical and mineralogical composition. It has a large amount of silica and contains great quantities of tridymite. The principal feldspar phenocrysts are labradorite, and although one finds hypersthene, augite is rare and olivine is completely absent. Feldspar microlites are very scarce; they are oligoclase and albite. Their lengths do not exceed 0.009 mm, and their thickness is less

* The dikes are numbered starting at Apano Meria and proceeding from west to east.

than .002 mm. These inclusions are from mafic rocks with relatively large grain sizes. They contain olivine but no hypersthene. The feldspar microlites have an average length of 0.02 mm and a diameter of 0.006 mm; they appear to be labradorite. Tridymite is completely absent.

The inclusions in dike no. 7 are even more remarkable than those in the preceding one. Some samples of this rock have the texture of a microbreccia, in which the grains reach dimensions of several tenths of a millimeter. Most of these fragments appear to come from the same material as the rock in which they are found. They do not differ from the latter except in the somewhat finer size of their mineral grains and in the variable abundance of globules of amorphous material that make up the groundmass, but what is most distinctive about them is the different fluidal alignment of the crystals. The orientation of bands of microlites differs abruptly from one fragment to the next. Thus they are bits of lava detached from their original setting after they had taken on this texture; they were subsequently caught up in the amorphous material that finally solidified. The rock in which they are found has a more uniform fluidal texture; the microlites in it form wavy bands that wrap around the inclusions, showing that the latter were solid while the matrix was still fluid. Some of the fragments in this rock have characteristic textures and chemical compositions similar to those of dike no. 47. Dike no. 7, which is an intermediate rock rich in tridymite, also contains debris of basic composition. These fragments, like those already mentioned, are enclosed in the fluidally banded material making up the main part of the dike.

In summary, when the dikes of Thera are examined under the microscope, all of them are seen to have a pronounced fluidal texture. Thus, the rocks must already have been laden with crystals before they stopped moving. These crystals were, for the most part, brought up from greater depths. They do not seem to have crystallized during the final stages of cooling and solidification of the rocks.

One objection can be raised against this interpretation. It is based on the following observation: the chemical and mineralogical composition of the dikes does not differ significantly in a vertical direction, but it has a notable lateral variation. The crystallinity declines toward the margins. Each dike has a more or less glassy selvage at its edges. The color of this marginal part is normally darker than that of the rock making up the central part of the dike. The thickness of some of these borders may be less than a centimeter, but in other cases it can reach as much as 30 centimeters. They are better developed in dikes of basic compositions than in rocks with greater silica contents. Owing to the rarity of crystals, the material of the outer margins is a brownish, often completely homogeneous tachylite. But one almost always finds that these selvages have wavy bands that are distinguished from one another by the intensities of their color. Quite often, one also sees fine opaque specks or very elongated, thin crystallites grouped in sheaves, star-like clusters, or oriented along a single direction. The textures are similar to those that are so typical of the scoriaceous slags of high-temperature furnaces. Finally, in addition to a few scattered normal crystals, one finds clear, anisotropic crystals in extremely thin lamellae forming parallelograms with more or less inclined edges.[2] These probably formed in place. The perfect integrity of their forms, the sharpness of their angles and edges, together with their very thin shapes make it unlikely that they were carried far from their point of origin. Thus, these selvages are essentially glasses containing delicate crystals that grew in place after the dike was emplaced. How can one explain their formation and profound difference from the material in the central parts of the dikes? If the magma was a viscous melt charged with numerous suspended crystals at the time it rose from greater depths, how can the selvages be explained? Why are the crystals that the magma normally carried almost completely absent? The existence of these glassy selvages implies that the phenocrysts were not contained in this part of the original magma.[3] Is this not simpler than assuming that the magma crystallized only after it was emplaced and cooled? If so, the selvages are easily explained: the cooling being more rapid, the lack of crystals would be a simple consequence of the form of the dike.

The problem this raises is a serious one, even though it may have a simple explanation. The formation of selvages is the result of the initial stages of a process that begins with the rise and eruption of the magma. The first injection of molten material encountering the cold walls of the fracture must instantly solidify as a crust. The filling of the interior of the dike follows later as the magma begins to erupt at the surface. Thus the selvages are a solidified layer coming from the upper part of the subsurface melting pot, and the interior of the dike comes from a deeper part of the same reservoir.

Is it not reasonable to suppose that a considerable difference of crystallinity comes from the different layers of the magma chamber that fed the dike? The amorphous groundmass of all the dikes is richer in silica than all the crystals it contains. This silica-rich composition, together with its vitreous state, ensures that it would be less dense than the crystals, and we can assume that the same would also be true when it was in the liquid state. It is likely, therefore, that in the interior of a magmatic reservoir, processes of liquid segregation were taking place in the heterogeneous magma, and these processes could have given rise to a variety of compositions, with the upper levels being richer in the silica-rich liquid and poorer in crystals.[4]

It is not surprising, therefore, that the material ejected first would be more glassy, while subsequent lavas coming from the same vent would be more crystal-rich.

In support of this explanation of lavas as consisting of a mixture of crystals suspended in a liquid with the composition of the glass we can cite the following facts. When one takes a sample of lava from a flow that appears to be totally molten and cools it rapidly in a way that causes it to solidify abruptly, one finds that the resulting solid product is just as crystalline as the slowly cooled lava from the same flow. The explanation for this must be that crystals begin to grow before the solidification, because if the crystals had grown only during solidification their development would be notably influenced by the rate of cooling. Rapid cooling must have the effect of inhibiting crystallization or at least reducing the number and sizes of the crystals.[5] But in this case, the differing rates of cooling do not change the crystallinity in any way. During each eruption of Vesuvius the guides make medallions that they sell to tourists. These souvenirs, which have the size of a five franc piece, are made by means of a pair of tongs that has a mold at the end of each branch. Viscous lava seized with this instrument is flattened and quickly solidified by contact with the iron tongs.

I tried this myself at Etna in 1865 by taking a small projection of lava and dropping it into a mixture of water and snow.

In every case, the quickly cooled lava has exactly the same microscopic texture as that found in the same material where it cooled in the interior of a thick flow.

Finally, I should add that in most volcanoes of basic composition, nature performs this same experiment on a grand scale. One sees in these volcanoes layers of lava that are very crystalline but quite thin. For example, the small summit cone of the large volcano on the island of Pico in the Azores is covered with tongues of lava that are at most a few centimeters thick but have remarkably abundant crystals. The rapid cooling of the thin sheets of lava at this altitude (about 2000 m.) seems to have done nothing to impede the growth of crystals. Thus the crystals in these lavas must have begun to form before the magma reached the surface.

If the microscopic examination of the lavas of Santorini demonstrates that the crystals were formed before the eruption of the assemblage that makes up the rock, this conclusion offers additional confirmation of another generally accepted principle that was formerly a subject of much controversy. It shows that all the crystals of these lavas nucleated and grew from the same material in which they are found today and that they are not the surviving crystals of an older rock that underwent partial melting. The majority of these crystals contain numerous microscopic inclusions of an amorphous material with the same color, homogeneous texture, or globular shapes as the groundmass in which the crystals are set. There is no doubt, therefore, that they grew from the amorphous matrix in which they are now found. They are not the residual debris from remelting of another rock.

Some of the inclusions are irregular while others have good polyhedral forms reflecting the symmetry of the crystals in which they are trapped. In the second case, their form tends to be more complex than that of their host. For example, in a feldspar cut perpendicular to its plane of symmetry and showing a hexagonal section determined by the intersections of the faces m, t, and g_1, it is not uncommon to find inclusions having the form of a dodecahedron corresponding to the faces h_1, m, t, g_2, and g_1. The angles of these dodecahedrons deviate little from 150 degrees with some differing by less than a few minutes of a degree, others by slightly more; the angles of the inclusions are ordinarily so well defined that they can be measured with more precision than those of the host crystal itself.

The polyhedral inclusions can be thought of as a kind of negative crystal or pseudocrystal. It seems logical, therefore, to attribute them to unequal growth of the crystal in which they are enclosed by filling faceted cavities in the surfaces of the faces as the crystal grew from the surrounding amorphous material and to their subsequent enclosure as growth advanced. This would account for their having forms closely related to that of the crystals in which they are found and would explain why most of the edges are parallel to those of the surrounding crystal. It also explains why they tend to be distributed in straight lines parallel to these faces.

Very few of the irregularly shaped inclusions have any recognizable orientation. Most often, they are elongated in a particular direction that is the same as that of their alignment and corresponds to the orientation of one of the faces of the enclosing crystal. As their angles become more accentuated and their sides less rounded, they grade into euhedral inclusions. On the other hand, it is difficult to distinguish them from the amorphous secondary products that penetrate the crystals along cleavages and have deposited irregularly shaped material. At times it is even impossible to decide whether one is simply dealing with a deposit that has infiltrated the crystal or whether one is looking at a more or less altered glassy inclusion. A true inclusion is normally characterized by sharp boundaries; it does not merge gradually into the surrounding material as the products of alteration commonly do.

In cases where one has difficulty in distinguishing these two possibilities, the best way to resolve the question is to crush the mineral to a grain size of about 0.05 mm and boil it for 2 or 3 hours in hydrochloric acid. The amorphous inclusions will resist the action of the acid, whereas the products of infiltration are completely or at least partly dissolved.

Some of the inclusions are completely homogeneous amorphous material of uniform color and no particular structure, but in the rocks we are considering here this is uncommon, especially in the case of inclusions in feldspars. Much of the amorphous material in them forms strands of uneven color. Some parts are colorless, others have various shades of brown. The brown part commonly takes on the form of rounded blebs similar to those in the glassy groundmass of the rock. These small globules have a brownish or purplish tint in their center and darker brown or even black margins. Some are scattered uniformly throughout the extent of the inclusion, in some cases sparsely, in others so densely that the entire inclusion is black and opaque. More commonly, they are distributed in strips, most of which have a certain regularity, especially in euhedral inclusions. Thus one finds them making up the entire central part of the inclusion, leaving only a narrow colorless border. In other cases they are scattered throughout all the inclusion except for a rounded space that one would interpret as a section through a gas cavity if the outline were not a shadowy line and the interior completely colorless. Many are in small strands that are terminated by straight lines, giving them the outlines of crystals. In one fairly common type the brown part with globules has a simple rectangular or sharply angular shape within a clear circular zone that is colorless and free of blebs. In another type that is even more common, the inclusions are rectangular, and granular material fills two opposite quadrants in such a way that the inclusions are separated into four parts of which two are colorless while the other two have a brownish color due to the small blebs.

It is rather common to see in a single thin section abundant small globules in the inclusions in feldspars, while there are none at all in those of the pyroxenes and olivines. It seems that the ferruginous material to which the globules owe their color has been extracted from the glassy inclusions in pyroxenes and olivines to become part of the compositions of the host minerals as well as the iron oxides that often accompany them. Particularly in the pyroxene, the glassy inclusions are almost completely colorless and contain no globules, but instead they have one or more crystalline grains of iron oxide, as though these grains have resulted from separation of the iron oxides that normally make up part of the amorphous material.

Many of the glassy inclusions in minerals of the dikes of Thera contain gas-filled cavities. These cavities resemble rounded bubbles with dark borders and clear centers. Their diameters rarely exceed one or two hundredth of a millimeter. The fact that they do not move when the thin section is heated shows that the inclusion in which they are set is solid and therefore glass.

Inclusions that are rich in small globules and also have a bubble of gas tend to have the globules tightly packed around the bubble. Many of the latter touch the wall of the bubble, which is so strongly colored that the bubble takes the form of an opaque black circle. It is rare to see several bubbles in the same inclusion.

Rocks that show evidence of having been chemically altered tend to have very few inclusions with gas bubbles. For example, whenever the thin section of one of the dikes of Thera contains calcite concretions, ferruginous deposits, or clots of tridymite, one can almost always expect the feldspars that surround them to have very few bubbles. Most likely, the water that was contained in this material infiltrated into the inclusions in the crystals, penetrated the bubbles, and deposited the material it contained in solution in a way that did not alter the shapes.

When one dissolves several grams of feldspar from the dikes of Thera in hydrofluoric acid, samples containing inclusions rich in bubbles give off no trace of gas, whereas feldspars from lavas of the recent eruptions in the center of the bay release small quantities of a mixture of gases, mainly hydrogen.*[6]

---

* The procedure one uses is as follows: small grains measuring about 0.01 mm in diameter are placed in a glass flask with a narrow neck and a capacity of 50 cc filled with boiling water and, at the same time, a small test tube filled with hydrofluoric acid is introduced taking care that

The similarity of older lavas of Thera to those of the recent eruption leads one to think that there are also gas bubbles in the inclusions of minerals in the dike rocks I have studied, but at the same time the experiments just mentioned show that the amount of gas is below detection.

When one washes powdered Santorini lavas in distilled water, the wash water always contains sodium chloride and traces of alkali sulfates in solution. Microscopic examination reveals no sign of these salts in the rocks. Nor does it provide any information on the source of the water that comes out of the rock when it is heated. On only one occasion have I seen in the olivines of dike 12 what I thought might be hydrous inclusions, but because the bubbles in the inclusions show little or no mobility when heated to high temperatures, I doubt whether this is sufficient proof of the presence of water in the minerals of any ancient or modern Santorini lavas I have studied until now.

All the lavas that I have crushed to a powder and treated with boiling nitric acid yield a liquid that gives a yellow precipitate when ammonium molybdate is added to it. This precipitate is characteristic of that obtained when testing for phosphorus in these rocks, even though the microscope shows no evidence of the presence of apatite.

These considerations are the result of general observations based on all the dikes of Thera. We come now to the important division of these rocks into two groups, and we note first of all the petrographic characteristics that support this distinction.

---

it remains vertical. The attack of the acid on the glass presents no problem. The test tube having been introduced, the flask is closed by a plug that is fitted with a glass tube of small-diameter (but not capillary) that is open at both ends and extends to the middle of the flask. One then tilts the flask gently while immersing the outer end of the tube in the plug in a bowl of water. The acid comes in contact with the feldspar and attacks it. The liquid in the flask expands from the heat of the reaction and part comes out; the gas that is produced collects in the flask, where it can then be recovered.

I have carried out this operation several times using 20 grams of feldspar each time. The feldspar separated from the rocks of dikes 5, 19, 32, and 62 gave no trace of gas. The same is true of feldspar from the recent lavas of the Isles of May. Feldspars from a lava of Giorgios gave off 0.6 cc, and those of two lavas of Aphroessa gave up, in one case 0.8 cc and in another 3.7 cc.

One can do this in yet another way.

On heating either the minerals of the lavas or the lavas themselves in a vacuum produced by a mercury pump, one always obtains water rich in hydrochloric acid and quantities of gas that are greater than those from the preceding method. I have tried this techniques several times using all imaginable precautions to avoid accidental contamination with foreign material. These experiments seem to prove that the rocks contain solid or liquid carbon-rich material in amounts that are too small to be detected under the microscope.

The rocks of the first group contain microlites of triclinic feldspar, normally with the substantial extinction angle of labradorite. In addition, there are iron oxides and augite which, on occasion, may be the dominant mineral. All the amorphous material is very granular. The larger crystals are anorthite, labradorite, augite, olivine, apatite, iron oxides, and sphene. In the more altered parts of the rocks, one finds patches of calcite, aragonite, and wispy deposits of limonite.

Rocks of the second type normally contain only microlites of triclinic feldspar (with the very small extinction angles of albite and oligoclase) and iron oxides. Rare grains of pyroxene are very widely dispersed. The feldspar microlites are normally individual crystals that are more widely separated and have shorter dimensions than those in the preceding group of rocks. The amorphous material tends to have fewer globules, and, as a result, its color is not so dark. The larger crystals are labradorite and very small amounts of oligoclase and sanidine, then augite, iron oxides, hypersthene, apatite, and accidental grains of olivine. In almost every thin section, one notices, in addition to the clusters of tridymite that fill vesicles, thin streaks of limonite.

Thus, the first type is characterized by the calcium-rich composition of its feldspars, by its abundance of olivine and corresponding rarity of iron oxides and hypersthene among the larger crystals, and by the presence of fine granular pyroxene. The reverse is true of rocks of the second group. The second type of rock is distinctly less basic than the first; this is clearly shown by the fact that it is almost completely devoid of olivine but is rich in tridymite. It is surprising, however, to see that it is this rock that tends to contain larger grains of iron oxides. If rocks of the first type are considered basic, those of the second type should be thought of as being more acidic.[7]

A great number of individual feldspars belonging to the two types we have just distinguished do not have the bands of birefringent colors normally seen under crossed nicols in crystals of the sixth system.[8] Some of these are simple crystals of a uniform color. Others have the binary twinning of orthoclase and are uniformly colored on both sides. But it would be a serious error to take as monoclinic all crystals lacking the characteristic birefringent bands of the triclinic system. As we shall see shortly, chemical analyses show this most convincingly, but the microscope, if properly used, can show it equally well. We know that monoclinic crystals cut parallel to their orthogonal axes, that is, on faces in the zone ph_1, have symmetrical forms and even have rectangular outlines. Moreover, these sections have rectangular cleavages parallel to their symmetry directions. And, finally, their extinction positions under crossed nicols are parallel to these directions of symmetry or, if they are rectangular, parallel to the sides of the rectangle.

When monoclinic feldspars are cut in sections other than those of the zone ph_1, these sections are not symmetrical, and cleavage traces are not rectangular, so the extinction direction is not parallel to both observed edges.*

In crystals belonging to the triclinic system, the form of the sections, the orientation of cleavage traces, and the extinction directions are all analogous to those just noted as characteristic of the latter case.

It can be concluded, therefore, that in microscopic sections of feldspar-bearing rocks it is important to determine the type of extinction in sections in the zone ph_1, because the form of the interference figures, orientation of cleavage traces, and type of extinction characteristic of the feldspars are best seen in that orientation.

Feldspar crystals in sections in which the cleavage traces are rectangular and extinction is parallel to one of the polarization directions are almost the only ones that can clearly be related to the monoclinic systems. Individual feldspars that are normally illuminated between crossed nicols and go to extinction in two directions that do not correspond to any of their sides can be considered to belong to either the fifth or sixth crystal systems. If the extinction directions of many of the crystals are parallel to their sides, it is likely that most of the sections that have inclined extinction belong to the monoclinic system. If they are rare, however, it is likely that there is a mixture consisting of considerable proportions of both triclinic and monoclinic feldspars. And finally, if all the extinction directions are inclined, triclinic crystals must be very dominant, and if monoclinic ones are present, they cannot be very numerous.

If the dike rocks of Thera are viewed in this sense, one is led to the following conclusions. Rocks of the first type (those rich in olivine) have scarcely any feldspars with multicolored bands of birefringence. Individual crystals with uniform birefringence have inclined extinction in all sections. Exceptions to this rule are extremely rare, and their parallel extinction is not always perfect. In addition, the sections are almost never symmetrical and their cleavage traces are not exactly rectangular. Thus one can conclude that in these lavas monoclinic feldspars are completely lacking or, at least, their presence is very doubtful.

---

* In the totally fortuitous case in which a crystal is accidentally cut perpendicular to one of the optic axes, there would be extinction in all orientations of the crystal.

In rocks of the second type (those with coarse iron oxides, veinlets of tridymite, and little or no olivine) one still finds feldspars with multicolored stripes under crossed nicols, but the feldspars in these rocks generally have uniform interference colors, and a few have parallel extinction. From that it can be concluded that the feldspars in these lavas are mainly triclinic, but they also include a few that are monoclinic.

It is rather common to see in these rocks Carlsbad twins, in which one half has an extinction direction oblique to the trace of the plane of twinning, while the other is parallel to it. The side with the oblique extinction commonly has bands of various interference colors, and in these cases the bands can be divided into two alternating groups, one with extinction oblique to the twin plane and the other parallel to it and always with the same orientation as the side with the uniform interference color. I conclude that the crystals are probably twinned according to two different systems, one monoclinic and the other triclinic.

The optical characteristics just pointed out are not easy to apply, for, according to the observations of Mr. des Cloizeaux, the extinction angle with respect to the edge mt is very small on faces in the zone ph_1 in feldspars very rich in silica. This is particularly true of oligoclase. So it is easy to confuse this case with the one in which a monoclinic feldspar is cut parallel to its orthogonal axis. In thin sections of rocks, the problem is made even more difficult by the fact that one is never sure of seeing a section that is precisely in the zone ph_1.

When one examines under crossed nicols platy feldspars extracted by means of a magnetic separator and views them resting on the face g_1 which tends to be the one on which they are flattened, one can easily see that many of the lamellae do not go to extinction in any direction. These are crystals that are twinned. One also sees that many of the untwinned crystals have an extinction angle of about 25 degrees with respect to the edge pg_1, as is the case for labradorite. The others have a very small extinction angle with respect to the same edge; these are either sanidine or oligoclase. But the optical methods available to us at this time permit no further distinction.

The fact that oligoclase is more abundant than sanidine is shown when the tests of Szabo and Boricky[9] are applied to feldspar lamellae that are flattened on g_1 and have an extinction direction parallel to pg_1.

As a rule, in samples from the dikes of Thera the number of crystals that remain without birefringent colors under crossed nicols and go to extinction parallel to one of the sides of their section becomes greater as the olivine and pyroxene in the rock become scarcer and tridymite more important. In other words, the abundance of these minerals relative to that of the feldspars seems to be related to the amount of silica in the rock.

The minerals in the two types of dike rocks of Thera have certain noteworthy differences in the proportions of their inclusions. Inclusions of amorphous material are more common and more granular, and the assemblages of feldspars are more irregular in the basic type than in the acidic. Gas bubbles are less common, and the amorphous material of the inclusions is more likely to show the effects of secondary alteration.

Inclusions of amorphous material in the pyroxene and olivine differ in the two types. In the olivine, they are few in number, colorless, and almost all lack bubbles of gas.

Crystalline inclusions are more common and more varied, especially in the feldspars of the acidic rocks than in the basic type. Apart from the pyroxenes with rounded granular forms and the apatite that is found in the feldspars of both types of rocks, one commonly sees iron oxides in the feldspars of the acidic rocks as well as long, colorless, birefringent prisms with oblique terminations that are probably triclinic feldspars picked up from the microlites in the groundmass.

Pyroxenes of the acidic rocks are also richer in inclusions of iron oxides and apatite than those of the basic rocks. The pyroxenes of the acidic dike no. 56 are particularly remarkable for the great quantities of apatite one sees in them and for the prisms with elongated hollow forms and cores of granular amorphous material.

Finally, hypersthene is seen only in rocks of acidic composition and seems to have a wide range of abundance.

In summary, a microscopic study of the dikes of Thera confirms the conclusions drawn from general observations based on other volcanic rocks and establishes the following general principles:

1. The crystals of the rocks grew from the molten amorphous matrix in which they are immersed and they encased particles as they solidified.

2. The crystals were formed before the magma stopped moving.

3. The iron oxides and apatite were the first minerals to crystallize from the molten magma; olivine was next in the order of formation, then, after a somewhat longer interval, monoclinic pyroxene, hypersthene, and, last of all, the feldspars.

4. The microscopic study shows that the dikes of Thera are of two types, each with the particular characteristics that were described in the preceding section.

Finally, the microscope provides not only evidence of the nature of the secondary minerals, such as hydrous iron oxides, calcite, aragonite, zeolites, chlorite, and tridymite, but it also yields interesting information on the mode of formation of the last of these minerals, which is not exactly secondary but a contemporary product of the eruption. It is easy to show that tridymite is not an essential component of the rock that has been formed from the molten magma in the manner that feldspar, pyroxene, olivine, apatite, and iron oxides have crystallized. It is not distributed uniformly throughout the rock in the same way as the other minerals but is found in small groups that form more or less distinct streaks. It is the result of silicification that in some instances pervades the entire rock and in others is limited to certain parts of it. In many thin sections, one observes strands through which the silica has penetrated. These parts are distinguished by the sharp outlines of the tiny globules contained in the amorphous material, by the absence of ferruginous deposits between the crystals, and even more by an absorption of reflected light. This latter characteristic is by far the most important. When part of a lava is penetrated by opaline deposits, it is enough to examine the thin section in reflected light to distinguish immediately the entire extent of its infiltration. One sees the silicified parts illuminated by the reflected light, and one can note that all the nests of tridymite are within these zones. Thus the siliceous infiltration and the tridymite that is one of its products are not derived directly from the chemical components of the original magma.

On the other hand, the silicification is not the result of late-stage processes following emplacement of the dike. It is not due to superficial water that penetrated through deep levels of the ground and carried in silica that it picked up along its path; nor is it from water coming from greater depths after the dike was completely solidified and cooled. In either one of these hypotheses, the siliceous deposits would be found throughout the entire length of the two vertical cracks produced by contraction at the contact between the dike and the adjacent host rock. These cracks would have served as channelways for the free circulation of silica-charged water, and, as a result their margins would have been the parts most strongly affected. In a word, the selvages should have been more silicified than the central parts of the dikes, whereas the reverse is what one sees. The silicification, therefore, occurred before the cracks bordering the dike had opened and the injected material still occupied the full width of the dike. The silica must have passed through the central part of the dike rather than the selvages, for the tridymite is more common in samples from the center of the dikes than in those from the margins.

It also follows that most of the free silica in the rock must have been introduced before the dike solidified, for it is evident that the siliceous fluids could not have moved through the long dimensions of rocks as compact as those in the central parts of the dikes. This central part would not remain the most favorable route for transporting fluids after the dike was solidified.

Silica can certainly be dissolved, for we can see that there are always notable amounts of it in the water of molten lava. The presence of water under these conditions is demonstrated, as we know, by the dense steam that is given off from very fluid lavas and from the explosions that take place in craters.

Lava in the molten state is clearly capable of transporting dissolved silica, for very few of the solid fragments of lava that it carries show any effect of silicification, while the material that surrounds them may be charged with deposits of opal and tridymite.

In the rocks of some of the dikes of Thera (notably in those of nos. 8, 49, and 51), tridymite or opal is found in numerous small vesicles, about one millimeter in diameter or in trains of vesicles that parallel the direction of flow for lengths of up to 5 millimeters. It is in these spaces that the dissolved silica was deposited. Water trapped in these narrow spaces and heated to red-hot temperatures must have had a strong corrosive effect on the surrounding rock and developed on a large scale features similar to those that have recently been the subject of remarkable experimental studies.

In general, the geometric form of the space occupied by tridymite indicates that the material from which it was deposited moved only slightly with respect to the igneous liquid making up the bulk of the rock. These spaces are drawn out in the direction of flow, but their elongation is normally very weak. The silica-rich aqueous solution was probably carried within the interior of molten mass; it participated in the same motion but, being constrained by the viscosity of the surrounding liquid, it had only a slight motion relative to its host.

The alteration caused by this superheated water and in turn by the siliceous deposits continued to act not only while the rock was still molten but even after it solidified, so long as the temperature was still sufficiently elevated. Some of the capillary spaces with diameters of at most 2 or 3 thousandths of a millimeter and considerable lengths (in some cases several millimeters) can scarcely be regarded as cracks produced by contraction during solidification; the opal that fills them must therefore have been deposited after they formed.

A microscopic examination of the lavas of Thera shows that these rocks were not erupted at a temperature higher than that at which they began to melt. We know, in fact, that without exception all of them contain, for example, crystals of feldspar. Thus, their temperature was below the melting temperature of this mineral, but they had just the right temperature for the amorphous matrix to be molten. The amount of this amorphous material differs from one dike to another, so it seems that the fluidity of the magma must have been greater in the lavas that contained large amounts of it than in those with large proportions of crystals. Moreover, this amorphous material does not have a constant composition; the more basic its composition, the more molten it was. Samples of basic rocks rich in amorphous material must therefore have had the most marked fluidity. This can be seen from their microscopic characteristics. The basic rocks are more homogeneous; crystals are always better formed in rocks that had the greatest fluidity. Many of these crystals are corroded and partly remelted, as though they had crossed through zones of different temperatures, but as a rule they are unbroken and show no sign of violent mechanical effects. Acidic rocks, on the other hand, are commonly heterogeneous and have fragments of other rocks that were suspended in a solid state. Their temperatures were not high enough to melt these inclusions. A few consist of a microbreccia that could not have had great mobility. Many of their crystals are broken and crushed as though they had been subjected to strong mechanical effects. The fracture surfaces of these broken crystals are clean; they show no sign of remelting. They are not corroded but simply broken and abraded at their edges.

We should also note that at the time of their eruption, the acidic lavas appear to have had a lower temperature than basic ones, for otherwise, it would be difficult to explain the presence of basic inclusions in an acidic lava that had a smaller proportion of melt.[10] If one assumes that these two types of lava come from a single underground magma chamber that had a uniform temperature, the basic lavas could only have reached the surface at a hotter temperature if they had greater mobility and ability to retain heat.[11]

The microscope also reveals interesting properties that the lavas had when they were discharged. It shows how the movement that was taking place when these features developed varied in intensity, not only from one dike to another but even in the same dike. One part of a given dike may record rapid flow and regular fluidity, while an adjacent zone may show evidence of irregular fluid motions with unequal currents moving at different rates and in various directions. One commonly finds streaks that seem to have lost all motion prior to solidification. Thus, in dike no. 52, zones of aligned feldspars are found adjacent to others in which the molten material was undisturbed long enough for the feldspars to arrange themselves in star-shaped groups that are remarkable for their forms, abundance, and arrangement.

Each of these groups of feldspars has the appearance of a small tuft of fine needles converging on a center. Their average diameter is about 0.02 mm. Their small microlites have thicknesses of about 0.001 mm. Another interesting aspect of this sample is its great numbers of very small rounded spaces filled with colorless amorphous material that can be seen in the middle of the groups of microlites. The regular forms of these spaces, the nature of the material that fills them, and the absence of brown globules of the kind that are common all around them suggest that they are vesicles filled with siliceous deposits and not random bits of the normal amorphous material found elsewhere in the rock. Some of these spaces are actually true vesicles with nothing in them. They are distributed in long, slightly sinuous streaks.

In the preceding pages we have examined the general results of a microscopic examination of the dikes of Thera. These results are impressive, for they illustrate how much can be learned through this type of study. The microscope not only reveals the principal mineralogic characteristics of the rocks of Thera and enables us to divide them into two groups, but it throws light on several interesting details of their origins. They illustrate the degree to which petrology depends on the use of this instrument. But no matter how skilled one is in the use of the microscope, petrography is not able to resolve all petrologic problems. The great majority of eruptive rocks are made up of essential minerals that include one or more members of the feldspar group. The microscope can give only incomplete information on the particular nature of these minerals that one finds in a given rock. It allows us to make only approximate identifications.

The great difficulty of distinguishing one feldspar from another by optical means alone tends to give one the impression that rocks never contain more than a single triclinic feldspar. This belief, which until recently was regarded as axiomatic, is erroneous. As already pointed out in the preceding chapter, studies of the lavas of the most recent eruption of Santorini have shown that several triclinic feldspars coexist in those rocks. The following observations clearly show that the same is true of the dikes of Thera.

The compositions of the triclinic feldspars making up most of the phenocrysts of each of the two classes of dike rocks have been deduced from their optical properties, but some control is needed to verify these determinations. I have had to resort to carefully considered chemical analyses to demonstrate as rigorously as possible the presence of two distinct varieties of feldspar. These tests have definitely resolved the question. The main triclinic feldspar found as phenocrysts in the basic dikes of Thera is definitely anorthite; in the more acidic dikes it is labradorite.

So, if the microscope is not capable of distinguishing the triclinic feldspars, must one consider labradorite and oligoclase to be distinct minerals of constant compositions, or should we consider them isomorphic chemical mixtures of albite and anorthite? This is one of the most serious questions of mineralogy. With the support of Professor Tschermak, the distinguished director of the mineralogical section at Vienna, des Cloizeaux and von Rath have recently carried out work that has an important bearing on this question. Chemical studies, crystallographic measurements, and studies of optical properties have all been brought to bear on the problem. Without presuming to offer a final decision, we can consider which of the possibilities is in best accord with the deductions drawn from our present work. The facts on which this is based are the result of chemical tests and analyses, aided by use of the microscope. Before laying them out in detail, it is essential to review briefly the historical background and the fundamental facts of the problem.

We know that all the minerals of the feldspar group are alumina silicates with monoxide bases, and that they share the common characteristic of a constant proportion of oxygens of these monoxide bases to those of the alumina. This ratio is 1 to 3. But the proportion of oxygen with alumina to that with silica can vary. In the monoclinic feldspars it is always 3 : 12, while in triclinic feldspars it drops from 3 : 12 to 3 : 4. This variation in the composition of the triclinic feldspars does not seem to be regular and continuous between the two extremes. Between the upper end-member, albite, which is characterized by the proportions 1 : 3 : 12, and the lower end-member, anorthite, with the proportions 1 : 3 : 4, labradorite has the proportions 1 : 3 : 6. And we might also include andesine, which has the proportions 1 : 3 : 8.

All the well-crystallized, unaltered, triclinic feldspars belong to these types. But in most cases, on analyzing triclinic feldspars separated from a variety of rocks, it was found that they have different ratios, such as 1 : 3 : 5 or 1 : 3 : 11, or even ratios in which the numbers for the oxygens combined with silica were not even integers. In some instances this complication was due, as Mr. Ch. Sainte-Claire Deville has shown, to alteration of the mineral or impurities, but was this always the case? Was the complexity of the formula always the result of accidental effects? Many mineralogists do not think so. Waltershausen and, after him, Rammelsberg and Schéerer have come out in favor of a gradual passage from one triclinic feldspar to the next. Delesse and Sterry Hunt are of the same opinion.[12] And Tschermak himself has given new evidence that the chemical composition and physical properties of any triclinic feldspar can be explained very well as a mixture of albite and anorthite in a molecular association. The only base in pure albite is sodium; pure anorthite has no monoxide base other than calcium.

All intermediate triclinic feldspars contain both sodium and calcium. As the composition of the feldspar approaches that of albite, one sees its specific gravity diminish and approach that of the pure mineral, while at the same time, it becomes richer in sodium and poorer in calcium. For a feldspar of a given silica content, one can immediately predict the proportions of sodium and calcium it should have by considering it a mixture of albite and anorthite.

Tschermak pointed out the regular variations in the development of cleavage parallel to t, in the angle of p on g_1, which is:

85°50' in an anorthite of Vesuvius
86°04' in a labradorite of Saint Paul
86°08' in oligoclase from Tvedestrand
86°29' in albite of Windisch Matrey

Von Rath has recently shown a continuity of the same order in the positions on g_1 of the junction of crystals twinned according to the pericline law. This line, which in albite is inclined from front to back with respect to the edge pg_1 (the crystal being viewed in the orientation in which it is normally drawn), is essentially parallel to this edge in oligoclase. It is inclined from rear to front in labradorite, and even more so in anorthite.

Des Cloizeaux has shown a similar continuity in the positions of the optic axes, except in the case of labradorite, which in this respect is an exception to the general rule. I should add here that the numerous analyses of triclinic feldspars carried out in recent years all tend to confirm the hypothesis of Tschermak. Moreover, we can confidently say that all our analyses of triclinic feldspars are in essential agreement with this hypothesis.

To explain the association of albite and anorthite and the replacement of one by the other in the same assemblage, Tschermak points out that the formulas of the two minerals can be written:

$$Ca\ Al\ Al\ Si_2\ O_8\ \text{anorthite}$$
$$Na\ Al\ Si\ Si_2\ O_8\ \text{albite (doubled formula)}$$

so that they differ only in the replacement of an atom of silica for one of alumina.[13]

Before attaching great importance to this comparison of two formulas that rests on an equivalence of two elements as different as silica and alumina, we should note that, according to those who have proposed this hypothesis, it is not simply a matter of isomorphism, as, for example, in the alkali-earth carbonates or alum. In their opinion, there is no continuity whatever in the series of triclinic feldspars. Members of this series certainly have markedly distinct forms that are favored under natural conditions, especially in the case of labradorite and oligoclase. Professor von Rath, who also fully accepts Tschermak's theory, believes that each triclinic feldspar is a combination in definite, proscribed proportions of a certain number of atoms of anorthite with a certain number of atoms of albite.[14] According to his interpretation, it is the result of atomic substitution and not a continuous molecular replacement. Labradorite, according to this hypothesis, would be composed of 3 parts anorthite and one part albite. This results in an atomic substitution, not a continuous molecular replacement. Labradorite would be composed of 3 parts anorthite and one of albite; andesine would be formed from one part albite combined with an equal part of anorthite; oligoclase would be a combination of 3 parts albite and one of anorthite, and its true formula would not correspond to the ratios of oxygens, 1 : 3 : 9, but 1 : 3 : 10. The other triclinic feldspars would result from combinations of albite and anorthite in more complex proportions. Thus we might find, for example, a combination of 6 atoms of albite with one of anorthite, 4 of albite with 3 of anorthite, and so forth. If one wishes to apply Tschermak's theory rigorously to all the triclinic feldspars for which analyses have been published so far, it would be necessary to assume associations of anorthite and albite in even more complex ratios. The multiplicity of names this would require, if the theory were totally accepted, would mean giving up the idea of recognizing distinct types in the series of triclinic feldspars and returning to the original idea of a continuous succession in the isomorphous substitution of albite and anorthite. One would then encounter the difficulty of explaining the frequency in nature of combinations corresponding to labradorite and oligoclase.

Throughout this discussion, we have spoken only of chemical combinations of albite and anorthite and have not mentioned the possibility that mechanical mixtures of these two feldspars could be responsible for intermediate compositions, such as labradorite and oligoclase and that the simplest chemical tests would reveal the presence of anorthite in an oligoclase even if it were a separate, physically interspersed substance. Tschermak and his disciples have dismissed this hypothesis, but I shall try to show that, in the case we are concerned with here, this explanation is closer to the truth.

The studies of feldspars in the dikes of Thera have established the following facts.

1. Labradorite is a well defined mineral species.

2. There is no feldspar intermediate between anorthite and labradorite.

3. The triclinic feldspars that appear to be intermediate between anorthite and labradorite are physical mixtures of anorthite, labradorite, and other triclinic feldspars richer in silica.

4. The triclinic feldspars reported to have ratios between 1 : 3 : 6 and 1 : 3 : 7 are mechanical mixtures of labradorite with other triclinic feldspars that are richer in silica.

5. Tschermak's rule is true in practice, but its precision can be attributed to several factors: first, the feldspars are mechanical mixtures; second, analyzed crystals contain inclusions of other crystals and amorphous material with compositions of feldspars that are richer in silica than their hosts; third, tridymite is commonly distributed throughout some of the feldspars augmenting the silica content of the feldspar analyses without changing the ratio of oxygens combined with alumina and monoxide bases.

These principles have been deduced from the following evidence. We have first verified the validity of Tschermak's rule by analyzing feldspars that were separated by means of an electromagnet and examined under the microscope in order to eliminate any conspicuous accidental impurities. The agreement between the results of these tests and the calculated values has been quite satisfactory as shown by the following table, which compares the results of chemical analyses with the closest composite feldspar calculated according to Tschermak's rule.*

---

* The dikes are numbered in order from west to east starting at the northwestern point of the island of Thera. (See plates XXXI and XXXII.)

Feldspars with silica contents between 45.4 and 46.5%

	Dike no. 4	Dike no. 5	Dike no. 30	Dike no. 34	Dike no. 31	Dike no. 20
Silica	45.8	45.7	46.4	45.6	46.1	46.5
Alumina	35.3	34.1	33.3	34.1	33.8	33.9
Lime	16.2	1.8	1.4	1.5	16.8	16.8
Magnesia	1.8	0.6	0.8	0.8	0.6	0.5
Soda	0.9	1.8	1.9	1.8	2.4	1.9
Potash	trace	0.0	0.2	0.2	0.3	0.4

Feldspars with silica contents between 45.4 and 46.5 % calculated according to Tschermak's rule.

	1 mole of albite 15 of anorthite	1 mole of albite 10 of anorthite
Silica	45.7	46.9
Alumina	35.2	34.4
Lime	1.8	16.8
Soda	1.3	1.9

Feldspars with silica contents between 47.0 and 49 %

	Dike no. 28	Dike no. 2
Silica	4.9	49.4
Alumina	32.6	32.4
Lime	16.7	14.2
Magnesia	1.1	0.6
Soda	1.7	3.0
Potash	trace	0.4

Feldspars with silica contents between 47.0 and 49.2 % Calculated according to Tschermak's rule.

	1 albite 8 anorthite	1 albite 6 anorthite
Silica	4.6	48.9
Alumina	33.9	33.0
Lime	16.2	15.2
Soda	2.3	2.9

Feldspars with silica contents between 50.0 and 55.5 %

	Dike no. 32	Dike no. 42	Dike no. 51	Dike no. 11	Dike no. 19	Dike no. 62	Dike no. 49	Dike no. 53
Silica	50.8	51.0	52.1	52.2	52.4	54.2	55.4	55.0
Alumina	30.9	30.1	29.5	29.3	29.0	28.3	28.6	28.6
Lime	14.0	13.1	14.1	12.6	13.6	11.0	10.2	11.3
Magnesia	0.7	0.6	0.6	trace	0.4	0.2	0.6	0.1
Soda	3.4	4.2	3.6	5.0	4.1	5.6	4.5	4.6
Potash	0.2	0.6	0.1	0.9	0.5	0.7	0.7	0.4

Feldspars with silica contents between 50.0 and 55.5 %
Calculated according to Tschermak's rule

	1 albite 4 anorthite	1 albite 3 anorthite	1 albite 2 anorthite
Silica	51.0	52.	55.2
Alumina	31.6	30.5	28.7
Lime	13.6	12.2	5.8
Soda	3.8	4.6	5.8

Proportions of oxygen

	Dike no. 4	Dike no. 5	Dike no. 30	Dike no. 34	Dike no. 31	Dike no. 20	Dike no. 28	Dike no. 2
Silica	24.32	24.37	24.75	24.32	24.59	24.80	25.55	26.41
Alumina	16.29	15.74	15.37	15.74	15.00	15.65	15.05	14.95
Lime	4.63	5.09	4.97	4.99	4.80	4.80	4.77	4.05
Magnesia	0.72	0.24	0.32	0.32	0.24	0.20	0.44	0.24
Soda	0.22	1.45	0.47	0.45	0.60	0.47	0.42	0.75
Potash	0.00	0.00	0.03	0.03	0.05	0.06	0.00	0.06
Sum	5.57	5.78	5.79	5.79	5.69	5.53	5.63	6.00

	Dike no. 32	Dike no. 42	Dike no. 51	Dike no. 11	Dike no. 19	Dike no. 62	Dike no. 49	Dike no. 53
Silica	27.09	27.20	27.79	27.84	27.95	28.91	29.55	29.33
Alumina	14.26	13.89	13.60	13.52	13.39	13.06	13.20	13.20
Lime	4.00	3.85	4.03	3.60	3.89	3.15	2.92	3.23
Magnesia	0.28	0.24	0.24	0.00	0.16	0.08	0.24	0.04
Soda	0.85	1.05	0.90	1.25	1.02	1.40	1.12	1.15
Potash	0.03	0.10	0.02	0.15	0.10	0.13	0.13	0.06
Sum	5.16	5.24	5.19	5.00	5.17	4.76	4.41	4.48

The following table shows the densities of the analyzed feldspars and their ratios of oxygen to monoxide bases, alumina, and silica.

Dike no. of the feldspar	Specific gravity	Observed ratio of oxygens		Calculated ratio of oxygens
4	2.792	1.02 : 3 : 4.50	1 albite, 15 anorthite	1 : 3 : 4.50
5	2.779	1.10 : 3 : 4.65	1 albite, 10 anorthite	1 : 3 : 4.73
34	2.782	1.10 : 3 : 4.64		
31	2.749	1.09 : 3 : 4.79	1 albite, 8 anorthite	1 : 3 : 4.89
30	2.723	1.13 : 3 : 4.83		
20	2.745	1.06 : 3 : 4.79		
26	2.707	1.12 : 3 : 5.09		
27	2.697	1.03 : 3 : 5.28	1 albite, 6 anorthite	1 : 3 : 5.14
32	2.713	1.08 : 3 : 5.70	1 albite, 4 anorthite	1 : 3 : 5.60
42	2.702	1.01 : 3 : 5.87	1 albite, 3 anorthite	1 : 3 : 6.00
51	2.686	1.14 : 3 : 6.12		
11	2.692	1.11 : 3 : 6.18		
19	2.675	1.16 : 3 : 6.26		
62	2.686	1.08 : 3 : 6.64	1 albite, 2 anorthite	1 : 3 : 6.67
49	2.694	1.00 : 3 : 6.72		
53	2.697	1.02 : 3 : 6.67		

In all these numerical results one sees a small excess of monoxide bases relative to the amount needed to satisfy the ratio of 1 to 3 between the oxygens of these bases and those of alumina. Such an excess may be due to several causes. Unless freshly prepared, the analytical reagents, such as nitric and hydrochloric acid, may contain impurities derived from their glass containers. The lime may also contain traces of magnesia. Very pure oxalic acid, ammonium nitrate, and oxalate must be kept free of contamination. All these products should be distilled to see that they leave no residue when calcined on a sheet of platinum.

A second source of error comes from the glass vessels in which filtration and the first stages of evaporation are carried out. The final evaporation is always made in platinum containers.

The source of the most serious errors is no doubt the presence in the crystals of inclusions of pyroxene, which augment the proportions of monoxide bases, especially magnesium.

I have made every effort to eliminate feldspar grains containing other material, but if I had to use only grains that are absolutely free of inclusions, I would have to reject all of my analyses, for not one feldspar grain in a hundred appears pure under the microscope. I have had to be content with rejecting the grains with the most inclusions and keeping only those with a relatively small number in order for the analysis to be affected as little as possible. On average this means eliminating between half and three-quarters of the grains by means of the electromagnet.

If we accept the results in the preceding table, the numbers provide a remarkable confirmation of Tschermak's theory. It is worth noting that this theory is based mainly on work of this kind. Crystals of feldspar from various rocks have been analyzed, and since many compositions were neither labradorite nor oligoclase but something in between, it was concluded that intermediate feldspars are in fact a combination of albite and anorthite.

It is very rare to be able to work with large, perfectly homogeneous crystals, but when this is possible they normally have the formulas of labradorite or oligoclase.

My own experience has been quite similar to that of the illustrious mineralogist who deduced the theory. I find that density decreases in direct proportion to the silica content and that lime decreases as the content of sodium increases. In short, the material I have analyzed gives results in conformity with it being a chemical combination of albite and anorthite. Knowing the silica content, one can use Tschermak's rule to calculate the corresponding amounts of lime and sodium making up their compositions.

This rule expresses two facts. In the present case, it represents very well the compositions of feldspars extracted from the rocks of Thera. If we assume for the moment the validity of this rule and the theory Tschermak has drawn from it, it is obvious that my results conflict with the view that the triclinic feldspars are associations *in fixed proportions* of albite and anorthite. They can be interpreted only as associations of these two minerals in a great range of proportions that can only be expressed by large numbers. This is inconsistent with the chemical relations normally seen in minerals.

It would seem, therefore, that one must fall back on an explanation based on the hypothesis of continuous isomorphic substitution as it was proposed by Mitscherlich, in which case it is difficult to explain the frequency of labradorite and oligoclase in the same terms as the other triclinic feldspars.[14]

Evidence based on chemical analyses that support Tschermak's theory assumes that all these analyses are of homogeneous material, but in the case of the present study I have been able to show that this is not necessarily true.

Thus, as we have already noted, the feldspar grains extracted from the dike rocks of Thera are mixtures of several mineral species, mainly anorthite and labradorite. The other feldspars associated with these in smaller amounts are richer in silica and are part of the upper range of the feldspar series.

Anorthite is very easy to identify. When a mixture of triclinic feldspars has been reduced to grains with a uniform diameter of about 0.05 mm, it is sufficient to boil them for half an hour in concentrated nitric or hydrochloric acid to attack all the grains of anorthite, while grains of other triclinic feldspars remain unaffected by the acid, especially nitric acid. After the grains have been boiled and dried, one notes that those of anorthite have been inflated and have turned milky white, while the others preserve their original size and transparency.

They can also be distinguished from one another by examining them under the microscope. Grains of anorthite will have become cloudy after being boiled in acid even though they may not lose their original form. They have more small cracks than before, and their surface is crossed by irregular scratches like those seen under the microscope in all gelatinous substances, such as those seen in many varieties of opal. Under crossed nicols, these same grains have lost all their birefringence, and in this way they stand in conspicuous contrast to the grains of other species of feldspar that, under the same conditions, still have their normal interference colors. In this way, one can see that anorthite is very predominant among the feldspars from dikes no.

4, 5, 34, 31, 30, and 20. Though slightly less abundant, it is also the main feldspar in dikes no. 28 and 2, and it is seen in smaller proportions in the feldspars of dikes no. 32, 42, and 19. It is completely absent from the mixtures of feldspar in dikes no. 51, 11, 53, 66, 49, and 61.

The difference in the effects of boiling acids (especially nitric acid) on anorthite and labradorite is so great that it is sufficient in itself to distinguish the feldspar that is most affected as anorthite.

I have nevertheless made every effort to obtain more positive evidence. I have separated grains of feldspar measuring 0.05 mm in diameter from dike no. 34 and boiled them in concentrated nitric acid for a quarter of an hour. After this time, the grains of anorthite were sufficiently altered to be easily distinguished; most lost all their normal birefringence in polarized light. Only the cores of a few grains still had their usual interference colors when observed under crossed nicols, but they could easily be distinguished from grains of other types of feldspar that had lost neither their fresh appearance nor their birefringence. With a little patience, they can be separated from one another under a magnification of 60 power.

Grains that were affected by acid were returned to the liquid in which they had been heated and analyzed by the normal method. This produced the following results.

		Oxygens	
Silica	43.6	23.3	
Alumina	36.5	16.8	
Lime	18.4	5.3	5.9
Magnesia	1.5	0.6	
Soda	trace		
	100.0		

The proportions of the oxygens are 1.05 : 3 : 4.14.

Feldspar grains separated from a sample of dike no. 2 were treated in the same way, and the composition of the altered feldspars was found to be as follows.

		Oxygens	
Silica	43.8	23.4	
Alumina	36.4	16.8	
Lime	19.0	5.4	
Magnesia	0.6	0.24	5.69
Soda	0.2	0.05	
	100.0		

The proportions of oxygens are 1.02 : 3 : 4.1.

Thus the feldspar that was most affected by the acid was anorthite.

The differences between these results and those of the formula for this mineral are considered to be within the margin of error of the analytical procedure. In the case of dikes no. 4, 5, 34, 31, 30, 20, 26, and 2, the proportion of separated feldspar grains that were not affected by acid shows that they consist almost entirely of triclinic feldspars. The presence of monoclinic feldspars is very doubtful.

We have already noted that most of the feldspars seen in thin sections of these basic lavas have the colored stripes characteristic of the triclinic system and that none of those with uniform interference colors and more or less symmetrical cross-sections have an extinction direction parallel to any of their long edges.

One can verify these same observations by examining a certain number of separate grains that have resisted the action of acids. The irregular form of fractures normally inhibits proper measurements, but cases in which it is possible to apply the test are common enough to show that monoclinic feldspars are, if not totally absent, extremely rare. I should add that the test procedures of Szabo and Boricky confirm this view by demonstrating that potassium is very scarce in these feldspar grains.

Having established this, it is clear that the triclinic feldspars associated with the anorthite in these basic lavas are not confined to labradorite.

In all the rocks, including even those of dikes no. 28 and 2, anorthite is so dominant that the presence of only labradorite in the part affected by acid can not account for the excess of silica found in the mixtures of feldspars from these rocks over that which would be appropriate for the formula for anorthite. In this regard, the small proportion that is not affected by acid must be rich enough in silica to form mixtures with anorthite that have as much as 49 percent of this component. I have been able to carry out only a partial analysis of the unaffected fraction of the mixture of feldspars separated from dike no. 2; I found 51 percent silica and noted in addition that this fraction was very rich in sodium. These observations confirm the conclusions based on the microscopic examination.

We should note at this point that the rocks having mixtures of feldspars in which anorthite is most dominant are the ones that are also richest in olivine and have little or no hypersthene or phenocrysts of iron oxides.

Thus the characteristics deduced from microscopic studies to be distinctive of the basic rocks forming the dikes of Thera are consistent with the results obtained from chemical studies.

We now proceed to the feldspars separated from dikes no. 32, 42, 51, 11, 19, 62, 49, 53, and 66. In these, anorthite is almost totally absent. One finds it only in the feldspars of dikes no. 32, 42, and 19. The first two of this latter group are the only ones in which the feldspars have compositions with proportions of oxygen intermediate between those of anorthite and labradorite.

But in this case we are also dealing with mixtures of feldspars in which triclinic forms are dominant. This can be seen from a microscopic examination in polarized light.

As we have already seen, we can also demonstrate that, here too, a triclinic feldspar is dominant and that this feldspar is labradorite.

There is little to be gained by using this chemical procedure as an unequivocal test for anorthite in the basic dikes of Thera, for the labradorite and other feldspars that are richer in silica are also only slightly impervious to acids. For that reason, we must resort to more indirect methods.

The microscope, together with chemical tests, has clearly shown the rarity or even the complete absence of anorthite in this group of mixed feldspars.

The same instrument has also revealed the presence of sanidine grains and even greater amounts of oligoclase in most of the mixtures, especially in those with high silica contents. The sanidine can be identified in sections belonging to the zone ph_1, which are symmetrical and have an extinction direction parallel to the long dimension of the crystals. (The grains also tend to be rectangular.) Oligoclase can be recognized by examining g_1 faces in which the extinction direction is parallel to the edge pg_1. We find in this way that, except for the samples from dikes no. 32, 19, and 42, in all the analyzed mixtures feldspar containing components that notably increase the silica contents of the mixture and an absence of feldspar components that make it lower than the silica content of labradorite. There can be no question, therefore, that grains of sanidine and oligoclase have raised the silica content of the mixtures, so that it closely approaches that of labradorite.

To verify this idea, I have carried out the following test on a mixture of feldspars separated from a sample of dike no. 66.

The feldspar in this rock was reduced to a uniform grain size of 0.05 mm in the normal way and then purified by hand picking under low magnification (60 power) and polarized light with crossed nicols. All grains with a uniform interference color or simple twins were rejected as possible monoclinic crystals, and grains with alternating parallel bands of interference colors were retained as the only crystals that were undoubtedly triclinic. When the remaining grains were analyzed, they gave the following results.

		Oxygens	
Silica	52.6	28.0	
Alumina	29.4	13.6	
Lime	12.9	3.7	
Magnesia	0.2	0.1	5.0
Soda	4.9	1.2	
Potash	trace		
	100.0		

The ratio of oxygens is 1.10 : 3 : 6.2.

If one compares the analysis of the mixture of feldspars from dike no. 66 with that of the same material after the separation described above, one is struck by the differences between the two sets of results.

There is a smaller proportion of silica in the sorted material than in the original mixture of feldspars; lime is about the same, but potash has disappeared. In short, apart from a slight increase in the amount of sodium that is probably within the precision of the analyses, the hand-picked material has a composition much closer to that of labradorite than the mixture of feldspars from which it was extracted.

Despite the care with which the feldspars were separated, the analysis indicates slightly more silica than the amount that labradorite should have, but this excess can easily be explained if one considers the glassy inclusions that are in almost all feldspar grains and if one takes into account the fact that, as will be shown later, such material has an elevated silica content. If there were still oligoclase grains in the labradorite, they would have the same effect.

For all these reasons, we believe that the main feldspar of these intermediate dike rocks of Thera is labradorite and that the excess of silica found in the analyses of feldspars taken directly from these rocks is attributable mainly to small amounts of sanidine and oligoclase mixed with the labradorite. The predominance of labradorite that characterizes these rocks is just as characteristic as the predominance of anorthite is of the more basic rocks.

Thus a series of analyses that at first seemed to offer new support for the theory of Tschermak, seems, on the contrary, to contradict the theory by providing a different explanation for the remarkable rule that was demonstrated by that eminent mineralogist. Instead of a series of isomorphic chemical components, we see nothing more than physical mixtures of specific feldspar species, mainly anorthite and labradorite.

A study of the pyroxenes and olivines in the dike rocks of Thera is of less interest than that of the feldspars, but I have undertaken it in order not to neglect anything that may serve to establish the mineralogical character of these rocks. Olivine is found in notable proportions only in the most basic rocks. Its composition presumably varies from one member of the group to another, as is clear from analyses of olivines separated from dikes no. 4 and 27, shown in the table below.

	Olivine in dike no. 4	Oxygens	
Silica	36.8	19.6	
FeO	25.8	5.7	
Lime	0.8	0.2	20.5
Magnesia	36.6	14.6	
	100.0		

Specific gravity = 3.627. Proportions of oxygens 1.04 : 4.

	Olivine in dike no. 7	Oxygens	
Silica	37.4	19.9	
FeO	25.0	5.5	19.78
Lime	0.5	0.2	
Magnesia	37.1	14.08	
	100.0		

Specific gravity = 3.614. Proportions of oxygens 1.01 : 4.

It is remarkable that these olivines, which as very close in composition to those in the anorthite-bearing blocks brought up in the lavas of the last eruption, differ notably from the olivine deposited in geodes by sublimation. They contain less silica, lime, and magnesia than the latter, but their iron content is much greater (on average 25.4 percent instead of 13.9).

Pyroxenes are found in labradorite-bearing rocks as well as in lavas with anorthite. It was interesting to see whether there were compositional differences between these two very different types and, if so, whether they are uniform within a given type of rock. To this end, I have separated and analyzed these minerals from two anorthite-bearing rocks from dikes 4 and 26 and three from rocks with labradorite taken from dikes 11, 19, and 53.

Hydrofluoric acid was used in making the separation. Few of the pyroxenes in dikes 4 and 26 are larger than 0.5 mm, so any other method of separation would be impractical. Concentrated hydrofluoric acid quickly dissolves all the aluminous minerals in the powdered rock and leaves a fine powder of crystalline pyroxene that need only be washed in flowing water while working the pulp with the finger to get rid of the small amount of gelatinous material that adheres to it. The dried powder was treated with a bar magnet to remove the iron oxides, and a brief hand picking under a magnifying lens completed the purification of the material for analysis.

The crystals extracted in this way are still intact; the faces are shiny and the edges sharp, so the acid seems not to have affected them. When mounted on a glass slide with Canada balsam they make a beautiful microscopic specimen. The analyses were carried out, as in the case of the feldspars, by means of the excellent method of Mr. Henri Sainte-Claire Deville.

	Pyroxene from dike no. 4	Oxygens		Pyroxene from dike no. 27	Oxygens	
Silica	51.7	2.6		49.4	26.3	
Alumina	1.5	0.7		2.1	1.0	
Fe_2O_3	1.6	0.5		0.9	0.3	
FeO	8.5	1.9		10.4	2.3	
Lime	22.3	6.4	14.1	23.7	6.8	14.5
Magnesia	14.4	5.8		13.5	5.4	
	100.0			100.0		

Proportions of oxygens	2 : 1.03	2 : 1.10
Specific gravity	3.367	3.384

	Pyroxene from dike no. 11	Oxygens		Pyroxene from dike no. 19	Oxygens		Pyroxene from dike no. 53	Oxygens	
Silica	50.5	26.8		49.5	26.4		50.2	28.8	
Alumina	3.8	1.7		2.9	1.3		4.3	1.9	
Fe_2O_3	0.4	0.1		0.7	0.2		0.9	0.3	
FeO	19.2	4.3		1.1	3.8		24.2	5.4	
Lime	6.6	1.9	13.9	12.6	3.6	14.2	7.3	2.1	12.6
Magnesia	18.9	7.6		16.7	6.7		12.3	4.9	
Soda	0.6	0.1		0.5	0.1		0.8	0.2	
	100.0			100.0			100.0		

Ratio of oxygens with silica to those with nonoxide cations	2 : 1.04	2 : 1.07	2 : 0.94
Specific gravity	3.438	3.417	3.452

One sees from the table above that the two types of rocks in the dikes of Thera have very different pyroxenes.[16] Those in the anorthite-bearing dikes are rich in lime and have little iron oxide. The reverse is true of the pyroxenes in the labradorite-bearing dikes. One also sees that the analyses are very similar to those of labradorite- and anorthite-bearing lavas of the recent eruption. There too, the calcium-rich pyroxene is associated with anorthite and the iron-rich one with labradorite. Thus, the association of minerals is consistent, despite the difference in age of the rocks and the intensity of the cataclysm that created the bay in the interval between the present eruption and the period in which the dikes were emplaced.*

---

* One could object that the blocks with anorthite erupted during the current activity could be fragments torn from older rocks under the volcano, but this does not invalidate our observations that the labradorite-bearing lavas form the bulk of the mass produced by the present eruption.

A microscopic examination of this material shows that there are also other important differences. In the anorthite-bearing rocks, the pyroxene is, as already noted, exclusively augite; in the labradorite-bearing ones it is a mixture of both hypersthene and augite in which hypersthene is much more abundant.

Thus, even though I tend to have faith in the analyses, I have felt it best to carry out additional ones, in order to learn the composition of the glassy material and the microlites it contains.

It is clear from these analyses that, without exception, the bulk compositions of all these dikes are richer in silica than the feldspar phenocrysts that are found in them.

Moreover, if one considers a single dike that has parts with differing proportions of microlites, one finds that a few of these microlites have quite different compositions; the parts with the most microlites tend to be less siliceous than the others. This is especially true in the case of basic dikes. One can conclude from this that in certain cases there are notable differences between the compositions of the microlites and the glassy material in which they are enclosed.

Several lines of evidence, such as the silica contents of the labradorite-bearing magmas, the fact that the extinction direction of their feldspar microlites is approximately parallel to one of the directions of polarization, and the similarity of the rocks to lavas of the recent eruption that contain microlites of albite and almost always oligoclase, support the view that the feldspar microlites in the groundmass of these rocks should be given as much importance as the large phenocrysts.

The question posed by anorthite-bearing lavas is more difficult. A certain number of the microlites in them have extinction directions almost parallel to their long axes, but more often the angle is greater and in some instances may be as large as 30 degrees. This indicates that the compositions of these feldspars are those of labradorite.

Thus, even though albite and oligoclase can be found among the microlites of these lavas, it is evident that labradorite is the dominant feldspar in the groundmass.

The analytical results that follow will enable the reader to relate my observations to those of other scientists who have dealt with the bulk compositions of the Santorini lavas.[17]

ANALYSES OF BULK ROCK COMPOSITIONS

Anorthite-bearing rocks

	Dike no. 4	Dike no. 34	Dike no. 27[18]	Dike no. 31	Dike no. 20
Silica	51.8	52.4	51.5	51.9	51.7
Alumina	20.1	21.3	24.5	22.2	22.4
Fe_2O_3	11.6	8.7	8.8	7.1	7.4
Lime	11.9	11.8	10.9	9.3	10.4
Magnesia	3.4	3.9	2.3	5.2	4.3
Soda	1.1	1.8	1.2	3.7	3.4
Potash	0.1	0.1	0.8	0.6	0.4

Labradorite-bearing rocks

	Dike no. 32	Dike no. 11	Dike no. 19	Dike no. 49	Dike no. 62	Dike no. 53
Silica	53.9	57.2	56.3	60.9	63.6	64.6
Alumina	25.6	19.1	18.5	21.6	20.2	18.7
Fe_2O_3	5.7	6.9	6.0	4.3	5.9	6.2
Lime	6.8	7.1	9.4	4.2	2.5	2.8
Magnesia	1.9	3.5	5.5	1.5	0.6	1.5
Soda	3.4	4.2	3.6	4.6	5.0	4.7
Potash	2.7	2.0	0.7	2.9	2.2	1.5

As noted earlier, all lavas for which compositions are given here are richer in silica than the phenocrysts of feldspar they contain, and all other minerals found as phenocrysts - pyroxene, olivine, and iron oxide - can only lower the silica content of the bulk rock. It follows, then, that the excess silica can only be attributed to the groundmass of the rock, i.e. the glass and microlites.

Magnesia is also present in much greater amounts in the rock as a whole than in the feldspar phenocrysts. This must be due to olivine and especially pyroxene which are present either as phenocrysts or as microlites.

Alumina, on the other hand, is always present in smaller amounts in the rock than in the feldspar that has grown in it. This is due to the relatively small amounts of this element in the iron-bearing minerals of the rock and, no doubt, to the compositions of the amorphous material and feldspar microlites that have compositions higher in the feldspar series and therefore less aluminous than the anorthite or labradorite phenocrysts.

We should note again that the amount of iron in these rocks diminishes as silica increases, even though the microscope shows that large phenocrysts of iron oxide are abundant in the most siliceous lavas. This apparent inconsistency must be due to the greater number of very small grains of iron oxide in anorthite-bearing rocks and to the abundance of olivine in the same rocks.

In addition to the sixteen dikes discussed above, I have studied in somewhat less detail other dikes with the same relations to the volcano.* I studied these under the microscope, and I have separated sufficient amounts of their feldspars to test them with acid and determine the degree to which they are altered by this reagent. I find that some of these feldspars have the characteristics of mixtures dominated by anorthite, while others have grains that are little affected by boiling nitric acid. The dikes in which anorthite is most important are also those with the most olivine. The dike rocks in which the feldspars are resistant to acid have little or no olivine, and one sees under the microscope that they contain phenocrysts of iron oxides. This confirms the division into two groups set up earlier on the basis of a few special samples that were the subject of more complete optical and chemical studies. Thus, the dikes of Thera can now be divided as follows.

1. Rocks with phenocrysts of anorthite and microlites of labradorite: Dikes no. 4, 5, 9, 10, 12, 13, 14, 16, 17, 20, 21, 27, 28, 30, 31, 32, 33, 34, 36, 37, 38, 39, 40, 41, 45, 50, 52, 53, 54, 55, 64.

2. Rocks with phenocrysts of labradorite and microlites of albite-oligoclase: Dikes no. 1, 6, 7, 8, 11, 15, 18, 19, 24, 25, 26, 35, 42, 43, 44, 46, 47, 49, 51, 57, 58, 59, 60, 61, 62, 63, 65, 66.

Let us now consider some of the geological deductions that can be drawn from field observations bearing on the age and distribution of dikes belonging to the two groups we have just set up. These deductions are as follows.

1. The oldest lavas of the two edifices, Megalo-Vouno and Mikro Profitis Ilias, are anorthite-bearing. These lava flows and intrusions are distinctly cut by both anorthite- and labradorite-bearing dikes.

It is worth noting that the same relations hold for the lavas of the most recent eruption. The lava of this eruption contains labradorite, and anorthite is found only in rounded blocks that were probably torn from the older rocks at greater depths by the rising labradorite-bearing magma.

2. Dikes with anorthite are found in both Mikro Profitis Ilias and Megalo-Vuono, but they are not numerous in the latter.

The reddish lava of trachytic appearance and platy texture that forms the base of Mikro Profitis Ilias differs in certain respects from the anorthite-bearing rocks of the dikes; it is richer in silica and sodium and has a conspicuous fissile texture, but its feldspar is readily attacked by acid and rich in lime, so there is little doubt that it is anorthite.[19] The composition of this lava is shown by the following analysis.

---

* A few dikes have not been studied owing to the difficulty of reaching them and to chance circumstances that did not permit sampling them.

Silica	56.0
Alumina	23.5
Fe_2O_3	5.3
Lime	6.7
Magnesia	2.6
Soda	5.5
Potash	0.4
	100.0

3. Dikes of the labradorite-bearing kind are found in both of these structures, but they are especially common in that of Mikro Profitis Ilias where they appear to have invaded and uplifted a major part of the edifice, especially in the summit region.

It is worth noting that the dike that is closest to Apano Meria (no. 1) is the one that is most remote from the others containing labradorite.

4. As shown in the earlier table, the two types of dikes commonly interfinger.

5. Eruptions of the two types of lava must have alternated with each other, for many labradorite-bearing dikes are cut by anorthite-bearing ones.

6. There seems to be no relationship between the mineralogical composition of the dikes and their orientation. Among those that strike NNW as well as those that have a trend more toward the NNE, one finds both dikes with anorthite as well of others containing labradorite. The only difference is that, in general, the rocks in NNW-trending dikes are brownish-yellow, whereas almost all those with a NNE trend are dark black. The dikes oriented NNW also tend to be thicker. They seem to have been more viscous than the others at the time they were emplaced, and they have been more strongly oxidized. The rocks are more altered and contain more ferruginous material. The labradorite-bearing rocks that follow this same trend also tend to contain more tridymite. Aqueous solutions appear to have played a somewhat more important role in the eruptions of their lavas.

We should also say a few words about the place that should be assigned to the series of modern lavas relative to the two types of rocks found as dikes in the northern part of Thera.

Before doing this, we should first consider the various systems of classification used today and determine which is best suited for dealing with the rocks we considered here. All the many schemes for classifying rocks are based on different assumptions and principles.

The oldest ones are based solely on physical properties. Color, density, and texture were the main criteria used in them.

Now that bulk compositions are available for more volcanic rocks from various localities, most recent classifications have been based on silica content rather than mineralogical composition. Volcanic rocks are thought to be products of magmas derived from subsurface sources of different chemical compositions, and this is said to account for the chemical differences of the lavas. I intend to explain why both of these approaches are invalid and should be rejected as criteria for classification.

The minerals of volcanic rocks are also useful for classification. Among the systems based on this principle, the most successful have been those emphasizing the nature of the alumina-silicate minerals, which are usually considered the principal mineralogical components of the rock.

In this way rocks having nepheline, leucite, and nosean have been distinguished from the feldspar-bearing rocks with which they were formerly confused. Extending this method further, one can separate the feldspathic rocks according to the nature of the feldspar they contain. When the feldspar species were still thought to be perfectly distinct minerals, there was no reason to question a division of volcanic rocks on this basis. The ferromagnesian minerals in the same rocks supported the same approach and also served to establish various subdivisions.

Now under strong attack, this method is left in disarray by the proposals of Professor Tschermak. If all triclinic feldspars are only isomorphic mixtures in all proportions of anorthite and albite, there is no reason to distinguish oligoclase from labradorite, and the only remaining division is between the monoclinic and triclinic feldspars; volcanic rocks cannot be divided on the basis of their plagioclase. Albite, anorthite, and the intermediate feldspars belong to a single group of rocks, regardless of which of these minerals is dominant, and divisions within this very large group of feldspar-bearing rocks can be made only on the basis of the other minerals found in them.

If Tschermak's rule is accepted, all the advantages of the method disappear. I believe, therefore, that a service has been rendered to science in showing how this rule should be interpreted and in furnishing new information in favor of maintaining labradorite as a well-defined species of feldspar.

Another objection, based on my personal experience, can be made to this kind of classification system. If several kinds of feldspar can exist together in the same volcanic rock, why should only one of them be used to place the rock in a particular niche? If a rock contains microlites of albite, oligoclase, and labradorite as well as phenocrysts of a monoclinic feldspar, labradorite, and anorthite, why should any one of these feldspars determine the rank of the rock? This question must be answered.

First, the true nature of the feldspar microlites is still uncertain, and there are so many difficulties in determining their compositions by chemical means that one cannot use them as a basis of classification. Second, among the feldspar phenocrysts, one is almost always so dominant over the others that its presence really determines the character of the rock.

While recognizing that a natural classification should take account of groundmass minerals as well as phenocrysts, I am inclined to adopt tentatively an artificial classification based on the nature of the leucocratic component that is dominant among the phenocrysts. If one accepts this view, one must distinguish not only rocks with leucite, nepheline, nosean, and monoclinic and triclinic feldspar but also those with anorthite, labradorite, and possibly andesine, depending on which mineral is the predominant phenocryst.

The method based on physical properties must be rejected completely. In the case of the rocks we are concerned with here, it leads us to consider some of the light-colored, platy rocks found in the dikes of Thera as phonolites, even though they are nothing more than labradorite-bearing lavas. It would result in grouping with the trachytes any light-colored, rough-textured lava in which olivine cannot be detected with a hand lens. Possible errors of this kind show the dangers of such a system of classification.

Classifications based on the average silica content of the rocks must also be rejected. In our case, they would lead to inclusion of labradorite-bearing rocks found in the dikes of Thera in the same category as sanidine-bearing trachytes that have very nearly the same silica content.

For these reasons, I shall use a classification based on the distinction of modern eruptive rocks according to the leucocratic mineral components that are most prominent among its phenocrysts. Accordingly, the dikes on Thera naturally fall into the two groups characterized by labradorite and anorthite. We should also note that the groundmass feldspar lends support to this division, since the groups to which the groundmass feldspars belong vary along with the phenocrysts of feldspar.

But in our case this division should be extended. It is essential to separate the labradorite-bearing rocks of Santorini from other common labradorite-bearing basalts or from certain other basic lavas, such as those of Etna.

The common occurrence of opal and tridymite and the rarity of olivine in the labradorite-bearing rocks of Thera are the main characteristics that will enable us to maintain this distinction.

Within the category of volcanic rocks with labradorite there is ample room to distinguish a relatively acidic group to which the labradorite lavas of Santorini belong and another group of labradorite-bearing rocks of the kind found on Etna. In dealing with the anorthite-bearing lavas, one should make a similar distinction and set aside as one subdivision the relatively felsic rocks of the type characterized by those of Santorini and another, more basic subdivision that would include certain rocks of Reunion that were recently described by Mr. Vélain.[20]

In addition, each of the acidic groups can be divided into two subgroups: one with greater acidity that, strictly speaking, constitutes the true acidic group, whereas the other has properties closer to those of the basic group and is best regarded as intermediate. Just as sanidine-bearing lavas have been divided into trachytes and rhyolites and lavas with oligoclase into andesites and dacites, depending on whether or not they have free silica, labradorite-bearing rocks can be divided into two corresponding groups. The labradorite lavas of Santorini belong to a subgroup lacking silica minerals. Tridymite cannot be considered exactly equivalent to the free silica of rhyolites and dacites; that mineral is not an essential component of these rocks.[21]

The anorthite-bearing lavas of Santorini are rich in olivine and normally have no opal or tridymite. If one must also divide the acidic anorthite lavas into two groups, one with and the other without free silica, the anorthite-bearing rocks of Thera would fall into the silica-poor category. A silica-rich subgroup is not unlikely, for studies of some of the anorthite-bearing dike rocks of Thera show that there is nothing to indicate that anorthite is incompatible with free silica.[22] The recent work of Mr. Vélain has provided very interesting examples of rocks of this kind.*

The rocks of dike no. 52 contain anorthite as well as tridymite, but it is significant that olivine is rare, whereas it normally is common in the anorthite-bearing dike rocks. The incompatibility of tridymite and olivine is illustrated by this example, but this incompatibility is far from absolute, for it is not uncommon to find in a single thin section of this rock, and even in the same field of view under the microscope, both of these minerals together.

In summary, some of the dike rocks of Thera belong to an intermediate group with labradorite, others to a group with anorthite, but, on the whole, the labradorite-bearing rocks, because they have an intermediate content of silica and a feldspar that falls in the middle of the triclinic series, should be considered rocks that are indeed intermediate, whereas all the rocks with anorthite are definitely basic, including even the more acidic varieties.

To complete this study, the following table provides a summary of the dikes along with notes on the geological relations in which they are found.

* We should note, however, that the silica in these rocks is generally in the form of tridymite.

Number	Width	Orientation	Phenocrysts	
1	1.0 m	N 5°W	labradorite	
2	3.0 m	N10°W	anorthite	Thin distinct selvages; rises very high
3	1.0 m	N20°E	anorthite	Rises very high
4	1.0 m	N30°E	anorthite	
5	0.8 m	N20°E	anorthite	
6	1.2 m	N30°W	labradorite	
7	0.6 m	N20°W	labradorite	
8	2.0 m	N40°W	labradorite	
9	0.3 m	N40°E	anorthite	
10	0.3 m	N40°E	anorthite	
11	2.0 m	N30°E	labradorite	
12	5.0 m	N30°W	anorthite	
13	0.2 m	N30°E	anorthite	
14	1.5 m	N40°E	anorthite	
15	3.0 m	N30°E	labradorite	
16	4.0 m	N30°W	anorthite	
17	0.4 m	M20°W	anorthite	
18	2.0 m	N20°E	labradorite	
19	6.0 m	N20°E	labradorite	Selvages showing their eastern faces; cuts dike no. 20, is cut by dike no. 21
20	1.0 m	N30°E	anorthite	
21	0.3 m	N30°E	anorthite	Bifurcates at base; cuts dikes 19 and 20
22				Small dikes cutting a ledge of lava
23				coming from dike no. 21
24	10.0 m	N45°W	labradorite	
25	6.0 m	N30°E	labradorite	Rises very high
26	4.0 m	N35°W	labradorite	Made up of three adjoining dikes
27	0.5 m	N30°E	anorthite	
28	1.5 m	N20°W	anorthite	
29	0.4, 0.2 m	N30°W		Three small dikes exposed over a limited distance
30	1.0 m	N30°W	anorthite	
30'	0.3 m			
31	1.0 m	N30°E	anorthite	Dikes 31 and 32 are adjacent
32	1.0 m	N30°E	anorthite	Dike no 32 has two distinct selvages
33	0.6 m	N20°E	anorthite	Dike with distinct selvages
34	0.3 m	N35°E	anorthite	Cuts dike no. 33
35	2.0 m	N30°E	labradorite	
36	0.3 m	N25°E	anorthite	
37	1.0 m	N25°E	anorthite	
38	0.4 m	N30°E	anorthite	
39	0.3 m	N30°E	anorthite	
40	0.8 m	N35°E	anorthite	
41	1.5 m	N40°E	anorthite	Exposed on its eastern face
42	8.0 m	N20°W	labradorite	Distinct selvages; yellow, fine partings
43	1.0 m	N25°W	labradorite	Distinct selvages; top tilted eastward from base
44	1.0 m	N30°W	labradorite	Cut by dike no. 43
45	9.3 m		anorthite	Curves toward the west
46	1.0 m	N20°W	labradorite	Curves toward the west; beside no. 4
47	1.2 m	N30°W	labradorite	Distinct selvages
48	0.3 m			
49	15.0 m		labradorite	Distinct selvages
50	1.5 m		anorthite	

Number	Width	Orientation	Phenocrysts	
51	2.0 m	N15°W	labradorite	
52	4.0 m	N40°W	anorthite	
53	1.2 m	N45°E	anorthite	Abundant tridymite although the rock is very basic
54	1.5 m	N20°E	anorthite	
55	0.3 m	N20°W	anorthite	
56	0.2 m	N10°E		
57	2.0 m	N30°E	labradorite	
58	2.0 m	N45°E	labradorite	
59	0.3 m	N45°E	labradorite	Bifurcates at the base, join with dike no. 60
60	3.5 m	N60°E	labradorite	
61	1.5 m	N65°E	labradorite	
62	2.0 m	N45°W	labradorite	Visible in the cliff and on the separated promontory
63	0.6 m	N45°W	labradorite	
64	2.5 m	N45°W	anorthite	
65	1.4 m	N65°E	labradorite	
66	1.8 m		labradorite	Distinct selvages

CHAPTER EIGHT

PETROGRAPHIC STUDY OF THE ROCKS IN THE SOUTHWESTERN PART OF THERA

In my earlier geological descriptions of the region around Akrotiri in the southwestern part of Thera I pointed out that this district was characterized by a great abundance of amphibole-bearing andesites. Some are compact rocks; others are loose scoria or tuffs. I also noted that there were a few basic rocks in the same region.

A microscopic examination of these rocks has led to observations that are sufficiently interesting to merit a special chapter of their own.

BASIC ROCKS OF THE AKROTIRI REGION

Among the basic rocks, we note first the lava forming the reefs in front of the harbor of Balos; this rock has a mineralogical composition substantially different from that of most other rocks of Santorini. Among the phenocrysts, one sees scarcely anything but olivine. This mineral has the form of rounded grains that are altered at their margins and along cracks running through them. Their diameters range from 0.2 to 1.0 mm. Feldspar and pyroxene are present only as microlite, but they have dimensions considerably greater than those normally seen in the lavas of Santorini. The feldspar microlites have an average length of about 0.3 mm and a width of 0.02 mm; they are simple crystals or twins joined along their long axes. Many have quite large extinction angles (30 to 35 degrees). The pyroxene microlites average about 0.02 mm across, and crystals of iron oxides have similar dimensions. Amorphous material is scarce and contains many small globules. The phenocrysts of feldspar are readily attacked by boiling acids; after boiling them in nitric acid for a quarter of an hour they lose all their birefringence under polarized light. The liquid produced by dissolving them shows a microscopic and chemical reaction with compounds of lime. The microlites, however, resist the action of acids. One can conclude from this that the phenocrysts are anorthite, and the microlites are labradorite. In short, these rocks have all the physical properties and mineralogical features of basalts. They represent a transition between the anorthite-bearing lavas of Santorini and true basalts.

The scoriaceous blocks that form the brownish-black cliffs at Balos are also composed of anorthite-bearing lava, but it is much less basic than the one just described; first, it is more feldspathic and, second, anorthite is not the only feldspar found as phenocrysts. Labradorite is also present in this form and in some specimens is very abundant. The phenocrysts of feldspar are for the most part intact and unbroken, but some are very altered, and the glassy inclusions have cracks filled with limonite identical to that making up much of the surrounding groundmass. The cleavages parallel to p and q_1 are very sharp, but many contain ferruginous material. Although most of the crystals are single individuals, many are multiply twinned according to the albite law. Most of these twins are relatively narrow.

A few crystals of pyroxene are found among the phenocrysts in this same rock, but olivine seems to be completely absent. The groundmass in which these phenocrysts are enclosed contains great numbers of microlites of labradorite and a few grains of pyroxene. Iron oxides are present as numerous small granules that have been converted almost entirely to limonite as a result of the strong alteration that affected the material when it was erupted.

This lava is almost in contact with the preceding one, strongly tempting one to view the two as products of the same volcanic event, but their marked mineralogical differences seem to rule this out. There must have been two distinct eruptions, one that formed the more basic rocks and another that produced a lava that was transitional between the anorthite- and labradorite-bearing types.

A third rock with basic characteristics can be seen high on the cliff at Balos. It is remarkable for its ropy surface and its distinctive appearance under the microscope. It has the general appearance of a dark brown trachyte filled with innumerable round vesicles that make up almost as much of the volume as the rock itself. The only phenocrysts are rare, poorly formed pyroxene crystals, a few well-developed crystals of labradorite, and only a few rare crystals of anorthite. Most of these two triclinic feldspars are simple individual crystals in which the two cleavages p and q_1 are well developed. One might easily mistake them for sanidine, but when seen in sections in the zone ph_1, their extinction angles are greater than 0 degrees. Those of labradorite resist the action of boiling nitric acid, while those of anorthite are readily attacked by this acid and the solution obtained in this way has an abundant precipitate when ammonium oxalate is added. The grains of anorthite that were treated with fluosilicic acid and then dried leave a gelatinous, white, translucent precipitate characteristic of calcium salts and, at the same time, the normal fluorosilicates of lime. After evaporation, one of the grains of labradorite produced a substance with properties that were similar but less pronounced and, in addition, the usual types of fluorosilicates of sodium and lime. The cubic forms of potassium fluorosilicates did not appear.

These tests offer a striking example of how one can use chemical reactions to examine feldspars that have been separated with an electromagnet. Phenocrysts are so scarce and small in the rocks high on the cliff of Balos that analyses, even at a qualitative level, of their feldspars seemed at first to be an insurmountable task. And yet, thanks to these procedures the problem could be solved without difficulty.

An interesting relationship revealed by a microscopic examination of this lava is the common presence of very crystalline fragments of a labradorite-bearing lava (lacking anorthite). The tiny patches of amorphous material in these inclusions are colorless and contain microlites of oligoclase and albite. They are totally different from the rest of the lava.

An analysis of the ropy lava of Balos gave the following results:

Silica..................................	57.2
Alumina............................	19.5
Fe_2O_3.............................	9.2
Magnesia.........................	3.1
Lime..................................	5.7
Soda.................................	5.2
Potash.............................	0.1
	100.0

In the village of Akrotiri itself, especially below the houses on the southern side of the town, one can find a small outcrop of black and brownish rock that closely resembles the rocks in the slopes of Balos. When examined under the microscope, the material making up this body is seen to have the same components as most of the rocks at Balos. It is a lava with phenocrysts of anorthite, microlites of labradorite, and notable amounts of pyroxene.

And finally, about 200 meters to the west of the village of Akrotiri up slope from a windmill there is a body of reddish scoria composed of a very similar rock. This locality must certainly have been an eruptive center that discharged not only lavas but large amounts of gas and acidic fumes.

The basic rocks that I have just described are not the only ones of this type in the region around Akrotiri. The promontory that sticks out on the southern side of Cape Mavrorachidi offers yet another example of a lava in which the silica content must have been relatively low. But its main phenocrysts of feldspar are not anorthite but labradorite. This particular rock makes up a thick mass of red and brown scoria that contrasts in form and color with the lighter-colored rocks it transects. Under the microscope, it is seen to contain several types of phenocrysts: (1) labradorite in thin, multiple twins, commonly with ferruginous inclusions, especially in their centers; (2) rare anorthite that can only be distinguished by separating the crystals and treating them with boiling nitric acid; and (3) augite in simple and twinned crystals that are crossed by cleavage traces and prominent, irregular fractures. A great number of these augite grains have lost their greenish tint and are almost as colorless under the microscope as crystals of olivine. They are quite resistant to the action of acids. The amorphous material is colorless but charged with small, dark granules of iron oxides or limonite and with occasional tiny crystals of altered pyroxene that in places are almost opaque. This amorphous material has an overall dark reddish-brown color and is seen to be colorless only in a few isolated spots. It also contains microlites of feldspar in simple crystals or binary twins that for the most part have very small extinction angles. Their dimensions have a wide range from one sample of the rock to another. It seems likely that almost all these feldspar microlites are of albite and oligoclase and that only a few are of labradorite.

ACIDIC ROCKS IN THE REGION AROUND AKROTIRI

Despite their large number of scattered exposures, the basic and intermediate rocks just described are actually rather exceptional in the immediate vicinity of Akrotiri. Most of the rocks found throughout this district are dense or tuffaceous silica-rich rocks that have hornblende as a very characteristic component. This latter mineral is not found in any of the rocks of the Kameni islands or those of the northern and eastern parts of Thera. It is absent from the lavas of Therasia and Aspronisi as well. In this region around Akrotiri, however, it is quite prominent. It gives the rocks of this southern part of Thera a very special character that further accentuates their distinctive mineralogical aspects.

If one adopts the petrographic classification scheme proposed by Professor Rosenbusch, the rocks of the Kamenis, Therasia, Aspronisi, and the northern part of Thera are augite andesites, whereas those of the southern part of Thera should be assigned to a group of amphibole andesites. Thus the petrographic criteria lend support to the stratigraphic observations and show that two very distinct series of volcanic rocks have been produced in the archipelago of Santorini.

The region around Akrotiri has been the focus of very intense silicification. Most of the rocks have been altered and permeated with various forms of silica. It was not just simple secondary processes that were responsible for these deposits, for the effects are seen in quite large masses of rock. In many places the silicification is both extensive and profound. All evidence indicates that it began simultaneously with the eruptions and continued long after. It is especially evident in the elevated and superficial parts of the terrain, as if abundant siliceous emanations had affected the surface layer. All along the escarpment between Loumaravi and Cape Mavrorachidi, for example, the slope is covered with a layer of quartz-rich debris littering all the ground surface, while in the bottoms of ravines and the low parts of the cliffs the most conspicuous phenomenon is the extensive alteration caused by acidic waters. The minerals have lost much of their birefringence and are impregnated with opal and zeolites. Chalcedony, however, is rare in these localities.

In the areas where rocks with free silica are prominent one still finds outcrops that appear to be unaltered and without any marked signs of silicification. I shall describe their mineralogical composition and their main textural varieties.

The most common type is a rough, gray rock, some parts of which are dense, bluish gray while others are porous and have a lighter tint. This rock forms a thick ledge that can be seen in the ravine that borders the western side of the escarpment just mentioned. Under a hand lens one can distinguish feldspar, hornblende, and augite in crystals that have diameters of about a tenth of a millimeter. The microscope reveals the following additional minerals: sanidine, oligoclase, labradorite, brown hornblende, and light green augite. The sanidine is in rectangular plates with parallel extinction. The labradorite is recognizable from its more or less rectangular sections in the zone ph_1 and its extinction angles of 25 to 30 degrees. It occurs both as single crystals and as binary or multiple twins. The probable presence of oligoclase is indicated by the large number of crystals with multiple twins that have extinction angles of less than 5 degrees. But of all the feldspars, labradorite appears to be the most common.

Along with the minerals just mentioned, one sees apatite and iron oxides, but both of these are relatively scarce. All these crystals are set in a colorless, amorphous matrix containing abundant small crystals in the form of colorless or pale yellow laths and irregular yellowish grains or globules. True microlites are rare, and as a result the groundmass is almost extinct under crossed nicols, but one can easily recognize a few microlites of several feldspar species. One rather common type has short, rectangular, rarely twinned crystals with parallel extinction. If they were not completely resistant to boiling nitric acid, one would be tempted to think they were nepheline, but since they are definitely feldspars, they must be sanidine. Another more numerous variety forms more elongated, very thin crystals that are either single individuals or binary twins and have an extinction direction parallel to their long axis; these are probably oligoclase. Another type has an extinction angle of as much as 18 degrees but normally less than 12 degrees; these crystals tend to be larger and are made up of binary or multiple twins. They are probably albite. Grains of titaniferous iron oxides are enclosed in all of the various components of the rock.

The feldspar phenocrysts are rich in glassy inclusions; some of these are single or multiple droplets, while others have irregular or polyhedral forms. Almost all the amorphous material of inclusions is brown, whereas the normal groundmass is colorless. This indicates that the inclusions were incorporated into the growing feldspars while the primitive magma was still in the reservoir and its various chemical constituents had not yet been partitioned into its separate mineral components.[1]

In some instances, the amorphous material of the groundmass is exceptionally rich in small brown globules, but all of these are poor in iron. Grains of pyroxene and iron oxides are almost completely absent, so that when the rock is crushed very little can be extracted with an electromagnet, and feldspars can be separated in this way from only a few special samples. It is quite difficult, therefore, to carry out chemical tests of the feldspars in this rock.

Nevertheless, deductions based on optical observations must still be confirmed by chemical tests. I have therefore felt obliged to carry out this sort of verification, at least for the feldspar phenocrysts.

With the electromagnet one can remove pyroxene, amphibole, and small amounts of iron oxides from the rock. One must then resort to tedious hand picking to separate the feldspars from the amorphous material and verify their birefringence under polarized light.

Having done this, it is necessary to sort out the monoclinic from the triclinic feldspars. I have been able to use only one difficult and purely qualitative method, which consists of examining under crossed nicols all the feldspar grains in the field of view and putting to one side all the grains with a uniform interference color* and to the other side those with triclinic bands. (Doubtful grains were rejected.) I then analyzed the two sets of powdered crystals and obtained the following results.

	Grains of uniform color	Proportions of oxygen		Grains with alternating stripes	Proportions of oxygen	
Silica	65.5	35.0		60.8	32.5	
Alumina	22.2	10.2		25.2	11.6	
Fe_2O_3	0.2			0.3		
Lime	3.4	1.0		7.8	2.2	
Magnesia	0.3	0.1	2.7	0.2	0.1	3.7
Potash	2.3	0.1		0.8	0.1	
Soda	6.0	1.5		4.9	1.2	
	100.0			100.0		

Ratio of Si : R' : R = 10.3 : 3 : 0.88   Ratio of Si : R' : R = 8.2 : 3 : 0.92

The analytical results for the first sample do not correspond exactly to the composition of a monoclinic feldspar and those in the second do not correspond to the composition of either oligoclase or labradorite. One can be sure, however, that (1) the amphibole andesite contains a significant proportion of monoclinic feldspar, and (2) the analyzed triclinic feldspars are a mixture of oligoclase and labradorite.

These results are in accord with those reached from optical observations, but we can also draw other conclusions.

First, one should note that neither of the two analyses is in agreement with Tschermak's rule; it is quite clear that we are not dealing with simple mixtures of different feldspars.

The first feldspar has less silica than is normally found in monoclinic feldspars; the ratio of oxygens for silica to those of trivalent and monovalent oxides is 10.3 : 3 : 0.88 instead of 12 : 3 : 1, showing that the monoclinic feldspar is mixed with single crystals or binary twins of triclinic feldspar.

Moreover, it was evident a priori that there had been mixing due to the method of sorting that was used. Almost all the crystals of feldspar are flattened along g_1 and, when they are removed from the matrix of a rock, they

---

* Naturally, the variations due to unequal thicknesses near the edges of crystals were ignored.

tend to lie on that face, so that the multiple albite twins cannot be detected under polarized light. It is difficult to be sure whether one is dealing with a crystal with multiple albite twins or simply with a binary twin following the Carlsbad Law.

Moreover, is it not surprising that the analysis of the feldspar that was thought to be monoclinic has so little of the monoxide cations? What should one make of the large proportion of sodium and calcium shown by the analysis? The general silicification of all the rocks in the Akrotiri region may explain these features if impurities have been introduced into what was really a triclinic sodium-calcium feldspar at the same time that the silica content was raised to a level close to that of orthoclase. Thus the proportion of mixed triclinic feldspars could be quite substantial.

Tests with fluosilicic acid following the method of Boricky[2] tend to support these deductions. I have subjected several grains to this type of test. In most cases this resulted in only fluorosilicates of sodium and calcium with white, gelatinous opal, but a few produced cubic crystals of the potassium fluorosilicate as well. Thus, the material that I analyzed was not just monoclinic feldspars but rather a mixture of this feldspar with free silica and triclinic feldspars.[3]

The highly sensitive and precise method of Szabo for determining the compositions of feldspars by means of the color of flames also shows that the feldspar that was thought to be monoclinic rarely has more potassium than is found in supposedly pure oligoclase.

The analysis of the triclinic feldspar in the second column seems to show that, despite their optical properties, the dominant triclinic feldspar in the rock is in fact oligoclase. But again, as in the case of the analysis in the first column, we see a deficiency of monoxide cations that seems to point to alteration of the feldspars by acidic water. One can infer, therefore, that these same solutions have deposited a certain amount of free silica, which has raised the silica content of the mixture and given it a composition close to that of oligoclase, whereas, in reality the dominant feldspar is actually labradorite.

The abundance of lime in the triclinic grains shown by quantitative analysis and by both the considerable proportion of calcium fluorosilicate as well as the opaque nature of the white, gelatinous material obtained when the grains were treated with hydrofluoric acid support the conclusion that the major part of the feldspar in the rock is labradorite. And finally, the tests made by the method of Szabo lend further support to this conclusion. Thus the optical properties, in conjunction with qualitative tests, modify the deductions that one would otherwise be tempted to draw from the seemingly more reliable chemical analysis. So one sees that the chemical analysis quite often resolves questions raised initially by optical observations, but at the same time the latter help in interpreting and verifying the chemical tests. These two methods act as controls on one another in mineralogical research. In some instances the results are in accord with one another, but in others one or the other may appear to be incorrect and the two methods seem contradictory.

I have carried out similar studies on other rocks that resembled those just described but had been more or less silicified. These investigations were designed to obtain more information on the nature of the feldspars found in these rocks. All the results I obtained resemble those I have outlined above; the main differences seem to be in the relative proportions of the monoclinic and triclinic feldspars. In certain samples (that came from Cape Akrotiri rather than the region of central Akrotiri) sanidine turned out to be relatively abundant, even though it is in all cases less common than the triclinic feldspars. Its chemical properties, which were determined qualitatively by the methods of Professors Boricky and Szabo, were consistent with the optical properties.

I should note here a practical technique I find helpful in identifying sanidine. Each of the crystals behaves as though it were composed of thin, randomly oriented lamellae, so that the extinction direction is not uniform throughout the entire section of the crystal. When one rotates the slide under crossed nicols, each grain, when in its position of maximum extinction, is partly shaded by a vaguely defined shadow that seems to pass over its entire surface. I have often found this characteristic very handy. When dealing with grains separated from a rock containing more than one type of feldspar, I have used this property to separate the monoclinic grains. In most cases, the qualitative tests have confirmed the accuracy of the method used to identify the crystals.

I should also mention another useful method that in some instances may be the only practical way of recognizing the presence of monoclinic feldspar. I have found it particularly helpful in identifying sanidine in rocks that also contain labradorite and oligoclase. Having crushed the rock and obtained a uniform grain size by means of a sieve with a mesh measuring 0.5 mm on a side, a certain number of intact feldspar crystals can be found in the resulting powder. These are sorted out under a magnifying glass and examined under a polarizing microscope. It is quite easy to see that most of these crystals have the forms of parallelograms in which the smaller angle is 63 degrees and the exposed face is g_1. On this face, the direction of the cleavage traces parallel to p identify the side of the parallelogram defined by the intersection pg_1. Mr. des Cloizeaux has shown that the extinction angles measured from this edge on face g_1 have the following values;

> For orthoclase.................. 4° to 6°
> - oligoclase.................... 1° to 3°
> - labradorite................... 26° to 28°

In the present case, precise measurements of these angles demonstrate beyond doubt the presence of three species of feldspar in the same rock.

On the face p of orthoclase and on the face passing through the edge pg_1 and perpendicular to g_1 (i.e. very close to p) of triclinic feldspars, the extinction angles measured from the edge pg_1, according to Mr. Des Cloizeaux, are:

> For the monoclinic feldspar.........0°
> - oligoclase.................... 2° to 3°
> - labradorite................... 10° to 14.5°

In the less common case in which feldspar grains are seen lying on the face p, one can also obtain part of these measurements. The visible faces of orthoclase are symmetrical, and those of the triclinic feldspars are essentially the same. The traces of the cleavage g_1 are parallel to the edge pg_1. Moreover, one often sees triclinic bands in oligoclase and labradorite. In the present case, by examining the p faces it has been quite easy to see the abundance of labradorite and verify the presence of a monoclinic feldspar in the same rock, although, unfortunately, the measurements are not precise enough to say how common oligoclase is. But this close examination has shown the important fact that the rocks of Akrotiri, despite their pale gray colors, rough textures, and abundant silica, are not trachytes but labradorite amphibolites (or, following the nomenclature of Professor Rosenbusch, amphibole andesites). The dominant feldspar phenocryst is labradorite with much smaller amounts of sanidine; phenocrysts of oligoclase are even less common.

The optical observations lead to still other conclusions. One interesting question is whether the monoclinic feldspars of the amphibole andesites have their optic axes in the plane perpendicular to g_1 as they are in the undeformed orthoclase of granites or in the plane of g_1 as in orthoclases raised to high temperatures (and deformed). Owing to the small size of the crystals, it was very difficult to find the orientation of the optic planes. Thanks to the work of Mr. Michel-Lévy and the curves he has determined for the extinction angles of feldspars in their principle orientations, especially in the zone g_1h_1 giving the different curves for orthoclase mentioned above, I have been able to verify that, among the numerous monoclinic feldspars I have examined, most belong to the variety in which the optic plane is perpendicular to the plane of symmetry. I will cite only two examples.

In one grain, two individuals are twinned according to the Carlsbad Law, and the two crystals are symmetrical. If we refer to the two faces visible in the twins as x and x', the angles of the two edges with the trace of the twin plane is about 72 degrees; the extinction angles on each side of this trace are 35 degrees. According to a table prepared by Mr. Thoulet[4] for the angles of the faces from the zone g_1h_1, one sees that the angle of 35° formed by the edges px and h_1x corresponds to a face in the zone g_1h_1 inclined at 49° on g_1. Moreover, the curves determined by Mr. Michel-Lévy indicate that on a given face in this orientation, the extinction angle is 35 degrees for undeformed orthoclase and 25 degrees for the deformed type.[5] The angle of 35 degrees for this grain indicates that we are dealing with the first of these two types.

The second example, which, like the first, is taken from a dense andesite at Akrotiri, is even more conclusive. The form of the twinned crystals is very similar to that of the preceding one. The angle of the edges px and $p'x'$ with the trace of the twin plane is 74 to 75 degrees. The extinction angles on each side do not differ by more than 3 degrees and average 38 degrees. According to the table of Mr. Thoulet, these sections correspond to an orientation of about 54 degrees with respect to g_1. For a face with this orientation, the curves of Mr. Michel-Lévy give an extinction angle of about 40 degrees for undeformed orthoclase and a maximum of 28 degrees for deformed orthoclase. So it is clear that the feldspar in the amphibole andesite of Akrotiri is a sanidine of the first of these two types.

In addition, one can conclude from these examples that the angle of the optic axes must be very small.

In the sanidines of some of the very siliceous pumiceous tuffs[6] of Akrotiri, the plane of the optic axes lies in the plane of symmetry. Thus this feldspar belongs to the so-called deformed variety of orthoclase. We know from the curves of Mr. Michel-Lévy that the extinction angles from h_1 increase in the zone g_1h_1 more rapidly for deformed orthoclase than for the other type. The measured extinction angles of certain monoclinic feldspar in the Akrotiri andesites increase very rapidly as the angle of the orientation increases. For example, I have observed in one example of binary twins that the two individual crystals have sections that are almost rectangular (the smaller angle of these faces being 88 degrees) and that in the twins the extinction angle from the trace of the twin plane is 34 degrees. According to the table of Mr. Thoulet, these faces make an angle of not more than 4 degrees with the face h_1, and the extinction angles of Mr. Michel-Lévy indicate a maximum of 15 degrees for undeformed orthoclase and notably larger angles for deformed orthoclase. Thus the Akrotiri sanidines must belong to the latter type.

In summary, it appears that the angles and orientation of the optic axes of sanidines in the siliceous rocks at Akrotiri are not uniform. In most samples, the sanidine is of the undeformed type that has a small optic angle (at most 15 to 20 degrees). This is especially true of samples taken from large bodies of massive rocks. Deformed sanidine has been seen in samples from certain varieties of highly altered pumiceous tuff. In these, the optic angle reaches 30 degrees.

It is difficult to find a rational explanation for these observations, but, without claiming to offer a definite solution to the problem, I will hazard the following guess. The rocks of Akrotiri, being of volcanic origin, have sanidines that formed before the magma was erupted, so the feldspars must have risen from greater depths. It is likely, therefore, that at the moment of eruption, the sanidines that had already been formed had their optic axes in a plane parallel to g_1. In rocks of thick bodies that cooled slowly the sanidine took on the properties corresponding to a stable molecular configuration in which the sanidines were undeformed. In contrast, the ejecta that cooled quickly preserved the molecular arrangement of high-temperature conditions, and the feldspars still have their optic axes in the plane of symmetry.[7]

The monoclinic feldspars in the amphibole andesites of Akrotiri normally are twinned according to the Carlsbad Law, but it is not unusual to find Baveno twins.

The triclinic feldspars commonly have multiple albite twins with numerous, very thin bands. Nevertheless, as I have already mentioned, these feldspars also have binary twins with large individuals in either groups of smaller bands or single twins. Baveno twins may also be seen in these feldspars, but almost all of these are combined with albite twins. In most cases, however, multiple twins are joined in such a way that four groups of bands alternate two by two along a line oblique to their long axis. In this case, it is a matter of two series of bands, joined according to the albite law, combined with two other series twinned according to the pericline law. Stelzner[8] was the first to point out that this combination is common in volcanic rocks, and Mr. Michel-Lévy has determined their optical properties. In all the feldspars, the intersection ph_1, which is the axis of rotation of pericline twins, coincides almost exactly with the pole to g_1, which is the axis of rotation of albite twins. For this reason, the extinction positions of the four series of bands in these two sets of twins are almost the same.

The combination of pericline twins with albite twins also differs from the combination of two groups of crystals according to the albite and Baveno laws in that the trace of the oblique intersection of the composition planes is not very noticeable in thin sections; it is neither as regular nor as continuous as it is in the combination of albite and Baveno twins.

Many of the samples of amphibole andesite that I have collected around Loumaravi scarcely differ from the one I have just described. The sizes and relative proportions of the constituent minerals are very similar, and most of the rocks also contain abundant deposits of tridymite and opal like those in the silicified varieties discussed earlier. Almost all of these rocks show evidence of marked fluidity.

Of the various types of unsilicified amphibole andesites found in the vicinity of Akrotiri one is worthy of special mention. It is found in thin layers low in the cliff on the east side of the landing at Balos. One could easily mistake this rock for a sandstone, for it is quite granular and easily disaggregated.

The color is bluish-gray. Under the hand lens one can distinguish white grains of feldspar, colorless, glassy grains making up most of the matrix, and a few crystals of amphibole, pyroxene, and iron oxides. The microscope shows the normal properties of these components more clearly. The amphibole is brown, the pyroxene pale green. A few crystals that are intact show their characteristic cleavages. The amphibole is dichroic, but the pyroxene is not noticeably so.

Sanidine is not abundant, but triclinic feldspars are common, mostly in albite twins. Both of these feldspars resist the action of boiling nitric acid. They contain many inclusions of brown glass with one or more bubbles. Thus, in terms of the large crystals, this variety of rock is perfectly identical to the preceding one, but it differs from it completely in the form of its groundmass. It has a more pronounced perlitic structure.

Aligned streaks of amorphous material are speckled with small crystals and the parts with fewer crystals are grooved with wide perlitic cracks. There are no microlites at all. When viewed under the microscope with the nicols crossed, a thin section of this rock is seen to be completely dark. But this is not true if the rock is disaggregated and passed through a sieve before examining it under crossed nicols, even if one has been very careful to select only grains with a diameter corresponding to the normal thickness of a thin section.

In this case each grain of the rock has the usual polarization characteristics of globular material of a gummy nature. Those that are nearly equidimensional spheres and very small are seen to be marked by a figure consisting of a black cross on a white background; those that are spherical and hence larger have the appearance of a black rectangle with a border of four white bands separated from each other by a narrow black strip. The appearance of these large latter grains is a modification of that of the smaller ones in that the center of the black cross is better developed. Irregular grains have figures with strange forms that depend on their shape as well as their basic structure. If one heats the grains to a dull red, some break down, while others retain their outward form but cease to react to polarized light. If one breaks one of them with a blow, the debris also ceases to show an effect under polarized light or does not show it to the same extent. Polishing a thin section seems to have the same effect, for it destroys the temper; this is why ordinary thin sections do not have the same interference colors in the parts composed of perlitic globules. But it is possible to obtain thin sections in which these globules still react to polarized light. To do this, one must avoid cutting into them by leaving the section at twice the normal thickness. If one takes the grinding beyond that, taking care to be very gentle, the globules can be cut without completely destroying their temper. Their centers will not respond to polarized light, but there will be a thin zone in their margin with bluish-white interference colors under crossed nicols.

To complete this discussion of the perlitic grains I will add that when one heats them to a bright-red they melt quickly to a viscous liquid that becomes inflated and turns into a very porous white pumice. The loss of weight is only a few parts in a thousand.

If one submits the pulverized rock to the action of concentrated hydrofluoric acid, the perlitic material dissolves easily with a strong increase of temperature, and one obtains a pretty white crystalline powder containing amphibole, pyroxene, and iron oxides, either singly or accompanying feldspars, depending on how long one has prolonged the process.

There is a remarkable difference between the composition of the glassy globules of this perlite and that of the principal feldspar in the rock. This feldspar is triclinic, and its extinction angle indicates that it is labradorite. As in the other amphibole andesites of the Akrotiri region, a monoclinic feldspar seems to be present but only in very small proportions.

If feldspar grains separated from this rock are treated with fluorosilicic acid and the product of their solution is evaporated according to the method of Professor Boricky, one obtains elongated, hexagonal prisms of sodium fluorosilicate, large oblique prisms of calcium or magnesium fluorosilicate, and swallow-tailed needles of calcium fluorosilicate. The white, gelatinous material coating the crystals is opaline. No cubic form resembling potassium fluorosilicate was distinguished.*

The glassy globules treated in the same way give different results. The gelatinous material is colorless, and cubes of potassium fluorosilicate are abundant; the sodium fluorosilicate is in short, hexagonal prisms lying flat or on their base. Calcium fluorosilicate occurs only in long swallow-tailed needles.

Analyses of the glassy globules and the feldspar associated with it gave the following results.

	Vitreous amorphous globules	Feldspar	Oxygen	
Silica	70.7	59.0	31.4	
Alumina	16.3	26.4	12.2	
Fe_2O_3	0.3	0.2		
Magnesia	0.4	0.1		
Lime	1.6	8.7	2.5	
Soda	7.2	4.9	1.2	3.8
Potash	3.5	0.7	0.1	
	100.0	100.0		

Ratio of amounts of oxygen 7.8 : 3 : 0.94

The analyzed glassy globules were perfectly free of all foreign material, so these results indicate their true composition. One notes that the figures for the feldspar correspond to no simple atomic formula and the silica content is too great for the labradorite composition obtained from optical properties. This anomaly has two causes: (1) the rock must contain small amounts of sanidine, and (2) the analyzed feldspar probably had a small coating of glass.

Despite these effects that tend to raise the silica content of the analyzed material, the feldspar is still seen to have more silica than oligoclase,

---

* This does not prove that sanidine was completely absent from these grains, for the treatment with hydrofluorosilicate acid does not ordinarily produce notable amounts of cations that are present in only small amounts.

while the silica content of the globules is substantially greater than that of monoclinic feldspars. It is interesting to note that the feldspar that crystallized from a magma with excess silica and large amounts of potassium was not sanidine but labradorite.

A sample of pumiceous tuff I collected from the northern part of the archipelago belongs to this same group of eruptive products. It, too, is an amphibole andesite with free silica, but it differs from the two examples just described in the nature of its amorphous material. It has a pumiceous texture with long, drawn-out filaments that have a brown or yellow tint and no trace of secondary alteration. As usual in pumice, this rock is made up of elongated vesicles, all of which are drawn out in the same general direction. Each is fringed with a corona of rounded, brown crystallites forming a bristled surface with weak birefringence. They result from the early stages of devitrification.

In summary, when the amphibole andesites of Akrotiri lack free silica they are very rich in microlites and grade into a pumice that is more or less charged with crystals. In all cases, the phenocrysts in these rocks are the same.

In addition, the amphibole andesites lacking silica minerals fall into three varieties that can be distinguished from one another by the nature of the groundmass in which the phenocrysts are set. The first has numerous microlites of albite-oligoclase; the second has tiny crystals and a perlitic texture; the third has nothing but an amorphous groundmass made up of stretched pumiceous filaments.

Each of these varieties has given rise to siliceous products and to liparites with various textures and forms. As a rule, the rocks that have been impregnated with silica have become harder and more coherent. They have acquired a finer grain and new colors of pink and green. Those with the most alteration resemble a compact grinding stone in their hardness and over-all appearance. The more siliceous they are, the fewer small crystals they have. In all cases, the microlites have very small dimensions and grade into somewhat larger crystals that, for the most part, are very thin laths of feldspar. Judging from their extinction angles, the microlites in these rocks have exactly the same characteristics as those in the andesites without free silica. The phenocrysts in them are also the same, and pyroxene is no less abundant relative to amphibole. The same is true of the proportion of triclinic feldspar relative to sanidine. These minerals also have the same inclusions as the rocks described earlier. The feldspars in particular are remarkable for the great numbers of gas bubbles in their glassy inclusions. It is not uncommon to find crystals of triclinic and monoclinic feldspar containing thirty or so inclusions of this kind, and in a few cases their numbers are even much greater.

In general, the phenocrysts have little alteration, even when the amorphous material is completely impregnated with free silica and the microlites have almost disappeared. In a few cases, however, the phenocrysts show signs of more or less advanced alteration. The feldspars and pyroxenes are almost always affected more than the hornblende. The feldspars are first crossed by thin, irregular cracks filled with opal, then the infiltration spreads, and under crossed nicols one commonly sees that nothing remains but a few plates of feldspathic material that still retains its optical properties under polarized light, while in normal light the feldspar crystal appears to have preserved its normal form and color. At times, the alteration goes even farther, and the feldspar appears to be infiltrated by quartz. The parts of the feldspar converted to quartz are easily recognized, especially in polarized light. The cleavage traces have either disappeared or have been greatly enlarged and have become more irregular. Under crossed nicols, their interference colors tend to be yellowish-white rather than the bluish-white of unaltered feldspar. The parts that have been infiltrated have rounded borders, so that they have patchy forms that may either extend over a wide section of the crystal or be concentrated in its peripheral zone. Some form only a thin, irregular zone confined to an area that follows the outline of the margins. In one instance, I found quartzose margins with a spheroidal form.

The pyroxene, on being altered, is transformed into chlorite. The hornblende, as already mentioned, is rarely altered. Its outer surface is clean, and its margins are free of the iron oxides that are normally so common in the hornblendes of amphibole andesites (especially in those of the Cantal). When it is altered, it loses its color and pleochroism. In one of the samples of siliceous andesite from Akrotiri, the amphiboles have three different forms, each of which still preserves the characteristic cleavages of this mineral. One variety is colorless and not perceptibly pleochroic; a second type is green and weakly pleochroic; and a third is brown and quite strongly pleochroic. The first two varieties seem to be modifications of the third.

The iron oxides, which are uncommon in the Akrotiri andesites without free silica, are completely absent from the rocks containing siliceous material, but in the latter they seem to have been responsible for the deposits of hematite and limonite that are often seen, particularly in the form of clear, red globules from 1 to 3 hundredths of a millimeter in diameter, or as dark red needle-like prisms.

The amorphous material in the siliceous andesites of Akrotiri has exactly the same appearance under the microscope as the glassy material in thin sections that have been exposed briefly to the action of hydrofluoric acid. It has lost none of its transparency but has become slightly cloudy and has the appearance of a gelatinous substance. The crystallites have disappeared, and various infiltration products, with or without color, are seen here and there. In some cases the microlites can no longer be seen, but ordinarily they can still be distinguished, even though they are profoundly altered.

Opal is abundant in most of these rocks. It is present in three main forms. The simplest variety lacks any regular shape and does not react to polarized light. It generally has many small cracks similar to those caused by desiccation of a hydrous substance. I shall refer to this type of opal as gelatinoid. It is essentially a colloid, but grades into the crystalline variety, tridymite. The transition comes about in the following way. The mosaic formed by cracking of the gelatinous opal becomes more regular, and the outlines of the grains tend to become polygonal. In a great many cases, it is impossible to decide whether one is looking at simple polyhedral cracks in the original gelatinous material or a mass of imbricated crystals. The tridymite in the amphibole andesites of Akrotiri rarely has the sharp hexagonal forms seen in the modern rocks of Santorini. It never has well-developed bluish-white hues under crossed nicols, even though it may seem to be almost the only constituent of the main groundmass of the rock.

The second variety of opal is common in what mineralogists refer to as hyalite (when it is found in association with chalcedony).[9] I shall add a few words here on the nature of hyalite. It is made up of more or less spherical grains, some isolated and some in clusters or encased in the gelatinous opal. It commonly lines the surface of vesicles, and in thin section can be seen to form botryoidal coatings on these cavities. Some of these are simple globules, but most are composed of concentric layers, often very numerous and unequal in thickness, which are also transparent and colorless but can be distinguished even in natural light. The globules of hyalitic opal are normally too small and too heterogeneous to show effects in polarized light, but when they are relatively voluminous with dimensions reaching 0.05 mm, it is rare for them not to show the same features in polarized light as those of quenched glass described in an earlier section, namely, a black cross that has four whitish quadrants and does not move when rotated under crossed nicols. This is what is observed whenever the globule is completely heterogeneous; that is to say the stress in the layers increases imperceptibly from the center to the periphery.

A more complex phenomenon is seen in a sample of hyalitic opal from a siliceous andesite of Akrotiri. The zonal structure of the spheroid stands out under crossed nicols owing to concentric layers that are almost colorless on their inner side and bright bluish-white on the outside. In addition, each layer is transversely heterogeneous and seems to be composed of small sectors as if it had begun to crystallize radially. This variety of opal grades visibly into chalcedony.

The third variety of opal, when in its simplest form, consists of radial growths with no birefringence at all. Because this type constitutes the most primitive form of spherulitic structure, I refer to it as spherulitic opal. In thin section, it is commonly seen in the middle of unaltered amorphous material or in coatings of opal that occupy colorless areas with such sharp boundaries that they look as though they were filling holes. By incorporating microscopic crystals of feldspar, these grade into siliceous spherulites; if they contain quartz, they grade, like the previous variety, into chalcedony.

In the amphibolitic varieties of andesite that still have many open spaces and rough cracks and in which silicification is only in its earliest stages, one sees no true quartz, either in the granular form or in distinct crystals, but most samples are thoroughly impregnated with silica and have surfaces that reflect light (what the Germans refer to as illumination) that attests to penetration of solutions throughout the entire fabric of the rock.

On the other hand, these same samples show the most extensive effects of zeolitization. The zeolites are distributed throughout the amorphous matrix of the rock and along the walls of vesicles. In the latter, the crystals point inward. There are at least two different species. One can identify faintly birefringent, rhombohedral crystals measuring about 0.01 to 0.03 mm across. These are probably chabazite. The second variety occurs as prisms that lack pointed terminations and have an average long dimension of 0.05 mm and a width of about 0.01 mm; almost all these crystals have a weak birefringence, which, in some cases, may be strong enough to produce bluish tints under crossed nicols when the direction of polarization is parallel to their long dimension. They are probably mesolite.[10] The third variety consists of exceptionally slender prisms (measuring about 0.05 by 0.001 mm) and arranged in rounded spheroidal clusters that are completely without birefringence.

The small knobs on which these products are deposited have a faint lavender tint when they are viewed under crossed nicols, and, what is even more common, they have a vaguely defined, nacreous white sheen like that seen on opal.

When these rocks are placed in boiling hydrochloric acid, the zeolites they contain dissolve and disappear. Moreover, the proportions of sodium chloride and calcium chloride that are formed from them are approximately those that should be found if one is dealing with a mixture of natrolite and chabazite. But the decomposition of these minerals was not as complete as one might expect. After boiling them in hydrochloric acid for two hours, it was still possible to make out their crystal forms, including even the pointed ends. If these minerals are indeed the zeolites they seem to be, one must conclude that the treatment with acid reduces them to a residual skeleton of silica.

Chemical tests indicate the presence in these same rocks of very small amounts of sulfuric acid. This suggests that the rocks should be considered alunites.[11]

In the more advanced stages of silicification, the amphibole andesites of Santorini are converted to compact rocks with an uneven fracture that produces a rough surface like that of a grinding stone. In outward appearance these rocks are identical to the more siliceous varieties of liparites and dacites. The color ranges from yellowish-white to dark violet. With a hand lens, one can distinguish a few crystals of amphibole and feldspar in most samples.

The microscope reveals that most of the minerals of amphibole andesites still retain their original forms. In some samples the crystals are remarkably well preserved. Pyroxene and amphibole have sharp outlines and uniform interference colors; the feldspars are intact and still preserve their cleavage and twinning. Beautiful examples of multiple albite and pericline twins can be found in certain specimens. In some instances, even the microlites and crystallites seem to be unaltered.

This apparent preservation of the crystals is seen in samples in which the amorphous material appears to be completely modified and infiltrated with opal and limonite; all the cavities are filled with opal and even chalcedony.

Along side these examples, however, one finds, even in the same rocks, examples in which phenocrysts of the essential minerals have been completely transformed to pseudomorphs. The feldspars have for the most part lost their birefringence, and the parts that have become isotropic are filled with opal.

The anisotropic parts are permeated with quartz which gives them a pale yellowish tint in polarized light.* Their edges are rounded and their cleavage is gone. The altered pyroxene contains chlorite and ferruginous material. The hornblende is more resistant, but it, too, shows signs of alteration in the form of discoloration and irregular patches of iron-rich material.

When the phenocrysts are altered, the microlites of feldspar are even more so. They are inflated, have lost their sharp edges, and many are curved and deformed as if they had been abraded. Some have completely lost their birefringence and could easily be mistaken for concretionary deposits, but many are only partly destroyed. A long blade with corroded edges may still remain birefringent in the central part of the grain, while the surrounding material is completely isotropic. These microlites behave exactly like those in recent lavas of Santorini when one treats them briefly with concentrated hydrofluoric acid. It is evident that in nature they have undergone the same kind of corrosion; the reactive agent was different but the effect was similar.

The absence of fragmented quartz in the silica-rich rocks is noteworthy, because in several samples (though admittedly only a few) the phenocrysts are broken as a result of the violent mechanical action to which they were subjected during extensive movement.

On the surface of these rocks one sees an abundance of various kinds of siliceous coatings, crusts of hyalite, chalcedony, and even crystals of quartz. These superficial deposits are especially common in the rocks on the southern slope of Loumaravi near the chapel of Saint John.

It is impossible to separate the silicified rocks of Akrotiri from the amphibole andesites of the same region or to divide them into two separate groups of rocks, but it is interesting to see whether one can find the place they would occupy in a classification system if one applied the criteria that are generally used.

The name *rhyolite* proposed by Richthofen[12] would not be suitable for them, because it implies not only an elevated silica content but much glass and a pronounced fluidal structure that these rocks do not possess.

The older term, *quartz-bearing trachytes*, can no longer be applied, for petrologists are now in general agreement that the name *trachyte* should be reserved for recent volcanic rocks containing sanidine; the dominant feldspar in the Akrotiri rocks is always labradorite.

The name *liparite*,[13] which is broadly applied to volcanic rocks containing free silica, would seem preferable, but it is clearly too general and

---

* This assumes, of course, that the section has the normal thickness.

vague, especially for a case in which it is possible to use more precise criteria for the nomenclature. Moreover, Professor Rosenbuch uses this classification to characterize in more detail recent sanidine-bearing volcanic rocks (which indicate the place held by liparites in his work).

I would prefer, therefore, the name *dacite* proposed by Doelter[14] for young volcanic rocks with plagioclase feldspar while recognizing that in certain samples monoclinic feldspar may be abundant, especially in samples with more silica (although this rule has exceptions). It will be recalled that the rocks of Santorini that are under consideration here differ from true dacites in that they lack quartz.

Pumiceous tuffs of the Akrotiri region

The amphibole andesite of Akrotiri has an equivalent pumiceous tuff that is very extensive and often difficult to distinguish from the andesites when the latter are strongly altered. The tuffs are composed of fine pumiceous debris, generally very altered, mixed with lithic fragments and broken crystals. These must be fragmental ejecta. The crystals are cracked and irregularly mixed except in the intact lumps of pumice found in some samples. There is little sign of extensive fluidal structure. The large crystals mixed with the fragmental debris include monoclinic feldspar, labradorite and oligoclase, hornblende, and, less commonly, pyroxene and iron oxides.

Most of the feldspars have corroded edges and are reduced to ragged plates. Triclinic feldspars with multiple twins are most strongly altered. Their twin lamellae are separated and dislocated, but more often their welding is more compact and their cleavage is no longer visible.

Pyroxene and even hornblende are commonly altered to chlorite or charged with reddish, ferruginous material, and the iron oxide minerals have gone to limonite. But for the most part it is the amorphous material that is most altered. As in the case of the siliceous andesites discussed earlier, it is reduced to a substance of gelatinous appearance in which one cannot distinguish the microlites, crystallites, or gas bubbles that are so common in the unaltered pumiceous parts.

The small fragments are cemented together by a colorless, amorphous incrustation speckled with exceptionally small, pale yellow grains. These grains differ from the usual small globules in the irregularity of their shapes. Numerous straw-yellow concretions are found in the cementing material. They form wormy bands, many of which are parallel to each other but most commonly are joined at all angles with the spaces between them filled with transparent amorphous material and zeolites identical to those described earlier.

Within the pumice lumps in this tuff there are accidental inclusions, some of which are still intact; even the amorphous material has escaped alteration. In these specimens one can see very well the pumiceous nature of the original material that makes up the tuff.

One of the best examples I have found comes from a sample collected on the northern flank of Mount Arkhangelos. The rock is grayish-white, has a rough surface, and is friable and porous. The hand lens reveals crystals of feldspar and hornblende. Under the microscope one recognizes the same minerals, along with smaller crystals of augite. These minerals are set in a brown, amorphous matrix with fluidal banding and large, elongated vesicles of the kind that is so common in pumices.

A great number of the feldspars in the sample are single crystals or binary Carlsbad twins. Some have rectangular cross-sections with extinction directions parallel to the sides. Judging from the appearance of these sections, the feldspar is probably sanidine. It is possible, however, that these crystals are oligoclase, for the extinction of this feldspar in the zone normal to g_1 has such a small angle (only 2 or 3 degrees) that it is impossible to distinguish it from sanidine in this orientation. But I should add that the case for sanidine is supported by the fact that, when treated with hydro-fluorosilicic acid, the feldspars develop no striations and produce notable amounts of potassium fluorosilicate. The feldspars with multiple twinning have the extinction angles of labradorite.

Some of the phenocrysts in the pumiceous tuff of Mount Arkhangelos are still unbroken, but a great number of the others are in small fragments. The amorphous material is rich in globules, which, instead of being uniformly rounded, have a bristly surface. (This peculiarity may be due to incipient alteration of the rock.)

The microscopic fragments one finds in the pumiceous tuffs of Akrotiri are normally of the two different kinds of amphibole andesite found in the same region. All the spherulitic and perlitic varieties seen in the rocks are also found among the fragments. Some samples also contain some of the silicified types of andesite, but it is very difficult to distinguish them from the tuff itself when the latter has undergone the same kind of alteration. White mica, older grains of quartz with fluid inclusions, and bits of micropegmatite are also rather common among the fragmental debris. These are probably derived from the underlying rocks or, more likely, from sand that covered the sea floor at the time of the eruption.

Almost all the pumiceous tuffs of Akrotiri are to some degree permeated with free silica. In general, the more this is developed the denser the rock. The most silicified tuffs grade into yellowish and greenish gray rocks with a rough fracture and an appearance identical to that of the dacites produced by silicification of the amphibole andesites.

The silica in these rocks takes various forms. In the hydrated state it is one of the varieties of opal; in the anhydrous form it has the characteristics of chalcedony and granular quartz. The chalcedony is found in rounded clusters, each of which is made up of a small number of separate radial masses. The sheaves that compose them are rarely well enough developed to have the optical properties normally associated with this variety of silica, but they react strongly to polarized light. The small crystals making up the sheaves go to extinction when the long axes making up the radial structure of the concretions are parallel to the polarizer. The chalcedony forms narrow bands that follow the contours of cavities within the rock or the solid fragments of which the rock is composed; the microscopic crystals of quartz are arranged perpendicular to the elongation of the bands.

The granular quartz is in small, irregular lamellae coating the other material making up the rock, but many of these lamellae have a marked tendency to form discrete crystals that terminate in points with straight edges. This kind of quartz is so abundant in some of the samples that it gives the rock an appearance reminiscent of that of certain types of ancient metamorphic rocks.

In the most silicified samples, some of the feldspars are totally altered. They are largely composed of opal, and their cleavage cracks are filled with chalcedony. All the original feldspar material appears to have disappeared.

The pumiceous tuffs of Akrotiri are much more altered at deep levels than in the upper parts of deposits. Sandy debris is also more common in the lower layers. The upper parts are very homogeneous, with coarse, unaltered components dominant over cement. The rock is granular, yellowish-white, and feels rough; that in the lower part is softer and has varied tints of yellow and green with varicolored patches due to the coherent sandy material.

All of the alteration of the amphibole andesites cannot have occurred after these rocks were emplaced; the same is true of the pumiceous tuffs. In both of these it is more likely that the silicification began at the moment of eruption, and in the case of the tuffs this is quite evident, for it should be born in mind that they had a submarine origin and that they resulted from agglomerations of explosively ejected pulverized material that fell into the water while still hot. But the strength of the metamorphic effects and, even more, the introduction of deposits show that the modifying effects continued long after the volcanic material was erupted. Very hot water or acidic steam attacked the lower parts of the deposits, dissolving some of the components, particularly silica. These components were transported in solution to the surface, where they came in contact with the atmosphere and were dropped.

It is in this way that the layers closest to the surface of the ground became the most silicified, so that in certain places around Akrotiri the rocks are covered with a thin coating of hyalite and quartz.

To conclude this discussion of the pumiceous tuff in the Akrotiri region, I give below the results of an analysis of one of these rocks. The tuff selected has an even gray color with minimal alteration; it is still pumiceous. All the feldspar crystals in the rocks were first removed by hand picking.

Heated to 100 degrees the rock lost 6.96 percent of its weight as water; between 100 degrees and a red heat it gave off another 4.40 percent for a total of 11.36 percent.

Composition of dehydrated material

Silica	73.2
Titania	3.2
Alumina	13.8
Fe_2O_3	2.4
Magnesia	1.1
Lime	0.6
Soda	4.9
Potash	2.1
	101.3

It is interesting to note that the silica content of the rock without its phenocrysts exceeds that of the feldspars it contains, especially that of labradorite, which is the most abundant of its feldspars. The elevated potassium and titania contents are also noteworthy.* It is curious to see that a feldspar of the sodium - calcium series, such as labradorite, has crystallized from a melt that is relatively rich in potassium and very poor in calcium.

Considering the essential composition of the magma, one would expect to find only sanidine or at most oligoclase. This example shows very well the problems presented by bulk analyses.

---

* This analysis was made by Mr. Thoulet, the preparator for inorganic natural history at the Collège of France.

Varieties of Acid Rocks from Cape Akrotiri

In addition to the normal types of amphibole andesites on Cape Akrotiri, several unusual varieties are worth special mention. I shall describe only three of these.

The first is a rough, pale-red rock composed of small grains the size of a pin head. Each of the grains is a siliceous spherulite. The structure of these spherulites is clearly visible, even with a lens of low magnification in natural light. Where a spherulite is cut through the middle, one can distinguish transparent filaments radiating from a point and fanning out in all directions. These filaments are so tightly packed against each other that even with a magnification of 300 one rarely sees the amorphous material that fills the space between them. Elongated ferruginous crystallites with poorly defined shapes are distributed here and there, sometimes in a random fashion but more often in rows. They are soluble in boiling hydrochloric acid.

If one examines the spherulites under low magnification and between crossed nicols, one sees that a certain number have scarcely any birefringence and that their interference colors change only slightly as the stage is rotated. Those that are composed of the largest crystals, however, show the greatest birefringence. They form sectors or, more commonly, parts of sectors in which the crystals making up the spherulite have a more or less uniform orientation. Although they rarely come to complete extinction, their direction of least birefringence is parallel to the radii of the spherulites. I have found no case in which the cross remains fixed as the stage is rotated. The individual crystals to which the spherulites owe most of their birefringence in polarized light normally have a radial orientation, but on closer examination one discovers that one of their axes is not uniformly oriented along the radius of the spherulite or exactly parallel to the longer axes of the crystals.

The explanation for this phenomenon becomes evident when the specimen is examined under a magnification of 600 power. In natural light, one sees that each radial filament is composed of a group of small, colorless crystals with a general elongation parallel to the radius. In the spherulites with the coarsest structure, these small crystals have diameters of up to 0.01 mm and lengths of 0.04 mm. They have pointed, oblique terminations. Several of them may be arranged end to end along the same radial direction and separated at their ends by a small amount of amorphous material. The section of each of these individual crystals normally takes the form of a small, elongated parallelogram in which the angles between the edges vary slightly from one crystal to the next along the same set of radial crystals.

When the section of the spherulite does not pass through its center, a number of the crystals are cut in a transverse direction. The sections obtained in this way are normally four-sided, but many are irregular and all have rounded edges.

Thus one must consider the constituent crystals of the spherulites as small prisms with lateral dihedral angles that are blunt, while those at the ends are sharply pointed.

Each of these small prisms, when seen under crossed nicols, has an extinction angle greater than zero degrees with respect to the edges of the long sides. The extinction angle may be as large as 30 degrees and commonly differs from one crystal to another along the same radius. In view of their optical characteristics, shape, and other properties, there is every reason to believe that these crystals are labradorite. It seems probable that, like the ordinary microlites in the same rock, they are elongated parallel to pg_1, but even though this may be the most likely interpretation, there is not enough positive proof to make it absolutely certain.

The spherulites must have been formed later than the amphibole and phenocrysts of feldspar that are common in the same rock, for they mold themselves around these minerals. The spherulites even developed after the microlites of feldspar (albite and oligoclase) that cross or limit their boundaries.

The amorphous material that cements the crystals of the spherulites is not opal, for the chemical analysis shows that the water content of the spherulites is very small. Nevertheless, a few pale yellow concretionary bands that extend between the spherulites have a globular hyalitic texture and lack birefringence. Moreover, one sees in a few rare cases small colorless spots that are undoubtedly made up of gelatinous, isotropic opal.

The division of certain spherulites into sectors that extinguish uniformly can be explained, as I have already noted, by a regular parallel arrangement of the individual crystals in each sector of a given spherulite. But in this case the regularity is far from perfect. It is important to note that the uniform extinction of these sectors is more apparent than real. Under high magnification one sees that each sector is actually composed of very large numbers of small crystals that have different extinction directions. For that reason, the extinction angles that one observes deviate only slightly from the average.

It is quite easy to separate all the components of the rock by means of the method of Mr. Thoulet[4] which utilizes solutions of bi-iodine of mercury in potassium iodide.

The feldspar separated from this rock has a specific gravity of 2.618 and the following chemical composition:

		Number of oxygens	
Silica	59.4	31.7	
Alumina	27.4	12.6	
Magnesia	1.4	0.5	
Lime	5.7	1.6	3.6
Soda	5.9	1.5	
Potash	0.2	0.0	
	100.0		

The optical properties of the analyzed feldspar grains show that they are made up of a mixture dominated by labradorite with lesser amounts of sanidine and oligoclase.

Spherulites separated in the same way have been analyzed for their bulk composition (it being impossible to separate them into their constituents). They have a specific gravity of 2.456 and the following chemical composition:

Silica	75.9
Alumina	14.2
Fe_2O_3	0.5
Magnesia	0.7
Lime	1.3
Soda	6.2
Potash	0.9
	100.0

One gram of the spherulitic material heated to 100 degrees lost 2 milligrams of water. When it was next heated for a quarter of an hour to a dark red heat it lost another 3 milligrams. Thus the water content is very slight.

As one would expect, the material is very siliceous. Its composition is quite different from that of the feldspar in the same rock. So these spherulites are not just feldspar crystals in a purely siliceous deposit.

A second notable variety of amphibole andesite is a perlite from Cape Akrotiri. The main material of the rock is a glassy, amorphous substance rich in crystallites and made up of perlitic globules with very fine sinuous outlines. Some of the crystallites are colorless, but most are ferruginous, opaque, and, since they alter to limonite, are probably composed of iron oxides. Each crystallite is a group of very small crystals arranged in a straight or curved line in the form of trichites.

But in some of the samples, the colorless crystallites are better developed and grade into microlites of albite. Broken globules of amorphous material produced by crushing the rock have no birefringence. This perlite is therefore different even though it is found in the same region and belongs to the same group of rocks.

The most unusual variety of amphibole andesites in the district of Cape Akrotiri is a dense, white liparite with the appearance of a breccia made up of small fragments. Under the hand lens it seems to be composed of small, milky white grains mixed with others that are glassy. The microscope reveals that it is made up of amorphous, transparent material drawn out in beautiful fluidal trails like those in obsidian, the space between them being filled with siliceous material. Some of these spaces are made up of more or less elongated vesicles, but others are stretched-out bands of very irregular dimensions. The material that fills them is opal that has two different forms. One of these is gelatinous; the other consists of more or less complete globules that, under crossed nicols, have the normal color and black cross of opal. The arrangement of these bands of opal with respect to the glassy material is such that on seeing them one is immediately inclined to regard them as fillings of the interstices formed by silica-rich solutions that permeated the obsidian breccia. These infiltrations are demonstrated by the presence in the bands of opal of small crystals that are distributed along their edges just like the floating material caught at the banks of a stream. Some of these small, transparent crystals are perfectly colorless; others have a light greenish tint. Their size ranges from 0.02 to 0.01 mm. Most have an octahedral form with an essentially rectangular base, but a few are prisms that also tend to be more or less rectangular and have basal terminations in which two or four additional faces have formed along the edges of the base.

In the case in which the form is octahedral, the extinction under crossed nicols occurs when the sides of the rectangle are parallel to the directions of polarization if this base is viewed on its face, but in other cases it is oblique with an angle that may be as great as 35 degrees.

Just as in the case in which the form is prismatic, the extinction may be parallel to the long edges of the prisms or it may be oblique with an angle of up to 35 degrees. These observations may be interpreted in the following two ways. Either the crystals belong to a single species that is monoclinic or they belong to two different species, one orthorhombic with mostly octahedral crystals, the other monoclinic with mostly prismatic crystals. But the general forms of these two kinds of crystals, their colors, and their transparency are so similar that it is difficult to believe that they are two different species, especially since there are all intermediate forms between the octahedra and the prisms with octahedral terminations.

Now if one adopts the hypothesis of a single crystalline species and considers these crystals as pyroxenes, as I believe to be the most advantageous approach, one must concede that the purely octahedral forms are related to pyroxenes lacking the faces g_1, h_1, and p_1 and showing only m faces and combinations of modifying facets on the base.

In this way the question can be resolved; mineralogical considerations do not contradict the pyroxene hypothesis in any way, and chemical reactions seem to lend further support to the idea. It is well known that pyroxenes tend to resist the action of hydrofluoric acid. When the crushed rock is treated with this acid, one sees that these crystals, whatever their forms, resist this reagent, whereas the glassy material and globules of opal are entirely dissolved.

These same crystals remain unaffected by boiling concentrated hydrochloric acid, so they cannot be olivine. Nevertheless, they are less resistant to hydrofluoric acid than the common augite of volcanic rocks, so it is likely that they are pyroxenes that are very rich in calcium and have less iron than common augite.[15]

Rocks of Mount Arkhangelos

The hill known as Mount Arkhangelos is made up of the same petrographic varieties as Loumaravi. The volcanic conglomerate one finds at the top has the same appearance under the microscope as certain types of Loumaravi andesites. The amorphous material swarms with feldspar microlites that are exceptionally slender (0.01 by 0.001 mm) and have very small extinction angles. The phenocrysts are monoclinic feldspars with abundant glassy inclusions, a triclinic feldspar (probably labradorite and oligoclase) that is equally rich in inclusions, hornblende, and iron oxides. The amorphous material is altered in many places and filled with spherulitic opal that has no birefringence. The microlites are albite or oligoclase. A conglomerate that is similar but composed of smaller, more altered fragments can be found at the northern base of the hill.

Rocks of the Obelisk

The same rocks are found in the steep promontory known as the Obelisk. The pumiceous tuff of the Obelisk is identical to that of Loumaravi; it is worth noting, however, that it tends to be much more altered, filled with isotropic concretions, and almost entirely lacking in microlites. The phenocrysts of feldspar are in small fragments. The only intact crystals of notable size are those of quartz, which have hexagonal forms with rounded edges and numerous fluid inclusions with mobile bubbles.

The rock at the base of the Obelisk and on the islets facing it is totally different from the amphibole andesites of Loumaravi and Mount Arkhangelos. Although many of the rocks are very altered and impregnated with chlorite, calcite, and limonite, they are almost identical to the modern lavas of the Kamenis.

The main feldspar phenocryst is labradorite. Relatively large microlites of albite (measuring 0.1 by 0.02 mm) are abundant; phenocrysts of pyroxene are common and well preserved, but hornblende is not seen. We must also note that between the blocks in the volcanic conglomerate there are deposits of quartz, calcium feldspar, and a zeolite which the scientists of the 1866 mission found and called *reissite*.[16] It is likely that formation of these lavas preceded that of the pumiceous tuff of the Obelisk and that they simply stand against them despite their apparent inclination. Some samples are tachylites without any microlites and various proportions of crystallites.

Rocks of Cape Mavro

Despite their proximity, the rocks forming the two capes, Akrotiri and Mavro, are quite different from one another. The dike of Cape Mavro, when seen under the microscope, is identical to the modern lavas of the Kamenis. It has phenocrysts of the same triclinic feldspar, pyroxene and iron oxides in a groundmass of albite, oligoclase, and iron oxides and brownish amorphous material.

The phenocrysts of triclinic feldspar are rich in inclusions of brown glass without gas bubbles. Sections through the zone ph_1 have maximum extinction angles of 25 degrees corresponding to labradorite. As the following analysis shows, they have the formula of this feldspar.

		Number of oxygens
Silica	56.7	30.2
Alumina	27.2	12.5
Iron oxide (Fe_2O_3)	0.2	
Magnesia	0.4	0.2
Lime	11.1	3.2 } 4.4
Potash	0.2	0.0
Soda	4.2	1.0
	100.0	

Ratio of oxygens: 7.24 : 3 : 1.05

Grains of feldspar extracted from the rock by means of an electromagnet resist very well the action of boiling nitric acid, but a very small number seem to be affected after a long treatment. The latter cannot be confidently considered to be anorthite.

The minerals in many of the lavas of Cape Mavro are altered, with the pyroxene transformed to chlorite and the feldspar replaced by chlorite and, more often, calcite. Much of the amorphous material has undergone similar alteration, and the iron oxides have been changed to limonite. And at the top of the cape, siliceous deposits are common but less so than at Cape Akrotiri.

Basic Rocks of Cape Akrotiri

It is interesting to note that, closely associated with the silica-rich rocks that make up almost all of Cape Akrotiri, a small ledge of violet colored lava can be seen at the base of the cliff close to the northwestern point. It is one of the better examples of an anorthite lava.

Phenocrysts of feldspar are rather rare; many are untwinned even though they are anorthite and are easily attacked by acids. Pyroxene, which is more common, tends to be surrounded by a corona of small, radially oriented crystals of titaniferous iron. Olivine is rare and severely altered. Feldspar microlites of relatively large dimensions (0.25 by 0.02 mm) have the composition of labradorite. In addition, the rock has many scattered grains of pyroxene and titaniferous iron oxides. The augite grains have a striking yellow color.

The body of reddish scoria that can be seen in the cliff east of Cape Akrotiri is also composed of anorthite-bearing lava very rich in olivine. Apart from that, the mineralogical characteristics are exactly the same as those seen in most of the anorthite-bearing lavas of Santorini. The reddish color is due to alteration of its iron-rich constituents, particularly pyroxene and small grains of iron oxides. The well-developed amorphous material is filled with reddish brown, translucent grains that have no birefringence. The sparse feldspar microlites are labradorite. The pyroxene phenocrysts are relatively well preserved.

CRYSTALLINE BLOCKS IN THE SCORIA OF THERA AND THERASIA

Among the rocks around Akrotiri, one of the most curious from the petrographic point of view is that of the coarse-grained blocks found at the base of the cliffs and in beds of cinders, particularly on the southern slope of

Thera between Cape Mavro and the Obelisk. Similar rocks can also be found on the shore of the bay near the harbor of Balos as well as along the base of the cliffs of Therasia. They are almost entirely crystalline and have no amorphous material like the glassy groundmass commonly found in modern volcanic products. The only isotropic substance one sees in them is opal, and even that is quite exceptional. The mineral species in them are monoclinic feldspar, labradorite, oligoclase, amphibole, diallage,[17] augite, biotite, apatite, iron oxides, quartz, chlorite, hematite, and opal. Most of these minerals make up the primary constituents of the original rock; chlorite and hematite seem to be the only secondary minerals. All the blocks collected from the various localities do not necessarily share all the minerals in this list, but those found in the cinders at Balos have all of them, though not always in the same proportions.

The dominant feldspars I found in the samples from this locality are monoclinic feldspar and especially oligoclase. Both of these form crystals up to 0.5 mm in length. They are clear and rich in glassy inclusions that have gas bubbles in them. In addition they have numerous inclusions of hornblende, biotite, pyroxene, iron oxides, and apatite. The oligoclase has many wide bands with a bluish interference color, and for the most part their extinction angle is close to zero with respect to the direction of elongation. The labradorite has pale or bluish gray interference colors. Almost all the grains are made up of many thin multiple twins with an extinction angle of less than 25 degrees from the composition planes. Combinations of albite and pericline twins are very common in this group of feldspars.

Hornblende is abundant and easily recognizable from its color, form, cleavage, dichroism, and extinction angle. In many cases, the longitudinal sections of this mineral tend to be iridescent under crossed nicols owing to the fact that each crystal is made up of a bundle of small prisms that are elongated in the same direction but turned in various orientations. The extinction direction of these small prisms has an angle of up to 15 degrees with respect to the long axis.

Certain elongated crystals with a fibrous form similar to that of the hornblendes just described have exceptionally little dichroism and have extinction angles of at least 35 to 40 degrees. They do not seem to have the cleavage of pyroxene. For that reason it seems appropriate to refer to them as diallage, even though they seem to lack the inclusions that are characteristic of that mineral.

Augite is common among the inclusions in the feldspars but is very rare among the phenocrysts of this rock. One should note, however, the presence of certain individual crystals of this mineral that are not dichroic and have a considerable extinction angle relative to the longitudinal edges. In addition, there are suggestions of two cleavages that are at nearly a right angle to each other.

The pyroxene and diallage are both pale green, whereas the amphibole is yellowish-green. The biotite has a brownish-yellow color. Its platy structure and the large difference of absorption of polarized light when it has different orientations under crossed nicols make it easy to recognize. Apatite, which is fairly common, has the characteristic features of that mineral.

The chlorite and hematite, which are equally easy to recognize, are products of alteration. The first is found in amphibole, the second in feldspar. The quartz is in irregular strands distributed through the space between the main crystals of the rock. Despite an intensive search with a very high-power lens and different temperatures, I have been unable to distinguish inclusions with mobile gas bubbles in the quartz. It seems likely, therefore, that the numerous inclusions with stationary bubbles that one sees are inclusions of glass. But in some of these one finds next to the bubble very small, colorless crystals with a cubic form and no birefringence. This would lead one to conclude that in these cases one is dealing with inclusions saturated with sodium chloride that has been precipitated as crystals. But for reasons that I shall explain later when discussing analogous inclusions in blocks found on Therasia, I doubt that this is the correct explanation. The inclusions formed by simple gas cavities are very numerous; in some instances they have the shape of a hexagonal double pyramid of quartz, but most are irregular. The crystalline inclusions appear to be the same as those seen in the feldspar, but in many cases it could be that some of the strands that look like quartz are actually irregular, corroded fragments of monoclinic feldspar. The absence of polygonal outlines or visible cleavage traces make it easy to confuse these two minerals, and the chances of error are increased by the fact that the feldspars in this particular sample are for the most part clear and perfectly transparent, just as the quartz is in the parts between the inclusions.

I have seen well-developed opal only in the form of granular, transparent encrustations on the surface of the rock or as dusty coatings composed of small, rounded grains measuring about 0.2 mm and showing under crossed nicols the normal cross of spheroidal glass.

Micropegmatites can be found deep in the interior of the rock, where they commonly occupy the interstices between crystals. They are just as well developed there as they are in the more attractive specimens of older rocks, such as porphyritic granites and kersantites. They are composed of small bits of quartz that tend to be aligned in parallel rows and show sections with the shapes of triangles or very elongated quadrilaterals.[18] All of the pockets of quartz grains are arranged in trails with a similar orientation. Most of them have very striking interference colors under crossed nicols and go to extinction at right angles to one another, but when a random section happens to cut through them in a direction that is close to the plane perpendicular to the optic axis of the quartz, their interference color becomes progressively less intense as the optic axis is closer to being perfectly centered. In that case, instead of elongated wedges one sees that the grains are parts of regular hexagons with a hollow interior of similar shape. From this, there can be no doubt that a certain number of these fragments resemble crystalline envelopes set in a substance identical to that which surrounds it.

The material between the wedges of quartz in these micropegmatites is birefringent and takes on bright colors under crossed nicols. It extinguishes in four directions that normally differ from those of the wedges of quartz and that often correspond to the extinction position of the crystals of feldspar in contact with them. But in some instances the wedges of quartz go to extinction at the same time as the intervening feldspathic material, which shows that the optic axis of the quartz is parallel to one of the axes of the feldspar. (This parallelism must be between the optic axis of the quartz and the axis of a monoclinic feldspar perpendicular to the plane of symmetry.) So, in certain cases that are not very rare, the wedges of quartz are set in material that remains dark between crossed nicols regardless of the orientation of the section. The most reasonable explanation that first comes to mind is that this substance is formed by feldspathic material cut perpendicular to one of the optic axes, but if one considers the great frequency of this phenomenon one must conclude that the material is really isotropic and that it is a glass similar to that in lavas or, more likely, opal. It is worth noting that this phenomenon has been observed mainly in samples with botryoidal opal of the kind discussed earlier.

Blocks of this same rock that I collected in the bed of ash near the base of the cliff on the southern coast of Thera between Cape Mavro and the Obelisk have all the minerals noted earlier except opal. Biotite tends to be very abundant. In certain samples, amphibole is missing, while, in addition to diallage, one sees augite with its distinctive cleavages and birefringence. In a few samples, not only is amphibole missing but the presence of diallage is questionable, while augite is exceptionally abundant. The blocks that are so rich in augite are also unusual in that the dominant feldspar in them is oligoclase. The crystals of this feldspar have multiple albite twins in which the narrow bands visible under crossed nicols commonly go to extinction at angles close to zero. Crystals of this oligoclase separated by means of the electromagnet resisted very well the action of boiling acids. Quartz forms irregular strips similar to those in the samples of amphibolite, but this mineral is less common than in the latter rocks. Micropegmatite is also much rarer and less well developed. Biotite and iron oxides are quite common, and apatite is fairly wide-spread. And finally, one can distinguish a few colorless, rounded grains that in natural light have all the superficial appearance of olivine. In polarized light, they have the characteristics of orthorhombic crystals, but the interference colors are not as strong as they would be in olivine of the same thickness. I can offer no opinion as to the identity of this mineral.

These samples, all of which are some variety of pyroxenite,[19] are completely lacking in interstitial glass, as are the samples of amphibolite found in the same localities. Micropegmatites are moderately well developed in them, and all their feldspars have many inclusions of iron oxides.

The ejected blocks one finds in the ash bed in the inner parts of the cliffs of Therasia have the same external appearance but are less friable. Under the microscope, one notes the same absence of interstitial glass and the presence of crystals of all the same minerals. But more micropegmatite is seen, while there is no biotite. Both diallage and augite are seen in well-developed forms. The diallage has a fine, fibrous structure and even the brown, linear inclusions that are characteristic of this mineral. The latter are aligned in two directions, one of which is parallel to the long dimensions of the grains while the other tends to be at an oblique angle. The diallage itself is not dichroic, but parts of the fibrous clusters within it have a marked dichroism, with the result that parts of the prisms of diallage may be dichroic. These dichroic parts are distributed in irregular strips.[20] In such a case, it is clear that these are replacement products that result in a transformation of the diallage into amphibole. The extinction behavior supports this interpretation.

Grains of iron oxides and hematite are wide spread in the form of inclusions. The monoclinic feldspar is riddled with inclusions of fine, parallel filaments that are distributed in bands separated by narrow spaces that are clear and unaltered, much as one often sees in the monoclinic feldspars of ancient rocks in which these minerals have undergone incipient alteration. The material making up these inclusions, which is easily dissolved in acids, is nothing but limonite that has infiltrated along cracks in the feldspar. Along with the monoclinic feldspar, one also finds much labradorite and a small amount of anorthite. The first of these two triclinic feldspars is finely striated and resists the action of acids; the second occurs in wide bands and has strong interference colors under crossed nicols. It is rapidly attacked by boiling nitric acid and loses all its birefringence. The most common inclusions in the feldspars are crystals, the most abundant of which are pyroxene and iron oxides, but they are also accompanied by a few inclusions of hematite and apatite.

Quartz forms rounded borders around the outlines of other crystals. It is rich in inclusions, particularly gas-filled cavities, but one also sees bubble-like inclusions. The material surrounding the bubbles and making up most of the inclusion does not react to polarized light, so it is probably glass; moreover, the bubble does not move spontaneously, even when the thin section is heated. In a great many cases one finds, in addition to the bubble, one or two transparent crystals with strong refraction and a light greenish tint. These crystals along with the bubble make up almost the entire inclusion. Most of the crystals form straight, short prisms that are almost cubic, but a few have more complex forms, occasionally containing very small grains of iron oxide. Their diameters do not exceed 0.005 mm. They rarely show signs of birefringence in polarized light. I would not care to speculate on what these crystals might be, but I believe I can safely say they are not composed of sodium chloride. Their light greenish color in natural light, their many facets, and their occasional weak birefringence seem to rule out this possibility.

The same ash layer contains equally common fragments of schist identical to those found on the flanks of Mount Profitis Ilias, particularly in the area around Gonia. I have also collected a few angular pieces of a rock that differs in many ways from the one making up the blocks described above. It is a gray rock composed of an amorphous groundmass in which there are a considerable number of hornblende crystals associated with a triclinic feldspar that is not affected by acids. The crystals of hornblende reach 3 millimeters in length and a millimeter in width. These rock fragments are probably a variety of amphibolitic trachyte ejected during submarine eruptions near Akrotiri.

The ash beds in which I found the rock fragments of greatest age are made up almost entirely of small pieces of pumice in which one can make out here and there a few crystals of augite, iron oxide, and triclinic feldspar. The last-named mineral has beautiful glassy inclusions and is not attacked by acids.

It has all the optical and chemical properties of labradorite.

Along the route up to Balos one can distinguish in the ash layer a bed of rounded concretions about the size of a hand lens. Under the microscope, these concretions have nothing notable in the way of crystalline material. One sees nothing in them but a more or less tightly packed agglomeration of the particles making up the ash. Each concretion is bounded by a brownish-yellow zone that loses its color on long exposure to acids. It is therefore likely that this color is due to the presence of small amounts of limonite that serves as the cement in the concretion and determines its shape.

CHAPTER NINE

CONSIDERATIONS ON THE ORIGIN OF THE ANCIENT PARTS OF SANTORINI (THERA, THERASIA, AND ASPRONISI)

The only ancient authors who wrote of the origins of the islands surrounding the bay of Santorini are Pliny and Apollonius of Rhodes. Both mentioned the appearance in the middle of the bay of an island to which the name Καλλιστη had been given.[1] Pliny gave no details of this event; he was content to note only that it took place in the fourth year of the 135th olympiad (237 B.C.). Apollonius included with his account an interesting legend and put the appearance of the island at an earlier date. He says that "Euphemos, in order to comply with the advice of the oracle, threw a lump of earth into the sea, and an island then rose that was called Kallisty and was the sacred nursemaid of the children of Euphemos." According to this account, these events took place at the time of the return of the Argonauts whose expedition is generally thought to have been completed during the fourth century before Christ.

Even though the accounts of the two authors may deal with factual matters, they may not refer to a local emergence during recent geological time as one might be led to assume. The archipelago of Santorini was not formed suddenly by ancient rocks emerging above water in a single event with instantaneous emersion of the submarine rocks; great numbers of eruptions obviously contributed to the formation of the islands of Thera and Therasia.

The many beds of ash, lapilli, scoria, and pumiceous breccia of various kinds that can be seen lying one upon another over a great extent of the cliffs around the bay show that the ground was formed in large part by a long series of volcanic explosions. Almost all these fragmental deposits were undoubtedly ejected into the air from vents that were more or less distant from one another. At the same time, however, most of the numerous layers of lava in cliff sections of these two islands, particularly around the northern part of the bay, are almost all products of subaerial eruptions; the parts of Thera forming the districts around Akrotiri and Profitis Ilias are the only ones that are clearly of submarine origin. Thus, Thera and Therasia, or, more properly, Kallisty, which was their first form, did not rise from the sea all at once.

Disregarding the erroneous explanations of Pliny and Apollonius, we can turn to the geologic history of Santorini and try to determine the mode of formation and age of the older rocks that make up the various parts of the archipelago.

The metamorphic rocks that form the large massif of Mt. Profitis Ilias and are seen in the face of the cliff of Athinios are the oldest visible rocks. They formed a small island before any of the eruptions laid down volcanic material around them. In most parts of this area they are exposed in disrupted beds that are very steeply dipping or even vertical. Because there can be no doubt that these were originally horizontally stratified marine deposits, it follows that they were subsequently deformed by powerful forces. This deformation most probably corresponds to the period when Pinde and Parnasse were uplifted near the end of the Miocene.[2] This happened before any volcanic activity had broken out in the surrounding region, for the marbles and schists that make up most of Profitis Ilias contain no traces of material of volcanic origin. No grain of pumice or fragment of lava is to be found in the marbles or phyllitic rocks. Moreover, there is a marked discordance between these rocks and the volcanic material that accumulated on and around them. The marbles of Profitis Ilias are almost vertical, whereas the tuffs of Akrotiri and the ash and scoria of the cliffs at Athinios and Cape Alonaki all have nearly horizontal layering. Many centuries had probably elapsed by the time the submarine eruptions of Akrotiri contributed the first volcanic material of Santorini.

The submarine origin of the pyroclastic beds and andesitic rocks of this region is demonstrated by the internal structure of the rocks making up these units. The frequency of perlitic cracks in the compact lavas of this district indicate that they were subjected to abrupt cooling of the kind that comes from sudden quenching in seawater. The lava flows are thick and short, and the material that forms them shows signs of strong alteration, but it is mainly in the white tuffs of this region that one sees the greatest amount of chemical alteration and deposits produced by water that had the components of these deposits in solution.

The most profound alteration was that of the material making up the tuffs. The hydrated minerals that form the cement binding the fragments together could only have been produced by the strong action of water at a temperature well above normal. It is possible, however, that this temperature was not much above 50 or 60 degrees, for it is in that range that Mr. Daubrée noted deposition in the bricks of Plombières of opal and zeolites very similar to those I have just described in the tuffs of Santorini.

The submarine origin of the Akrotiri tuffs is further shown by the presence in them of small grains of quartz, almost all of microscopic size. These rounded quartz grains are rich in fluid inclusions with gas bubbles. They are grains of sand from the seafloor that were entrained with the volcanic ejecta and incorporated into the tuff.

The ejections tended to be turbulent. Initially, it is true, they produced rather regular beds, but the stratification is weakly developed in the central parts of the deposit. The tuffs laid down at the end are also distinguished by the regularity of their bedding, which splits into thin plates and shows that the moderated strength of the ejections produced fine particles like those of ordinary volcanic ash.[3]

The substantial amounts of pumiceous tuff and compact andesitic rocks in the Akrotiri region probably came from three main vents corresponding to the promontories of Loumaravi, the Arkhangelos, and Cape Akrotiri. Each produced a separate group of rocks. The most important of these, that of Loumaravi, was separated by a wide interval from the limestones and schists of Athinios and Profitis Ilias and were filled with fragmental deposits of subaerial origin. The second body, that of Mt. Arkhangelos, is joined at its base to the first, but its upper parts are clearly separated. And finally, both of these units were separated from the rocks of Cape Akrotiri by a low area that, after receiving a rather thin layer of submarine debris was in turn covered by several beds of ash and scoria of subaerial origin.

The substantial thickness of each of these three bodies and their limited lateral extent is explained by their mineralogical and chemical nature and by the conditions under which they were erupted. The rocks are very silica-rich and refractory; they are very crystal-rich with many microlites and siliceous concretions, so they were probably very viscous at the time of eruption. The sudden contact with seawater must have cooled the magma and caused it to solidify immediately, so that it remained close to its source. The tuffs that were ejected rose into a rather thick layer of water, and, because the force of their eruption was absorbed by the resistance of the water, fragments of small size accumulated close to the vent.

The entire region around Akrotiri was under water. The sea even covered the peak of Mt. Loumaravi though only by a few meters.

The shallow depth of water at this point is shown by the planing off of the summit, by the ripple marks one can see there, and by the strong stratification that was developed in the pumiceous tuff. All these phenomena can be explained as the result of wave action. We know that the superficial levels of the sea are easily set in motion by the wind but that the motion is not transmitted beyond a very shallow depth.

Thus, not only were the eruptions under water, but the ejecta that piled up around the vent must never have breached the surface of the sea. I have placed special emphasis on the evidence for this conclusion, because pumiceous tuffs are not always produced in this way. Even though the bottom of a crater may be below sea level, the material ejected from it can accumulate to great thicknesses and reach a level well above the water. It is only necessary for the rim of the crater to be incomplete, with an opening on one side, for the water to gain easy access to the interior. Several tuff cones on the various islands of the Azores appear to have been formed under conditions of this kind. They are composed of material that is obviously of submarine origin, for it contains more or less well-preserved marine fossils, but the crater rim has a structure that is exactly the same as that of craters formed subaerially. The stratification of the ash beds is very sharp. Each layer has a double inclination with a steep slope toward the interior of the crater and a more gentle one toward the exterior. Such a structure could never have this shape if it were formed under water. The action of waves and marine currents would not have permitted the stratification to take on such a regular form.

No craters of the kind found in the Azores are seen in any part of Santorini. One must conclude, therefore, that the deposits of pumiceous tuff near Akrotiri must be made up of local material that was entirely of submarine origin.

The fossils one finds in the pumiceous tuff around Akrotiri provide conclusive proof of the way the deposits were laid down. They show that the underlying rocks of this region are made up of material that originated under water.

To add more evidence, we can recall briefly the nature of the outcrops of organic debris and the particular characteristics they present.[4]

1. These remains must all, without exception, be identified as coastal forms. The marine community to which they belonged inhabited very shallow water (at most about ten meters deep).

2. One finds them at very different elevations starting at 50 meters and extending to 174 meters above sea level, i.e., over a vertical range of 124 meters.

3. They are concentrated in restricted areas between layers that have no fossils. In places, there is nothing but small pieces distributed in thin irregular beds. This is the usual form in which almost all these fossils are found, but one of the deposits at an elevation of 140 meters on Mt. Arkhan-

gelos has different characteristics. The fossils deposited in a narrow, horizontal, semi-circular zone are intact, despite the fact that they belong to a species that is large and fragile. The creatures that produced them must certainly have lived in the same location where their remains are found today.*

4. The shells are embedded in the pumiceous tuff or attached to ledges of the same rock.

Several consequences arise from these facts. First of all, we see that the ejecta making up the pumiceous tuff were produced before the fossils were deposited. The marine life lived on a surface composed of volcanic material mixed with grains of quartz. It formed a large mass of pumiceous tuff on the surface of which the marine mollusks had scavenged.

The normal alteration and wear of the fossils found on the west flank of Mt. Loumaravi would lead one to believe that these organic debris were ejected along with the pumice. But this hypothesis is undermined by the fact that the fossils on Mt. Arkhangelos and at Balos are intact and still in place where they lived. The three exposures in close proximity to one another have a common base and could not have been formed under very different conditions. But how can we reconcile this with the very different degrees of preservation of the fossils? Among the explanations that come to mind the following seem most reasonable.

The fossiliferous beds of Loumaravi, like those of Mt. Arkhangelos and Balos, are, as we have noted, characterized by shallow-water species, so they were formed at a time when the surface of the sea was only slightly above the beds. The shells of Loumaravi are broken and rounded, because they belong to species that lived near the shore line of gently sloping, exposed beaches. Those of Mt. Arkhangelos are intact even though they are near the same levels, because they were deposited on the northern flank of a promontory at the entrance to a small cove and were protected from the direct impact of waves. The point of Arkhangelos protected their southern side, which was the direction of open water where waves broke. But apart from this minor difference, both must occupy today the same places they did originally. Thus neither of them was torn from the subsurface and mixed with the pumice.

The fossils also justify the conclusion to which we were led by examining the mineral components of the rocks; they show that the sea

---

* These may also have been the conditions where the fossils were deposited in the large body of rock west of Balos, but since this inaccessible site could not be observed at close hand, this is only my opinion.

formerly stood at a much higher level than it does today. A similar explanation probably accounts for the deposits at Balos.

An uplift of about 162 meters would have been required to bring the deepest shell beds above water and at least 212 meters to bring the summit of Loumaravi to its present elevation.[5] The ground movement was slow and sporadic. The submarine part of Akrotiri was exposed by a series of discontinuous uplifts, and, as a result, it gradually displaced the shore line in a horizontal as well as a vertical direction. From this came the great number of deposits of varied assemblages of fossils at the levels where one sees them today. The shell beds represent the various positions of sea level during stable periods, while the intermediate levels correspond to periods of active uplift.

Was this ground movement confined to this particular locality or did it affect a wide area? It is impossible to respond a priori to such a question, for volcanic regions are known to be subject to both types of phenomena. One sees local changes of the ground level as well as others that can be produced by general causes. The large-scale movements that upset sedimentary terranes have affected volcanic terranes as well, and we have a great many authentic examples of local ground movements in the immediate vicinity of craters. I should cite a few of these to show the common characteristics of these manifestations.

I witnessed phenomena of this kind at Vesuvius in 1861 and at Etna in 1865. The 1861 eruption of Vesuvius affected an entire area of several square kilometers around Torre del Greco and ruined the villa of that name just as though there had been a violent earthquake. Several hundred meters of the coast line had risen about a meter. At Etna in 1865, the effects of the bulging of the ground could be followed for a distance of about two kilometers.

At Santorini, it has manifested itself in remarkable local uplifts during the course of the two latest eruptions. At the beginning of the eruption of 1707, an uplift of the sea floor caused an island of pumice to appear that has been known for a century and a half as White Island.[6] From 1866 to 1870, the ground formed by the lavas of the preceding event underwent curious up and down movements, the details of which I have already described in earlier chapters.

Now if an eruption such as that of 1707 can lead to a part of the sea floor producing an island 15 to 20 meters high, one can appreciate the possibility of a much more important emergence under similar conditions.

Thus, the uplift that raised the summit of Mt. Loumaravi to an elevation of 212 meters can be considered a much larger analog of the submarine bulging to which White Island owed its origin in the last century.

In the present case, however, it is likely that the uplift was much more extensive and that the effects noted on a large scale at Santorini are only a variation on what was a more general phenomenon. We see on the shores of the archipelago and on other islands of this sea numerous beds of Pliocene fossils that are now perched at considerable heights. At Milos, for example, deposits of this kind are very common. Most often, they take the form of wide-spread beds that may be either horizontal or inclined and in some cases are continuous but in others are broken by faults. In a few rare instances, the deposits are reduced to small, conical mounds composed of thin horizontal beds standing on a base of volcanic rocks. There are particularly good examples of these curious geological features in the area between the two bays that cut into the eastern coast of the island of Metelin and on the flanks of the mountains in the interior of the island of Milos.

The fossils identified in these deposits include mollusks dating from all stages of the Pliocene epoch, but species belonging to the Upper Pliocene are by far the most numerous. There can be no doubt that at the end of that time an important uplift took place in a wide region that included not only the Grecian archipelago and the lands of the surrounding areas but almost the entire Mediterranean basin. Santorini could not have escaped an effect of this extent, so it is reasonable to attribute the emergence of Thera to this great geological event. The fact that most of the stratified deposits around Akrotiri, including even those with rounded and abraded shells, were deposited in horizontal or gently inclined beds lends support to this interpretation. The beds are not displaced, as they certainly would be if the uplift had been in a vertical sense and limited to a very restricted horizontal extent.

Thus, the most likely hypothesis is that the southern part of Thera became emergent as a result of an uplift that affected a wide region and that this emergence took place during a unique geologic episode.[7]

In all the preceeding discussion of the submarine origin of the Akrotiri region, we have been concerned only with the tuffs of Mt. Loumaravi and the andesitic rocks of Mt. Arkhangelos and Cape Akrotiri. These rocks are characterized by labradorite, sanidine, oligoclase, and abundant hornblende. Pyroxene is less common, iron oxides are rare, and olivine is totally absent. The rocks that are massive have a light gray tint. The fact that they have a marked tendency to be perlitic shows that, as we have already noted, they were quickly consolidated as they would be if discharged under the sea. The dark-colored rock that forms Cape Mavro, although its composition is very different, seems to have reached the surface under similar conditions. It consists of the same material as the lighter-colored rock that forms the small islets near the southern coast of Thera as well as the small body of columnar-jointed rocks on the shore. All these places show signs of physical and chemical action due to contact with a large mass of water while they were still incandescent. The breccia forming the main mass of Mavro is composed of strongly welded agglomerate with a cement composed of volcanic ash and impregnated with calcite, zeolites, opal, and chalcedony. There are no open spaces of the kind found between blocks in breccias of subaerial origin. The alteration is also more pronounced than in most eruptive breccias formed subaerially. It is also noteworthy that silicification is as well developed as in the nearby tuffs and massive andesites.

The radial joints that are so common in the rocks of the southern coast of Thera and the small offshore islands seem to indicate an abrupt cooling of the rocks and, although they are not definitive proof, they suggest that this lava was erupted under seawater.* It is likely, therefore, that both types of product are of submarine origin.

The labradorite-bearing lavas of Cape Mavro and the columnar structures on the southern coast of Thera were formed after the tuffs, breccias, and flows of trachyte at Cape Akrotiri, Mt. Arkhangelos, and Loumaravi, for one finds in the conglomerate of Mavro numerous fragments of gray andesitic rocks identical to the ones at these various localities, while the reverse is not seen, even though both were raised above sea level by the same uplift.

After the uplift, the center of volcanic activity moved a short distance northward into the area that is now the central part of the gulf. It formed a large central cone composed mainly of loose material and volcanic ejecta. From this subaerial edifice came the beds of lapilli, pumice, and ash that make up almost the entire section of cliffs in the southern and southeastern parts of Thera. There must have been a huge crater, for otherwise it would be impossible to explain the great extent and uniformity of these deposits.

The crater wall was continuous from the southwest around the south to southeast; the mantle of ejecta supported by the slopes of Profitis Ilias and

---

* Basic lavas commonly have columnar jointing even where they were formed subaerially, but acid or intermediate lavas like the modern ones of Santorini seem to develop this structure when they cool on contact with water.

the submarine rocks of Akrotiri formed a great rampart, so one should not be surprised to find no lava flows of importance exposed in the cliffs of this part of Thera.

If one can judge from comparisons with other large, central-vent volcanoes, the slopes of the cone must have formed a high Somma-like ridge extending from the southwest to the southeast and concave toward the north.[8] In a horizontal plan it represented almost half the circumference of a great belt open toward the north.

The elevation of the central part must have varied with time. It is certain that, at least on some occasions, seawater had easy access to the interior of the volcano, for otherwise it would be impossible to explain the banks of conglomerate exposed along the route up to Thira. These thick beds are composed essentially of blocks of all sizes, commonly quite large, and generally worn and rounded but randomly distributed and held together by a deeply altered, reddish tuff. The distribution of the blocks shows that they were ejected explosively and are not just erosional debris. At the same time, the alteration and compact nature of their cement show that they were affected by acidic solutions, but the alteration of stratified parts of this conglomerate is not so great that one would interpret it as the result of a submarine eruption. Moreover, the brick-red color of the rock seems to indicate the kind of oxidation produced by exposure to the atmosphere.[9] There can be no doubt, therefore, that most of the beds making up the southern cliff of Thera are composed of material ejected under dry conditions or through very shallow water. They are composed of pumice, ash, and lapilli that water has neither cemented together nor notably altered. Almost all this material that is so varied in appearance and structure must have come from the same vent. It has horizontal continuity with only local irregularities of minor extent. Thus, a single vent seems to have ejected, either alternately or at the same time, fragmental material that was altered to differing degrees by the action of water.

The first eruption tore from the rocks underlying the volcano a certain amount of crystalline debris with a mineralogical character that could only be from a much earlier geological period, probably the beginning of the Tertiary. As a result, if one assumes that these products brought up by volcanic means took various forms at the surface, just as lavas do, one must conclude that Santorini had been the site of eruptive phenomena before the opening of the vents that produced the pumiceous tuffs and andesites of Akrotiri. If so, there must have been a sizable amount of magma under Santorini, particularly around Akrotiri and Therasia, at the same time that the Tertiary euphotides, ophites, and granitic rocks were formed on the island of Elba.[10]

To summarize, the observed relations can be explained very well if one supposes that all the large pyroclastic deposits of Santorini are the products of multiple eruptions that produced a large crater, in the middle of which was a wide depression that from time to time was invaded by the sea but lost its water during eruptions.

If one accepts this hypothesis, all the fragmental units exposed in the cliffs of southern and southeastern Thera can be considered parts of a single large, more or less conical structure, truncated at an elevation of about 500 to 600 meters and having a large crater with a wide breach on its northern side. The central part of this crater was a low area that communicated with the sea.[11]

The cone was incomplete on the north, no doubt because the sea was deeper on that side, and the material ejected by the volcano disappeared under the water as fast as it fell.

The extent of this opening can be delineated by a straight line running from the promontory of Skaros on the island of Thera to the tip of Cape Tripiti on the island of Therasia.

During the period that the volcano was active and ejecta from the central vent produced the major part of the material in the cliffs of southern and southeastern Thera, secondary eruptions occurred on several occasions on the outer slopes of the main cone. These produced lava flows and domes or even local deposits of scoria. It is phenomena of this kind that must have been responsible for units such as the small body of lava at sea level near Therma as well as the flows seen in the lower third of the cliff near Cape Alonaki and the red scoria visible at a high level in the cliff at the southeastern angle of the bay.

Following formation of the large breached crater just described the large eruptions stopped temporarily, and the central vent seems to have entered a period of repose. Eruptions were then confined to centers situated a few hundred meters north of the opening of the large crater. Sizable parasitic cones were formed at the present sites of Megalo Vouno and Kokkino Vouno, and close to the point where Mikro Profitis Ilias now stands. Voluminous masses of lava (mainly anorthite-bearing), thin beds of scoria, and ash of various colors contributed to the growth of these secondary cones. The predominance of lavas over fragmental material indicates that explosive eruptions were less intense.

From the beginning of its growth, the highest point of Megalo Vouno was located more or less where the summit is now. About a third of the way up a cliff section below Megalo Vouno the ash beds slope toward the bay. The center of the parasitic cone was probably close-by but slightly to the north. The ash layers exposed in the cliff covered the lower southern flank of this parasitic cone.

The summit of the hill forming Mikro Profitis Ilias was probably some distance west of the cliff that today dominates the heights of this name. Substantial flows of lava form the base, but above these solid rocks one finds a mantle of ash layers that dip in opposing directions toward the north and south over the massive lavas.

Megalo Vouno had attained about half its height when eruptions of great violence took place about a kilometer toward the west near the present location of the channel between Thera and Therasia. Ash beds and pumiceous scoria form an extensive layer that can now be seen at the base of the cliff in areas close to the two islands.

On Therasia these light-colored ledges thicken in the direction of the channel separating this island from the shore of Apano Meria. This, together with the southward dip of the layers, shows that the vent from which the material came must have been somewhere close to the point I have just indicated.

Along the cliffs of Thera these same deposits can be seen to thin in the direction away from the edge of the channel of Apano Meria and to rise on the flanks of Megalo Vouno.

The main product of the eruptions from the parasitic cone was a thick bed of red scoria that reaches its maximum thickness near the point of Apano Meria. The same unit maintains its characteristic form as far as the base of the cliff of Therasia; it is very thick in the northern part of the island and becomes progressively thinner toward the south. This red scoria, as well as the underlying lighter-colored beds, are products of violent pyroclastic eruptions in which gases, particularly hydrochloric acid and water vapor, played a significant role. The phenomena that preceded the eruption of this material were quite different from those that preceded the formation of Mikro Profitis Ilias and Megalo Vouno. The final result of their accumulation was a cone that was composed almost entirely of fragmental debris and was probably breached toward the north-northwest in a direction very close to that of the elongation of the channel of Apano Meria.

Eruption of this unit does not seem to have interrupted the activity at the two cones of Megalo Vouno and Mikro Profitis Ilias, or if there was any decline of their activity it seems to have been only momentary. After activity at the crater of Apano Meria came to an end, eruptions must have continued at these two parasitic cones, for Megalo Vouno and Mikro Profitis Ilias continued to grow. The pyroclastic ejecta were not abundant, but the discharge of lavas was on a large scale. The lower levels of the cliff sections just above the basal rocks underlying these centers expose a series of thin sheets of lava most of which are separated by thin layers of scoria.

The lavas of Megalo Vouno flowed mainly toward the west and north. Those that flowed west spread over a gentle slope formed by the cone of Apano Meria and continued for considerable distances from their source. One of the flows, which can be seen in the upper part of the cliff, starts near the chapel of Stavros close to the highest point of Megalo Vouno and descends as far as the village of Apano Meria. Toward the north, the flows are shorter and steeper (up to 35 degrees and in a few cases as much as 45 degrees). To the east, the flows were obstructed by small cones and accumulations of scoria.

The most important of these bodies is the one that today has the greatest relief, Kokkino Vouno. It can be considered a subsidiary vent of the larger, nearby cone of Megalo Vouno. Kokkino Vouno must have remained active much later than Megalo Vouno, for its ejecta cover those of the latter. It also rests on the products of Mikro Profitis Ilias. The last eruptions of Kokkino Vouno ejected great quantities of ash and lapilli. The black and dark gray color of this material shows that water vapor and hydrochloric acid played only a minor role in the explosive eruptions. When these two agents come in contact with lava at high temperatures, they attack it and give it a more or less bright-red color. Hydrochloric acid reacts with the rocks to produce chlorides of iron which break down in the presence of air and water vapor to form hydrated iron oxides. The alteration takes place not only at the surface of the rocks but extends a certain depth into their interior, where it produces, particularly at the expense of iron oxides, irregular platy deposits of limonite and hematite. It also permeates the rock with opal derived from the break-down of silicate minerals. These modifications, which the microscope reveals in some of the scoriaceous red lavas of Santorini, are identical to those produced when the same rocks are treated artificially with air, water vapor, and hydrochloric acid at high temperatures. The lapilli of Kokkino Vouno were not ejected under conditions of this kind. They were erupted along with inert gases, such as nitrogen, or perhaps even with reducing agents, such as hydrogen or compounds of carbon.

Mikro Profitis Ilias was not built by a single volcanic vent. The ravines that cut its northeastern slopes expose bodies of scoria produced by near-by eruptions and other forms of emplacement. The upper parts of this group are made up of ash and scoria overlain by a thick slab of lava that dips toward the east at a considerable angle. Thus, it is clear that during the period it was active the vent that discharged most of this material was situated at some distance inward from the present cliff face and away from the interior of the bay.

In front of the opening of the large cone of ash, pumice, and scoria, the remains of which are seen today in the cliff faces of southern and southeastern Thera, there was a chain of smaller parasitic cones that completed the circumference of the large central crater on this side. The crater had the form of a vast cirque breached on the east and west and was surrounded on all other sides by escarpments at least 400 meters high.

The oldest of the lavas that make up Megalo Vouno and Mikro Profitis Ilias are anorthite-bearing, but the younger ones contain labradorite. These two types of lava appeared alternately. The conduits through which they rose can be seen in the cliff section. The magma that filled them solidified to form the sixty-six dikes described in an earlier chapter. Forty-nine of these belonged to Megalo Vouno and seventeen to Mikro Profitis Ilias.

This was the form of the volcano at the time the underground reservoirs were being replenished and discharging magma through the central vent of the large crater. There were no strong explosions or large outpourings of ejecta during this period but only repeated flows of lava. The material reaching the surface in a more or less completely molten state first flooded the crater floor and gradually raised its elevation. Held in on all sides, either by the walls of the large central cone itself or by the parasitic cones situated in front of the vent on the northern side, the lavas could flow out only through low gaps on the west and east. Those that flowed through the larger opening on the west spread to form the flows one now sees in the cliffs of Therasia. Those that went toward the east made up the numerous flows that one sees superimposed on one another in the section of the cliff of Thera between Mikro Profitis Ilias and the path up to Thira. Where they accumulated in the greatest thickness they formed the present summit of Merovigli.

These lavas, having essentially the same chemical and mineralogical composition as those discharged during the later eruptions that produced the Kamenis, were also just as viscous. None of them formed long flows; they all stopped at a distance from the center of the bay that did not exceed five kilometers. Their fronts were made up of talus slopes, one behind the other, like the steps on a stairway. The superposition of these lavas with their high flow fronts accounts for the steep inclination of the slopes on the eastern side of Thera and the western side of Therasia. The lava flows erupted in the interior of the bay during the eruptions of 1707 and 1866 have exactly the same appearance.

It is quite probable that a long period including several eruptive episodes was required to fill the large central crater and produce the numerous flows seen along the interior cliffs of Thera and Therasia. It is possible, however, that a single eruption lasting several years could have accomplished this effect. The last eruption, which lasted four years, accumulated lava in parts of the bay where the depth of the water was previously 200 meters and produced a pile of flows with a thickness of about 100 meters. When one considers these facts, it is easy to see that a slightly stronger and longer volcanic episode would, if sustained, have been able to produce the lava flows seen in the western and eastern cliffs.

Whatever the duration or nature of the activity may have been during the period before the lavas were exposed, there can be little doubt that the end result was to fill the large central crater almost completely. The main vent must have been surrounded by a pile of pyroclastic material that had the form of a cone with a crater near its center. The relations of this interior cone to the walls of the great cirque that had been formed inside the older edifice would have been similar to that of the present cone of Vesuvius and the surrounding ridge of Somma. Between these two structures there was a semicircular low area similar to that of the Atrio del Cavallo, which extends from the cone of Vesuvius to the base of Monte Somma. The Atrio of Santorini, as well as the ridge that surrounded it had the form of a semicircle open toward the north. It was probably wide and could not have been very deep relative to the height of the escarpment that surrounded it.

A much more pronounced depression forming a prolongation of the northeastern extremity of the atrium developed between the lavas of the central crater and the parasitic cone of Megalo Vouno. It ended in a deeply incised ravine that opened out toward the northwest, where the channel now lies between Thera and the tip of Apano Meria. It crossed parts of the body of red scoria which, as shown earlier, was present in this area. Evidence of the former presence of such a depression is found in the rocks at the northern end of Therasia. A large, localized mass of pumice contains rounded blocks that have a size and form that can be explained very well if one postulates that they were deposited at the inner end of a narrow ravine that descended from high on the volcano and collected alluvial debris that was carried into it by rain water.

From what we know of the configuration of volcanoes that have a central cone of this kind with a concentric circular ridge, and even more from what we have learned from the eruptions of the Kamenis about the relations of the modern cones to the lava flows they produced, we are justified in concluding that the central cone of Santorini was not very large and did not stand very high above the lavas that came from it. In reconstructing a mental image of this cone, it would be a mistake to project the outer slopes of Thera and Therasia upward until they meet at a height of about 2000 meters. It is much more likely that the elevation was a good bit lower.

The vent of the summit cone was probably no higher than the present summit of Merovigli, for if the flows of lava that can be seen in the upper parts of the eastern cliffs of Thera had their sources at an elevation of at least 400 meters above sea level, it is easy to show that they could not have come from much higher. Whatever the orientation of the cliff section cutting these flows and whatever their distance from their source, they are all horizontal or gently inclined. Steeply dipping flows are seen only on the northern and northeastern slopes of Thera, where they were caused by the parasitic cones of Megalo Vouno and Kokkino Vouno. All the lavas coming from the central crater are remarkable for their gentle slope, not only on Thera but also on Therasia. Where their extremities reach ground that today has a considerable slope, as it does, for example, near Vourvoulos, one sees that, as we noted earlier, many of them end in a set of tiers that are stepped back one from another.

Thus the central dome was almost completely filled by horizontal lavas and at its top had a wide, rough plateau that was cut by many irregularities of the kind one always sees, even in the most uniform fields of lava. The great Caldeira of Terceira is a striking example of what may be a similar plateau. This caldera is a wide, almost round cirque surrounded on the south and west by a raised belt of felsic volcanic rocks and confined on the north by a line of scoria cones and trachytic spires. It has a floor made up of basaltic lavas that overflowed through the interval that remained open on the eastern side. The adjoining flows that cover the surface form a kind of cavernous floor that is flat overall but broken by many small irregularities. The scoria cones from the foot of which the lavas emerged to form the floor of the Caldeira of Terceira form part of the rim, while at Santorini the point of emission of the lavas must have been closer to the center as it is, for example, at Vesuvius.

The plateau that capped the dome of Santorini was certainly surmounted by a small terminal cone around the principal vent of the volcano.

The lavas that came from this source had chemical and mineralogical properties that caused them to accumulate over the vent instead of spreading quickly as flat sheets in the way basaltic lavas usually do. These rocks, having piled up in a scoriaceous mass of moderate size, formed a rocky plug standing above the surrounding lava field. No part of the summit cone was composed of fragmental material, lapilli, or ash like that of most large volcanic structures; gases and steam appear to have played little role in the eruptions that led to the filling of the large cavity, which, as we noted earlier, was surrounded by a ridge of pumice and ash. Together with the parasitic cones on the northern side, this formed the crest of the volcano. One sees very little loose material between the lavas making up the cliffs of Therasia or Merovigli. Were it not for a few layers of ash and lapilli and a very small number of thin beds of scoria, there would be no evidence of pyroclastic ejecta whatever. In short, the composition of the eruptive products of the central cone must have been very similar to that responsible for the surface form of the structures making up the present Kameni islands. Ordinarily, volcanic cones composed mainly of pyroclastic ejecta are the only ones that attain great heights; explosive debris has a much greater tendency than lava flows to accumulate within a restricted range of the vent. A volcano that erupts without violent explosions or sudden discharges of volatiles, ash, or lapilli is much more likely to build a moderately large dome, especially if it discharges acidic lavas, but it will not produce a cone that would be distinguished for its height or the regularity and steepness of its slopes as is the case in volcanic regions characterized by intense explosion. Thus we have good reason to believe that the cone that crowned the central plateau of the great volcano of Santorini had only a modest height.

It is quite possible that this summit feature could have been a double rather than a single cone. One can see that the flows that spread toward the east from the cone of Merovigli did not originate at the same point as those that flowed in the other direction toward Therasia. But we should also note that these two groups of lavas have identical compositions and a very similar appearance. There is no reason why one must postulate that the volcano had a double central vent. The observed relations do not exclude other possibilities, and, in the absence of further proof, one is not justified in assuming that several cones crowned the high plateau of the volcano.

Around the same time that lavas were filling the large cirque of Santorini, secondary vents were active on the flanks of the volcano. The three eruptive centers of Megalo Vouno, Kokkino Vouno, and Mikro Profitis Ilias were still active and continued to add to the elevation of their summits. The four dikes of Therasia added flows of lava to those coming from sources at the central vent. Other new vents opened in the region around Akrotiri to form an almost continuous line from Balos through the present site of the village of Akrotiri to Cape Mavrorachidi. In the same region but more to the west, an eruption produced a large mass of brown scoria that can now be seen in the cliff not far from Cape Akrotiri. The lavas of the Akrotiri region are remarkably basic with a dark color and large amounts of scoriaceous material. Some have a corded surface. They were very fluid at the time of eruption and were crossed by many fractures that released volatile components.

The lavas that flowed from the central vent of the volcano in a direction corresponding to the present position of Merovigli filled a wide valley between Mikro Profitis Ilias and the thick accumulation of fragmental material on which the present town of Thira now stands. Exposures in the cliff section show that they completely filled this gap. It is evident that waters that collected in the ring-like atrium of the volcano ceased to flow out in this direction. Instead, they escaped in the opposite direction between the western edge of the circular valley and the series of superimposed flows that moved toward Therasia. They descended through one or more ravines and came out near the present site of Aspronisi in the area between that island and Cape Tripiti on Therasia. The lavas that flowed toward Therasia barred their way toward the northwest.

There is no doubt that torrential streams descending in a southwesterly direction produced the conglomerate seen just below the pumice on Aspronisi. The small depression in the beds making up this island may also have been produced in this way. One can confidently conclude that it corresponds to a low place in the old ground surface eroded by water flowing in this same direction.

A valley on the outer part of the large cone of Santorini beyond the Somma ridge that enclosed it on the south was remarkable for its location and for the way in which it originated. It had an unusual effect on the make-up of the Santorini archipelago as it exists today.

This east-west trending valley descended from a point in front of Athinios near the present position of Cape Alonaki and reached the sea between Cape Akrotiri and the island of Aspronisi. It has already been described by von Fritsch in his memoir on the older parts of Santorini.

In order to show that this was the configuration of the topography at that time, von Fritsch invoked the following reasoning. The depression must have already existed, he says, for otherwise lavas coming from the heights of the volcano would have spread over the region of southern Thera between Akrotiri and Mt. Profitis Ilias, as they did over other parts of this island and Therasia. It was this valley that deflected the lavas from their course. For the same reason, torrential waters descending from the upper slopes were diverted and did not erode a deep ravine on the Akrotiri peninsula as they did, for example, on the eastern slope of Thera. They did not cut into the rocks exposed in the cliff section as they would certainly have done if they had not been intercepted by the valley. As a final piece of evidence in support of this hypothesis, the author cites the bayward slope of the rocks bordering the inner cliffs in the western part of the Akrotiri district. The dip of these beds is attributable primarily to the same conditions that were responsible for the erosional topography.

All these arguments do not have equal weight. The first can be questioned, because the lavas coming from the central crater could have been stopped, as already suggested, by the escarpment of a Somma ridge that was convex toward the south. But the last two lines of evidence invoked in favor of the hypothesis are decisive. Regardless of the form of the central volcano of Santorini before the bay was formed, it is certain that alluvial streams should have flowed down the length of the eastern slopes. As a consequence, ravines would have been cut into the flanks of the mountain between Mt. Profitis Ilias and the village of Akrotiri had the streams not found an easier route toward the west. It is equally certain that after the volcanic rocks of submarine origin in the area around Akrotiri were uplifted they formed a partly emergent area that was elongated in an east-west direction. Loose material from the large central cone collected against this barrier. Ejecta from the main cone partly covered the mass of submarine material and maintained its relief. During their growth, the cone and the adjacent hill that was tangent to its base continued to have about the same relations, and the postulated valley formed a depression between them. Since the configuration of the cone and hill is beyond question, the existence of a depression between them is equally certain.

In the area south and southwest of the village of Akrotiri, the ground is cut by deep, wide furrows. The main ones can be seen on each side of the low ridge that begins at Loumaravi and ends at the point of Mavrorachidi. These ravines separate older ridges that owe their distribution and form to growth of Loumaravi and the Arkhangelos and to the flows that came from these centers. Their relief has since been accentuated by erosion.

Most of the ravines on the northern and northeastern slope of Thera were formed under similar conditions. Some are cut into the older volcanic rocks of the island, others extend only to the base of the younger rocks. Some were refilled during the pumice eruption and reopened by later fluvial erosion. Rains are now rare at Santorini, but they occasionally give rise to great torrents. Even in the narrowest part of Thera near the northeastern part of the island, storms occasionally produce very swift, voluminous streams. Considerable erosion takes place in a few minutes, and drainage courses become deeply incised. It is difficult, therefore, to distinguish ravines formed after deposition of the upper pumice from those formed earlier. Close examination of the distribution of the deposits provides some evidence relevant to this question, but a detailed study is required in each individual case.

When one observes the upper edges of the cliffs around the bay, one is struck by the unbroken continuity of lavas and layers of fragmental material underlying the pumice; one sees no erosional irregularities that have been filled with pumice. This can be explained if one considers how the drainage was controlled by the three large depressions that came down from the heights of the volcano and reached the sea near the present locations of the three channels connecting the bay of Santorini with the surrounding sea.

The depression corresponding to the channel of Apano Meria was deeply entrenched, probably through most of its length but certainly in its lower section. It was very steep and had a tortuous course, first from east to west in the section between Megalo Vouno and the central dome, and then turning from the south-southeast toward the north-northwest when the present channel of Apano Meria was formed.

The channel directed toward Aspronisi or, more properly, between Therasia and Aspronisi was even steeper and was probably very sinuous in its upper part, but in its lower section, it joined the valley between the southern slope of the large crater and the northern base of the hill at Akrotiri.

This valley constituted the third outlet by which water flowed from the central heights of the volcano. The area seems to have been a favorable one for vegetation, for planting was more feasible here than on the upper slopes where the scoria from recent eruptions was either sterile or supported only trees. The outermost peripheral zone of the island was very narrow, had abrupt slopes, and must have differed little from the central region in the type of vegetation is supported. Thus, the most habitable part of the island that could provide the most resources for humans was that formed by the basin between Akrotiri and Therasia. It is precisely on the two sides of this valley that one finds the settlements of the people who inhabited the island before the great pumice eruption and formation of the bay.

In one of the preceding chapters we described in detail not only the dwellings of this ancient population but also its various utensils, especially its pottery, which was remarkable both for the beauty of its form and for its abundance and diversity.

In summary, there was a time when all Santorini was a single large island with a high dome in its center. Its summit was probably barren and covered with scoria, but the upper slopes were wooded and its southern half was cut by wide fertile valleys. Although the topography was irregular, the island had the very appropriate name Στογγύλη or *round*, by which the ancient historians designated it. It was also referred to as Καλλίστη, *very beautiful*, for the vegetation that covered the upper slopes and the rich cultivation of its lower parts. These two names, both of which are totally inconsistent with present conditions, seem to be a legendary memory of what existed before the cataclysmic event that led to formation of the bay.

It need scarcely be mentioned that the basic features of the picture developed in the foregoing pages, if essentially correct, can be used to evaluate certain secondary consequences. The bay of Santorini can not have the same age as the rocks surrounding it. Anyone who considers the height and near-verticality of the cliffs, the nature of the surrounding rocks, the depth of the water measured in soundings, or the shape of the basin that forms its bottom will immediately reject the notion that there has always been a belt of islands separated from one another or connected in only a few parts of their circumference. So, without dwelling on this untenable idea, we can accept the former existence of a large, round volcanic island, the origin and development of which have already been described, and proceed to the more serious matter of the factors involved in the formation of the wide marine basin that today occupies what was once the central heights of the island of Santorini.

Among the hypotheses proposed to explain the bay, one that has played a particularly important role is known as the theory of craters of elevation. Proposed in 1802 by L. von Buch,[12] it has been supported by von Humbolt and Élie de Beaumont. Vigorously attacked by Constant Prévost, Poulett Scrope, Darwin, Virlet d'Aoust, and Lyell, it became the topic of a lively controversy which, for thirty years, has been of major interest to the scientific world and has divided geologists into two camps.

According to this theory, there are two types of craters. One, called craters of eruption, is formed by accumulations of lava and, especially, ejecta around the orifice from which this material comes. The other type, known as craters of elevation, owes its formation to bulging of the surface as the result of deep-seated, upward-directed forces that cause failure of the superficial rocks that are under great tension due to the internal pressure. The opening of a crater of elevation is analogous to formation of star-shaped fractures in a plate of glass that has been struck by a central blow; the upward-directed stress produces radial cracks around a central focus. Violent explosions then enlarge the space where the converging cracks intersect and in this way produce a circular depression surrounded on the inside by a steeply inclined escarpment and on the outside by a surface that slopes gently away from the rim. The sedimentary beds or the stratiform eruptive rocks, which, on average, were originally flat-lying, will be displaced from their original horizontal position; they will be uplifted around the central part of the rim and in places will be breached by vertical fractures that increase in width upward.

The immediate cause of the uplift would in some cases be an intrusion of solid rocks acting like a wedge against the under side of the surficial beds, but more often it would be the force transmitted by molten magma or by a sudden expansion of a large amount of gas that escapes from a deep-seated igneous mass.

Almost all the circular volcanic depressions that are considered good examples of craters of elevation have large dimensions; these include the calderas of the Canaries and Azores, the Somma rim of Vesuvius, the Val del Bove of Etna, and the great craters of Java. Some are circular, others elliptical, but most are as wide as the bay of Santorini and several are even larger.

A peculiarity shared by craters of this kind is that they are accompanied by one or more eruptive centers that may be located either inside the crater or at some point beyond the rim. The well-developed craters of elevation that are considered ideal examples of these great natural features are those in which a single pyroclastic cone stands in the middle of a large circular enclosure composed of a steep wall made up of rocks that may be volcanic but could equally well be plutonic or sedimentary. If they are of volcanic origin, they may or may not be identical to the rocks of the central peak that they surround.

The interior of a crater of elevation may either be dry, as are those of Vesuvius and Etna, or invaded by the sea, as it is in the case of Barren Island, Saint Paul, and Santorini.

These, then, are the essential features of the classical theory of L. von Buch. It is unlikely that this idea would have survived the death of its author had not Élie de Beaumont moderated it and added support in the form of very strong arguments.[13] It was he who showed that the central volcanic cones owe their form to explosive ejections. According to Leopold von Buch, von Humbolt, and even Dufrénoy, the central peaks of the main volcanoes were formed by uplift in the same way as the ramparts surrounding them. They maintain that one must distinguish three phases in the development of a complete volcano: first, inflation of the crust; second, outbreak and production of a crater; and, third, appearance near the center of the crater of a cone or dome with a vent penetrating to the earth's interior. Only after it has passed through all three stages is the form of the volcano complete. The uplift may stop after one of the first two phases or in the middle of the third; so that a dome may form without developing a vent. In this way, volcanoes take on different forms. Puy de Dôme, according to this view, was produced by inflation but without a subsequent outbreak; the island of Palma owes its origin to uplift that ended in the second stage; in the Cantal, a great cirque opened and the phonolite of Grioux came out in the center of the crater of elevation. At Vesuvius, Etna, Tenerife, etc. the volcanoes have passed through all three stages. The cone of Vesuvius,* von Buch says, had its origin in the great eruption of A.D. 79; "Vesuvius did not exist before that time; the cone we now see rose in the middle of the crater of elevation of Mt. Somma, and the southern and southwestern walls of this great mountain must have given way to the new outlet that continues to be one of the most active volcanoes in the world. The volcano emerged after already being formed in the earth's interior; it did not rise by accumulation of successive flows of lava. On the contrary, its elevation since that time has only decreased." Etna was also formed by "sudden uplift," and "one cannot fail to see that it was fully formed at the time it was born."†

These quotations illustrate how the theory of craters of elevation has since been amended by Élie de Beaumont. No geologist who has seen Vesuvius, Etna, or Tenerife would disagree with his statement that "the central cone of Vesuvius, the peak of Tenerife, and the upper cone of Etna developed in the same way as the parasitic cones on the flanks of Etna."‡

Moreover, there is no lack of examples of central cones that developed in less than a century in the interior of large cirques that von Buch and his followers considered craters of elevation, and among these, the new cone of Santorini may be the most complete and closely observed example. In all

---

* De la nature des phénomènes volcaniques dans les iles Canaries, p. 312, translation by Boulanger, Paris, 1836.
† L. von Buch, work already cited.[14]
‡ Poulett Scrope, Note on the formation of volcanic cones, §9.[15]

these cases, it has been possible to see that the new cones were composed of material ejected from their orifice and that ground movements played little if any role in their formation.

Nevertheless, even if we disregard these extreme ideas held by L. von Buch, we should still consider the theory of craters of elevation as it was developed by Élie de Beaumont and discuss each of the arguments he advanced in its favor. We will then be able to see how well they can be applied to the bay of Santorini.

The arguments invoked in support of this theory can be grouped according to two types of observation. The first of these is based on the configuration and structure of the strata seen in the walls surrounding craters of elevation; the other is based on historic examples of volcanic uplift. The first type forms the foundation of the theory, for it includes the basic arguments that have most often been challenged and are most tenaciously defended.

The strata of the rim of a crater of elevation are said to be inclined in all directions outward from the center, regardless of whether the rocks are sedimentary or volcanic. If they are sedimentary, they could not have originally been inclined as they are now. As shown by the regularity of their stratification and the positions of pebbles and fossil shells that one commonly finds in them, they must have been laid down horizontally. The way they are tilted is evidence of an upward-directed force and consequent rotation from their original position. If they are volcanic and have resulted from solidification of a molten mass, they are in thick, wide, and continuous sheets covering a great surface area. Compact and rarely vesicular, they have all the characteristics that molten lava acquires when it solidifies on flat ground. A voluminous flow can have a tight, coherent structure, a flat surface, and columnar jointing perpendicular to its base and top only if it spreads over a flat surface or ponds in a confined depression. "Lavas that have solidified on the inclined flanks of volcanoes leave only narrow trails of scoriaceous material that results from rapid solidification of their crust on the slopes of the cones that erupted them. If the lava in these flows is abundant, it separates itself from an irregular crust that cools by exposure to the atmosphere; it takes on the distinctive characteristics caused by rapid cooling except in situations where, even though the lava may still be liquid, it can no longer flow. It then spreads in wide flows that have the textures and forms of basaltic sheets only where they have crossed a flat surface around the base of a cone."* Volcanic cones are too steep-sided for any notable amounts of their eruptive products, whether they be fluid lavas or fragmental ejecta, to remain on their flanks. Whenever a lava has a slope of more than 3 degrees, it develops a broken, chaotic, rubbly form; if its slope exceeds 10 degrees, it leaves only a narrow trail of incoherent scoria. "The properties of lavas that have remained for long in a soft malleable state before solidifying have such properties that their course can be determined only by examining the trails they leave behind much in the way that melt-water leaves blocks of ice on the slopes down which it has flowed. Thus, one can readily explain why the structure of these rocks left on the surface by a large flow vary according to the slope of the ground; the way in which they develop is entirely dependent on the slope angle."†

In conclusion, Élie de Beaumont, on the basis of these observations, maintained that a lava takes on the appearance and form of basaltic sheets only when it solidifies on a surface with a slope of less than 3 degrees. He says that "the rubbly flows that are roughest, most disordered, and most difficult to cross on foot are those laid down on slopes of between 3 and 5 degrees, apparently because at this angle the crust of the lava could attain such a thickness without having already lost much of its velocity, so that the contest between these two competing factors reaches its greatest magnitude."

Applying these principles to the Val del Bove of Etna, Élie de Beaumont concluded that the beds that make up this huge depression have been displaced from their original positions. Since this reasoning can be applied to all the large, circular volcanic depressions regarded as craters of elevation, and particularly to Santorini, I think it best to reproduce the original text word for word: "None of these beds, which show consistent evidence of having solidified on more or less uniform slopes, have taken on a rubbly form and they have been even less likely to leave behind a trail of detached patches of scoria. The scoriaceous crust that formed on each of the flows in Val del Bove is normally covered either by lapilli, commonly very fine, or by angular fragments of scoria of moderate size. Nowhere on the surface does one find large, chaotically distributed blocks of the kind developed on rubbly lavas that flow and finally solidify; nowhere does one see

---

* Dufrénoy and E. de Beaumont, *Memoire sur les groupes du Cantal et du Mont Dore, et sur les soulèvements auxquel ces montagnes doivent leur relief actuel* (Ann des Mines, 1836, p. 535).

† Dufrénoy and E. de Beaumont, *Memoire sur les groupes du Cantal, etc.*, p. 536.

the upper crust of the flow broken, cracked, or turned up vertically as happens so often in rubbly flows. Nowhere does one see the entire base of the structure up-arched in a way that reveals a hollow space below it, as is observed so often in the rubbly flows of Etna and Vesuvius wherever a vertical section can be seen. Layered lavas of the Val del Bove differ in certain ways from modern flows that spread over almost flat ground where their motion becomes very slow and they soon stop of their own accord. I have yet to observe any that, judging from sections I have seen, appear comparable to a flow that ran any distance down a slope of more than 3 degrees .... For that reason, it is evident that those layers that have changed orientation are the ones that today are steeply inclined."*

Élie de Beaumont stressed with equal force the following argument based on the considerable width of the sheets of lava making up the walls of craters of elevation: "The simple fact that the ancient lavas have great widths in transverse sections tends to show that none of those that now have a significant inclination could have flowed down the present slope, for on such a slope the fractures that are filled by dikes of any length or orientation would have had ends at very unequal elevation, so that all the lava would have escaped from the lowest point and would have left a narrow trail following a course down the steepest slope, as happens when modern lateral eruptions break out on substantial slopes."†

The parallelism of the two faces of each sheet serves as the basis of an essential point of the argument made by Élie de Beaumont. "One can easily imagine that material of this nature (ejected products) falling on a more or less horizontal surface would produce a layer of nearly uniform thickness over a certain area, but if the surface on which the rain of ash accumulates had a considerable inclination, say 25 to 30 degrees, such a uniform thickness would be less likely .... The material making up the fragmental layers at the Val del Bove varies considerably from one layer to the next and spans the entire range of possibilities, and since the strength of their impact must also have varied from one eruption to another, there must have been a great number that could not have remained stable at their angle of repose of 27 degrees. Ejecta falling on a slope with this angle would have rolled toward the base of the slope where the material would accumulate in a greater thickness, while

---

* Élie de Beaumont, *Recherches sur la structure et l'origine du mont Etna*, chap. IV, p. 538.
† Élie de Beaumont, same work, p. 509.

on the slope itself, almost nothing would remain. In any case, the middle and upper parts of the slope could not remain covered with a constant thickness of accumulated ejecta; the lower part would be buried under a much greater thickness of the same material. This would result in a lack of uniformity in the thickness which in certain orientations would be very noticeable."*

Élie de Beaumont also points out that, owing to the slope of the terrain, lavas discharged by the various dikes visible in the Val del Bove would tend to flow out only in the down-slope direction. They would not be as numerous in one direction as in the other. This would lead to irregularities that are not seen in the sequence of layers. He says that "the dikes discharge their products in all directions, and this implies that the ground surface around the vents was originally horizontal."

Etna, which Élie de Beaumont has taken as one of the type examples of cones produced by uplift, is distinguished, in his view, by a characteristic form that differs from that of cones produced by accumulation of eruptive material. Instead of continuous, straight talus slopes steeply inclined over their entire extent, this volcano has two very different parts, (1) an outer part with gentle, even slopes and (2) a central mound with an average slope of 30 degrees and a small summit cone. Thus there is a break in slope on the flanks of the mountain. "The gentle slopes around the base were produced by deposition, but the abrupt, irregular bulges of the central part are caused by uplift . . . . This central bulge is due entirely to a central core that forms the main mass of the mountain."†

Having developed these arguments in support of the theory of craters of elevation, Élie de Beaumont attacks the idea that volcanoes are formed solely by accumulated ejected material and tries to show how the amounts of material discharged by Etna over the centuries are trivial compared to the amount required to form the entire mountain. The ash and scoria that have accumulated in the past two thousand years on the upper plateau of Etna have raised the surface only 1.25 meters. This is the amount measured in an excavation near Torre del Filosofo, a small, ancient ruin built on the platform that supports the upper cones of Etna. "Thus the river Nile is more effective in burying the monuments of Thebes and Memphis with alluvium than Etna is in burying the Torre del Filosofo under volcanic debris."‡

* Élie de Beaumont, *Recherches sur la structure et l'origine du mont Etna*, p. 512.
† Élie de Beaumont, same work, Chap III, p. 613 and 614.
‡ Élie de Beaumont, same work, Chap III, p. 594.

For all the objections to his ideas, Élie de Beaumont had serious responses that were often difficult to refute. Lyell objected that one finds on Etna thick slabs of lava with an almost basaltic density and that these flows are very young and spread over fairly steep slopes. Élie de Beaumont replied that the compactness of some parts shows that the lavas discharged on a daily basis from Etna are in fact capable of taking on the massive character of basalts under favorable cooling conditions, but that this characteristic is exceptional. In the face of observations made by the English scientist, he held fast to his earlier assertion that "a slope covered with basalt is just as obviously due to movement of the earth's crust as one covered by lacustrine limestone deposited from the waters of a lagoon, and that a cone covered with basalt is necessarily a cone of elevation."

Pointing out that the presence of compact lavas in modern flows is inconsistent with the theory of craters of elevation, his critics also confronted him with the objection that vesicular lavas are present in the walls of these craters. It was said that if these flows had moved horizontally and solidified in this position they would still be compact like basalts and never scoriaceous. There would be no reason why the innumerable vesicular cavities of the lavas should be elongated in directions consistently divergent with respect to the center of the volcano as though the molten material had descended from the heights of a central cone. Lyell[16] stated that "the strength of this argument, which was developed chiefly by Virlet d'Aoust, will be appreciated by those who understand that gas bubbles enclosed in a moving liquid take an oval form in which the direction of their major axis always coincides with that of the flow direction."

But Élie de Beaumont did not let this objection pass without a response. Here is his reply: "As for the objection that M. Virlet deduced from the very elongated bubbles in some of the trachytic rocks in the escarpment of Santorini, it has no foundation, because by that reasoning only a trachytic lava would be very flat, it must have spread over older ground, and its vesicles must have been elongated as the cells of a porous material are when it has been flattened. Nothing proves that they flowed on a slope inclined in the direction that they indicate today."

It was objected that if, as the theory of L. von Buch holds, the walls of craters of elevation result from radial fracturing of the earth's crust, they must have always been broken by deep tensional fractures. On the contrary, it is said, observations show that in many places the faces of these walls are continuous and without any perceptible breaks. The long cliffs of Thera, which form the northern, eastern, and southern rim of the bay of Santorini are good examples of the kind that is cited in opposition to the theory of craters of elevation. Scarcely had this objection been raised than Élie de Beaumont responded to the challenge and published in collaboration with Dufrénoy a lengthy memoir in which he calculated the total amount of fractures and joints that would be produced by a radial uplift of a given height. He found that at Tenerife and Palma this total had a maximum initial value of only 1/70 of the total circumference of the cirque, even if one considers the visible trachytic dikes in the escarpment as fractures that were later filled. For Santorini, one finds that the theoretical total would be at most 1/250 of the circumference or 38 meters, even when one takes into account the substantial depth of the bay and assumes the uplift began at the lowest central point. So this shows, according to Élie de Beaumont, that the results "do not indicate at all that one should find on the surface of Santorini the numerous valleys and ravines seen at Palma or even Tenerife. It is likely that the small sum of fractures is accounted for in joints or fissures of narrow width." Thus, he says, he has disposed of the objections that Virlet raised against von Buch's ideas that Thera, Therasia, and Aspronisi are the remains of a crater of elevation.

Partisans of the theory of L. von Buch invoke in its support the fact that in the centers of craters of elevation one often sees enormous domes of trachytic rocks that rise within the crater as they do, for example, at Astoni in the Phlegrean Fields and in the middle of the Val del Bove at Etna. They also stress the characteristic situation of phonolitic bodies of Sanadoire and of Tuiliere at Mont Dore and the domes of the same rock in the center of the massif of Cantal. But Élie de Beaumont has treated the last example with reservations. "We do not know," he says, "whether these bodies of phonolite have been uplifted, as in the case of Santorini, at a time when the areas where they are found had already taken their present form or whether they appeared at a time when Cantal had already acquired its present relief."* But the views of the master have not always been followed by all of his disciples. The latter, favoring the uncompromising ideas of L. von Buch, considered these eruptive rocks as having been extruded upward in a very viscous or solid state and as having caused, by their direct pressure, the break-up of the shallow basement

---

* Élie de Beaumont and Dufrénoy, *Sur les group du Cantal et du Mont Dore et sur les soulèvements auxquels ces montagnes doivent leur relief actuel* (Ann. des Mines, 1836, p. 564).

rocks. From this, they concluded that these centers revealed the direct causes and evidence of uplift. They regarded the domes in the middle of large cirques as clear proof of the mode of formation and a demonstration of the validity of the theory.

Having outlined the arguments in favor of the hypothesis of craters of elevation based on observations of phenomena of a purely geological nature, we must now consider the reasoning based on recorded historical events.

From the most remote times of ancient history, many new craters have come into being. It is beyond question that a certain number formed slowly and gradually by accumulating various kinds of volcanic ejecta around an eruptive vent. But there are others that have appeared during extremely violent events; in a few hours, cones, some of considerable size, have risen, and great cavities have opened in the upper levels of pre-existing volcanic chimneys. In the latter case, the phenomena have often been of such intensity that it seems reasonable that they correspond to gigantic explosions of an unusual kind and, because of this, such cataclysms have been thought to demonstrate the theory of L. von Buch, which requires a powerful manifestation of this kind.

The peaks of Timor, Cape Urcu, Carguairazo, the volcano Cosiquina in Central America, and Gelung-Gung and Popandayang in Java are well known examples of volcanic cones that, in the space of a few hours, have almost entirely disappeared leaving an immense void exactly like the craters of elevation. Before the catastrophe of Popandayang, the summit of this mountain had an elevation of 2,109 meters. After the terrible explosions, nothing but its base was left, and 40 villages had been buried under debris. Where the center of the mountain once stood, a cavity had been excavated 24 kilometers in length long and 10 wide. How, ask the supporters of the theory of craters of elevation, can one explain these facts without admitting that these mountains had suddenly collapsed into an abyss beneath them? What could be more reasonable than to conclude that each of these cones had an enormous cavity with a shell that suddenly failed so that the cavity underneath them was suddenly opened? This hypothesis accounts for both the sudden nature of the phenomena and for the size and shape of the resulting depression.

But the reasoning proposed by the supporters of L. von Buch's ideas does not end there. They cite two historic facts as irrefutable examples of the recent formation of craters of elevation. The first is the appearance of Monte

Nuovo in 1538 at the edge of the bay of Baia near Naples; the second is that of Jorullo in Mexico near the city of Arca in 1759.

Monte Nuovo is a cone 134 meters high with a base of two and a half kilometers. It has a crater 128 meters deep and is composed of scoria, lapilli, and ash. It is wrong, according to L. von Buch, to think it was formed by an eruption and made up of loose material, such as scoria and pumice. Solid beds of uplifted tuff are visible around the crater, and only the outer surface is composed of scoria.*

In support of this opinion, Dufrénoy reports a passage from a book by the Italian doctor Porzio[†] which reports that in 1538 the ground where Monte Nuovo now stands was suddenly inflated and that an explosion broke out from the underlying space. Here is the passage in question: "Following earthquakes that lasted two days and two nights, the sea withdrew about 200 meters and people living in the area could collect great quantities of fish along the shore; they noticed several springs of fresh water that were discharging there. Then on the third day of October (29 September by the new calendar), they saw a large area, near Lake Averne and between the base of Monte Barbaro and the sea, rise and suddenly become a growing mountain. Then at two o'clock in the night this great mass of ground broke open with a loud sound and vomited large amounts of flames, stones, pumice, and ash."

A person living at that time, Pietro Giacomo di Toledo, has left a detailed description of the phenomenon. A sentence in his account has been taken as supporting Porzio's version. He states that "violent earthquakes continued for two years; on 27 and 28 of September, they continued day and night; the plain between Lake Averne and the sea was raised and several crevices opened there." These words describe the first stages of formation of Monte Nuovo, and the inflation and outburst described by Porzio.

The other historic example cited in favor of the theory of craters of elevation, that of Jorullo, owes its renown to the great authority von Humbolt, who made it known. According to this distinguished naturalist, a hill measuring 168 meters at its center grew in a single night in the middle of what had been a plain covered with sugarcane and indigo and crossed by two streams. The middle was covered with thousands of small, fuming cones and was tilted toward the surrounding plain at an angle of about 6 degrees. Six large buttes aligned in the direction of the volcanoes of Colima and Popocatepetl stood

---

*   L. von Buch, *Phén. volcaniques*, p. 347. Paris, 1836.

†   Porzio, *Opera omnia, medica, phil. et math. in unum collata*, 1736. Cited by Dufrénoy in his memoirs describing the geology of France, v. IV, p. 274.

in the middle. The highest, named Jorullo, rose more than 500 meters above the plain. The convex surface of the protrusion dominated by these buttes was riddled with innumerable crevices from which burning gases were still being given off at the time of von Humbolt's visit. Although the temperature in these fissures had declined from year to year, it was still hot enough in 1780 to light a cigar at a depth of a few inches. The surface of the protrusion resonated under one's feet as if one were walking over a hollow space. Part of the courses of the two small streams that formerly crossed the plain had disappeared since the eruption. They went underground at the eastern edge of the mound and reappeared near the western side as hot springs.

These various peculiarities led von Humbolt to conclude that Jorullo owed its origin to uplift, that the surficial layer of ground covered a cavity, and that a more violent volcanic force completed the work already begun by breaking through the ground and producing a large crater. Jorullo was thus a cone of uplift arrested in the first stage of its development.

Such were the arguments on which L. von Buch and the geologists of his school built their theory for the normal formation of large volcanoes. In the summary I have just given, I have developed at length each of the lines of reasoning they invoked in an effort to give each its full extent and value.

It is time now to consider what is really known in this matter.

First of all, is it true that it is impossible for lavas to solidify on slopes of more than 6 degrees and form compact layers? Observations show quite clearly that they can; all the adversaries of von Buch's ideas have cited examples. Constant Prévost, Lyell, and Poulett Scrope have even used these as the basis for their arguments. Poulett Scrope says that he saw on Vesuvius in 1819 and 1820 thick currents of lava solidify on slopes of 33 degrees. Lyell points to observations of the same kind during the eruption of Etna in 1855.

To be sure, the accounts of these observers might be questioned because of the ardor they show in discussing this question, but they are not the only ones to cite such examples. Geologists who have long been indifferent to the struggle, and even supporters of the theory, have felt obliged to indicate a certain number of cases in which they had opportunities to note the solidification of lavas on fairly steep slopes. On several occasions since the beginning of this century lavas on the flanks of Vesuvius, Etna, and the volcano of Reunion have been seen to solidify on slopes of 10 to 30 degrees.

According to Junghuhn, several of the large volcanoes of Java offer examples of large sheets of lava that stopped and solidified on ground with very steep slopes.

The lava of the 1809 eruption at Saint Georges in the Azores archipelago forms an elongated layer descending from the uppermost parts of the island and has a slope of 35 degrees near the place where it reaches the sea.

The cone of Pico in the same archipelago occupies the western part of a mantle of mammillated lavas, and yet the slope is so steep that it is difficult to climb. The average inclination on the upper flank of the volcano is about 35 degrees and in places it is so abrupt that one has to use the irregularities of the rocks in order to get up it.

At Vesuvius in 1855 and 1872 and at Etna in 1865, currents of lava were seen cascading into ravines from high embankments that were built up when lava poured over them and suddenly solidified.

In the Sandwich islands, Dana observed on the flanks of Mauna Loa a number of pinnacles of lava that were 12 to 15 meters in height and had been produced by vertical jets of very fluid lava that were subjected to very rapid cooling.

At Santorini, the last eruption produced currents of lava that, mainly on the southeastern side, spread toward the sea on slopes of 20 to 25 degrees. Although the upper parts of these flows were made up of loose blocks, their interior is composed of compact, glassy lava that can be seen very clearly in longitudinal sections.

Examples of similar lavas can be seen among the older products of the Kameni volcanoes, particularly among those that were erupted from Nea Kameni in 1707 and others on the island of Palaea Kameni. The prehistoric rocks that make up much of the island of Thera provide other good examples of lava flows that solidified on slopes with considerable inclinations. One can see this particularly well on the northern slope of the cone of Megalo Vouno. Several of the flows exposed there are inclined at angles that are as steep as 15 to 20 degrees.

Extinct volcanoes provide similar examples, even though the evidence may be less clear. No one would question the origin or mode of formation of the thick flows that, in the Auvergne, descend from such peaks as Gravenoire, Nugère, and Pariou. All along the lengths of these flows, inclinations of 10 to 30 degrees are quite common. In short, there is no volcanic region, either ancient or modern, where one cannot find continuous sheets of lavas of substantial thickness that have come to rest at angles considerably steeper than that which Élie de Beaumont set as a maximum limit beyond which they should have left nothing but fragmental products to mark their paths.

The authors of the theory of craters of elevation have themselves cited a great number of facts that contradict their statements denying the possibility for lavas to solidify on steep slopes as coherent flows. Leopold von Buch and Élie de Beaumont have both conceded that certain lavas have solidified on the steep slopes of Tenerife. Moreover, the first of these two great geologists, in describing the volcanic island of Lanzarote, pointed out thick basaltic lavas that, "like black glaciers," descend from the crater to the base of the cone formed during the eruption of 1730. Von Humbolt, in the same work in which he described the formation of Jorullo as the result of uplift, says that when he visited the volcano, he could only reach the summit of the mountain by climbing a great cascade of solidified lava that came from the crater and descended to the plain in the form of a large promontory inclined at an angle of 35 degrees.

Examples of thick, continuous flows that solidified on very steep slopes are both numerous and incontestable, but one must recognize, nevertheless, that presently active volcanoes rarely produce extensive flows in large sheets of the kind formed in ancient outpourings of basalt. Modern flows tend to be narrow; when the lavas from an eruption are spread over a broad surface, one almost always notes that several currents of moderate width have been juxtaposed. The great width of certain recent outpourings of lava results from a combination of several narrow flows that have spread side by side in close proximity. In active volcanoes of our time, thick, compact, and continuous units are formed only when the lava flows across a horizontal surface with very little slope - in other words, under very exceptional circumstances. So if one sees in the walls of a so-called crater of elevation beds of lava that are steeply inclined and, at the same time, have a compact structure and continuous longitudinal sections over wide distances, one should conclude that in fact these beds must have spread out at the source on a surface that was close to horizontal, and it must be admitted that the intervention of strong mechanical effects are required to explain their present situation.

The aphorism posed by Élie de Beaumont to the effect that "any cone covered with basalt is necessarily a cone of elevation" is correct in designating, under the name basalt, not a rock of such and such mineralogical composition but a flow that has spread, like certain basalts, *in wide, thick, and compact sheets.* But that is not the crux of the problem. The argument must address the question of whether compact, wide, and continuous sheets of basalt are found in the walls of craters of elevation. A response based on actual observations seems at first to support Leopold von Buch's theory, for in all large cirques considered to be craters of elevation, if one stands in its center and surveys the surrounding ramparts, one is immediately struck by the apparent regularity and continuity of the layers exposed in the cliffs. Horizontal sections through lava flows, piled up in great numbers, seem as though they can be traced around almost the entire circumference of the inner walls of the crater. The layers visible in the escarpment of the Somma ridge at Vesuvius, as well as those that enclose the bottom of the Val del Bove at Etna or the amphitheater of the great caldera of Palma in the Canaries, have been cited as excellent examples. One could also mention the layered rocks of the great calderas of the Azores, but above all, one should certainly invoke the example of the lava flows that are so numerous and regular in appearance all along the inner cliffs facing the bay of Santorini.

What should one make of this? Are these impressions reliable, and how should one interpret them? These are the questions we have to resolve.

Let us say at the outset that in most cases where it is possible to make careful observations, it is seen that the continuity of the layers is only apparent. It has been noted that a unit that seemed perfectly regular and bounded by parallel surfaces over great lateral distances is in reality composed of narrow sections aligned side-by-side. In short, the relations seen in modern lava flows of great width is also what is observed in the layered rocks in the walls of so-called craters of elevation. At Monte Somma, in the bottom of the Val del Bove, and in the calderas of the Azores, this seemingly parallel stratification has been recognized long ago by Lyell and Poulett Scrope, but it is at Santorini that one can see the evidence most clearly. Certain layers of lava seem to form unbroken ledges for distances of several kilometers, but when examined more closely they are seen to have many discontinuities. Each one is composed of relatively short sections that abut against one another, adapting the shape of their margins to that of their neighbor. The apparent parallelism stems from the difficulty of detecting these junctions between separate flows. More often than not, the steepness of the escarpment makes it impossible to examine the relations of the rocks at close range, and when one sees them from a distance, only rarely is it possible to distinguish individual units that are closely joined to one another and tend to have the same physical appearance. Weathering and erosion of these lavas also tend to hide the discontinuities of layers, because talus, which is usually more abundant close to the junctions, conceals the true relations.

The walls of the Val del Bove have long been one of the main elements of this controversy. In the lateral parts of this vast depression, one sees a series of many ledges of steeply dipping lava that seem to have remained parallel over great distances. The dip of the beds varies from 20 to 35 degrees; it is not the same at all levels along the same vertical line. The ledges that occupy the highest positions in the order of superposition are also those that have the maximum inclination. Thus the parallelism is far from being as perfect as it might be if all the beds had been originally horizontal and of equal thickness and if a single force had tilted them from their original position and raised them in a way that gave them their present slope. Moreover, the apparent continuity of these lavas over long distances is not real. At the bottom of the cul-de-sac of the Val del Bove, where the lavas appear to be cut crosswise, their discontinuity is such that it seems to be the result of violent dislocations. On the lateral walls of the same embayment, I tried to follow by eye some of the ledges that seemed to continue for the greatest distance, but I always found that they disappeared at various distances, always rather short, from their apparent point of origin. Rarely can one follow them for a distance of more than two or three hundred meters.

In addition, even if the observations had indicated a great extent, it would not have offered any solid argument in favor of the theory of craters of elevation. Actually, if one starts with the opposite theory, one will note that the ledges of lava in the lateral parts of the Val del Bove represent longitudinal cuts through flows coming from the central part of the volcano and following the natural slope of the ground, so that they must have resulted from accumulations of piled-up volcanic debris. It is quite probable, if not certain, that some of these beds, if they had been formed in this way, must have had a length of several kilometers, not unlike modern flows. Their longitudinal sections would expose them continuously over very great lengths. If such a thing is not observed, it is doubtless because the flows that have created these layers of superimposed lavas were narrow and they did not follow directions that were perfectly rectilinear or parallel to the escarpment of the Val del Bove.

The argument in favor of Leopold von Buch's theory based on the continuity of lava flows would really be convincing only if this continuity could be demonstrated in layers cut in two directions, one following a vertical plane passing through the axis of the mountain and another along a second vertical plain perpendicular to the first. Now if one can occasionally follow banks of continuous lava for extended distances in the first direction, finding a comparable section in the second direction would be next to impossible. So it is not layers like those of typical basaltic flow that cover the so-called cones of elevation but many narrow, thin flows like those of modern eruptions. Thus an examination of the true facts wipes out the argument based on the structure and disposition of deposits that contribute to the walls of craters of elevation. We should note that this argument is the key to the theory we are now discussing. Élie de Beaumont devoted the most eloquent pages of his renowned memoir on Etna to supporting and developing it. In denying it any credible value, we leave the theory of Leopold von Buch with no more support than that to which even its partisans attached little importance.

But we should consider whether the other reasons are equally vulnerable to being upset and overturned.

What value, for example, must one assign to the argument based on the discontinuity of the slopes of certain craters of elevation? Let us take the classic example of Etna with its lateral talus and its rounded middle section. It is not possible, according to Élie de Beaumont, that the products of the central vent could ever produce this enormous mass, for they are of far too little importance. Despite the repeated eruptions from the summit crater of Etna that have occurred since Roman times, the upper plateau of this volcano has been raised by only an insignificant amount, as can be seen from the ruins of the Torre del Filosopho. But the explanation for this is very simple. What do the two thousand years that have elapsed since Roman times amount to compared to the immensity of geologic time? Even if the central crater of Etna had maintained the same average level of activity since it began, and even if the eruptive products that give the central part of the volcano its added elevation of 1.5 meters accumulated over a period of two thousand years, we would simply have to conclude that it took about 2 to 3 million years to produce the central mass, and such a lapse of time is not at all impossible or even unusual, even though Etna only became an active volcano in relatively recent geologic time. But there is no need to suppose that the volcano dates from such ancient times, for there is nothing to indicate that it has always had the same level of activity. On the contrary, it is likely that during its early stages, it was the center of more intense activity, and eruptions must have been stronger in the central region than they are today, for it is reasonable to assume that the production of parasitic cones scattered over the flanks has become more common as the central vent of the volcano has become higher and thereby offers greater resistance to the rise and eruption of magma.

One can see in the walls of Val del Bove that dikes are more abundant in the upper elevations than they are anywhere else on the volcano. What are these essentially vertical dikes if not fractures through which magma has risen from the interior of the volcano to pour out on the surface? No ordinary lava is ever fluid enough to spread as a flat sheet like water in a basin, and even if it were perfectly fluid near the source from which it is discharged, it would soon become viscous enough to pile up in place. As for pyroclastic ejecta, such as ash, lapilli, and volcanic bombs, they also have a strong tendency to accumulate around the eruptive vent. If they are voluminous and heavy, wind has very little effect on them. They always fall within a short distance from the vent from which they came. If the beds are thin and light, they still accumulate within a narrow radius when the air is calm and especially if they have absorbed rain water. Thus one can state that all places that have been centers of volcanic eruptions for even a limited time must take on the configuration of a more or less rounded structure.

The diameter of such a body depends principally on how abundant the material is and on the average degree of fluidity the magma had at the time it was ejected. It depends also on the mineralogical composition and chemistry, as well as on the atmospheric conditions under which each eruption takes place. In short, anything that tends to accelerate the rate of solidification of the molten magma naturally tends to reduce the radius within which it spreads and, as a result, makes the slopes of any body formed by several discharges of this kind more abrupt. So the discontinuity of the slopes of Etna result, not from a circular disturbance affecting older beds that were originally horizontal and later elevated around the center, but more to the central dome-like shape produced by piling up of material ejected from the interior of the volcano. The lateral talus owes its origin to discharges of lavas during eruptions that led to formation of parasitic cones.

Let us now consider the argument based on the presence of more or less voluminous isolated masses in the middle of craters of elevation. What importance should be attached to them? Are the rocks of trachytic appearance that make up the bodies of Musara, Capra, and Finocchio in the Val de Bove, the phonolitic cones of Grioux, Grionau, and the Usclade in the middle of the Cantal really the result of central surges that have controlled the outbreak from the ground and the development of craters of elevation? Nothing we see in them supports such an interpretation. The trachytic rocks of the Val del Bove differ in only minor respects from the common lavas of Etna, and they have the same mineralogical composition. There is no reason, therefore, to consider them older than most of the material that contributes to the walls of Val del Bove. And besides, there is the question of how they could have contributed to formation of such a cavity. How on earth could bodies of such insignificant size contribute effectively to formation of a ring in the middle of which they are scarcely distinguishable when one is standing on the highest parts of the surrounding wall? They are cut by dikes like those in the walls surrounding the depression, and nothing indicates that the latter have been displaced in any notable way from either verticality or their original orientation. The only debris one sees in these bodies is fragments from underlying rocks that fortuitously escaped the powerful forces that created the Val del Bove; there are no fragments that were torn from much deeper levels and carried up in the same event that caused the ground to be breached.

As for the phonolites of the Cantal, they are mineralogically different from most of the surrounding rocks, but their role in breaking through to the surface is unclear. Even if we assume that the central massif of the Cantal has always had the normal form of craters of elevation (which seems doubtful), the phonolitic bodies of Grioux, Griounaux and Usclade would be quite small compared to the size of a cavity that would have had a diameter corresponding to the distance between Plomb du Cantal and Puy de Chavaroche. At most, it could have played only a minor role in forming the enclosure, and in fact probably had no effect whatever. The trachytic rocks in the immediate area north of Grioux form regular beds; nothing indicates that they have ever been displaced from their original position and even less that they have undergone an upheaval comparable to what they would have undergone if a cone like that of Grioux had suddenly intruded them.

The supporters of the theory of craters of elevation have also cited in support of their ideas the rounded forms of a great number of trachytic domes that they consider voluminous chambers that were formerly inflated but failed to break out. This hypothesis cannot be justified. The typical shapes of these bodies are entirely a function of the chemical and mineralogical compositions of their lavas. Lavas that are very siliceous and rich in feldspars, especially the more silica-rich members of that series of minerals, and poor in ferromagnesian silicates, are not very fluid. It is not surprising, therefore, that at the time of eruption, such a lava was inconspicuous and formed a thick mass of limited extent. Moreover, many trachytic domes are formed, not just of voluminous blocks piled one on the other like an extrusion of lava, but in most cases they are composed in large part of beds of ash, lapilli, and scoria. Trachytic domes of the lower regions of the Auvergne generally come up in the middle of brecciated scoria cones, the eruptive origin of which is beyond doubt. Thus, the Puy de Goule and Petit Sarcouy are formed of scoria and surround the trachytic dome of Grand Sarcouy as satellites around that center. It is impossible to see there anything but local piles of acidic lavas of the kinds that have erupted along with abundant emissions of steam and gas and ejections of ash and lapilli.

The trachytic domes of the Andes are also composed in part of pyroclastic material. One sees there thick layers of ash and lapilli and thick masses of pumice. Their flanks have long, radial trails of clinkery lavas, with surfaces consisting of incoherent blocks and the same general appearance as rubbly lavas of more basic composition. In addition, some of these domes have had eruptions that are usually characterized by explosions due to sudden release of gas and steam, ash, and other fragmental material, and by abundant flows of scoriaceous or glassy flows. The volcano Cotopaxi began to erupt during the night of 14 September 1853 and discharged down its southwestern flank a current of molten lava that from a distance had the appearance of a long trail of fire. A similar event was observed on the same volcano by Reiss and Stübel[17] in December 1873, and these scientists state that they observed flows of youthful lava on a great number of other Ecuadorean volcanoes. So it is certain that most of the trachytic or andesitic domes have been formed in a way that excludes the possibility of simple uplift.

Finally, the growth of the body that heralded the beginning of the eruption of Santorini in 1866 illustrates very well the conditions of formation of certain types of bulbous bodies of trachyte that were formerly thought to have originated by uplift. This body was produced, from the outset, not by explosive eruptions but by quiet eruptions. The lavas were discharged in a viscous state, solidified almost as soon as they emerged, and accumulated close to their source.

Thus the composition of trachytic domes offers no serious argument in favor of the theory of craters of elevation.

Contrary to what has been said, the nature of deep ravines on the flanks of so-called cones of elevation offers no better support for Leopold von Buch's theory. While some of these fissures are produced by faulting, far more often they are simply ravines cut by surface waters into the loose material mantling the surface of the volcano. They become wider down-slope, because the erosive power of the water becomes stronger toward the base. Had they been produced by the mechanisms entailed in this theory, their widths would become greater upward.

We have seen how Élie de Beaumont responded to the argument put forward by Virlet d'Aoust on the basis of the fractures that should be present in the cliffs of Santorini if these cliffs really formed a belt around a cirque produced by uplift. The open sections of water that separate the islands of

Thera, Therasia, and Aspronisi are quite wide enough to conform to the theory of Leopold von Buch. The depressions of the Val del Bove and the caldera of Palma also meet the requirements of his theory, but how does one explain continuous walls around a crater that is said to be caused by uplift? How could Monte Nuovo, for example, have been formed by uplift when the upper rim of its crater has the form of a complete circle? The caldera of Seitie Cidades on the island of San Miguel in the Azores furnishes even better evidence of this kind. It is a circular ring with a diameter of 5 kilometers, and continuous, unbroken walls that rise 200 or 300 meters above the interior. It is impossible to imagine that a violent underground shock could have raised and spread the walls of such a depression without breaking their continuity and producing deep notches in them.

A cone with a round, unbroken central crater cannot be considered the result of radial fractures or as having been raised by a sudden violent force directed upward and acting on a solid horizontal layer.

To end this review of arguments based on the topographic forms of volcanoes, we must examine those based on the relations of dikes to the lavas they cut. According to Élie de Beaumont, if the ground was already tilted when the lavas poured out on the surface the flows would have moved mainly in the direction of the slope and would have produced a greater number of superimposed layers on that side. The equal and regular distribution of layers of lavas around each dike would thus be proof that the ground remained horizontal until the moment of uplift. This argument, like one discussed earlier, rests on the assumption that the layers making up the wall of a crater are regularly stratified, but we have already pointed out that this regularity is only apparent. It is rarely possible to confirm with certainty the connection of a dike to the lava it discharged, and in the cases in which the continuity with a dike can be demonstrated, one observes instead that the discharge has for the most part gone toward one side. It is impossible to confirm that the lavas from dikes always spread equally in all directions from the orifice.

It is true that in the lateral walls of Val del Bove the number of lava flows does not appear to increase toward the base of the slope, but that only means that many of these lava stopped a short distance from their source. If flows maintain more or less constant transverse dimensions all along their course, they tend to pile one upon the other in the upper parts of the volcano, but they separate and abut laterally against each other in their lower parts. In this way the divergence of lava flows will have the effect of compensating for the great number of individual units and causing them to have a more or less constant thickness.

In summary, the observed facts on which the theory of craters of elevation rests are either of questionable validity or subject to interpretations other than those proposed by the supporters of the theory.

Arguments based on phenomena that accompanied the formation of Monte Nuovo or Jorullo are even less convincing than those that are based on the observed form and structure of the so-called cones of elevation.

Modern geologists have completely revised von Humbolt's account of the origin of Jorullo. The results of geological investigations, as well as information collected with care from local inhabitants, have now been thoroughly discussed and evaluated. The event took place under perfectly natural conditions that were completely compatible with those that prevail during ordinary eruptions. After numerous earthquakes that for several years had devastated the country, the ground opened along an almost straight fissure extending from north to south. Ash and scoria ejected from this crack accumulated mainly around the places where the activity was strongest. In this way, six cones were formed along a single line, and the largest of these came to be known as Jorullo. This same fissure discharged strong currents of very viscous basaltic lava that spread all around the cones and produced a great elliptical plateau that was thick in the center and thinned toward the edges. Along the length of the flows, the surface of the lava quickly formed large cracks, along with many scattered fumaroles that reproduced on a small scale many of the same phenomena seen in the large cones situated along the main fissure. At each of these points, the discharge of gas continued along with small amounts of semi-fluid lava. The rapid solidification of this viscous ejecta caused it to pile up around the orifices where gas was being released. The prominent steep-sided cones that von Humbolt described have been called *hornitos*. The gaping crater of Jorullo was the source of the last flow of lava that was so viscous that it consolidated in the form of a rampart on the flanks of the cone. At the end of the eruption, a great quantity of black cinders came from the craters and accumulated in thick beds on the surface of the immediately preceding flow of lava covering the hornitos and giving them the form of small domes. Torrential rains accompanied and followed these phenomena. The ash, under the influence of the water and acidic gases, became altered and agglutinated into concretions with concentric layers. Incandescent lava surrounded dwellings that had been built in the middle of the former plain close to the point where Jorullo was to rise. It did not enter them, however. They still survive today with their vertical walls and horizontal floors still preserving the rectilinear form of the original buildings. One sees no sign of distortion. Trees planted around them also remain standing in their original positions. Would a large, sudden uplift of the kind advocated by von Humbolt have failed to affect the planting fields and buildings that were so close to the place where it occurred? Is it not clear that all these facts show that Jorullo was produced by normal volcanic processes?

The two accounts of Antonio Falconi and Giacomo di Toledo, both of whom were witnesses of the formation of Monte Nuovo in 1538, do not seem to indicate anything that would justify an application of Leopold von Buch's theory to this event. Here is what they tell us. Falconi says that "stones and ash were thrown out with a force like that of large artillery and in such quantities that it seemed they would cover the whole earth. After four days, the deposits in the valley between Monte Barbaro and Lake Averne formed a mountain as high as Monte Barbaro itself." Giacomo di Toledo, after describing the earthquakes that had devastated Calabria for several years, relates in these terms the appearance of Monte Nuovo. "Around two in the night, the ground opened near the lake and revealed a formidable vent from which fire, smoke, stones, and muddy ash escaped with great fury. A sound like that of the strongest thunder accompanied the tearing open of the ground. . . . The stones reached a height about equal to that of an arrow from a crossbow then fell back either on the margins or into the interior of the discharging vent. The mud had the color of ash. Very fluid at first, it thickened little by little and was ejected in such quantities that, along with the stones, it formed a thousand-foot-high mountain in less than twelve hours."

It is true that these two authors speak of a withdrawal of the sea over a space of about two hundred paces, which may imply an uplift of the ground, but when one considers the gentle slope of the beach, the uplift seems to have been very minor. This is especially true if one considers that the fall of ash and lapilli around the volcano must have contributed to raising the surface of the ground. Ground movements similar to those that could reasonably be expected at Monte Nuovo are commonly noted around other volcanic centers. At Santorini I have observed the very remarkable changes described in another part of this book. At Vesuvius in 1861, I witnessed an uplift of 150 cm around the shores of Torre del Greco and an elevation of the center of the town that was associated with destruction comparable to that of an earthquake. In the immediate vicinity of Monte Nuovo, the upheaval of the ground could not have been much greater than in these other instances. Dufrénoy points out that all around that area, on the shores of Lake Averne, Roman ruins such as the temples of Apollo and Pluto, do not appear to have suffered from this supposed uplift. "The walls that still remain have preserved their vertical positions, and the arches over them are in the same state as in other structures near the shores of the bay. The long gallery leading to the grotto of Sibyl on the opposite side of Lake Averne also remained perfectly horizontal. The only possible change was the floor of the chamber in which Sybyl rendered oracles, which is now covered by a few inches of water. This could only mean that there was a slight change in the level of Lake Averne." So one can conclude from this, as Lyell did, that if Monte Nuovo was formed in 1583 by an elevation of the original ground level, "one must at least recognize that the adjacent ground was not affected, and this does not accord with what the hypothesis leads us to expect." The same author also draws our attention to the fact that not only was there no uplift of the town of Tripergola, which was situated very close to the place where Monte Nuovo rose, but that the town disappeared under a blanket of muddy ash produced during the eruption.

We recall that Monte Nuovo had a completely circular form, had neither tension fractures nor breaches, and that is was not crossed by any apparent dikes, as it certainly would have been had several eruptions taken place in its center. The layers of which it is formed consist mainly of ash and lapilli tuffs quite unlike those of the Campanian tuff. As Leopold von Buch has himself said, in the middle of beds of fragmental material one notices, among other things, thin lenses that were probably formed by flattening of volcanic bombs that were still in a plastic state when they fell. The various pyroclastic beds one sees all along the inner side of the crater are regularly arranged, with nothing in their form or distribution to indicate that there was a violent disruption. At the bottom of the crater one can see no rock that could have been any part of an intrusion during formation of the volcano.

Thus everything shows that Monte Nuovo was formed by accumulation of ejecta that fell around the eruptive vent and that it did not result from inflation of older rocks to form a cavity that was breached at its summit.

We still must mention here another argument against craters of elevation that Mr. Rames has developed with remarkable vigor.* "It is quite extraordinary," he says, "to see these great geologists who advocate the theory of Leopold von Buch refusing, in the case of Monte Nuovo and other small cones of the Phlegrean Fields, to accept an eruptive origin while admitting that the cones of the lower Auvergne, like those on the flanks of Etna and the Island of Lanzarote were formed by eruptions. The distribution of material and the inclination of the beds are very similar in all these cones, and the only

---

* *Études sur les volcans*, by J-B. Rames. Savy, ed., Paris, 1866.

way they differ is that the former have been built of ash and fine trachytic pumice, the latter of scoria and basaltic ash. And on what basis do these geologists establish these different origins among volcanoes that are in all respects so similar? If any criterion exists that permits one to separate cones formed by uplift from cones formed by eruptions, it is hardly one to inspire confidence. In effect, von Buch and Humbolt affirm that the Somma of Vesuvius, the peak of Tenerife, and all of Etna as we see them today are due to the uplift they have undergone. Élie de Beaumont, for his part, declares that Somma and the core of Etna have been uplifted but that Vesuvius, the peak of Tenerife, and the cone of Etna come from eruptions just as the parasitic cones on the latter do, and Dufrénoy attributes all of these - Somma, Vesuvius, and its parasitic cones - to uplift. Similar differences in the evaluations of the best known volcanoes do not speak well of the strict mathematical rigor in which these scientists present their theory."

The same author also points to the profound difference between the forms of beds making up the so-called cones of elevation and those of uplifted and dislocated sedimentary rocks that make up great mountain ranges. The latter show clear evidence of the powerful forces to which they have been subjected following deposition and consolidation, but in general these characteristics are not seen in the layered volcanic rocks exposed in the walls of so-called craters of elevation. The latter have not been folded or turned vertically; their inclination is always moderate and similar to that of the lava flows of modern volcanoes.

"How," asks Mr. Rames, "does one explain the instinct of subterranean forces that leads them, in forming craters of elevation, to manifest themselves in places occupied by horizontally layered volcanic rocks and to know how to moderate their action to the degree that they always turn up beds at a very modest angle?"

The discussion into which we have just entered is a summary of the important debate that has long divided geologists into two opposed camps. At present, the struggle has subsided, and the theory of Leopold von Buch has been abandoned by most geologists concerned with the study of volcanoes. For them, it is now only a historical legend worthy of respect because of the imposing authority of the great men who have seen in it the expression of a scientific truth. But in view of the magnitude of the debate and the renown of the scientists engaged in it, we should still set out here in a brief summary the special reasons why it has no application to Santorini.

The cliffs that surround the bay of Santorini are composed, as we have already noted, of metamorphosed sedimentary beds, products of submarine eruptions, and various types of material coming from subaerial vents. An inflation that formed an open chamber and subsequently burst out through the roof cannot explain the relationships of any of these deposits. In fact, it is quite inconsistent with the observed facts. This can be shown if we review each of the three types of deposits that are involved.

The metamorphosed sedimentary beds are strongly warped into steep positions in all directions; though highly varied, the average direction is NNW and the direction of tilt is ESE. The dip is 25 degrees at the shore of the bay at Athinios, but it ranges from 45 to 60 degrees on the slopes of Profitis Ilias and Messa Vouno. These beds were, of course, deposited horizontally in the manner of all sedimentary deposits, then disrupted and elevated, but the uplift that displaced them from their original positions was not local; it was due to regional effects that were felt not only at Santorini but throughout the Cyclades and neighboring parts of the mainland. There is no relationship whatever between the form of the bay of Santorini and the direction and amount of tilt of the metamorphic rocks of Profitis Ilias, and there is nothing to indicate that their orientation is in any way a consequence of formation of the bay.

At Athinios, the layers of marble and schists exposed along the route up the cliff are not turned up uniformly to tilt toward the south or southwest, as they should be if they were deformed according to the theory of Leopold von Buch. Moreover, their inclination is much more gentle than it is in places farther to the east, but according to the theory, they should be at least as steep.

The deposits of submarine origin that make up most of the western part of Thera beyond the village of Akrotiri are regularly stratified in many places. In some places, as at Mt. Arkhangelos, they are horizontal, but in others they are tilted toward the coast and in others toward the bay in the form of an outward, dipping saddle-like structure. This is especially noticeable in the area between Loumaravi and Cape Akrotiri. Both of these orientations are equally inconsistent with the theory of L. von Buch, which holds that the beds should be uniformly tilted in a direction away from the bay.

The subaerial volcanic rocks do not fit the theory any better. In every locality around the bay where these deposits can be seen, they are tilted in different directions.

Most often, they dip away from the side next to the bay, but their inclination in this direction tends to be gentle, even when the slope of the ground is very steep, as it is on the outer slopes of the islands. Their surface, which is like the steps of a stairway, is often almost horizontal, while the slope of the ground where they end is considerably less than that of the stairway itself.

Although the ground surface on the outer slopes of the islands is inclined in a way that is consistent with L. von Buch's theory, one cannot necessarily conclude from this that each of its constituent layers is also inclined in the same way. In a vertical section perpendicular to the face of the cliff, one normally notes that the lower layers are almost horizontal, while the upper ones are strongly tilted toward the exterior. It follows that all the layers cannot have originally been horizontal, for they would remain parallel when uplifted.

But that is not the only problem. Not only does one see subaerial volcanic rocks still horizontal despite the supposed uplift, but one sees some that are exposed in slopes toward the bay, contrary to what the theory of L. von Buch would indicate. In the lower and middle parts of Megalo Vouno and at the base of Mikro Profitis Ilias, one can make out layers of ash and lava that plunge inward toward the bay.

The orientation of the numerous dikes that can be seen in the same part of the cliff is also inconsistent with the theory of craters of elevation. Most of the dikes have the form of vertical walls and their directions are between northwest and northeast. A few meters away, one sees other dikes that are oriented at 90 degrees to the first. A powerful uplift affecting a very restricted area, as the theory of L. von Buch proposes, would necessarily alter the verticality of at least some of these dikes. While those that are essentially vertical in the face of the cliff would maintain this orientation, those that are nearly parallel to the face of the cliff should be tilted by an amount that would depend on their orientation. No relations of this kind are seen.

An uplift would also have the effect of lifting rocks formed under water and making them emergent. All along the length of the cliff one should see differences in the structures of rocks formed under water and those that have always been some distance above sea level, but such differences are not seen. The lower layers have all the characteristics of subaerial rocks that are found in the upper units.

If one applies to Santorini the ideas proposed by L. von Buch to account for the surficial pumice deposits on Vesuvius and Tenerife, one would have to conclude that the mantle of pumice that covers all parts of the Santorini archipelago have been spread over the surface of the islands as a result of explosions immediately following the uplift. But underneath these pumice deposits, we find human dwellings that, according to the hypothesis we are considering, would have to have been constructed before the upheaval of the ground by the inflation and subsequent outbreak. These dwellings would have been displaced from their original positions along with the ground on which they were built; the walls would have lost their verticality. But today they are still perfectly straight, and nothing about them indicates that they have been moved.

Thus, an examination of Santorini shows that the conditions are inconsistent with the theory of craters of elevation.

Many other hypotheses have been proposed to account for the formation of large, circular volcanic depressions. Most of these have quickly passed into oblivion, and there is no need here to mention more than a few of the main ones.

Some geologists have thought that the explanation lies in rapid erosion by rain water. In their view, the caldera of Palma and the Val del Bove of Etna owe their origin mainly to gullying by rain. In these volcanoes, there would have been an original central crater of moderate size at the summit of a cone composed of ejecta. Gullying would soon have cut ravines into the slopes of the cone and one of these would breach the rim and, by forming an outlet for the central crater, would be enlarged and cut more deeply than the others. As it grew, it would serve as a channel to remove erosional debris along with the great amounts of rain water, so that the breach would quickly be accentuated. It would eventually become so large that it would no longer be controlled by the form of the original crater. In this way the enclosures would become cul-de-sacs as they are now. The position of the original crater would correspond to the bottom and the part excavated by water to the rest of the rim.

It is clear that such an explanation cannot be offered for great volcanic depressions like the caldera of Palma and Val del Bove. It cannot possibly explain their rims, which are still intact. The crater of Sete Cidades on the island of San Miguel and that of the island of Graciosa could never have been cut by water, for their rims are still complete.*

But even where large crater rims, like those of the Caldera of Palma and the Val del Bove have large openings, rain water cannot have played more than a very minor role in forming the depressions. The crest at the tops of the surrounding walls is very sharp and slopes outward. Thin streams of water falling on the Piano del Lago in Val del Bove, for example, have only

---

* The crater of Sete Cidades has two lakes without outlets; in Graciosa, rain water is concentrated in a small pond or falls into a cavern where it forms a small basin.

a weak effect, and, almost as soon as it falls, the rain disappears into the rocks. On all the surfaces covered with pumiceous scoria, one sees no marks left by the passage of running water, and ravines carved out by water are seen only at the lowest levels of slopes below the opening of Val del Bove where chestnut trees grow.

In the Atrio del Cavallo on Vesuvius and in the great amphitheater of Vulcano, one of the Eolian islands, no running stream of any size is to be seen. Certainly, in the case of Santorini, erosion can scarcely be invoked as the agent responsible for excavation of the bay. The bay is surrounded by a narrow, emergent ridge with slopes that are predominantly toward the outer sides of the islands. To imagine that there formerly existed a great elevated area outside the present site of the bay, higher than the present cliffs of Thera and Therasia and capable of diverting great amounts of water in the direction of the bay, would be pure conjecture. Nowhere does one see extensive deposits of rounded boulders and gravel, such as those seen wherever torrential waters are capable of carving out valleys several kilometers in width and several hundreds of meters in depth. Moreover, the deepest part of the bay is in its center, and running water could not excavate a valley with such a shape.

The sea is equally incapable of producing such a depression. The action of waves extends only to quite shallow depths and has the effect of leveling submarine surfaces rather than increasing their irregularities. As for deep ocean currents, they can, under certain conditions, sweep the bottom of the sea and produce channels, but it is impossible to imagine that in a very enclosed space, such as that of the bay, they could have excavated a depression 400 meters deep.

It is therefore a mistake, in my opinion, to call on water to play an important role in forming large volcanic cirques. At Santorini, in particular, any external effects cutting back the cliffs would only diminish the depth of the bay and would not tend to enlarge it.

The hypothesis that, in recent times, has been most favored by geologists is that explosions were the sole agent responsible for formation of so-called craters of elevation. The sudden release from erupting volcanoes of gases compressed under high pressure or of water vapor at high-temperatures has such powerful mechanical effects that one is quite easily led to overlook all other possible effects. Ordinary volcanic explosions can project blocks of enormous size and weight, and they can eject immense amounts of volcanic debris in the form of ash. Is it not reasonable, then, to imagine that, in great cataclysms, forces of this kind might be even more powerful? And if so, why should we not assume that they have broken through the shallow ground where the eruptions took place, reduced it to fragments, and ejected it over a wide area and in this way forming a wide, funnel-shaped cavity? In other places, one can find authentic, recent examples of very large craters that have been formed in very short times during explosive eruptions of horrible violence.

One of the most famous eruptions of this kind is that of Cosiguina in the Republic of El Salvador. Between the 20th and 23rd of February, 1835, the upper part of this volcano was removed by explosions, leaving only a truncated cone with a crater about 3 kilometers in diameter. The quantity of rocky debris thrown out was enormous. The town of La Union 60 kilometers away was covered by a layer of ash 13 centimeters thick, and the town of San Miguel 30 kilometers more distant received deposits of the same material with a thickness of 10 centimeters. Thirty kilometers to the south, the ashfall reached a thickness of 3 meters. The rain of ash fell over a distance of 1000 kilometers around the volcano. Very close to the site of the eruption, the volume of ejecta was so great that a virgin forest disappeared under the debris. Strips of pumice and two floating islands were formed in the adjacent sea. The total mass of ejecta from this volcano was estimated to have been about 3,000 cubic kilometers.[18]

The example of Popandayan on the island of Java is no less remarkable. On the 11th of August 1772, this mountain was the scene of a terrible eruption. Its upper part was removed by the explosions and in its place one now sees a wide depression open to the northeast and surrounded on the other sides by sharp ridges of impressive height. The hole has a strong resemblance to the Val del Bove.[19]

In 1815, the volcano Tambora on the island of Sumatra had an equally violent eruption. Ash was carried as far as 450 kilometers and at a distance of 60 kilometers houses were crushed under its weight. The crater of Tambora is 12 kilometers long, 7.5 wide, and 533 meters deep. Its horizontal dimensions differ little from those of the bay of Santorini, but its depth is greater.[20]

The greatest depth of the bay of Santorini is now 390 meters and the maximum height of the cliffs is 360 meters. Frequent landslides reduce both of these figures, and, as is normal elsewhere, there was probably a summit higher than the present cliff at the time the bay was formed, so the depth of the crater of Santorini must have been at least the sum of these two numbers or 750 meters. Since the bay has a length of 11 kilometers and a width of 7.5, the volume of debris removed by the geological event responsible for the bay can be placed at about 60 cubic kilometers. On the other hand, Mr. von Fritsch[21] concluded from an examination of the pumiceous tuff covering all of Santorini, that one tenth consists of older rocks, and the remaining nine-tenths of pumiceous fragments. Thus, according to this hypothesis, the explosions that created the bay would have to have ejected a total volume of solid material equal to about 600 cubic kilometers.

If one accepts these numbers and compares them with those just cited for Cosiguina, Popandayang, and Tambora, one sees that explosions capable of producing a crater-like cavity like that of the bay of Santorini are not at all impossible.

But several objections cast serious doubts on any hypothesis that holds that explosions were the unique agents responsible for the great depression lying between the present islands of Thera, Therasia, and Aspronisi.

First of all, if one compares the relative amount of pumice and debris derived from older rocks in the pumiceous tuff of Santorini, one easily sees that the relation is far from that represented by the figures just cited. The proportion of fragments of older rocks, instead of making up one-tenth of the tuff, does not appear to be more than one-hundredth. This means that it would not be 600 cubic kilometers of material that one would have to assume was ejected but 6000, and the eruption would have been much more intense than those of the volcanoes mentioned earlier. It would have to cover with ejecta an area of at least comparable size. One should therefore find today a layer of volcanic debris on the surface of the metamorphic rocks not only on nearby islands such as Nios and Sikinos but also on more distant islands, such as Paros, Naxos, and Syrna to the north and Crete to the south. No such surficial deposits have been observed, even on islands such as Nios and Sikinos.[22] The volume ejected during the eruption that formed the bay of Santorini must therefore have been relatively moderate. It does not seem to have been as great or as extensive as those responsible for the burial of Herculaneum and Pompeii. So the deposits left by these explosions are out of proportion to the scale that would be required if they corresponded to a total ejected volume of several thousand cubic kilometers. Even if the ejecta were relatively less important, where can one look to find the 600 cubic kilometers of volcanic debris resulting from the destruction of the central part of the large island of which Thera, Therasia, and Aspronisi represent the remains?

In the second place, if the excavation of the bay had been due solely to great explosions, the pumiceous tuff left by this cataclysm would have had a very inhomogeneous composition. The fragments of older rocks should be found almost exclusively in the first material to be erupted. One should see a deposit composed almost entirely of lithic blocks coming from the pile of lava that formerly covered the area of the bay between Merovigli and Therasia. Nothing of the kind is observed; the lowest parts of the pumiceous tuff have scarcely any more blocks of lava in them than the upper layers, and at the base of the tuff there is no notable amount of breccia made up of fragments of lava. The distribution of dense, lithic material in the pumiceous tuff is very uneven. Enormous blocks are seen at all levels. It is impossible to imagine that the older rocks occupying the site of the present bay had been broken and thrown out by explosions in the opening phase and that subsequent explosions ejected pumice mixed with only the debris that had fallen back into the crater.

The hypothesis that violent explosions were the sole agent responsible for the bay of Santorini, although attractive in its simplicity and the reality of some of the phenomena it postulates, cannot be accepted, for it conflicts with what we know about the pumiceous tuff covering the present islands. The only remaining explanation for the bay of Santorini is that a collapse preceded, accompanied, and followed the explosions.

This hypothesis accounts for the abrupt form of the cliffs around the bay, the extent and depth of the depression, and the small amount of older rocks found in the pumiceous tuff. It is also consistent with observations made at other craters in active eruption. On many occasions, when molten lava is visible in these craters, one sees that it is subject to frequent variations of its level. Sometimes it is seen to rise as high as the edge of the crater, and a few days later one finds that the surface has fallen back by notable amounts and that this happens without significant amounts of explosive discharge or any escape of lava from lateral vents.

Observations made by Spallanzani at Santorini and repeated countless times at other volcanoes, particularly at Mauna Loa in the Sandwich islands,[23] show on a very reduced scale how the phenomenon takes place. If the molten lava in an open crater can rise and descend so suddenly, why could not a similar rise and fall occur out of sight below the solid crust? It is reasonable to suppose that there could be a sudden withdrawal whenever a mass of molten lava, contained in the interior of a volcanic mountain, finds a lateral outlet at a relatively low level. The cavern of Graciosa in the Azores is striking proof that this is possible. This cavity unquestionably owes its origin to a phenomenon of this kind. It has the form of a vast vaulted chamber with a floor steeply inclined from northwest to southeast. Its largest diameter is about 200 meters and the smallest 100 to 120 meters. The single span of the ceiling rises 30 meters above the middle of the floor. To reach the cavern, one descends through a narrow, irregular crack. The bottom of the cavern where the slope is steepest is about 80 meters below the ground surface outside. The form of this opening can be explained only as the result of deep-seated lavas draining and leaving the shallower ones still in their original condition. The removal of the lava can reasonably be attributed to an outflow of underground magma, for most of the upper part of the fissure through which one reaches the cavern is filled with lava that has solidified there in the form of dikes. Elsewhere this same fissure must still communicate with the underground magma chamber, for it gives off hot, acidic gases. If fresh lavas flow through it some time in the future and it again serves as an outlet for high-pressure gases to become the focus of a new eruption, it is almost certain that the roof will not be strong enough to withstand the stresses on the walls of the fissure when volatiles are suddenly released and the shocks produced by repeated surges of magma. These effects will cause the roof to break up, become partly remelted, and be thrown out along with the fresh molten material. The mass of older rocks discharged under these conditions will necessarily be much less important than if the cavern had not already existed. At the same time, once the explosions have removed the roof rocks, the space occupied by the former cavern will form a deep crater with abrupt walls not unlike those of a miniature crater of elevation.

In the preceding lines I have tried to offer an explanation for the small amount of older debris in the pumiceous tuff of Santorini and for the steep cliffs surrounding the bay. To do this, I have had to call upon a series of three different phenomena: first, the reduction of the uppermost parts of the ground to a relatively thin crust prior to the eruption; second, the intervention of a sudden collapses; and, third, the triggering of violent explosions.

The first of these phenomena explains the scarcity of older debris in the pumice; the second explains the formation of the bay and the steepness of the cliffs; and the third accounts for the explosive eruptions.

The inadequacy of explosive phenomena alone to account for the observed facts has already drawn the attention of Vogelsang.[24] This is what led him to postulate that before the eruption an excess of heat rose from great depths to the upper levels of the crust, remelting a notable part of the near-surface rocks in a way that left only a thin, solid crust covering the layer of molten lava. This idea has the disadvantage of being only an unsupported hypothesis, but at the same time, it cannot be challenged by any solid evidence, for the only serious reason one can find to oppose it rests on the difficulty with which fluid lavas melt solid substances immersed in them. The lavas of ordinary eruptions that reach the surface bearing disseminated crystals in an amorphous matrix remain liquid by virtue of their heat, but, in addition, they commonly carry numerous solid fragments, either of the same rock or pieces of foreign rocks. So one can raise the objection that the temperature in the subterranean chamber, at least in its upper parts, is not hot enough to melt the older, solid lavas in its walls. But this objection might easily be rejected. Even if the lavas produced in most eruptions are very crystal-rich, it is not rare to see outpourings composed almost entirely of amorphous material, and that is exactly the case in the eruption that produced the bay of Santorini. The pumice ejected in the course of that eruption has very few crystals but is made up mostly of amorphous material. It is likely that the temperature of the source from which it came was too high to permit abundant crystallization. From this one can understand that a lava in contact with such a liquid could be partially fused by it; its amorphous part could become liquid while its crystals would also be partly melted or at least disaggregated and strewn about in the igneous fluid. So Vogelsang's theory is not impossible nor even unreasonable, but I still do not accept it, because it does not rest on positive evidence or any tangible observations. It is interesting, nevertheless, because it shows how the eminent geologist who put it forward felt compelled to find an explanation for the scarcity of fragments of older rocks in the immense quantities of pumice erupted by the volcano.

Several other hypotheses can be proposed to account for a collapse. Let us consider the main ones.

1. The natural compression of the unconsolidated beds of scoria that make up much of Santorini could, it is said, lead to a gradual settling that would cause a slow subsidence of all its central parts. Several objections to this hypothesis can be raised. The form of the steep cliffs that make up the walls surrounding the bay is incompatible with the idea of a slow subsidence, for the observed kind of rupture could be produced only by a violent event. Gradual settling of the ground would have a regional effect and would not be limited to a restricted area distinctly circumscribed in the way that the bay of Santorini is. There is no reason why the subsidence would be limited to the central part of the original island that is now occupied by the bay and present islands. The effects of such settling would be very marked in the remaining parts of the large original island and would still be visible along the cliffs of

Thera and Therasia, but nothing of this kind is seen. To cause enough settling of volcanic rocks to produce 750 m of subsidence of the ground, it would be necessary to have a section of rock with a total thickness of several thousands of meters. The proximity of the phyllitic rocks of Profitis Ilias and the notable amounts of schist and limestone fragments brought up by lavas or explosive ejecta near the center of the bay show that the metamorphic basement rocks are to be found at shallow depths.

2. Could subterranean solution, by either marine or pluvial waters, have removed some of the underlying rocks to produce the cavity that is thought to have existed before the collapse? Again, the response must be negative. It is true that there are localities, most notably in the Ionian islands off the coast of Cephalonia, where marine currents penetrate the ground and reappear mixed with fresh water after following an underground course, but this is exceptional and observed only in stratified calcareous rocks where the disturbance of the ground has opened cracks through which water could move and form channels. These conduits can enlarge themselves or at least maintain open passages at the margins of the calcareous rocks if the water has carbonic acid in solution. In volcanic rocks like those of the original island of Santorini, there could also have been, as on Nea Kameni, underground channels through which marine waters could move. But the layers of lava do not normally have enough thickness or continuity for the channels to attain large dimensions. Moreover, if such channels existed and the water were charged with carbonic acid, the process of altering volcanic rocks makes them more compact rather than cavernous. It is true that it dissolves some of the elements, but it brings in others and introduces oxygen and carbonic acid, and deposits them in the rock in various proportions.

As for currents of fresh water, their action could only be insignificant in an island with an area as limited as that of the original Santorini, especially if one thinks of the force the currents would have to have to penetrate and produce cavities as large as the bay. They could not have much effect except in the part of the ground above sea level.

There is no reason to suppose that the site of the present bay of Santorini formerly had thick deposits of substances such as rock salt or gypsum that the seawater could have dissolved and carried away. Not only would such a hypothesis be purely gratuitous, but it is inconsistent with what is indicated by examinations of the material in the present cliffs. Observing the eruptive components of the high escarpments of Merovigli and Therasia, it is impossible to imagine that their section was ever occupied by rocks other than volcanic ones.

Finally, if slow solution by water had caused a collapse of the ground, would one not expect to find in the escarpments traces of these phenomena, such as hollowed-out rocks with voluminous cavities and rounded, irregular outlines?

In summary, the formation of the bay of Santorini can be attributed only to a cataclysm of great violence. The origin of this event was deep-seated, and its manifestations were of the same intensity as those that normally occur in erupting volcanic centers. The explosions were the main agent involved, and if we adopt the premise that they were accompanied by a collapse, it is to explain two phenomena that the explosions alone cannot account for: first, the steep slopes of the escarpments and, second, the small amount of older rocks found in the pumice.

SUMMARY

The descriptive parts have such an important place in this book that it seems useful for me to summarize in a few pages the essential information brought to light by this work, as well as the main deductions developed from this new knowledge. In the interest of greater clarity, I have chosen not to reproduce the chapters of the text in rigorous order.

I

The recent eruption, like most volcanic phenomena of this kind, has given rise to two kinds of products. First, volatile material expelled in the form of gas and vapor that condenses easily under moderate conditions, and second, lavas discharged in a state of incomplete fusion or ejected as fragmental debris or scoria.

The gases that were emitted varied widely in composition during the course of the eruption. At the beginning, when the lavas first rose above sea level, they escaped intact as if poured from a container, without being oxidized by the air. Mixtures of gases, which were rich in combustible components, particularly hydrogen, ignited on contact with the incandescent rocks and produced flames in the central parts of blocks. For the first time, hydrogen gases collected from an active volcano have demonstrated that, contrary to the views held by most geologists, true flames may be present in volcanic vents.[1]

Thorough study of these gases has proven that in certain cases the hydrogen that was collected came from dissociation of water into its two constituent elements.[2]

Finally, as the eruption progressed and the lavas piled up in greater amounts, chemical analyses have shown that there was a steady variation of the compositions of the mixtures of gases emanating from the same source, until the combustible elements in them finally disappeared completely. In the end, the volcano had lost its primitive marine character; air penetrated mainly through the interstices between blocks forming a cone that rose above the level of the sea. When this happened, combustion of the hydrogen gases took place out of view in the interior of this mass of lava.

In addition to free hydrogen and methane, the emanations of Santorini have been shown to include the common volcanic gases, chlorine, sulfur dioxide, carbon dioxide, hydrogen sulfide, and nitrogen; the first two of these components have been found exclusively at points of emission where temperatures are above 100 degrees, the other three are found in all the fumaroles and appear wherever the temperature is only slightly above that of the atmosphere. The pattern of variation of emanations is therefore the same as that observed in purely subaerial volcanoes.

II

Water played an important role in various phases of the eruption. In the vapor state, it figured in the emission of volatile material, regardless of temperature, and must be considered the immediate cause of explosions.

Liquid water was the main component of hot springs with temperatures that varied according to conditions of the sea. This curious relationship has a rational explanation. The water of the springs can be thought of as being discharged at the orifice of a subsurface conduit that has the form of a siphon, one arm of which received cold water from the ocean while the other was heated by contact with hot rocks and diluted by gases from the volcano.

Iron chloride, expelled as a vapor, accompanied hydrochloric acid, but one of the volatile products that is normally very abundant among the products of volcanoes, ammonium chloride, is notably lacking in the emanations of Santorini. This tends to support the opinion of workers who consider this component to be of organic origin and think it is carried by the atmosphere above fumaroles that give off hydrochloric acid. At Santorini, the scarcity of nearby cultivated fields may very well explain the rarity of ammonia in the air.[3]

Finally, in the parts of the volcano where the incandescence was strong and in central eruptive vents themselves, spectroscopic measurements revealed the presence of volatilized salts of sodium and potassium.[4]

Later, after the eruptive phases ended, these salts which were deposited around the orifices of fumaroles could be collected and analyzed. Chlorides and sulfates of sodium and potassium as well as carbonates of sodium and magnesium have been identified among the components of these deposits.

Evaporation of seawater leaves a residue which, when heated, forms sublimates of analogous composition. These results are therefore of special interest, because they lend support to the interpretation of seawater as an active agent of the explosive eruptions.

At places where alkaline salts were volatilized, spectral analyses and chemical tests have demonstrated the simultaneous presence of water vapor, iron chloride, free hydrogen, etc.; in other words, all the components of volcanic fumaroles. This fact justifies the conclusions previously drawn from studies of Etna, namely that the hottest fumaroles give off all the volatile components of volcanoes at the same time and that fumaroles of lower temperatures are systematically depleted as their temperatures become too low for volatilization of the eruptive material.

Thus one finds that the chemical variations of fumaroles have a rational explanation and that the compositions of their mixtures are related to temperature.

III

Among the volatilized substances in volcanic conduits, there are some that are susceptible to reacting with each other and producing stable compounds. The hydrated oxides of iron, hematite, sulfur, free sulfuric acid, alum, and calcium sulfate found around fumaroles are produced in this way. Our understanding of the origin of these substances is well founded.

The same is not true, however, of certain crystallized silicates that are generally formed under different conditions. Though formed from elements that one normally thinks of as stable, they are found in cracks near the surface of volcanic rocks under conditions where they can only have been deposited as the result of volatilization.

Silicate crystals formed in this way have been collected at Santorini on the interior surfaces of tubular cavities resembling fulgurites.[5] These tubes have evidently been channels for gases of very high temperature. The crystals deposited under these conditions include anorthite, sphene, and the pyroxenes, fassaite and augite.[6]

Other crystalline silicates occur as accidental inclusions in the same lavas. These minerals come from altered blocks of limestone torn from the basement and carried up in the lavas.[7] The main minerals are anorthite, augite, fassaite, sphene, melanite garnet, and especially wollastonite.

Quartz and mica-schists, like limestone, occur as xenoliths in the Santorini lavas, but they seem to have suffered no detectable alteration of their main constituents, despite the high temperatures to which they have been raised.

V

Mineralogical studies of the 1866 lavas present great difficulties. The constituent minerals are very small and closely intergrown. Extraction of large crystals of feldspar by simple hand-picking was an extremely difficult procedure, and for other minerals in the rock it was simply impossible.

The highly varied orientation of the feldspar in thin section and the common absence of the multiple intergrowths so characteristic of the triclinic feldspars render their identification uncertain.

SUMMARY

At the time I started my research, the only known determinative methods were the qualitative ones of Szabo and Boricky,[8] which were based, in the first case, on the colors the feldspars give to flames, and, in the second, on the crystalline form of the fluosilicates obtained by treating these minerals with hydrofluoric acid. These methods are practical only for isolated grains of feldspar; their use under normal conditions assumes a common identity for all the individual feldspars of a rock, which in most instances is incorrect. Finally, when one can work with homogeneous material or at least when one is aware of the heterogeneity, there is no substitute for a quantitative analysis.

Under these conditions, I have been forced to search for new ways of separating the minerals of the rocks, and this has led me to devise two new procedures that, though difficult, are applicable to the recent lavas of Santorini. The first, based on the use of a strong electromagnet, permits a separation of feldspar; the second, based on use of concentrated hydrofluoric acid, enables one to separate the iron oxides and ferromagnesian silicates.[9]

VI

Crystallization of the minerals in the lavas of Santorini took place in two distinct stages. During the first, they grew to lengths that commonly reached 0.5 mm and average dimensions in excess of 0.05 mm. During the second period, the crystals reached notably smaller dimensions. The smaller *microlites*, which crystallized in this second stage, surround and wrap around the *large crystals*.[10]

Before the microscope became widely available for the study of rocks, only *phenocrysts* were known. The groundmass enveloping the crystals that can be seen with a magnifying lens was considered to be a uniform material lacking any crystallinity. Actually, swarms of *microlites* abound in this material. Their discovery has been one of the major achievements of mineralogical microscopy.

I have undertaken the first chemical study of these microscopic crystals and have provided information on their composition. I have also been able to gain useful data on their optical properties. In so doing, I have shown that most feldspar microlites of a rock belong to a higher order of feldspars than the large crystals with which they are associated.

The optical crystallographic work of Michel Lévy has confirmed my results and has provided additional information on the properties of the microlites.

The minerals that occur as microlites in the average lava of 1866 are feldspar and titaniferous iron oxides. The feldspar is in small crystals elongated along the pg_1 axis; the iron oxide is in granular crystals.

The dominant feldspar among the phenocrysts is labradorite, but that is not the only feldspar in the rock that has this form; crystals of anorthite, as well as small amounts of oligoclase and sanidine, are also found as phenocrysts. The microlitic feldspar is albite mixed with a substantial amount of oligoclase.[11]

Finally, the whole mass is held together by a glassy matrix that represents the last residue of crystallization - the part of the rock responsible for its fluidity at the time when its constituent minerals had started to crystallize. This amorphous material has a composition very similar to that of albitic feldspar but is richer in silica and potassium. It is curious to see a substance with the composition of feldspar and silica making up the last remaining liquid of the rock.[12]

The order in which the minerals crystallized to form the rock was as follows:

1. phenocrysts of iron oxides
2. apatite
3. ferromagnesian silicates (augite and hypersthene)
4. phenocrysts of feldspar
5. granular microlites of titaniferous iron oxide
6. feldspar microlites

VII

The common lava of the 1866 eruption brought up xenoliths derived from lavas with compositions very different from that of their host. In some of these blocks the labradorite phenocrysts typical of the normal lavas are in large part replaced by oligoclase, and, at the same time, hypersthene is seen to be much more abundant than augite. In another much less common type, the feldspar phenocrysts are represented almost exclusively by anorthite; hypersthene is absent, augite is very well developed, and olivine is present. As will be shown later, this second type makes up an important part of the ancient lavas of Santorini.

VIII

Almost all the essential minerals of the 1866 lava contain foreign material. Some of these microscopic inclusions are crystalline, while others are composed of an amorphous substance. The feldspar phenocrysts, for example, contain crystals of pyroxene, iron oxides, apatite, or even triclinic feldspars.

The inclusions of glass are bits of the ambient material that surrounded the crystals when they were growing. A gas bubble is trapped in almost every glassy inclusion. The form, structure, and distribution of inclusions in the interior of crystals provide precise information on the manner in which the latter formed in the lava. Recognition of their nature and characteristics is one of the finest discoveries revealed by the modern microscope.[13]

To what was already known about microscopic inclusions, I have added a few observations on the distribution of the colored material that is spread throughout them and in many cases follows the laws of some sort of crystalographic symmetry. This crude ordering of chemical elements in an originally homogeneous magma constitutes the first stage of speciation of minerals in the rock. The small particles that separate in this way, usually referred to by the general term *crystallites*, offer very interesting prospects for study.

The research that I have been able to carry out on the nature of the substance filling the bubbles in vitreous inclusions seems to establish that this substance is not simply gaseous. They contain very small quantities of a material with properties similar to those of organic matter.

IX

On repeated occasions, clouds of ash were ejected during the last eruption. This volcanic ash is from the same magma as the lavas, but it was disrupted by the sudden escape of gas and steam that rose through it while it was still more or less fluid. The condition of the ash and its crystallinity are a consequence of the state of the magma at the moment the ash was formed. The more fluid the magma the more pumiceous the ash.

If the lava was already charged with crystals, especially microlites, so too was the ash. This has been true of most of the ash of Santorini. When discharging lavas so crystalline that they were almost solid, the ashes were also rich in crystals, particularly microlites. Moreover, the minerals they contain are identical to those of the lava.

An important observation should be noted here. Whatever the crystallinity of a lava may be, the ash that comes from it is always richer in glassy material. This is due, no doubt, to the gases and steam that precede ash eruptions, passing preferentially through the more fluid parts of the lava, so that the ash must correspond more closely to the mineralogical character of the liquid fraction of the material from which it is derived.

We should note that at the moment ash is ejected, the phenocrysts are already formed within the still-fluid amorphous material, so it is not surprising that a great many of those found in the ash are completely separated from the of the material. Their ejection into the atmosphere has led to a separation from the liquid that enclosed them.

X

Extraordinary ground movements took place at the site of the recent eruptions. Though confined to an area around the principal eruptive centers of the volcano, they have noticeably modified the ground level.

Phenomena of this kind have long attracted the attention of geologists, just as they have mine. Many of the illustrations in the section at the end of this volume faithfully represent the most visible modifications of the ground resulting from these movements. The elevation changes, which have always been slow and gradual, were more pronounced at the beginning of the eruption than during the later periods of volcanic activity. The main effect has been one of subsidence. The former pier of Nea Kameni built at the southeastern point of this islet has subsided and the sea has reached different heights in the houses that bordered it. In the southwestern part of the same islet, uplift and subsidence followed one another, alternating according to the corresponding variations of various eruptive phenomena such as opening of new fissures, release of gas, discharges of hot springwater, and notable increases of the ground temperature.

The volcanic ground movements are local and weak; they have nothing in common with the great cataclysms that disturb sedimentary beds and give rise to mountain ranges. The recently proposed interrelationship between these two different types of phenomena cannot, therefore, be sustained. At Santorini the contrast is striking; there one sees, essentially side by side, the effects of both. Lavas of subsurface origin have undergone only limited ground movement; in contrast, sedimentary rocks and submarine volcanic material in the southern part of Thera have been elevated to great heights by a broad uplift extending to all the neighboring parts of the Mediterranean basin.

XI

In addition to the phenomena that will be discussed in a later section, the complete development of the eruption involved:

1. opening of the ground
2. formation of a cone and crater
3. discharge of lava flows

Santorini has been the scene of all these manifestations during the course of the last eruption. Their detailed study is the main topic of the second chapter of this work. Topographic modifications brought about by the eruption are otherwise consigned to the maps and the various plates that appear at the end of the volume. The ground did not open along the entire length of a gaping fissure as is generally observed in eruptions of subaerial volcanoes. But the two active vents designated by the names Giorgios and Aphroessa are linked by parallel cracks running between them in a way that, in this case, conforms to Gemellaro's rule which holds that all volcanic eruptions break out along the length of more or less rectilinear fissures.

At the beginning of the eruption, the discharge of lava was the most conspicuous phenomenon; the emission of blocky material went on silently, and it was only after several days that explosions and ejections of incandescent material began. As a result, in the beginning the newly formed structures had no resemblance to the familiar kinds of volcanic cones. They were simply rocky mounds, lacking a crater and resembling old trachytic domes. Little by little, however, ejections became more common; ash and scoria covered the surface of the rocks; more violent explosions dug into the summit of the main cone and produced vents with the forms of craters. Thus a mass of incoherent rock of trachytic appearance - a *homogeneous cumulo-volcano*, the term used by one of the witnesses of these initial phenomena - gradually transformed itself into a cone covered with a blanket of scoria and ash like that of most volcanoes formed by basic lavas.

This observation is one of great importance because of the deductions it allows one to make concerning the geology of eruptive rocks. It shows, in effect, how arbitrary is the marked distinction that geologists have recently made between trachytic bodies and lava cones and craters. This absolute distinction, generally considered fundamental, should be abandoned or, at least, greatly qualified.

Instead, we should search for another explanation to explain the variations in the form of the cone produced by Santorini's most recent eruption. The discharge of semisolid rocks and the escape of volatile fluids took place from the same orifices; the two types of products must therefore contribute to formation of the central cone. The lavas, in particular, have piled up there because of their low degree of melting, which necessarily causes them to accumulate in place. Because this moderate degree of melting is a natural consequence of the chemical and mineralogical composition of the rocks, it is not surprising that one sees its effects so clearly in the older trachytic rocks.

The emersion of the Isles of May offers an interesting example of submarine discharge of lavas; it is a remarkable case of a submarine eruption that ended shortly after it began. The glassy character and fragmentation of the rocks are note-worthy, but no other aspect of their structure or composition distinguishes them from those erupted under subaerial conditions. Fragmental ejecta provide a better distinction between the products of submarine and subaerial eruptions than do lavas.

XII

Eruptions similar to those of 1866 have taken place in the bay of Santorini since the beginning of historic times and have given rise to the islands known as the Kamenis. The purpose of the first chapter of this work is to establish the locations and exact dates of their appearance and to discuss all available historical records, particularly the detailed accounts of the most recent events. Geologists have a strong tendency to take as pure legend any reports of ancient observers that deviate from the pattern of common geologic phenomena. I have tried to avoid this pitfall and to rectify the erroneous views of my forerunners.

XIII

The great catastrophe responsible for the bay of Santorini occurred in prehistoric times, for it has not been mentioned by early writers, but a civilized population with artistic tastes were witnesses and victims of this event. The excavations of which I have been one of the instigators and the petrographic methods I have applied to the pottery that has been unearthed have yielded much information on the ancient inhabitants of the original island. The people were workers and fishermen. They raised herds of goats and sheep, cultivated grain, made flour and olive oil, wove cloth, and fished with nets. They lived in buildings with shaped wooden structures and masonry of trimmed stone. They made vases with strange designs and a distinctive style. Most of their tools were made of stone, usually lava but also of glass or shaped obsidian. They used gold and probably copper, but these metals were very rare. There was an abundance of wood on the island, whereas now there is only a single type of tree (a palm) in the entire archipelago.[14] Viticulture, which today is the only form of agriculture, seems to have been unknown at that time.

Microscopic study of thin sections of various samples of pottery found in the buried buildings has made it possible to deduce the sources of clay used to make their pottery.[15] The information obtained in this way supports the geologic evidence that Santorini formerly had a valley opening to the sea where the large bay is today.[16]

XIV

The islands of Thera, Therasia, and Aspronisi are the remains of the large island that existed before the present bay was formed. They are made up of three kinds of rocks:
 1. metamorphic rocks (marbles and mica schists)
 2. subaerial volcanic rocks
 3. submarine volcanic rocks

The topography and detailed stratigraphy of these units, which are the subject of one of the chapters of this work, are illustrated with maps and figures at the end of the volume.

In particular, a series of diagrams provides useful information on the cliffs surrounding the bay.

XV

The products of subaerial volcanism are the only ones that can be seen on most of the islands of Santorini. They consist mainly of dense lavas, scoria, pumice, and dikes. The steep cliffs of northern and northeastern Thera provide especially good sections of the dikes and facilitate collection of good samples.

Samples obtained in this way have been closely examined by the same chemical and microscopic methods used in studying the lavas of the recent eruption. The main results of this study are as follows.

 1. In each of these rocks there are always several feldspars that differ from the phenocrysts, but in all cases one of these feldspars is greatly predominant over the others.

 2. The dominant feldspar of phenocrysts is always triclinic, in some cases labradorite, in others anorthite.

 3. Depending on whether the most abundant feldspar of phenocrysts is labradorite or anorthite, the mineralogical association of components of the rock varies accordingly.

 4. Labradorite phenocrysts are regularly accompanied by hypersthene,

augite, and iron oxides, while albite and oligoclase, as well as iron-titanium oxides, occur as microlites. In addition, tridymite and opal are remarkably common.

5. As anorthite increases so too do augite, iron oxide (which is less abundant than in labradorite-bearing samples), and phenocrysts of olivine. Labradorite, augite, and titaniferous iron oxides account for most of the microlites.

The feldspar phenocrysts rarely have the multiple twins that are characteristic of the triclinic system, but even where they have simple or binary twins, an optical examination of separated grains by the methods of Szabo and Boricky, shows that in most cases they are labradorite or anorthite, occasionally oligoclase, and rarely sanidine.

It has been known for some time that in a few ancient igneous rocks, especially granites, the monoclinic feldspar is normally accompanied by a triclinic feldspar (oligoclase), but in volcanic rocks, it is said without proof that only a single feldspar is found in any given volcanic rock. This widely accepted opinion is not in accord with the facts outlined here; the presence of different feldspars in the same eruptive rock must now be considered the normal case.

The studies just discussed have provided other results that are helpful in resolving one of the most serious, long-standing problems of petrography, namely, the identities of the triclinic feldspars. Are all the feldspars simple isomorphic chemical mixtures of albite and anorthite? Must labradorite and oligoclase be dropped from the nomenclature of mineralogy, as proposed by Professor Tschermak's well-known theory (which is only an amplification of the earlier ideas of Sterry Hunt)? That is the basic question. Analyses of feldspars separated from various rocks gave a number of compositions intermediate between those that characterize the feldspar species known formerly. Moreover, their proportions of sodium and calcium are in accord with the silica contents of the analyzed material.[17]

The possibility that more than one feldspar can occur in a single rock seemed inadmissible, and yet that is the case in a great number of rocks collected from dikes on Thera. It is quite easy to demonstrate the coexistence of labradorite and anorthite. These facts explain what appeared to be verifications of Tschermak's rule, but at the same time they definitely upset the theory that was based on it.

The chapter dedicated to a study of the dikes of Thera also includes new information on tridymite and, in particular, on its mode of formation in volcanic rocks. This crystalline variety of silica was formed at the time of eruption under conditions where water vapor was trapped in the lavas.

The mantle of pumice that covers the entire archipelago of Santorini is related to the lavas with labradorite phenocrysts. The phenocrysts and microlites have the same composition as those in the lavas, but they are less abundant. Microlites are especially rare, suggesting that the pumice eruptions took place when they were just beginning to develop. On the other hand, the large amounts of glass account for their elevated silica content.

XVI

The submarine products of Santorini are found exclusively in the southern parts of the island of Thera in the area around Akrotiri. A few lavas with mineralogical compositions identical to those of the subaerial lavas can be seen among the submarine material, but most of the latter belong to a completely different type. They are felsic rocks. While labradorite is still abundant among the phenocrysts, it is accompanied by large proportions of more evolved types of feldspars. In addition, hornblende is the characteristic ferromagnesian mineral; augite is subordinate, and the presence of hypersthene is doubtful. Iron oxides are rare, and olivine is completely lacking. The groundmass is made up of oligoclase, albite, and sanidine. This type of rock is so distinct from the subaerial lavas of Santorini that the two types can be assigned to different lithologic families within Professor Rosenbusch's remarkable classification scheme, namely the amphibole- and augite-andesites.

But the extent of silica enrichment in the submarine rocks of Santorini does not end there. Most are charged with free silica, mainly in the form of opal. The appearance of these silica-saturated rocks is highly varied. They can be either massive or scoriaceous. Some that are impregnated with opal and zeolites are as hard as grinding stones.

One finds here curious perlitic and spherulitic varieties of glassy rocks, the structure and mineralogical compositions of which have been the subject of close study. The feldspars that occur as phenocrysts cannot be extracted magnetically owing to the scarcity of iron in the glass and silicates that surround them, but they are easily separated by the Thoulet method based on their different densities in concentrated solutions of mercury bi-iodide and potassium iodide.

XVII

The ash found at the base of the cliffs of Therasia and Thera contains numerous blocks brought up from the underlying basement series. All these blocks are crystalline rocks quite unlike the sedimentary and volcanic units exposed at the surface. Lacking vitreous material and microlites, they are aggregates of crystals with granitoid textures. Various feldspars, hornblende, and augite are associated with diallage and quartz; the latter are commonly combined with feldspar to form micropegmatite. In their petrographic character these rocks resemble gabbros, kersantites, diabase, and serpentine. I consider them part of the early Tertiary granitoid series.[18]

XVIII

The submarine origin of most of the rocks of the southern part of Thera is shown by the presence of remains of marine animals in the trass[19] of this region. The mollusks and zoophytes that these remains come from are characteristic of the late Pliocene.

The fossil occurrences occupy restricted parts of different levels. In some, the shells are abraded and overturned; in others they are intact and still in place.

All these pieces of evidence testify to an important uplift that exposed the southern part of Santorini. This ground movement has been slow and more or less contemporary with that which had such a notable effect on the Mediterranean basin at the end of the Tertiary period.

XIX

What deductions should one draw regarding the general classification of the rocks of Santorini, and what place should these rocks occupy in such a classification? Before answering this two-fold question, it should be emphasized that classifications of rocks cannot have the same degree of rigor or precision that is customary in the natural sciences. The best established rock-types are gradational into one another and are not as well defined as species. Following an ingenious scheme put forward by Professor Rosenbusch, one can compare their relationships to that of a knotted cord in which the names correspond to the thickened parts. It would be a mistake to believe that one can neatly separate rocks into a certain number of well-defined groups; a natural petrologic classification should simply indicate the main rocks and show how they are linked by intermediate compositions. This method differs little from the more conventional ones used in more precise classifications. It is still a matter of subordinating secondary characteristics, establishing successive divisions within the larger assemblage, and defining a gradational scale of various factors that are considered in the order of their importance. Chief among these factors are those that are essential to the classification of rocks, such as mode of formation, age, and mineralogical assemblage. This method depends on the chemical compositions of the constituent minerals, their structures, and associations. The relative importance of these factors has been perceived in different ways; one after another has been given priority according to the individual author's particular view and the nature of his work.

When one considers rocks in their geological context as the constituent elements of large parts of the earth and not simply as aggregates of minerals of particular compositions, their mode of formation becomes the most important consideration. It is this factor that necessarily governs their most consistent and distinctive compositional variations and accounts for the principal characteristics listed above. The structure of a rock is more closely linked to the way it was formed than to its chemical or mineralogical composition.

Thus, the basic division between eruptive and sedimentary rocks or, what amounts to the same thing, between massive and stratified rocks is perfectly justified.

Another important characteristic, from the geological point of view, would seem to be the age of the rock, for in the case of sedimentary rocks, it has long been recognized by all geologists that age is the essential basis for classification of that group of rocks. But it is not quite the same in the case of igneous rocks. Michel Lévy was the first to emphasize the relations of age, structure, and mineralogical composition. He deserves credit for bringing to our attention the remarkable reappearance, during the Tertiary period, of various types of rocks known from older formations. This piece of evidence enabled Professor Rosenbusch to separate pre-nummilitic[20] igneous rocks from younger ones. In addition, following up on the work of Michel Lévy, it seems that the division could now be extended farther, because subdivisions similar to those used to classify sedimentary rocks can be deduced from age considerations in dealing with pre-nummilitic and post-nummilitic eruptive rocks.

Consider next the use of mineralogical features as a basis for subdividing rocks. Characteristics of this kind, often mistakenly ignored in dealing with sedimentary rocks, may be carried too far in the case of igneous rocks. Nevertheless, they are so useful, especially when the petrographic relations of the essential minerals are taken into account, that they deserve our close attention.

If characteristics stemming from the mode of formation and age are used as a basis for a general petrologic classification and other intrinsic characteristics, despite their importance, are assigned only a third rank, we can consider how these principles can be used to classify igneous rocks. In the great majority of rocks, two sorts of silicate minerals are seen: a colorless one composed of silica, alumina, and alkalies and belonging to the family of feldspars and closely related minerals; the other dark ferromagnesian silicates that are normally brown or dark green. In general, the first tend to be more constant in a given rock and, hence, more distinctive. They should, therefore, be used in preference to the second group, which serves mainly to place the rock in a more extended subdivision. When orthoclase, albite, oligoclase, labradorite, and anorthite were recently taken into account as diagnostic mineral species, it became possible to distinguish within the pre-Paleogene and post-Paleogene rocks five groups, each characterized by one of these feldspars. Since then, Tschermak's ideas about the composition of the feldspars have been widely accepted and all the triclinic feldspars are generally considered to be chemically related in all proportions. It is no longer possible to make any division based on feldspar, except between rocks with orthoclase and those with only plagioclase. Most post-Paleogene volcanic rocks have been called trachytes or andesites depending on whether they belong to the first or the second of these two groups.

Examination of the Santorini rocks shows that the four established types of triclinic feldspars have a real identity. Beyond that, it is evident that, in a complete classification, one must reorganize the divisions of the rocks according to the four triclinic feldspars.

Two objections to this procedure seem to arise from the study of Santorini rocks.

The first is based on the simultaneous presence of several triclinic feldspars as phenocrysts in the same rock. One might wonder why one of these should determine the place of the rock in a general classification rather than the other.

The answer is simple. In almost every case, one of the phenocryst feldspars is greatly dominant over the others, and can serve to characterize the rock. Moreover, the objection has no more validity than the one that can be raised in the case of rocks that contain both orthoclase and plagioclase, which, as we know, is quite common and has never been questioned.

A second, more serious objection results from the fact that the composition of the groundmass feldspar is usually different from that of the phenocrysts, and either one or the other must be used to identify the rock. Again, the mode of formation must be considered; a rock that has crystallized in two separate stages must be separated from one that has crystallized during only a single period, and in the latter case it is necessary to distinguish whether the rock is lacking in either phenocrysts or microlites. Rocks with a crystalline groundmass fall into two major categories. In one, the extinction angle of the feldspar is negligible, or very close to the axis of elongation; in the other, it can exceed 20 degrees. If this difference is generally valid, it could serve as a way of subdividing rocks according to their feldspar phenocrysts.

As already noted, the ferromagnesian silicates provide a subsidiary means of characterizing igneous rocks, but one must also consider the presence or absence of uncrystallized melt or siliceous glass between the crystals. And finally, the presence or absence of free silica on the one hand or abundant olivine on the other makes it possible to establish subgroups of acid or basic rocks.[21]

Using a petrologic classification based on these principles, the volcanic rocks of Santorini fall readily into a natural order. Consider first the *amphibole andesites* (as defined by Rosenbusch) of the Akrotiri district:

1. Their age is known to be late Pliocene.
2. They had two stages of crystallization.
3. The microlites have parallel extinction.
4. Labradorite is the dominant phenocryst.
5. Hornblende is the most abundant magnesian silicate. Some samples contain no free silica, while others are strongly silicified.
6. The interstitial material is amorphous or just beginning to crystallize.

The *augite andesites* (as defined by Rosenbusch), which comprise the modern lavas as well as the older submarine rocks, belong to two groups.

I. Those with phenocrysts of labradorite have the following characteristics:
   1. They are Quaternary or Recent in age.
   2. They had two stages of crystallization.
   3. Their microlites have parallel extinction.
   4. Labradorite is the dominant phenocryst.
   5. Hypersthene and augite are the magnesian bisilicates.
   6. Silica is commonly present as tridymite.
   7. The rock is essentially glassy.

II. Lavas with anorthite phenocrysts are characterized as follows:
   1. Both the massive lavas and modern blocky flows are Quaternary in age.
   2. They have had two stages of crystallization.
   3. The microlites have extinction angles greater than 20 degrees.
   4. Anorthite is the dominant phenocryst.
   5. The magnesian bisilicate is augite.

6. Olivine is common.
7. The magma was largely uncrystallized.

Applying the same criteria to the granitoid blocks found in the oldest volcanic debris, we note that:

1. These rocks are older than at least late Pliocene.
2. They have had two stages of crystallization, probably much closer to each other than in the case of the andesites just described.
3. The last melt to crystallize formed graphic micropegmatite.
4. Some of the feldspar phenocrysts are labradorite, others oligoclase associated with small amounts of orthoclase. In the labradorite group, the amount of micropegmatite is vanishingly small.[22]
5. The magnesian silicates are pyroxene, diallage,[23] and amphibole.
6. Free silica is very abundant. It has the form of quartz, both in the micropegmatite and in separate grains. It also occurs as opal.
7. There is no glassy material except in inclusions.

XX

In the last chapter, which is devoted to the history of the formation of Santorini, I have sought to determine the order of appearance of the various rocks forming the islands enclosing the present bay.

Since an ancestral island composed of marble and mica schists had submarine volcanic rocks on it, considerable uplift must have occurred. When the eruptions became subaerial, they spread abundant debris from several vents and produced a large island the slopes of which were soon covered with trees. A fertile valley in the southwestern part of the island was cultivated, while the summit region remained covered with lava.[24]

A violent collapse, accompanied and followed by formidable explosions of pumice, created the bay.[25] Finally, the eruptions that have taken place in historic times have built the Kameni islands.

The craters of elevation theory cannot be applied to Santorini.

I have laid out the arguments of the partisans of this famous theory, attempting as best I could to convey the enthusiasm with which the idea has been attacked and defended.

The reasons for supporting it differ greatly in merit. For example, arguments based on the order of events are seen to fail under the weight of evidence, but those based purely on geological relations have been sustained

SUMMARY

much longer. The most important of the latter are the essential keystones of the theory. They can be summarized in the two following assertions.

1. In the walls of a so-called crater of elevation, layers of compact lava of uniform thickness and exposed in long, transverse sections resemble modern basaltic flows that have flowed over flat, horizontal ground.

2. During modern eruptions, all lavas laid down on slopes of more than 6 degrees have been scoriaceous and discontinuous.

These two arguments, already rejected by the scientific opponents of L. von Buch and Élie de Beaumont on the basis of studies of Vesuvius, Etna, and several other volcanoes, are shown to be equally invalid for Santorini.

Examination of this volcano reveals two additional facts that decisively oppose the theory. The walls of the so-called crater of elevation of Santorini expose vertical dikes that are oriented perpendicular to each other. Could a central uplift and outward tilting have occurred without changing the verticality of either one of these sets of dikes oriented at right angles?

Second, other exposures in these same walls show superimposed lavas with various inclinations. Some are tilted, in accord with the theory of L. von Buch, but others are gently inclined in the opposite direction. Such orientations are inconsistent with the idea of a central uplift affecting all the beds, for such a hypothesis requires that layers that are now horizontal were originally strongly inclined inward toward the central axis of the volcano; their discharge would have been in defiance of the laws of gravity and inconsistent with all that is known about modern eruptive rocks.

The craters of elevation theory must therefore be decisively abandoned; it can no longer be considered as more than part of the noble debris that advancing science leaves strewn along its path.

Subsequent Geological Studies

Following publication of Fouqué's monograph in 1879, the islands attracted the interest of few geologists until a new eruptive episode began in 1925. A major collaborative study undertaken by Hans Reck, Neuman van Padang, and their co-workers (1936) greatly clarified the structure of the islands and their relations to the Aegean arc. This work provided the first comprehensive interpretation of the geological evolution of the Aegean region. In addition, their detailed studies of the pumice units resulted in a much clearer picture of the caldera-forming eruptions, primarily by pointing out that the pumice mantling the islands has three distinct parts: the basal unstratified "Rose Pumice", a cross-bedded middle part, and an upper, chaotic layer devoid of bedding.

A provocative paper published by Spyridon Marinatos (1939) just after the outbreak of World War II proposed that the Bronze Age eruption of Santorini was responsible for the sudden demise of the Minoan civilization of Crete. Though received at first with skepticism, Marinatos' hypothesis was viewed with much greater interest when extensive layers of youthful pumice were found on other islands and the floor of the eastern Mediterranean (Mellis, 1954; Ninkovich and Heezen, 1965; Watkins et al., 1978; Keller, 1980, 1981). Although little if any Minoan ash was found on Crete (Pichler and Schiering, 1977), the resulting controversy triggered a surge of investigations by geologists and archaeologists from many nations.

The impressive studies by Hans Pichler and his co-workers included the first detailed geological map (Pichler and Kussmaul, 1969-70, included with this volume). Comprehensive petrologic studies by Nicholls (1971), Pichler and Kussmaul (1980), Huijsmans (1985, 1988), Druitt (1985), and Druitt et al. (1989) traced the evolution of the magmas through the entire geologic history of the volcano. Druitt and his co-workers (1989) divided the magmatic evolution of the volcano into two major cycles each of which began with eruption of mafic to intermediate magmas and terminated in major rhyodacitic eruptions.

Marinatos' work provided an equally strong impetus to archaeological studies. Under his direction, the exploratory trenching begun at Akrotiri in 1967 rapidly developed into systematic excavations that have given us a better appreciation of the scale and sophistication of the Minoan culture on Santorini.

Pre-Minoan Volcanism

A wealth of new geochronological information has yielded a much clearer picture of the early history of the islands. The record of explosive volcanism in the Aegean Arc has been summarized by Fytikas et al. (1984) and Keller et al., (1990); that of Santorini has been compiled by Druitt et al. (1989).

The volcano stands on an uplifted block of Triassic limestones and Triassic to Early Tertiary schists and phyllites (Papastamatiou, 1958; Tartaris, 1963; Davis and Bastas, 1980; Skarpelis and Liati, 1990). These rocks, which now form the high ridges of Mt. Profitis Ilias, Mesa Vouno, and Gavrilos, appear to have been emergent long before volcanism began around the end of the Pliocene. Two structural lineaments in the basement series, one trending northeast and a weaker one toward the northwest, have controlled the distribution of vents and form of the caldera (Hoskins and Edgerton, 1971; Budetta et al., 1984; Heiken and McCoy, 1984).

The oldest volcanic rocks consist mainly of submarine breccias of hornblende dacites and andesites erupted from sources in the vicinity of Akrotiri Peninsula and Cape Therma between 0.63 and 1.59 million years ago. A second volcano, known as Thera, rose near the southeastern corner of the present bay and eventually joined Akrotiri and the nonvolcanic island of Profitis Ilias to form a single island. Another small island was formed by the composite cones of Mikro Profitis Ilias and Megalo Vouno at the northern end of what is now the main island of Thera. Activity then shifted to a new center, known as Skaros, near the village of Merovigli. The dark andesitic lavas and tuffs of Skaros are overlain by widespread phreatomagmatic tuffs that are believed to have been erupted from another center near the present island of Therasia. The products of these scattered vents eventually coalesced to form the ancestral island, Stronghyle.

Nicholls (1971) divided the volcanic rocks of Santorini into four series, each erupted from a separate set of vents but sharing a similar calc-alkaline character. Huijsmans (1985) has demonstrated that the compositional evolution from high-alumina basalts through andesites and dacites to rhyolite can be explained as the result of crystal fractionation of the minerals found as phenocrysts and in gabbroic inclusions. The smooth evolution along this trend was modified somewhat by periodic mixing with new influxes of more basic magma.

The major early eruptions seem to have been separated by intervals of quiet lasting tens or even hundreds of thousands of years. The first large caldera was formed at the time of a large pumice eruption about 100,000 years ago (Seward et al., 1980). Another pumice eruption occurred about 79,000 years ago, and the caldera is believed to have collapsed a second time about 18,000 years ago.

The last major eruption prior to the Minoan event occurred about 18,000 years ago (Pichler and Friedrich, 1976). It produced a distinctive reddish rock, known as the Cape Riva ignimbrite, which was used in many of the buildings at Akrotiri. A mature soil horizon developed during the following 15,000 years of repose (Friedrich et al., 1977). Plant remains found in the soil layer (Lacroix, 1896; Schuster, 1936; Friedrich et al., 1977) indicate that the vegetation and climatic conditions differed little from those of the present. Fouqué concluded that Stronghyle, the name used for pre-Minoan Santorini, was a single island with a broad central valley open to the south. Pichler and Friedrich (1978, 1980) showed that this valley was at least partly submerged to form a bay. Although Heiken and McCoy (1984) concurred with this interpretation, they postulated a much larger bay they believed to be an earlier caldera. Friedrich and his co-workers (1988) proposed that the bay contained a small island, not unlike the present Neae Kameni. Although opinions still differ as to the exact geometry, most recent reconstructions have followed this general form (e. g. Erikson et al, 1990; Druitt and Francaviglia, 1992; Friedrich, 1994).

The Minoan Eruption

Judging from a combination of geological and archeological evidence, Fouqué (1879, p. 129-131) placed the major pre-historic Minoan eruption between 2000 and 1500 BC. More recent estimates, based on correlations with Egyptian chronology, placed the event between 1500 and 1550 BC (Hubberten et al., 1990; for discussions of the archaeological evidence see Aiken, 1988; Manning 1988). This differs from radiocarbon dates of about 1625 BC for carbonized material from Thera (Betancourt, 1987; Betancourt and Michael, 1987), tree-ring chronology giving 1626 BC (LaMarche and Hirschboeck, 1984; Kuniholm et al., 1996), and a mean date of 1645 BC for eruption products in cores of the Greenland ice cap (Johnson et al., 1992;

Zielinski et al., 1994). While the conflict with archaeological evidence has yet to be resolved, the new dating places the eruption nearer the earlier dates (Betancourt, 1987, 1990; Davis, 1992).

Although it is generally thought that there was no written record of the Minoan eruption, Greene (1992) has argued that Hesiod's *Theogony* preserves eye-witness accounts of the eruption.

The sequence of events leading up to the Minoan eruption has been worked out from detailed studies of the pumice deposits and their relations to archaeological evidence found at Akrotiri. The eruption was preceded by moderate tremors that probaby led to evacuation of the island, then by more severe earthquakes that caused extensive damage to buildings. At least some of the inhabitants returned and began to clear the debris of damaged structures, but they fled again during the opening stages of the eruption (Doumas, 1990).

The eruption opened with mild phreatic explosions that left a thin layer of fine grey lithic ash that is preserved beneath the pumice in a few scattered localities (Heiken and McCoy, 1990). The main eruption has been divided into four phases. The first produced a vertical plinian column that laid down the unit that Neuman van Padang (1936) referred to as the "Rose Pumice". Judging from the variations of thickness and grain size, the vent must have been in the vicinity of the present Kameni Islands (Pichler and Kussmaul, 1972; Bond and Sparks, 1976). Because the pumice produced by this phase has a maximum thickness of about 7 meters in the vicinity of Thira and thins toward the north, south, and west, the wind direction was probably toward the east. Although much of the caldera was filled with sea water at this time (Friedrich *et al.*, 1988), water was excluded from the vent during the initial eruptions of this phase, possibly because the eruptive center was on the pre-Kameni island. Later, however, activity shifted to a vent in the eastern-most part of the present bay. The pumice produced by these eruptions has a maximum thickness of 3 to 4 meters near Merovigli. The number of lithic fragments increases upward indicating that the vent was becoming larger, and a similar increase in the proportion of andesitic scoria suggests that the eruption was tapping deeper levels of the reservoir.

As the magnitude of the eruption increased, the vent grew toward the southwest and into the flooded interior of the island. Keller and Ninkovich (1972) and Pichler (1973) pointed out that the pumice of this second stage includes spectacular base-surge deposits laid down by pumice flows and characterized by undulating, dune-like bedding. Slightly thicker than the underlying Rose Pumice, it seems to have been produced by repeated pulses, some of which were separated by brief intervals of repose. Finally, the climactic phase of the eruption came from multiple vents associated with the collapse. The pumice contains a much greater proportion of glassy inclusions and fragments of older rocks. It had an original thickness of about 60 meters in the central part of Thera. The fourth and final stage produced ignimbrites from vents beyond the rim of the deepening caldera.

In 1980, Pichler and Friedrich pointed out that early estimates of the volume of pumice were much too large, and since that time all estimated volumes have been much smaller. Pyle (1990), calculated a total volume of 28 to 29 km^3 (dense rock equivalent) of which 41 percent was ignimbrite and 59 ash fall. More recent estimates are as small as 13 to 18 km^3, 60 percent of which was produced in the final phase. These new estimates are close to the volume of the northern part of the caldera.

Post-Minoan Eruptions

The historical record is confused by a lack of agreement on when the various intra-caldera islands were formed. Strabo reported that an island called Hiera (or Iera) was formed in 197 BC. Fouqué thought this to be the now-buried, submarine feature he called Banco. Other workers, notably Reck (1936), concluded that it was not Banco but Palaea Kameni, which Fouqué believed to be the island formed in 46 AD and referred to in ancient accounts as Thia. Later workers concluded that Thia was formed in 19 AD and was destroyed by wave erosion.

Recent work has failed to define the age of the Ayios Nikolaos lava on Palaea Kameni. Although it is normally assumed to have followed the large pumice eruption of AD 726, Pichler and Kussmaul relate it to the Aproessa lavas of 1866. The fact that it appears to be banked against a fault scarp formed in 1457 would support a younger age.

All but one of the post-Minoan eruptions have been near the center of the bay along a northeast-trending regional structure extending from the Christiana Islands outside the caldera in the southwest to Kolombo volcano near the northeast coast of Thera. The linear feature stands out clearly in the regional gravity field (Budetta et al. 1984).

The submarine eruption of 1650 was located at the northeastern extremity of this structure about 6.5 kilometers from Cape Kolombos. A clearer view of the eruption has emerged from the discovery by Lambros (1885) of an unpublished eye-witness account and from detailed bathymetry showing that the submarine volcano has a small caldera (Akylas, 1925). The magnitude of the tsunami has become more apparent from studies of buried ruins that were exhumed when the wave carried away the thick cover of pumice and revealed the underlying surface at Perissa and Kamari (Friedrich, 1994).

Georgalas (1962) has provided an excellent summary of activity since 1870, as well as an extensive bibliography for the entire eruptive history of Santorini. The multi-volume reports of three international conferences sponsored by the Thera Foundation in 1969, 1978, and 1989 comprise a major part of the research completed in the last few decades. Druitt and his co-workers have recently submitted for publication a comprehensive volume on Santorini, and an English version of Walter Friedrich's popular book *Feuer im Meer* should appear shortly.

Eruptive History of Santorini

The following summary is based mainly on compilations by Georgalas, 1962, and Druitt et al., 1989. Ages of pre-Minoan eruptions are given in thousands of year (ka).

645-360 ka   Eruptions of hornblende-bearing rhyodacite in the vicinity of Akrotiri. Later eruptions (520-350 ka) from cones at Capes Balos, Kokkinopetra, and Mavrorachidi covered the earlier siliceous rocks with andesitic scoria and lava. During this same period (530-439 ka), andesitic lavas and scoria were erupted from Peristeria Volcano at Megalo Vouno in northern Thera.

360-180 ka   Eruptions of andesite and dacite from several cones, including those near Cape Alai (360-230 ka), Cape Therma, and Cape Alonaki (~225 ka).

203±24 ka   Eruption of Lower Pumice 1. Rhyodacitic plinian airfall and ignimbrite deposits. Erupted from a vent near the present Kameni Islands

~80 ka   Eruption of Lower Pumice 2. Similar to Lower Pumice 1. Followed by caldera collapse.

~170 ka   Andesitic lavas (172 ka) at Cape Simandiri and pumice-fall deposits and ignimbrites of about the same age at Cape Thera and Therasia

~100 ka   Middle Pumice sequence of wide-spread Plinian airfall deposits, lithic breccias, and ignimbrites followed by eruptions of agglomerates and scoria from a source north of the present Kameni Islands.

76-54 ka   Formation of the lower part of Megalo Vouno and Kokkino Vouno cinder cones. Eruption of widespread pumice and pyroclastic surge deposits of Vourvoulos, followed by andesitic agglomerates and scoria (Upper Scoria 1). The tuff ring near

Cape Kolombos at the northeast corner of Thera is on the same alignment and may date from the same eruptive episode. Incremental caldera collapse probably took place about this time.

~70 ka Construction of the Skaros lava shield consisting of andesitic domes overlain by more mafic andesites and basalts. An age of 67 ± 9 ka has been obtained from this sequence.

~55 ka A large eruption of the andesitic scoria, probably from the Skaros volcano (Upper scoria 2).

21-18 ka Formation of the dacitic and andesitic lava domes of northern Santorini and eruption of the upper lavas of Megalo Vouno. Cape Riva eruptions of air-fall pumice and ignimbrites. Caldera collapse.

~1628 BC Minoan eruption. In a succession of four eruptive phases, about 30 km^3 of pumice were discharged from a vent near the present island of Nea Kameni. Localized collapse gave the caldera its present form.

197 BC Formation of Hiera, the first island to emerge in the bay. Although the evidence is ambiguous, Hiera was most likely the now-submerged reef of Banco northeast of Nea Kameni. See p. 8-9 for Fouqué's discussion and Friedrich (1994) for a summary of classical sources.

?19 AD According to Pliny, another island, called Thia, rose in the bay at this time. Fouqué (1879, p. 6) concluded that this was an error and that this event was probably the eruption of 46 AD.

46 AD A dome rose within the bay to form a second island called Thia. Most modern authorities agree with Fouqué's conclusion that this eruption was responsible for the main part of Palaea Kameni.

726 AD A large eruption of pumice from an uncertain source, possibly

a new vent at the northern end of Palaea Kameni. A lava flow at Ayios Nikolaos has been assigned to this event, but this too is uncertain.

1457　The faulting responsible for the steep fault scarp on the northeastern side of Palea Kameni has been attributed to earthquakes near the end of this year. The only known record is an inscription on a wall in the ofrmer fortress at Skaros. It has long since disappeared but the text was recorded by Kircher (1665) and is given by Fouqué on page 11. It is uncertain whether there was any volcanic activity at this time.

1570-73　A new dome rose in the bay forming the island of Mikra Kameni.

1650　26 Sept. to 6 Dec. Following a series of strong earthquakes, explosive eruptions occurred about 6.5 km northeast of Cape Kolombos forming a small island that soon disappeared. 50 persons on Thera and the entire crew of a passing ship diedfrom toxic fumes. The dome that remained when the eruption ended rises 300 m above the sea floor to a summit that is now 19 m below the surface.

1707-11　23 May 1707 to 11 Sept. 1711. A new dome rose between Palaea and Mikra Kameni to form the island of Nea Kameni.

1866-70　On the 26th of January, 1866, under water explosions broke out in Voulcano Bay on the SE side of Nea Kameni and a dome, named Georgios, emerged from the water on 4 February 1866. A second dome, Aphroessa, appeared west of Georgios on 13 February. Lava continued to flow from both domes until that of Aphroessa stopped on 17 May 1866; Georgios continued to grow until the end of the eruption on 15 October 1870. A third dome, called Reka, began to grow next to Aphroessa on 10 March 1866 but was later covered by the lavas of Georgios. A fourth center, known as the Iles of May, was reported by Fouqué, but was attributed by Reiss and Stübel to a submarine lava from Aphroessa.

1925-26   The volcano had only solfataric activity between 1870 and 1925 when, on the 11th of August, the first of a series of explosions took place in the channel between Mikra Kameni, Nea Kameni, and Georgios. A lava dome, named Dafni started at a depth of 19 meters and rose above the water on 12 August 1925 to unite Mikra and Nea Kameni in a single island to which Georgalas (1962) gave the name Neae Kamenae. By the time the activity came to an end in January of 1926, a new dome, Dafni, had reached a height of nearly 90 meters and covered a large part of the island with over a hundred million cubic meters of new lava.

1928   On 23 January 1928, after 20 months of quiet, another new vent opened 300 meters south-south-east of that of Dafni. The slow discharge of lava was accompanied by mild explosions, and when activity ended on 17 March another dome, Naftilos, had been formed. It rose about 15 meters above its base and had a volume of 40,000 cubic meters.

1939-41   On the 20th of August, 1939, submarine explosions began in what was formerly St. Georgios Bay on the western side of New Kameni. Two small domes, Triton A and Triton B, soon emerged and mild explosions formed a funnel-shaped vent to which was given the name Warathron. In late September a new vent broke out near the western base of Nea Kameni. A dome, named Kténas, eventually grew within the crater to a height of about 12 meters, and lava from the same source covered the Triton domes and filled what remained of St. Georgios Bay. The Warathon vent, which had ceased to erupt at the beginning of September, returned to activity on the 13th of November and four days later began to produce a dome, now called Fouqué, which, by the time it ended in mid November of 1940, covered the summit of Nea Kameni, the Kténas dome, and parts of Georgios and Aphroessa. In July of 1940, another new vent opened in the low area between the Fouqué lavas and the slopes of Georgios and produced three small dome, called Smith A, Smith B, and Reck. The first two ceased to grow after about two months, but Reck continued until mid October by which time it was partly destroyed by explosions. On 24 November, lava again began to flow from a new vent on the eastern slope of the Fouqué dome and continued until the beginning of July 1941, by which time it had built the 125 meter-high dome that came to be called Niki.

1950 On 10 January 1950, after a repose of eight and a half years, explosive eruptions broke out on the cone of Georgios and continued until 2 February. A small dome, called Liatsikas, formed in the new vent, and two lava flows moved short distances down the northern and southern flanks. Since that time only solfataric activity has been observed.

Translator's Notes

The text given here follows the original in all respects, but Fouqué's two introductory sections have been consolidated, and the Summary, which he placed near the beginning now follows the last chapter. In the interest of economy, some of the plates have been omitted, but the original numbering has been retained. Except for the Introduction and Summary the pages numbers closely match those of the original.

Many place names on Santorini have more than one English spelling. In most instances, I have kept the spelling consistent with that on the geological map of Pichler and Kussmaul that accompanies this volume. The only exceptions to this rule are Kameni and Palaea, for which I have adopted what seems to be the most common English spellings in current use. The name Neae Kameni is used for the present island formed by the union of Nea Kameni, Mikra Kameni, Giorgios, and Dafni.

Introduction

1. Léonce Élie de Beaumont (1798-1874), one of the most respected and influential geologists of his time, was a strong advocate of the theory of "craters of elevation" that Fouqué's work on Santorini effectively demolished. He died five years before the original version of this book was published.

2. Charles Sainte-Claire Deville (1814-1876) developed techniques for collecting and analyzing volcanic gases and was the first to offer a comprehensive theory explaining the distribution and compositional relations of volatile emissions from active and dormant volcanoes (Sainte-Claire Deville, 1857). His brother, Henri, introduced Fouqué to geology, coauthored his first publication, and arranged for him to go to Vesuvius with Charles in 1861.

Chapter 1. Historical Accounts of the Formation of the Kameni Islands

1. Philippson (1896) believed that all the islands produced by these eruptions subsequently disappeared and that none corresponds to Palaea Kameni. Druitt et al. (in press) follow Fouqué's interpretation.

2. A church of this name still exists, but it has probably been rebuilt.

3. On the geological map of Pichler and Kussmaul these lavas are indicated as unit K_3 which would make them products of the eruption of 1866. Druitt et al. (in press) believe that they were erupted shortly after the explosive eruption of 726 and the faulting responsible for the steep cliff on the northeast side of the island. See note 6.

4. Fouqué recognized that large eruptions tend to be preceded and followed by longer intervals of repose. A statistical analysis of well-dated eruptions by Simkin et al. (1981) has shown this correlation to be valid.

5. See Friedrich (1994) for a discussion of this account.

6. My colleague, Dr. Mary Jaeger, to whom I am indebted for translating this difficult latin text, points out that the Romans counted inclusively, including the dates at each end of a sequence, so that the seventh day before the Calends of December would be the 25th of November. In the same way, in line 4, 2 is added to 1455 to give the year 1457.

> Great-hearted Franciscus Crispus, surely an offspring of heros,
> you see the destruction that has shown us wonders,
> with fourteen hundred and fifty-five of Christ's years
> slipping by, and adding two to that number,
> Seven days before the Kalends of December, with a great rumble,
> immense Theresinus tore enormous rocks from Kameni,
> and with a groan, a crag appeared from the depths of the waves
> bringing with it a great and memorable portent.

Fouqué seems to be the only person to cite this event. If he is correct in relating it to the faulting of Palaea Kameni, the lava flow at Saint Nicolas Cove (Ayios Nikolaos) discussed in the preceding section must have been erupted later, for it seem to post-date the faulting.

7. The waves that struck the east coast of Thera came far inland and washed away much of the Minoan pumice around Kamari and Perissa. At the latter place it revealed the ruins of a post-Minoan settlement that had been buried by alluvial deposits coming from the slopes of Mesavouno. The tsunamis produced by the Minoan eruption carried pumice to elevations of 20 meters on the island of Ios and dropped pumice above the shorelines of Anatolia. It seems unlikely, however, that they could have reached the palace of Knossos, which stood at an elevation of 60 meters on the island of Crete.

8. The effects described here resemble those caused by the gas clouds released during the eruptions of Diëng, Java, in 1979, and Lake Nyos, Cameroons, in 1986.

9. Detailed bathymetry shows that the cone of Kolombos rises from a depth of 280 m to within 20 m of sea level. The floor of the summit depression is 512 m below sea level. During the eruption of 1650, it produced tsunamis that resembled those caused by the Krakatoan eruption of 1883 when waves hitting the coasts of Sumatra and Java took the lives of 90,000 persons. The waves circled the earth twice.

10. A cubit is defined as the length of the forearm from the elbow to the tip of the middle finger. It is normally taken as 18 inches.

11. Numerous examples are known in which vegetation has been rejuvenated in the years immediately following a large ash eruption that supplies fresh nutrients to the soil.

Chapter 2. The Eruption of 1866

1. De Cigalla with is various alternate spellings is a common Greek name, Derica'kka. The person referred to here lived on Santorini in the nineteenth century. He is credited with reintroducing the tamarisk tree, which grew on the island before the Minoan eruption (Friedrich, 1994).

2. The sources of this and the following reports can be found in the list of general references at the end of this volume.

3. A similar red stain can be seen today in the small inlet on the northeastern coast of Palaea Kameni. A stain of this kind was observed on the water of Lake Nyos after the eruption of 1986. It probably had a similar origin.

4. Zinc melts at 420#C.

5. Fouqué refers to carbon dioxide as carbonic acid.

6. This technique of thrusting a rod into lava has been used to measure viscosities in the field. For example, Einarsson (1949) used it to estimate the viscosity of lavas erupted from

NOTES ON PAGES 65 THROUGH 98

Hekla in 1947-48. He calibrated the method by measuring the time and effort required to drive the "penetrometer" into a substance, such as asphalt, for which the viscosity is known over a range of temperatures. The technique is further described by Gauthier (1973) and by Pinkerton and Sparks (1978).

7. Sainte-Claire Deville's (1857) classification of fumarolic emissions was based mainly on the compositions of gases at different temperatures. At the highest temperatures, the assemblage includes H_2, CO, CH_4, and metallic chlorides and fluorides. As temperatures decline, hydrogen and carbon monoxide are replaced by CO_2 and H_2S, and at the lowest temperatures only H_2O remains. This sequence is seen in a given locality as activity declines with time and, at a given time, with increasing distance outward from a volcanic center.

8. Lead acetate is used to test for hydrogen sulfide.

9. It is unlikely that much hydrogen could be produced in this way.

10. The orientation of the magnetic anomalies was probably the result of a dike connecting the two eruptive centers.

11. These seem to have been among the first analyses of volcanic fumes by remote spectrographic techniques.

12. Incandescence has been observed under water where lavas have entered the sea along the coast of Hawaii. Fouqué's explanation of the insulating effect of steam is essential correct.

Chapter 3. Prehistoric Structures of Santorini
Most of the notes on archaeological topics in this chapter were supplied by Professor J. L. Davis and Mr. J. M. L. Newhard, University of Cincinnati.

1. More recent excavations indicate that the population of Santorini fled before the eruption reached its cataclysmic stage (Heiken and McCoy, 1990; Sparks and Wilson, 1990). Fouqué's view was based on the discovery of a human skeleton mentioned in a later paragraph. Ironically, after more than a century of further excavations, no further evidence of this kind has been uncovered.

2. Pozzolana is a siliceous tuff or ash used in making a type of cement that is especially resistant to seawater.

3. Other contemporary references to the discoveries on Therasia include Bory de Saint-Vincent (1833), Lenormant (1866), and Virlet d'Aoust (1866).

4. Fouqué is probably referring to Skala Phira, a landing area directly below the modern town of Thira.

5. Cylindrical masonry structures of this kind have also been found in the Pthellos Quarries south of the town of Thira and on Christiana, but in both cases they should be earlier than those found on Therasia, (i.e. of the earlier Middle Bronze Age). For the Phtellos remains, see Doumas (1973) and Marthari (1982). For the Christiana remains, see Doumas (1976).

6. The use of rubble, mud, and wood in these walls appears to be identical to practices found in excavations at Akrotiri. For additional details concerning Theran construction practices, see Palyvou (1988).

7. It is very improbable that any roof had an arch, cf. Shaw (1977).

8. Other instances of columns supporting roofs are found at Akrotiri, primarily in the middle of second-floor rooms. See Palyvou (1990a).

9. Some paleobotanists give Minos Kalokairinos, who dug at Knossos in 1878, credit for first identifying botanical remains in archaeological excavations. See Hansen (1991). While

Kalokairinos worked in 1878 and published in 1899 and 1900, Fouqué worked in 1867 and published his account in 1879. For a summary of the past and present flora and fauna of Santorini see Friedrich (1994, Chap. 5).

10. This "saw of flint" is almost certainly as example of a well-known type of denticulate, commonly made of rhyolitic obsidian. See Torrence (1986a, p. 95), and Blitzer (1992).

11. This "adjacent courtyard" is strange; houses at Akrotiri are not of the house and compound variety.

12. In her dissertation, C. Palyvou (1988, 1990a and b) categorized the 85 windows of Akrotiri into four groups. The windows described here would probably fall into the "Type B" category, having a long vertical axis and an opening with an average area of 1.00 m^2. They are found on both the first and second floors of buildings.

13. The use of ashlar masonry on the corners of buildings is well attested at Akrotiri. See Palyvou (1988).

14. This use of ashlar masonry at corners is familiar in Theran and Minoan architecture (see Palyvou, 1980). The cylindrical hole is presumably for a dowel (Shaw, 1971). The letters or numerals were plausibly mason's marks, such as those found in Minoan architecture and their presence at Therasia is significant. For the use of mason's marks at Akrotiri, see the discussion following Palyvou's presentation at the Thera Conference of 1990.

15. This must be the earliest recognition of paleosols on the island, at least for archaeological purposes. Fouqué's observations have recently been amplified by Limbrey (1990) and Davidson (1978).

16. Another example of this style of construction is found in room Delta 1a at Akrotiri.

17. Room R is not indicated in Fouqué's diagram.

18. Petruso (1992) includes an overview of weight systems of the prehistoric Aegean. Lead was most often preferred over stone, although stone weights have been found in Late Bronze Age contexts. The stones found by Fouqué may have been weights, since their weights are rough multiples of the standard Aegean unit of 61 grams.

19. This container must be what has been called a "Syro-Palestinian" tripod mortar, although it is now certain that they were manufactured locally in the Aegean. Numerous examples have been illustrated from the Akrotiri excavations. See also Runnels (1981) for a discussion of the volcanic stone used to make these.

20. See note 15.

21. The blades described here are what today are called "prismatic blades," and are characteristic of Bronze Age assemblages in the Aegean. The triangular point could in theory be of Neolithic age, since there are Neolithic finds from Akrotiri, but it is more likely Late Cycladic, particularly if it had a tang and a hollow base. On the typology of Cycladic lithic implements, see Cherry and Torrence (1984, p. 12-25), and Torrence (1986).

22. Televantou (1984) discusses in detail the physical aspects of jewelry from Akrotiri and its representation in art.

23. Not all of the pottery at Akrotiri was wheel-made.

24. The description here refers to what is now known as a "beaked jug" or "bird jug," because of the beaked appearance of the spout. For an illustration of the jug referred to in this passage, see Fouqué (1867) and Friedrich (1994, p. 133.)

25. A brief description of the excavations by Mamet and Gorceix (1870a and b) can be found in Doumas (1983).

26. The pottery from the early excavations of Fouqué, Mamet, and Gorceix was reexamined by Renaudin (1922). At that time, the collection was in the French School at Athens.

27. This must be what today is called an oval-mouthed amphora. The backward-bent spout (rejete´ en arrie`re) is probably another reference to a "beaked jug" of some sort.

28. Not all the pieces described here fit the standard unit of 61 grams described by Petruso (1992). The stone balls probably were not weights, for, as Petruso points out, nearly all certain weights so far discovered on Thera have at least one flat side, so that the piece can rest firmly on a balance pan.

29. More recent work at Balos is described by Davis and Cherry (1990).

30. Jones (1987) discusses storage practices in the Aegean region. The small pea Fouqué calls arakas is a young form of faba, which is still raised on Santorini today.

31. The filters are technically called "rhyta" today. See Renaudin (1922, fig. 11) for a photograph of the three undecorated rhyta. The decorations of circles and bands are probably what are known as "circle patterns." See Furumark (1941).

32. In modern terminology, the elliptical vessels are called kymboi. The excavations at Phylakopi uncovered other examples of these large basins (Atkinson et al., 1904), fig. 112-115, plate XXX.

33. Friedrich et al., (in press) have reexamined the clay used in pottery and agree that it certainly came from Santorini, probably from part of the Loumaravi complex on Akrotiri. The actual source may now be under water.

34. The wall paintings mentioned here are the well-known Keftiu representations discussed by E. and J. A. Sakellarakis (1984) and Strange (1980).

35. On obsidian sources, see Renfrew et al. (1965, 1968). Compare the small saw with those described by Torrence (1986). On the acquisition of obsidian from Melos, see Torrence (supra n. 21). Fouqué was wrong in thinking that this commerce was limited to the "Stone Age." The use of obsidian for chipped stone tools is known even in the Classical period (Runnels, 1982).

36. These objects, which are now known to have been loom weights, are found in great numbers at Akrotiri. They were used to pull the warping threads (which hung from a beam) tight as cloth was woven. See Barber (1991) for an excellent discussion of Aegean weaving practices.

37. Studies of the soil layer under the Minoan pumice have revealed little evidence that trees were abundant except in drainage channel where they were sheltered from the wind and could find more water (Friedrich, 1994).

38. Clear evidence has been found in the excavations at Akrotiri that an earthquake preceded the eruption. See note 1 for this same chapter.

39. Considering the limited evidence at his disposal, Fouqué's estimate is remarkably close to the date (in the late 17th century BC) generally accepted today. For a discussion of the date of the eruption and its archaeological context see Davis (1992).

40. Phoenician colonization in the 15th century BC is now considered unlikely (Coldstream, 1982).

41. It is now thought that the prehistoric population of the islands originated in Caria (southwestern Asia Minor). See Broodbank and Strasser (1991).

Chapter 4. Description of the Present State of the Kamenis and the Two Underwater Cones in the Bay of Santorini

1. Although the present Kameni islands were formed in historic time, it is now thought that there was an ancestral Kameni at the time of the Minoan eruption. Thera, Therasia, and Aspronisi seem to have constituted a single island since the early Quaternary. For a discussion of the ages of the rocks of these islands, see Druitt et al., (1989).

2. The cliff referred to here must be the one on the northeastern side of Palaea Kameni. As the accompanying map of Pichler and Kussmaul shows, that side is clearly bounded by a fault but is protected from the direct action of the sea.

3. The word calanque is used is many parts of the Mediterranean region for a cove or creek.

4. Leycester published his account of Santorini in 1850, not 1859, fifteen years after that of Forbes (1844). For more recent information on the fossil evidence for uplift, see Seidenkrantz (1989) and Friedrich (1994).

5. It is unfortunate that this mass of pumice has been buried by younger lavas and can no longer be examined, for, as Fouqué reasoned, it could help clarify the nature of the volcano and its eruptive history. If the pumice was indeed a product of the Minoan eruption, then there must have been deep water in this part of the present bay, but the pumice could well have been produced from a vent near the northern end of Palaea Kameni during the eruption of A.D. 726. An analysis would resolve the matter.

6. Fouqué recognized that the rate of advance of lava flows is a function, not only of viscosity and slope angle, but also of the rate of discharge. Walker (1973) showed that this factor is more important than all others combined.

7. The illustrations of these changes, shown in the plates at the end of this volume, are among the most dramatic examples of groundlevel changes associated with volcanism.

8. Owing to an inadvertent drafting error, the submarine cone described here is not shown on the geological map that accompanies this volume.

9. A detailed bathymetric survey (Hoskins and Edgerton, 1971) showed that this is a regional tectonic trend and a major structural element of the volcano (Heiken and McCoy, 1984).

Chapter 5. Chemical Study of the Products of the Recent Eruption of Santorini

1. Fouqué was a leader in the development of the techniques of magnetic separation.

2. A carbon-zinc battery invented by R. W. von Bunsen in 1841.

3. It is not clear whether Fouqué recognized the significance of his observation that the last interstitial liquid to crystallize has the composition of alkali feldspar and silica and corresponds to that of highly differentiated magmas, such as rhyolites and granites. He did recognize, however, that silica-rich liquids are slow to crystallize.

4. The amount of CaO in the analysis (10.49%) indicates that the sample contained a small amount of other material, possibly augite, feldspar, or glass. The small inclusions that Fouqué mentions below may be exsolved Ca-rich pyroxene.

5. Andesine and bytownite were not generally recognized at the time this was written.

6. As noted earlier (note 3) the glassy residue has the composition of alkali feldspar and silica. In this case, it is found in a more evolved rock containing iron-rich hypersthene.

7. The excess alumina and low silica suggest that the analysis was poor. Alumina and silica are the most difficult components to determine accurately by the classical wet methods used at that time.

8. Fouqué later identified this mineral as melilite. See Fouqué (1890a).

9. This interpretation is in accord with the work of Nicholls (1971).

10. The "sixth crystal system" is that of triclinic minerals, such as plagioclase.

11. Possibly a product of devitrification of the glass.

12. The methods for collecting and analyzing samples described here were quite satisfactory, even by more modern standards. The methods used to determine H_2S, CO_2, and O_2 are similar to those used today. The measurements of CH_2 and H_2 were adequate for most purposes but less precise than those of the other components (R. Symonds, pers. comm., 1996).

13. Gases that rises through water tend to lose H_2O, SO_2, HCl and HF to the water (R. Symonds, pers. comm., 1996).

14. The unusual amounts of hydrogen Fouqué reports are largely an artifact of the incomplete analysis. If water and other important components were included, the amount of hydrogen would not differ greatly from that in other volcanic gases. Nevertheless, it is true that the amount of H_2 relative to CO_2 tends to increase with time as magma is degassed and the contribution of meteoric steam become relatively more important (R. Symonds, pers. comm., 1996).

15. Only a very small part of the hydrogen and oxygen could come from dissociation of water.

16. This interpretation is based on the belief that the intermediate plagioclases are closely intergrown mixtures of albite and anorthite; the concept of solid solution with continuous paired substitution of Na-Si for Ca-Al had not yet been recognized. Although Fouqué realized that the extinction angle of oligoclase was very small, he seems not to have been aware of the systematic relations between the composition of plagioclase and extinction angles. It would be another fifteen years before his colleague, Michel-Lévy, would publish his well-known determinative curves (1894, 1896).

17. The techniques referred to here are based on flame and fused-bead tests developed by Jozsef Szabo (1876) and on Emanuel Boricky's (1877) method based on etching with fluosilicic acid (H_2SiF_6).

18. Fouqué recognized "eolian differentiation" as responsible for spatial variations in the grain-size and composition of ash from a single eruption. In the preceding paragraph he makes an important distinction between two main types of ash: lithic and juvenile. The former is produced by low-temperature fragmentation of older rocks and the latter by vesiculation of fresh molten magma.

Chapter 6. Description of the Older Parts of the Santorini Archipelago

1. This cape is designated Ayios Nikólaos on the accompanying map of Pichler and Kussmaul. The town of Apano Meria is now called Oia.

2. Some of the elevations given by Fouqué differ slightly from those on more recent maps. In most instances the difference is less than 5 meters. See pages 245 and 246 for his comments on the reliability of the barometric measurements.

3. The thickness of the pumice at this locality has been reduced to 30 to 40 meters as a

result of quarrying and erosion. The aragonite blocks are probably the stromatolites discussed in a later section. (See note 22 below.)

4. The principles outlined in this discussion of the role of water in oxidation are essentially those held today the only difference being that water is now known to dissociate, not to hydrogen and oxygen, but to H^+ and OH^-.

5. These submarine breccias were related to viscous dacitic lavas that built a cluster of domes in the Akrotiri area. The growth of these domes was interrupted by periodic explosive eruptions, many of which were phreatomagmatic (Pichler and Kussmaul, 1980).

6. Although Fouqué is correct in associating the pisolites, or, more properly, accretionary lapilli, with eruptions through water, they are formed in the wet eruption column, not in agitated water.

7. The term pyroxene, as it is used here, refers only to augite; hypersthene was not considered a pyroxene at the time this was written. Only after Warren and Modell (1930) showed that its crystal structure is very similar to that of the calcium-rich pyroxenes was hypersthene fully accepted into the pyroxene family.

8. The "lower pumiceous tuff" Fouqué describes here is the unit Neumann van Padang (1936) called the Lower Pumice (Bu for "Unterer Bimstein"). It actually consists of two parts, Bu-1 and Bu-2, with ages of about 180,000 and 203,000 years respectively (Druitt et al., in press). It is especially conspicuous in the section below the town of Thira, where it has a combined thickness of 70 meters (Günther and Pichler, 1973). Both units are products of eruptions that opened with a pumice fall and culminated in pumice flows. Judging from the variations of thickness and grain size, the source of Bu-1 was probably in the southern part of the present island of Thera, whereas the source of Bu-2 was more to the north in the area of today's Kameni islands. The Bu series is related to an early caldera. Judging from its thickness and wide distribution, the Bu eruption could have been much stronger than the Minoan one.

Günther (1972) used the name Middle Pumice (Bm for "Mittlerer Bimstein") for a third layer between the other two. It is seen in the caldera wall below the town of Thira where the black, glassy dacite is displaced by a fault. Although it has a strong lateral variation, it can be followed for several kilometers. Below Merovigli, the Bm layer is a strongly welded, massive red airfall tuff, but below Thira it is pitch black. Within a few kilometers southward, it changes color and structure from brownish welded pumice to bright-white pumice. This indicates that the eruption center of the Bm layer must have been within the present caldera in the vicinity of Merovigli.

9. The locality referred to by Fouqué as Cape Lazaret is now called Cape Katothira. It is located a short distance south of the landing place below the town of Thira.

10. The metamorphic complex, locally referred to as the Cycladean Massif, is made up of Triassic to Tertiary sediments of the former Tethys Sea that extended from southern Europe to as far east as Indonesia. Greywackes, shales, and limestones were folded, metamorphosed, and uplifted during the Early and Middle Cenozoic, when the African and Indian plates collided with Eurasia. They are now exposed in the mountain chains extending from the Alps through the Himalayas to southeastern Asia. Subsequent to the work of Fouqué, the rocks have been studied by Phillipson (1896), Neumann van Padang (1936), Papastamatiou (1958), Tataris (1963), Davis and Bastas (1978), Murad and Hubberten (1975), Skarpelis and Liati (1990), and Skarpelis et al. (1992). The metamorphic rocks between Athinios and Cape Plaka are intruded by a late Miocene granite for which Skarpelis et al. (1992) measured an age of 9.5 million years.

The granitic intrusion was also encountered at a depth of 252 meters in a boring at Megalokhorion (Skarpelis and Liati, 1990).

11. It is now thought that only the pumice of the first phase of the Minoan eruption (Bo-1) was deposited at high elevations. The air-fall deposits of the plinian phase were followed by base surge eruptions and pumice flows that passed through low elevations of the caldera rim.

12. The mineral deposits are found mainly in veins of pyrophyllite, magnetite, chalcopyrite, galena, and talc within the skarns adjacent to a granitic intrusion (see note 10 above). Some of these minerals were used as pigment for frescoes on the walls of buildings at Akrotiri and the talc was used on pottery (Friedrich, 1994). The deposits were mined for lead, zinc, and silver at the beginning of this century (Murad and Hubberten, 1975).

13. Fouqué is probably referring to the ruins of ancient Thera on Mesa Vouno, a southern spur of Profitis Ilias. Excavations by F. Hiller von Gaetringen (1899-1904) between 1895 and 1902 uncovered remains ranging in age from the early classical period to the third century A.D.

14. The deformation and uplift of the metamorphic series took place mainly during Oligocene to Early Miocene time, well before the beginning of Late Cenozoic volcanism.

15. The section between Cape Athinios and Cape Plaka exposes part of a belt of blueschist-facies rocks that extends throughout much of the Cyclades islands.

16. These rocks are not shown on the accompanying map.

17. Fouqué used the term trass for this type of pumiceous tuff that resembles possolan.

18. Reported by von Fritsch (1871). See Quenstedt (1936), Tataris (1963), and Seidenkrantz (1989) for more recent studies.

19. This is probably the Cape Riva ignimbrite, dated at 16,000 BC. It is shown on the accompanying geologic map of Pichler and Kussmaul as a very thin layer below the Minoan pumice. Although Fouqué was responsible for introducing the term nuée ardente into geological literature (Hooker,1965), the origin of ignimbrites had not yet been recognized. Although he called the rock a conglomerate, he clearly recognized its igneous origin.

20. Although details of the Minoan eruption are still being debated, volcanologists agree that three or four successive phases took place over a considerable interval of time, certainly years and possibly decades.

21. Fouqué seems to have been puzzled by the crude layering of the pumice and the rounding of the mafic xenoliths. At that time, he believed that all of the pumice had been ejected at high angles and had no way of recognizing deposits of low-angle blast, the importance of which only became apparent after Moore (1965) described the 1965 base-surge eruptions of Taal volcano. Radially directed ring-shaped density flows expand swiftly from the base of the eruption column and leave dune-like deposits with intricate cross-bedding.

22. The stromatolites offer clear evidence that marine water occupied what is now the northern part of the bay. Analyses of their carbon isotopic ratios shows that this area was underlain by a magmatic source of carbon. Their carbon-14 ages range between about 15 and 23 thousand years (Eriksen et al., 1990; Friedrich, 1994).

Chapter 7. Petrographic Study of the Dikes in the Northern Part of Thera

1. Although it is true that the products of each episode are distinctive, Huijsmans (1985) found that silica contents tend to decline in most eruptive cycles, the main exception being the cycle preceding the Minoan eruption in which the trend was one of increasing silica. He found no discernible variations in the post-caldera lavas.

2. What Fouqué describes here could be cross-sections of hollow crystals that grow during rapid cooling. They are common in basalts that have been rapidly quenched in water or wet sediments.

3. Fouqué's explanation for the lack of phenocrysts in the chilled margins, though probably correct, is not the only possibility. Dispersive shear stresses in the margins of flowing magma ("Bagnold effect") can cause crystals to be swept into the interior, where the velocity profile is more uniform.

4. In these two remarkable paragraphs, Fouqué's proposes the now widely accepted concept of a zoned magma in which a silica-rich liquid overlies a more basic magma with crystals that have been concentrated by gravitational segregation. Similar relations have been found elsewhere, particularly in large, caldera-forming eruptions.

5. These few sentences are a succinct statement of the principles governing the crystallization of silicate melts as they are understood today. Fouqué recognized that the prior presence of nuclei enables the crystals to grow, even during rapid cooling. See Higgins (1996) for a detailed study of the sizes and shapes of crystals in the lavas of Santorini.

6. Bubbles of gas in fluid inclusions were described by Sorby in 1858, but Fouqué was the first to separate and analyze them.

7. The two groups of rocks defined here correspond to the major types of magmas, one that precipitates olivine and another that precipitates a silica mineral. In later sections (p. 313 and 332) he notes that hypersthene is found only in the latter type, and on page 331 he states that calcium-poor, iron-rich monoclinic pyroxenes that we would now call pigeonite are found only in rocks without olivine. Fouqué recognized that the second group was more silica rich ("acidic") and elsewhere (p. 316 and note 9) he notes that such magmas have lower temperatures, but he does not seem to have linked these differences to crystal fractionation. Taken as a whole, the rocks follow a normal calc-alkaline trend of silica enrichment and depletion of iron.

8. Fouqué seems to be referring to the polysynthetic albite twinning that is so common in feldspars of the sixth (triclinic) system. The binary twins mentioned in the next sentences are probably Carlsbad twins, which, though characteristic of monoclinic feldspars, are also quite common in plagioclase.

9. See note 17, chapter 5.

10. Fouqué's correctly deduced that mafic magmas were hotter and less viscous than felsic ones. This seems to have been the first recognition that the temperatures of such magmas differ in a systematic way (Harker, 1909, p. 186).

11. I have not translated Fouqué's term "conductibilité ... pour la chaleur" as "thermal conductivity," for he states that the mafic magmas are hotter when they reach the surface; if they had a greater thermal conductivity the reverse would be true. Instead, he seems to mean that, being hotter, they have a great capacity for retaining and carrying heat.

12. Although these authors recognized the compositional continuity of the plagioclase feldspars, they considered intermediate compositions mechanical mixtures of the two end-members, anorthite and albite.

13. Although they recognized the sympathetic variations of Ca, Na, Al and Si, none of the mineralogists studying the plagioclases seems to have viewed them in terms of paired substitution of Na and Si for Ca and Al.

14. As written, this sentence is misleading, because Fouqué uses the term "atom" for a unit of albite or anorthite in a "molecule" of plagioclase when he seems to be thinking of something like our present concept of a unit cell. In the following sentences, I have translated "atom" as "part".

15. Fouqué was clearly troubled by these conflicting lines of evidence. It was not until he and Michel-Lévy completed their study of the feldspars about ten years later that he arrived at the modern view of a solid solution series.

16. This Ca-poor, Fe-rich pyroxene is clearly pigeonite, but the mineral was not recognized as a distinct pyroxene until 1900, when A. N. Winchell described and named it in his doctoral thesis at the University of Paris.

17. Fouqué was not the first person to analyze the rocks of Santorini. Starting with Abich in 1841, almost every geologist that came to the island offered new information on the composition of the rocks. In the following paragraph, Fouqué notes that none of the minerals contain as much silica as the total rock and concludes that the excess silica must therefore be in the groundmass. It is curious that he does not follow this reasoning one step further and state that removal of crystals would cause the remaining liquid to evolve to more silica-rich compositions. A few paragraphs later, he notes that iron decreases as silica increases, as Bowen would argue in his debate with Fenner fifty years later, but he is puzzled by the greater abundance of coarse oxide grains in the iron-poor rocks without recognizing the significance of his observation that these are phenocrysts and not part of the groundmass.

18. In the original text, this sample is labeled no. 26, but in the table at the end of the chapter, the dike with this number is said to be labradorite-bearing. In his personal copy of the monograph, H. S. Washington, who later studied the same rocks, changed the number to 27. (See note 22 below.)

19. Johannsen (1939, vol. 3, p. 347) expressed doubt that the plagioclase in this rock could be anorthite, because the analysis of the rock shows substantial amounts of sodium. Huijsmans (1985, p. 101-107), who studied the plagioclases in lavas of this same section, reported bimodal populations and stated that many of the crystals show resorption and other evidence of reaction. He found plagioclases with as much as 88 percent anorthite. These calcium-rich feldspars could not have crystallized from a melt of the composition Fouqué reports, but they could be xenocrysts.

20. Charles Vélain (1875, 1878) was working on the lavas of Piton de la Fournaise about the same time as Fouqué was working on Santorini.

21. Fouqué considered quartz more diagnostic than tridymite, because, as he notes on page 314, he found that much of the latter is secondary.

22. Washington (1897, p. 368) mentions lavas of this kind at Santorini and coined the name santorinite for anorthite-bearing rocks with more than 60 percent SiO_2. Such rocks have been found in other island arcs, notably the Antilles and Japan.

Chapter 8. Petrographic Study of the Rocks in the Southwestern Part of Thera

1. This brief statement shows that Fouqué saw the effects of crystallization on the composition of the remaining liquid but did not seem to recognize its significance as a mechanism of differentiation.

2. See note 17, chapter 5.

3. Note that in his search for an explanation for these irregular compositions Fouqué was willing to call for "mixtures" of different minerals, even to the extent of mixing monoclinic and triclinic feldspars. The analysis seems to show too little lime and too much silica and alumina, indicating that, as Fouqué suspected, the sample was affected by the hydrothermal alteration that he says affected this region.

4. Thoulet was an analyst who worked with Fouqué. See foot note on page 365. In 1879, he published a brief note describing the heavy liquids he used for mineral separations.

5. It was long thought that microcline was a mechanically deformed variety of orthoclase.

6. Fouqué referred to these tuffs as "trass." (See note 15, Chap. 6.) In the following discussion, the term is translated as "pumiceous tuff."

7. This statement indicates that Fouqué recognized that the structure of feldspars is reordered on going from a high- to a low-temperature state.

8. Alfred Wilhelm Stelzner (1840-1895) is best known for recognizing melilite basalts as distinctive igneous rocks.

9. Hyalite is a clear, colorless form of opal.

10. In the original text, this mineral is referred to as "mesotype," but this must be an error. Fouqué probably meant mesolite, the name used for a type of zeolites intermediate between natrolite and scolecite.

11. Pierre A. Cordier (1777-1862) gave the name alunite to hydrothermally altered, feldspathic igneous rocks containing large amounts of alum, but today the name is used for a hydrated sulfate with the composition $KAl_3(SO_4)_2(OH)_6$.

12. The name rhyolite was coined by von Richthofen (1860, p. 156) for a volcanic rock equivalent to granite.

13. Roth (1861, p. xxxiv) coined the name liparite for rocks of Lipari that are essentially the same as those to which von Richthofen (1860) had just given the name rhyolite.

14. According to Johannsen (1932), the name dacite was proposed by Guido Stache (in F. von Hauer and G. Stache, 1863, Geologie Siebenbürgens, Vienna, p. 71-72) for a volcanic equivalent of quartz diorite. The characteristic feldspar was said to be oligoclase. If a potassium feldspar were present, the rock would be a trachyte.

15. Fouqué (1890) later identified this mineral as anhydrite. It is probably derived from Neogene marine sediments beneath Santorini.

16. The name reissite was coined by von Fritsch in 1870 for a zeolite similar to epistilbite but with small amounts of potassium and sodium.

17. Fouqué lists diallage as a component distinct from augite. In this instance, it may be uralite, an altered form of augite.

18. This must be what is referred to today as graphic quartz.

19. The name pyroxenite was coined independently in the same year, 1857, by H. Coquand (Traité des roches, Paris, pp. 114-118) and F. Senft (Die Feldspaten, Breslau, pp. 42 and 150).

20. This is probably uralite, an amphibole formed by low-grade metamorphism of Ca-rich pyroxenes.

Chapter 9. Considerations on the Origin of the Ancient Parts of Santorini (Thera, Therasia, and Asrponisi)

1. Green (1992) considers a passage in Hesiod's Theogony an eye-witness account of the

Minoan eruption, but this interpretation is not widely shared.

2. See note 10, chapter 6, on the deformation and uplift of the metamorphic series.

3. See Pichler and Kussmaul (1980), Friedrich (1994), and Druitt et al. (in press) for summaries of more recent studies of the early submarine volcanism in the Akrotiri region.

4. The age of the fossils is placed by Seidenkrantz (1989) at about 2 millon years, certainly not less than 1.6 million. The beds in which they are found contain volcanic debris eroded from a nearby emergent source, so one must conclude that the volcanic island already existed at that time (Friedrich, 1994).

5. The amount of uplift of the fossiliferous beds is estimated to have been about 200 meters (Seidenkrantz, 1989; Friedrich, 1994).

6. See pages 146 to 152 for an account and discussion of the appearance of White Island.

7. Regional uplift occurred during the early and middle Cenozoic (see note 10, chapter 6). The metamorphic rocks of Santorini were already above sea level in Pliocene time, and subaerial volcanism began about 3 million years ago. Friedrich (1994) has recently provided an excellent summary of these events.

8. The arcuate ridge known as Somma is a remnant of the rim of the caldera formed by the A.D. 79 eruption of Vesuvius. The name is now used for any ridge of this kind.

9. Fouqué is probably referring to the Cape Riva ignimbrite. It is interesting to see how he attempted to explain these rocks at a time when their true origin was not understood.

10. The names euphotide and ophite refer to gabbro and dolerite.

11. The depression in the central part of Santorini before the Minoan eruption was more than a crater; it is now thought to have been a caldera slightly smaller than the present one (Pichler and Friedrich, 1980; Heiken and McCoy, 1984; Friedrich et al., 1988; Druitt and Francaviglia, 1990).

12. Christian Leopold von Buch (1774-1853) is best known for his role in disproving the Neptunists theories of his professor at Freiberg, Abraham Gottlob Werner (1749-1817). Although his reputation was somewhat tarnished by his "craters of elevation" theory (von Buch, 1802, 1818), he was widely respected as one of the great geologists of the 19th century. The works of Prévost, Poulett Scrope, Darwin, Virlet d'Aoust, and Lyell mentioned in the following sentence can be found in the list of references.

13. Elie de Beaumont (1836).

14. Von Buch (1818).

15. George Poulett Scrope (1797-1876), one of the leading volcanologists of his time, is best known for his study of the volcanoes of the Auvergne (1825). His work (1859, 1862) did much to discredit the theory of craters of elevation.

16. Charles Lyell (1797-1875) was a geologist of wide-ranging authority. His classic Principles of Geology, published in 1830, was the first comprehensive book on the earth sciences. Several versions of his paper refuting the craters of elevation theory were published, including one in English (Lyell, 1858) and another in German (Lyell, 1859).

17. Reiss and Stübel (1868).

18. The volume of ejecta from Cosiguina (which is not in El Salvador but in northeastern Nicaragua) was less than 10 cubic kilometers, but the crater is indeed a true caldera (Williams, 1952).

19. Although the eruption of Papandajan in 1772 resulted in collapse of the northeastern part of the volcano, it did not produce a distinct caldera (Neumann van Padang, 1929, 1951).

20. The "crater" produced by the 1812 eruption of Tambora is generally considered a caldera (Neumann van Padang, 1951).

21. Von Fritsch (1871).

22. Pumice has been found on many Aegean islands, including Nios, Kos, Sikinos, Rhodes, Trianda, as well as in western Turkey (Keller, 1980, 1981; Sullivan, 1988; Pyle, 1990). Little if any Minoan pumice has been found on Crete (Pichler and Schiering, 1977).

23. The modern name of the Sandwich Islands is Hawaii. During periods when a lava lake has been present in the pit of Halamaumau, the level of the magma has been observed to rise and fall by tens or even hundreds of meters in a matter of days.

24. Vogelsang (1867) proposed that the dearth of older rocks in the eruptive debris was the result of large-scale melting by the fresh magma. Fouqué saw that solid rocks can be in thermal equilibrium with a liquid of about the same composition and that extra heat is required to melt them.

Summary

1. This is one of the few cases in which true flames have been recorded in volcanic vents; the combustion seems to have resulted from a combination of unusually large amounts of hydrogen with atmospheric oxygen.

2. It is doubtful whether significant amounts of hydrogen can be formed by dissociation of water at the temperatures observed in volcanic eruptions, but it could result from oxidation of iron or other reduced components in sediments under the volcano.

3. More recent work has confirmed that most of the ammonia and methane in volcanic gases is derived from low-temperature organic sources when meteoric water circulates through sediments (Robert Symonds, pers. comm., 1995)

4. This is one of the first reported applications of spectrographic techniques to the analysis of volcanic gases.

5. Fulgurites are columns of glassy rock formed when lightning strikes soil or loose sand and fuses it.

6. Prior to this, it was not realized that high-temperature silicates can be deposited from a vapor phase.

7. More recent work (Nicholls, 1971b) confirms this interpretation.

8. See note 17, Chapter 5.

9. Fouqué pioneered the development of magnetic separation.

10. This is the first time that a compositional distinction was noted between the large crystals that grow in magmas prior to eruption and those of the groundmass that crystallized during rapid cooling at or near the surface. The term phenocrysts, introduced ten years later by Iddings (1889), is used in this translation wherever this was clearly the meaning of Fouqué's term large crystals.

11. In showing that the compositions of groundmass crystals are more evolved than those of phenocrysts, Fouqué demonstrated that the latter had grown from a liquid of more primitive composition, which could have been either the original magma in which they are now found or a different source from which they were mechanically separated. It is not clear, however, whether he appreciated the full significance of this difference.

12. Fouqué's observation that the last remaining liquid had the composition of alkali feldspar and silica seems to be the first recognition of a "granitic" residuum in the groundmass

of igneous rocks. Fouqué's use of the term amorphous material for glass is retained in the translation.

13. In 1858, Sorby described small bubbles of liquid and gas in crystals at a time when many influential geologists believed that igneous rocks were deposited from the sea and that the fluids in these bubbles would have the composition of seawater.

14. More recent work has shown that the island had only a few trees, mainly tamarisk, olive, and dwarf palm, which were concentrated in low areas near stream beds where they had water and shelter from the wind.

15. The use of the petrographic microscope to study pottery was one of Fouqué's most important contributions to archaeology.

16. The valley deduced by Fouqué is now thought to have been an ancestral caldera (Pichler and Friedrich, 1980; Heiken and McCoy, 1984; Friedrich et al., 1988; Druitt and Francaviglia, 1990).

17. In 1864, Tschermak proposed a classification of feldspars that treated the six plagioclases as separate minerals rather than a continuous isomorphic series. He thought that the composition of andesine had a very narrow range occupying only a small part of the compositional gap between oligoclase and labradorite. By demonstrating an unbroken range of Na-Ca and Si-Al ratios, Fouqué confirmed the view of Hunt (1854, 1855) and others that the plagioclases form a continuous compositional spectrum, but he thought the intermediate compositions were mechanical mixtures of two isomorphic end-members, albite and anorthite.

18. While many of these coarse-grained xenoliths are derived from the underlying metamorphic rocks, recent work has shown that some are products of deep-seated crystallization of the Santorini magma (Nicholls, 1971a).

19. Trass, a pumiceous tuff resembling pozzolan, is used in the production of hydraulic cement.

20. The "nummilitic" age corresponds to what would now be called Paleocene. Fouqué's rejection of age as a basis for classifying igneous rocks was a break with the standard practice of his day.

21. The terms "acid" and "basic", widely used at that time to distinguish between felsic, silica-rich and mafic, silica-poor rocks, is no longer favored. The more fundamental distinction Fouqué made between rocks with olivine and those with one or more silica minerals is the basis of the modern division based on saturation or under-saturation with silica.

22. The term micropegmatite used here for intergrowths of quartz and alkali feldspar has been discarded in favor of granophyre.

23. The dark-colored form of augite, diallage, is distinguished by conspicuous parting. At the time this was written, it was thought to be a distinct mineral.

24. The "valley in the southwestern part of the island" is now known to have been a large bay similar to the present caldera. As already pointed out in note 14, the island had few trees.

25. Although Fouqué was not the first geologist to propose collapse as the main mode of formation of calderas, he was the first to provide convincing geological evidence to discredit the alternative theory of "craters of elevation."

References

Abich, H., 1841, *Geologische Beobachtungen über die vulkanischen Erscheinungen und Bildungen in Unter- un Mittel Italien*. Braunschweig, 131 + xi pp.

Akylas, V., 1925, *Volcanoes and Thera Island*, Athens (in Greek)

Aiken, M. J., 1988, The Minoan eruption of Thera, Santorini: A reassessment of the radiocarbon dates. In: R. E. Jones and H. W. Catling, eds., *New Aspects of Archeological Science in Greece (British School at Athens, Occasional Paper 3 of the Fitch Laboratory)*, 19-24.

Atkinson, T., R. Bosanquet, C. Edgar, A. Evans, D. Hogarth, D. McKenzie, C. Smith, and F. Welch, 1904, *Excavations at Phylakopi in Melos*, London.

Barber, E., 1991, *Prehistoric Textiles: The Development of Cloth in the Neolithic and Bronze Ages with Special Reference to the Aegean*, Princeton.

Betancourt, P. P., 1987, Dating the Aegean Late Bronze Age with radio-carbon, *Archaeometry*, **29**: 45-49.

Betancourt, P. P., 1990, High chronology or low chronology: The archaeological evidence. In: D. A. Hardy, J. Keller, V. P. Galanopoulos, N. C. Flemming, and T. H. Druitt, eds., *Thera and the Aegean World III*, **3**: 19-23.

Betancourt, P. P., and H. N. Michael, 1987, Dating the Aegean Late Bronze Age with Radiocarbon: Addendum. *Archeometry*, **29**: 212-213.

Blitzer, H., 1992, The chipped stone, ground stone, and worked bone industries. In: W. A. McDonald and N. Wilkie, eds., *Excavations at Nichoria in Southwestern Greece, Vol. II: The Bronze Age Occupation*, Minneapolis, 712-756.

Bond, A., and R. S. J. Sparks, 1976, The Minoan eruption of Santorini, Greece. *Jour. Geol. Soc. London*, **132**: 1-16.

Boricky, E., 1877, *Elemente einer neuen chemisch-mikroskopischen Mineral- und Gesteins-analyse*. Archiv d. naturw. Landesdurchforschung von Böhmen. III, 5.

Bory de Saint-Vincent, M., 1833, *Expedition Scientifique de Morée, section des sciences physiques, II.2*, Strasbourg.

Broodbank, C., and T. Strasser, 1991, Migrant farmer and the Neolithic colonization of Crete. *Antiquity*, **65**: 233-245.

Buch, L. von, 1802, *Geognostische Beobachtungen auf Reisen durch Deutschland und Italien*, vol. 1, Berlin.

Buch, L. von, 1818, Über die Zusammensetzung der basaltischen Inseln und über Erhebungskrater. *Abhandl. Akad. Wiss. Berlin*, 1818-1819.

Budetta, G., D. Condarelli, M. Fytikas, N. Kolios, G. Pascale, A. Rapolla, and E. Pinna, 1984, Geophysical prospecting on the Santorini islands. *Bull. Volc.* **47**: 447-466.

Cherry, J., and R. Torrence, 1984, The typology and chronology of chipped stone assemblages in the Prehistoric Cyclades. In: R. L. N. Barber and J. A. MacGillivray, eds., *The Prehistoric Cyclades*, Edinburgh.

Coldstream, J. N., 1982, Greeks and Phoenicians in the Aegean. In: H. Niemeyer, ed., *Phönizier im Westen*, Mainz.

Dana, J. D, 1890, *Characteristics of Volcanoes*, New York, 399 pp.

Darwin, C., 1844, *Geological Observations on the Volcanic Islands Visited During the Voyage of H. M. S. Beagle*, London.

Davidson, D., 1978, Aegean soils during the second millenium BC, with reference to Thera. In: C. Doumas, ed., *Thera and the Aegean World I*, 726-739.

Davis, E. N., and C. Bastas, 1980, Petrology and geochemistry of the metamorphic system of Santorini. In: C. Doumas, ed., *Thera and the Aegean World II*, 61-79.

Davis, J. L., 1992, Review of Aegean prehistory, I: The islands of the Aegean. *Amer. Jour. Arch.*, **96**: 699-736.

Davis, J. L., and J. F. Cherry, 1990, Spatial and temporal uniformitarianism in Late Cycladic I: Perspectives from Kea and Miulos on the prehistory of Akrotiri. In: D. A. Hardy, J. Keller, V. P. Galanopoulos, N. C. Flemming, and T. H. Druitt, eds., *Thera and the Aegean World III*, **I**: 185-200.

Doumas, C., 1973, Phtellos: Ysterominoïki Ia pimeniki; egkatastasis para ta Phira Thiras. *Archaiologike Ephemeris*, 161-166.

Doumas, C. 1976, Protokykladiki keramiki apo ta Christiana Thiras. *Archaiologike Ephemeris*, 1-11.

Doumas, C., 1983, *Thera: Pompeii of the Aegean*, Thames and Hudson, London. 168 p.

Druitt, T. H., 1985, Vent evolution and lag breccia foundation during the Cape Riva eruption of Santorini, Greece. *Jour. Geol.*, **93**: 439-454.

Druitt, T. H., and V. Francaviglia, 1992, Caldera formation on Santorini and the physiogeography of the islands in the Late Bronze Age. *Bull. Volc.*, **54**: 484-493.

Druitt, T. H., R. A. Mellors, D. M. Pyle, and R. S. J. Sparks, 1989, Explosive volcanism on Santorini, Greece. *Geol. Mag.*, **126**: 95-126.

Druitt, T. H., L. Edwards, M. Davies, M. A. Lanphere, R. Mellors, D. Pyle, R. S. J. Sparks, and B. Barreiro, in press, Santorini Volcano. *Mem. Geol. Soc. London*

Dufrenoy, P. A., 1834, Sur les terrains volcaniques des environs de Naples, *Mém. pour servir à une description Geol. de France*. **IV**: 227-420

Einarsson, T., 1949, The eruption of Hekla, 1947-1948, Pt. IV, 3: The flowing lava. *Soc. Sci. Islandica*.

Élie de Beaumont, L. 1836. Recherches sur la structure et l'origine du Mont Etna. *Ann. Mines*, **9**: 175-216, 575-630; **10**: 351-370, 507-576.

Élie de Beaumont, L., and P. A. Dufrenoy, 1833, Mémoire sur les groupes du Mont Dore et du Cantal et sur les soulèvement auxquels des montagnes doivent leur relief. *Bull. Soc. Géol. de France*, **3**: 205-274.

Eriksen, U., W. L. Friedrich, B. Buchardt, H. Tauber, and M. S. Thomsen, 1990, The Stronghyle caldera: Geological, palaeontological and stable isotope evidence from radiocarbon dated stromatolites from Santorini. In: D. A. Hardy, ed., *Thera and the Aegean World III*, **2**: 139-150.

Forbes, E., 1844, Report on the mollusca and radiata of the Aegean Sea and on their distribution considered as bearing on geology. *Rept. Mtg. Brit. Assoc. Adv. Sci.*, 1843, 130-193.

Fouqué, F., 1867, Premier rapport sur une mission scientifique à l'île Santorin. *Archives des Missions Scientifiques*, 2e ser., **4**: 224-252.

Fouqué, F., 1879, *Santorin et ses éruptions*. Masson, 440 pp.

Fouqué, F., 1889, Sur le bleu égyptien ou vestorien. *Bull. Soc. Min. Fr.*, **13**: 325.

Fouqué, F., 1890, Revision de quelques mineraux de Santorin. *Bull. Soc. Min.*, **13**: 245-251.

Fouqué, F., and H. Gorceix, 1869, Étude chimique de plusieur des gaz à éléments

combustibles de l'àtalie centrale. *Compt. Rend. Acad. Sci.*, **69**: 946-950.
Fouqué, F., and A. Michel-Lévy, 1882, *Synthèse des minéraux at des roches*. Masson, Paris, 423 pp.
Fouqué, F., and A. Michel-Lévy, 1886a, Mesure de la vitesse de propagation des vibrations dans le sol. *Comptes Rendus Acad. Sci.*, **102**: 237.
Fouqué, F., and A. Michel-Lévy, 1886b, Expériences sur la vitesse de propagation des vibrations dans le sol. *Comptes Rendus Acad. Sci.*, **102**: 1290.
Friedrich, W. L., 1994, *Feuer im Meer*. Spektrum Akademischer Verlag, Heidelberg, 256 pp.
Friedrich, W. L., U. Eriksen, H. Tauber, J. Heinemeir, N. Rud, M. S. Thomsen, and B. Buchardt, 1988, Existence of a water-filled caldera prior to the Minoan eruption of Santorini. *Nat. Wiss.*, **75**: 567-569.
Friedrich, W. L., H. Pichler, and S. Kussmaul, 1977, Quaternary pyroclastics from Santorini, Greece and their significance for the Mediterranean palaeoclimate. *Bull. Geol. Soc. Denmark*, **26**: 27-39.
Friedrich, W. L., M-S. Seidenkrantz, and O. B. Nielsen, in press, A reconstruction of the ring island, natural resources and clay deposits from the Akrotiri excavation. *Jour. Geol. Soc. London*
Fritsch, K. von, 1871, Geologische Beschreibung des Ringgebirges von Santorin. *Zeit. d. geol. Gesll.*, **23**: 125-209.
Furumark, A., 1941, *Mycenaen Pottery*, vol. 1: *Analysis and Classification*. Stockholm.
Fytikas, M., F. Innocenti, P. Manatti, R. Mazzuoli, A. Peccerillo, and L. Villari, 1984, Tertiary to Quaternary evolution of volcanism in the Aegean region. *Geol. Soc. London, Spec. Publ.* 5: 687-699.
Fytikas, M, N. Kolios, and G. Vougioukalakis, 1990, Post-Minoan activity of the Santorini volcano. Volcanic hazard and risk, forecasting possibilities. In: D. A. Hardy, J. Keller, V. P. Galanopoulos, N. C. Flemming, and T. H. Druitt, eds., *Thera and the Aegean World III*, **2**: 183-198.
Gauthier, F., 1973, Field and laboratory studies of the rheology of Mount Etna lava. *Phil. Trans. Rot. Soc. London, A*, **274**: 83-98.
Georgalas, G. C., 1962, *Catalogue of the Active Volcanoes of the World including Solfatara Fields. Part XII Greece*, Int. Assoc. Volc., 40 pp.
Gorceix, H., 1874a, Phénomènes volcaniques de Nisyros. *Compt. Rend. Acad. Sci.*, **78**: 444-446.
Gorceix, H., 1874b, Aperçu géologique sur l'île de Kos. *Compt. Rend. Acad. Sci.*, **78**: 565-568.
Gorceix, H., 1874, Sur l'étude des fumarolles de Nisyros et quelquesuns des produits de l'éruption de 1873. *Compt. Rend. Acad. Sci.*, **78**: 1309-1311.
Gorceix, H., and C. Mamet, 1874, Constructions de l'époque antéhistorique, découvertes à Santorin. *Compt. Rend. Acad. Sci.*, **73**: 476-478.
Greene, M. T., 1992, *Natural Knowledge in Preclassical Antiquity*, Johns Hopkins University Press, 182 pp.
Günther, D., and H. Pichler, 1973, Die Obere and Untere Bimsstein-Folge auf Santorin. *N. Jb. Geol. Paläont. Mh.*, **7**: 394-415.
Hansen, J., 1991, *The Palaoethnobotony of Franchthi Cave*, Bloomington
Harker, A., 1909, *The Natural History of Igneous Rocks*, London, 384 pp.

Heiken, G., and F. McCoy Jr., 1984, Caldera development during the Minoan eruption, Thira, Cyclades, Greece. *Jour. Geoph. Res.*, **89**: 8441-8462.

Heiken, G., and F. McCoy Jr., 1990, Precursory activity to the Minoan eruption. In: D. A. Hardy, J. Keller, V. P. Galanopoulos, N. C. Flemming, and T. H. Druitt, eds., *Thera and the Aegean World III*, **2**: 79-87.

Higgins, M. D., 1996, Magma dynamics beneath Kameni volcano, Thera, Greece, as revealed by crystal size and shape measurements. *Jour. Volc. Geoth. Res.*, **70**: 37-48.

Hiller von Gaertringen, F., 1899-1904, *Thera. Untersuchungen, Vermessungen un Ausgrabungen in den Jahren 1895-1902*. Berlin, Verlag G. Reimer, 4 vols.

Hooker, M., 1965, The origin of the volcanological concept *nuée ardente*. *ISI*, **56**: 401-407.

Hoskins, H., and H. E. Edgerton, 1971, Normal-incidence 5 and 6 kHz sonar profiles delineate sea floor rock and sediment types of Thera, Greece. *Acta Int. Sci. Cong. Volcano Thera, Athens*, **1**: 325-365.

Hubberten, H-W., M. Bruns, M. Calamiotou, C. Apotolakis, S., Filippakis, and A. Grimanis, 1990, Radiocarbon dates from the Akrotiri excavations. In: D. A. Hardy, J. Keller, V. P. Galanopoulos, N. C. Flemming, and T. H. Druitt, eds., *Thera and the Aegean World III*, **3**: 179-187.

Huijsmans, J. P. P., 1985, Calc-alkaline lavas from the volcanic complex of Santorini, Aegean Sea, Greece. *Geol. Ultraiectina, Rijksuniversiteit to Utrecht*, 41, 316 pp.

Huijsmans, J. P. P., M. Barton, and V. J. M. Salter, 1988, Geochemistry and evolution of the calc-alkaline volcanic complex of Santorini, Aegean Sea, Greece. *Jour. Volc. Geoth. Res.* **34**: 283-306.

Humbolt, A. von, 1824, *Uber den Bau und die Wirkung der Vulkane*, Berlin.

Hunt, T. S., 1854, Illustrations of chemical homology. *Amer. Jour. Sci.*, **18**: 269-271.

Hunt, T. S., 1855, Examinations of some feldspathic rocks. *Phil. Mag.*, ser. 4, **9**: 345-363.

Iddings, J. P., 1889, The crystallization of igneous rocks, *Bull. Phil. Soc. Washington*, **11**: 65-113.

Johannsen, A., 1931-1939, *A Descriptive Petrography of the Igneous Rocks*. 4 vols., 2nd ed., University of Chicago Press.

Johnson, S. J., H. B. Clausen, W. Dansgaard, K. Fuhrer, N. Gundestrup, C. U. Hammer, P. Iversen, J. Jouzel, B. Stauffer, and J. P. Steffensen, 1992, Irregular glacial interstadials recorded in a new Greenland ice core. *Nature*, **359**: 311-313.

Jones, G., 1987, Agricultural practices in Greek prehistory. *Ann. Brit. Sch. Athens*, **82**: 115-123.

Keller, J., 1980, Prehistoric pumice tephra on Aegean islands. In: C. G. Doumas, ed., *Thera and the Aegean World II*, 49-56.

Keller, J., 1981, Quaternary tephrachronology in the Mediterranean region. In: S. Self and R. S. J. Sparks, eds., NATO Advanced Studies Institute, ser. C, **75**: 227-244.

Keller, J., and D. Ninkovich, 1972, Tephra-Lagen in der Ägäis. *Z. Deutsch. Geol. Ges.*, **123**: 579-588.

Keller, J., Th. Rehren, and E. Stadlbauer, 1990, Explosive volcanism in the Hellenic Arc: A summary and review. In: D. A. Hardy, J. Keller, V. P. Galanopoulos, N. C. Flemming, and T. H. Druitt, eds., *Thera and the Aegean World III*, **3**: 13-26.

Kircher, A., 1665, *Mundus subterraneus*, Book IV, Amsterdam

Kraft, M., 1991, *Les Feux de la Terre - Histoires de Volcans*. Gallimard, Paris, 208 pp.

Kténas, C. A., 1926, L'éruption du volcan des Kamenis (Santorin) en 1925. I. *Bull. Volc.*, **3**: 3-64.

Kténas, C. A., 1927, L'éruption du volcan des Kamenis (Santorin) en 1925. II. *Bull. Volc.*, **4**: 7-46.

Kténas, C. A., and P. Kokkoros, 1929, Le dome parasitaire de 1928 et l'évolution du volcan des Kamenis (Santorin). *Bull. Volc.*, **5**: 87-97.

Kuniholm, P. I., B. Kromer, S. W. Manning, M. Newton, C. E. Latini, and M. J. Bruce, 1996, Anatolian tree rings and the absolute chronology of the eastern Mediterranean. 2220-718 B. C. *Nature*, **381**: 780-783.

Lacroix, A., 1896, Sur la découverte d'un gisement d'empreintes végétales dans les cendres volcaniques anciennes de l'île de Phira (Santorin). *Compt. Rend. Acad. Sci.*, 123: 656-659.

Lacroix, A., 1904, *La Montagne Pelée et ses éruptions*. Masson, Paris, 662 pp.

LaMarche, V. C., and K. K. Hirschboeck, 1984, Frost rings in trees as records of major volcanic eruptions. *Nature*, **307**: 121-126.

Lambros, S., 1885, Unpublished narration on the eruption near the Thera Island in 1650. *Bull. Hist. Ethn. Soc. Greece*, 2 vols. Athens.

Lenormant, F., 1866, Découverte de constructions antéhistorique dans l'île de Therasia, *Revue Archéologique*, 2e série, **14**: 423-432.

Leycester, E. M., 1850, Some accounts of the volcanic group of Santorini or Thera, once called Calliste, or the Most Beautiful. *Jour. Royal Geograph. Soc.*, **20**: 1-38

Liatsikas, N., 1942, Der polyzentrische Ausbruch des Santorin Vulkans 1939 -1941. *Prakt. Acad. Athènes*, **17**: 30-36.

Liatsikas, N., and G. Georgalas, 1928, Strukturelle Unterschiede der primären und sekondären Lavastrom des Dafni Vulkans. *Centr. Bl. f. Min. Geol. u. Pal. Jg. 1928*, Abt. B. Nr. **6**, 337-342.

Limbrey, S., 1990, Soil studies at Akrotiri. In: *Thera and the Aegean World III*, **2**: 377-382.

Lyell, C., 1858, On the structure of lavas which have consolidated on steep slopes; with remarks on the mode of origin of Mount Etna and on the theory of "Craters of Elevation.", *Phil. Trans. Royal Soc., London*, 703-786.

Lyell, C., 1859, Über die auf steilgeneigter Unterlage erstarrten Laven des Ätna und überdie Erhebungskratere. Mit Zusätzen und Änderungen des Verfassers übertragen von Herrn Roth, aus Philos. Transact. for 1858, 148. Part 2. *Zeitschr. d. Deutsch. Geol. Ges.*, **11**: 149-250.

Mamet, C., 1874, *De Insula Thera*, Insulis (E. Thorin), Paris.

Mamet, C., and H. Gorceix, 1870, Recherches et fouilles faites à Théra (Santorin). *Bull. Ecole Fran. Athènes*, **9**: 183-191; **10**: 199-203.

Manning, S., 1988, The Bronze Age eruption of Thera: Absolute dating, Aegean chronology and Mediterranean cultural relations. *Jour. Medit. Arch.*, **1**: 17-82.

Marinatos, S., 1939, The volcanic destruction of Crete. *Antiquity*, **13**: 429-439.

Marthari, M., 1982, Anaskafi sti thesi Phtellos Thiras (excavations at Phtellos on Thera during 1980). *Archaiologika analekta ex Athenon*, **15**: 86-100.

McBirney, A. R., 1990, An historical note on the origin of calderas. *Jour. Volc. Geoth. Res.*, **42**: 303-306.

Mellis, O., 1954, Volcanic ash horizons in deep-sea sediments from the eastern Mediterranean.

Deep-Sea Res., **2**: 89-92.

Michel-Lévy, M., 1894, *Etudes sur le détermination des feldspaths dans les plaques minces au point de vue de la classification des roches.* Librairie Polytechnique, Baudry & Cie, Paris, 71 pp.

Michel-Lévy, M., 1896, *Etudes sur le détermination des feldspaths dans les plaques minces (deuxième fascicule) sur l'éclairement commun des plagioclases zonés.* Librairie Polytechnique, Baudry & Cie, Paris, 73-109.

Michel-Lévy, M., 1905, Notice sur F. Fouqué. *Bull. Soc. Fran. Mineral.*, **28**: 1-19.

Morgan, L., 1988, *The miniature wall paintings of Thera. A study of Aegean culture and iconography.* Cambridge Univ. Press.

Murad, E., and H. W. Hubberten, 1975, Sulfide mineralization in phyllites from the island of Thera, Santorini archipelago, Greece. *N. Jb. Miner. Mh.*, **7**: 300-308.

Neumann van Padang, M., 1929, *Goenoeng Papandajan*, Exc. Guide Fourth Pac. Sci. Cong., Java.

Neumann van Padang, M., 1936, Die Geschichte des Vulkanismus Santorins von ihren Anfängen bis zum zerstöenden Bimssteinbruch um die Mitte des 2. Jahrhausend v. Chr. In: H. Reck, *Santorin - Der Werdegang eines Inselvulkans und sein Ausbruch 1925-1928.* **1**: 1 -72.

Neumann van Padang, M., 1951, *Catalogue of the Active Volcanoes of the World, Part I Indonesia*, Internat. Volc. Assoc., Naples, 271 pp.

Nicholls, I. A., 1971a, Petrology of Santorini volcano, Greece. *Jour Petrol.*, **12**: 67-119.

Nicholls, I. A., 1971b, Calcareous inclusions in lavas and agglomerates of Santorini volcano. *Contr. Min. Petrol.*, **30**: 261-276.

Nicholls, I. A., 1971c, Santorini volcano, Greece - tectonic and petrochemical relationships with volcanics of the Aegean region. *Tectonophysics*, **11**: 337-385.

Nicholls, I. A., 1978, Primary basaltic magmas for the pre-caldera volcanic rocks of Santorini. in: C. Doumas, ed., *Thera and the Aegean World I*, 109-120.

Ninkovitch, D., and B. C. Heezen, 1965, Santorini tephra. *Proc. 17th Symp. Colston Res. Soc.*, London, 413-453.

Palyvou, C., 1988, *Akrotiri Thiras: Oikodimikai technai kai morphologika stoicheia stin Ysterokykadiki architektoniki (Akrotiri, Thera: Building Techniques and Morphology in Late Cycladic Architecture).* Athens Polytechnic University.

Palyvou, C., 1990a, Architectural design at Late Cycladic Akrotiri. In: D. A. Hardy, C. Doumas, J. A. Sakellarakis, and P. M. Warren, eds., *Thera and the Aegean World III*, **1**: 44-55.

Palyvou, C., 1990b, Observations sur 85 fenêtres du Cycladique Récent à Théra, *L'Habitat égéen préhistorique*, (Athens), 190-139.

Papastamatiou, J., 1958, Sur l'age des calcaire cristillins de l'île de Thera (Santorin). *Bull. Geol. Soc. Greece*, **3**: 104-113.

Pègues, M. L'Abbé, 1842, *Histoire et Phénomènes du volcan et des iles volcanoiques de Santorin.* Paris (Imprimerie Royale)

Petruso, K. M., 1992, *Keos VIII: The Balance Weights*, Mainz.

Phillipson, A., 1896, Die Inselgruppe von Thera. In: H. von Gaertringen, ed., *Thera Untersuchungen, Vermessungen und Ausgrabungen in den jahren 1895-1902*, vol. **1**, Berlin.

Pichler, H., 1973, "Base surge" - Ablagerungen auf Santorin. *Naturwissen-schaften*, **60**: 198.

Pichler, H. and W. L. Friedrich, 1976, Radiocarbon ages of Santorini volcano. *Nature*, **262**: 373-374.
Pichler, H. and W. L. Friedrich, 1980, Mechanism of the Minoan Eruption of Santorini. In: C. G. Doumas, ed., *Thera and the Aegean World II*, 15-30.
Pichler, H., and S. Kussmaul, 1972, The calc-alkaline volcanic rocks of the Santorini group (Aegean Sea, Greece). *N. Jb. Miner. Abh.*, **116**: 268-307.
Pichler, H., and S. Kussmaul, 1980, Comments on the geological map of the Santorini islands (with coloured geological map 1:20,000 and two profiles). In: C. G. Doumas, ed., *Thera and the Aegean World II*, 414-427.
Pichler, H., and W. Schiering, 1977, The Thera eruption and Late Minoan-IB destruction of Crete. *Nature*, **267**: 819-822.
Pinkerton, H., and R. S. J. Sparks, 1978, Field measurements of the rheology of flowing lava. *Nature*, **276**: 383-385.
Poulett Scrope, J., 1859, On the mode of formation of volcanic cones and craters. *Quart. Jour. Geol. Soc.*, **15**: 505-549.
Poulett Scrope, J., 1862, *Volcanoes. The Character of Their Phenomena, Their Share in the Structure and Composition of the Surface of the Globe, and Their Relation to its Internal Forces*, 2nd ed., London, 490 pp.
Prévost, C., 1855, Consideration générales et questions sur les éruptions volcaniques. *Compt. Rend. Acad. Sci.*, **44**: 866-876.
Pyle, D. M., 1990, New estimates for the volume of the Minoan eruption. In: D. A. Hardy, J. Keller, V. P. Galanopoulos, N. C. Flemming, and T. H. Druitt, eds., *Thera and the Aegean World III*, **2**: 113-121.
Quenstedt, W., 1936, Tertiäre und quartäre Mollusken von Santorin. In: H. Reck, ed., *Santorin - der Werdegang eines Inselvulkans und sein Ausbruch 1925-1928*. Verlag Von Dietrich Reimer, Berlin, 3 vols.
Rames, J-B., 1866, *Etudes sur les volcans*, Paris.
Reck, H., ed., 1936, *Santorin - Der Werdegang eines Inselvulkans und sein Ausbruch 1925-1928*. Verlag Von Dietrich Reimer, Berlin, 3 vols.
Rehak, P., 1996, Aegean breechcloths, kilts, and the Keftiu paintings. *Amer. Jour. Arch.*, **100**: 35-52.
Reiss, W., and A. Stübel, 1868, *Geschichte und Beschreibung der vulkanischen Ausbrüche bei Santorin von der ältesten Zeit bis auf die Gegenwart*. Heidelberg.
Renaudin, L., 1922, Vases de Thera. *Bulletin de Correspondence Hellénique*, **46**: 113-159.
Renfrew, C., J. Cann, and J. Dixon, 1965, Obsidian in the Aegean. *Ann. Brit. Sch. Athens*, **60**: 225-247.
Renfrew, C., J. Cann, and J. Dixon, 1968, Further analysis of Near Eastern obsidians. *Proc. Prehist. Soc.*, **24**: 319-331.
Richthofen, F. von, 1860, Studien aus den ungarisch-siebenbürgischen Trachytgebirgen. *Jahrb. K. K. Geol. Reichsanst.*, Wien, **11**: 153-278.
Reiss, W., and A. Stübel, 1868, *Geschichte und Beschreibung der vulkanischen Ausbrüche bei Santorin von der ältesten Zeit bis auf die Gegenwart*. Heidelberg.
Ross, L., 1840, *Reisen auf den griechischen Inseln des ägäischen Meeres*, J. G. Cotta, Stuttgart and Tübingen. Reprinted in 1912 by Max Niemeyer, Neudruck Halle.
Roth, J., 1861, *Gesteinsanalysen*, Berlin

Runnels, C. N., 1981, *A Diachronic Study and Economic Analysis of Millstones from Argolid, Greece*. Indiana University Press.

Runnels, C. N., 1982, Flaked stone artifacts in Greece during the historical period. *Jour. Field Arch.*, **9**: 363-373.

Sainte-Claire Deville, C., 1857, Memoire sur les emanations volcaniques. *Bull. Soc. Géol. France*, ser. 2, **14**: 254-279.

Sakellarakis, E. and J. A., 1984, The Keftiu and Minoan Thalassocracy, *The Minoan Thalassocracy*, Stockholm.

Seebach, K. von, 1867, *Der Vulkan von Santorin*, in: *Sammlung gemein-verständlicher wissenschaftlicher Voträge*, 2nd serie, Heft 38, Berlin.

Seebach, K. von, 1868, *Ueber den Vulkan von Santorin und die Eruption von 1866*. Gottingen, in: *Abhandlungen der Physicalischen Classe der Koniglichen Gesellschaft der Wissenschaften zu Göttingen*, **13**: 3-81.

Seidenkrantz, M. S., 1989, Foraminiferfauna fra Akrotirihalven. *Georapporter*, **11**: 22-25.

Shaw, J., 1971, Minoan architecture: Materials and techniques, *Annuario della Scuola Archeologica di Atene e della Missioni italione in Oriente*, **49**, Rome.

Shaw, J., 1977, New evidence for Aegean roof construction from Bronze Age Thera. *Amer. Jour. Arch.*, **81**: 229-233.

Simkin, T., and R. S. Fiske, 1983, *Krakatau 1883 - The Volcanic Eruption and Its Effects*. Smithsonian Institution Press, Washington, D.C., 464 pp.

Simkin, T., L. Siebert, L. McClelland, D. Bridge, C. Newhall, and J. H. Latter, 1981, *Volcanoes of the World*, Hutchinson Ross, 233 pp.

Skarpelis, N., K. Kyriakopoulos, and I. Villa, 1992, Occurrence and $^{40}Ar/^{39}Ar$ dating of a granite in Thera (Santorini, Greece). *Geol. Rund.* **81**: 729-735.

Skarpelis, N., and A. Liati, 1990, The prevolcanic basement of Thera at Athenios: Metamorphism, plutonism and mineralization. In: D. A. Hardy, J. Keller, V. P. Galanopoulos, N. C. Flemming, and T. H. Druitt, eds., *Thera and the Aegean World III*, **2**: 172-182.

Sorby, H. C., 1858, On the microscopical structure of crystals indicating the origin of minerals and rocks. *Quart. Jour. Geol. Soc.*, **14**: 453-500.

Sparks, S. R. J., and H. Sigurdsson, 1977, Magma mixing: a mechanism for triggering acid explosive eruptions. *Nature*, **167**: 315-318

Sparks, R. S. J. and C. J. N. Wilson, 1990, The Minoan deposits: A review of their characteristics and interpretation. In: D. A. Hardy, J. Keller, V. P. Galanopoulos, N. C. Flemming, and T. H. Druitt, eds., *Thera and the Aegean World III*, **2**: 89-98.

Strange, J., 1980, *Kaphtor, Keftiu*, Leiden

Stübel, A., 1868, *Die supra- und submarinen Gebirge von Santorin in photographischen Nachbildungen der an Ort und Stelle gefertigten Reliefkarten*, Leipzig.

Szabo, J., 1876, Ueber eine neue Methode die Feldspate auch in Gesteinen zu bestimen. *Sajat Kiadasa*, Budapest.

Tataris, A. A., 1963, The Eocene in the semi-metamorphosed basement of Thera Island. *Bull. Geol. Soc. Greece*, **6**: 232-238.

Televantou, C., 1984, Kosmimata apo tin proïstoriki Thira, *Archaiologike Ephemeris*, 14-54.

Thoulet, J., 1879, Note sur un nouveau procédé pour prendre la densité de minéraux en fragments très petits. *Bull. Soc. Mineral. France*, **2**: 189-191.

Tissot, C., 1991, Ferdinand Fouqué, vulcanologue (1823-1904). *Rev. de L'Avranchin et du Pays Granville*, **68**: 497-516.

Tschermak, G. 1864, Chemisch-mineralogische Studien. I. Die Feldspathgruppe. *Sitzb. Akad. Wiss. Wien*, **50**: 566-613.

Torrence, R., 1986a, *Production and Exchange of Stone Tools*, Cambridge.

Torrence, R., 1986b, Chipped Stone. In: J. L. Davis, *Keos V: Ayia Irini, Period V*, Mainz, 90-95.

Vélain, C., 1875, Sur la constitution géologique du massif volcanique de l'Ile de la Réunion. *Compt. Rend. Acad. Sci.*, **80**: 497, 900.

Vélain, C., 1878, Description géologique de la presqu'île d'Aden, d l'île de la Réunion, des îles Saint-Paul et Amsterdam. Thesis presented to the Faculty of Sciences. Paris, p. 49, 211.

Verbeek, R. D. M., 1885, *Krakatau*. Batavia, 546 pp.

Virlet d'Aoust, P. T., 1838, Note sur les volcans de Santorin et de Milo. *Bull. Soc. Géol. France*, **9**: 168-176.

Virlet d'Aoust, P. T., 1866a, Histoire des Kaimenis, ou îles volcaniques nouvelles du golfe de Santorin, dans l'archipel de la Grèce. *Les Mondes*, **11**: 350-357, 476-484.

Virlet d'Aoust, 1866b, Coup d'oeil général sur la topographie et la géologie du Mexique et de l'Amérique Central, *Bull. Soc. Géol. France*, **23**: 14-50.

Vogelsang, H., 1867, *Philosophie der Geologie und mikroskopische Gesteinsstudien*, Bonn.

Vogelsang, H. P., 1872, Ueber die Systematik der Gesteinslehre und die Entheilung gemengten Silikatgesteine. *Zeit. deutsch. geol. Ges.*, **24**: 507-544.

Walker, G. P. L., 1973, Lengths of lava flows. *Phil. Trans. Roy. Soc. Lond.*, A, **274**: 107-118.

Warren, B. E., and D. I. Modell, 1930, Structure of enstatite $MgSiO_3$, *Zeit. f. Krist.*, **75**: 1-14.

Washington, H. S., 1897, Italian petrological sketches. V. Summary & Conclusions. *Jour. Geol.*, **5**: 349-377,

Washington, H. S., 1926a, The eruption of Santorini in 1925. *Jour. Washington Acad. Sci.*, **16**: 1-7.

Washington, H. S., 1926b, The Santorini eruption of 1925. *Bull. Geol. Soc. Amer.*, **37**: 349-384.

Watkins, N. D., R. S. J. Sparks, H.Sigurdsson, T. C. Huang, A. Federman, S. Carey, and D. Ninkovich, 1978, Volume and extent of the Minoan tephra from Santorini volcano: New evidence from deep-sea cores. *Nature*, **271**: 122-126.

Williams, H., 1941, Calderas and their origin. *Univ. Calif. Publ. Geol. Sci.*, **25**: 239-346.

Williams, H., 1952, The great eruption of Cosequina, Nicaragua, in 1835. *Univ. Calif. Publ. Geol. Sci.*, **29**: 21-46.

Zielinski, G. A., P. A. Mayewski, L. D. Meeker, S. Whitlow, M. S. Twickler, M. Morrison, D. A. Meese, A. J. Gow, and R. B. Alley, 1994, Record of volcanism since 7000 B. C. from the GISP2 Greenland ice core and implications for the volcano-climate system. *Science*, **264**: 948-952.

Index

Accretionary lapilli 263, 378, 419
Age of the Minoan eruption 129-130
Akrotiri, amphibole andesites 341, 345, 347, 350-364, 368-371
    description of region 276-286
    discovery of ruins 104
    excavations 104-123
    felsic rocks 344-372
    pumiceous tuffs 362-365
Alafouzos, excavations on Therasia 96, 99
Alignment of vents 186-187
Alteration, effects of 255, 315-316, 357, 380, 387
Alunite 360
Ammianus Marcellinus 3, 5
Ammonia 435
Ammonium chloride 435
Amphibole andesites, see andesites
Analyses, bulk rock 343, 365
    feldspar 197-198, 201, 202, 206, 212, 297, 322, 326, 329, 347, 368, 371
    garnet 209
    glass 296, 355
    olivine 202, 203, 205, 330
    pyroxene 195-196, 202-205, 209, 212, 296, 331
    wollastonite 208
Ancestral island 126, 129, 298, 387-398, 450
Andes, volcanoes of 417
Andesite, amphibole-bearing 341, 344, 347, 350-364, 368-371, 445, 449
Antonio Falconi, account of formation of Monte Nuovo 420
Aphroessa 46, 48-50, 52-53, 56-64, 68-75, 77-79, 81-87, 157-159, 167-172, 180, 187-188, 191, 201, 212, 226-227, 231-232, 309
Apollonius of Rhodes 379
Arkhangelos, rocks of 363, 370, 381, 383, 385-386
Ash, effects on vegetation and animals 30-31, 81, 82, 89
    eruption of 1866-67 233-234
    formation of 81, 234, 439
Aspronisi, description of 291-292
Atmospheric pressure, effect of 50, 88, 179, 189
Aurelius Victor 6
Auvergne, ancient volcanoes 410, 416
Azores 382, 399, 410, 412, 418, 425, 429

Balos 341-343, 353, 373, 378, 383-384, 395
Banco 7-9, 11, 182-184
Barometric pressure, see atmospheric pressure
Bay of Santorini, see also caldera
    form and bathymetry 235-243
    prior to the Minoan eruption 379, 396-398
Blocks in pumice 248, 293, 387, 392, 426, 429
Bones, animal 103, 114, 120-121
    human in ruins 99
Bombs, volcanic, types of 79-80
Boricky 234, 312, 327, 348, 349, 355, 444
Bory Saint-Vincenet 95
Buch, L. von 398-401, 405-409, 411, 413-414, 417-418, 421-424, 451
Buildings, prehistoric, construction of 97-98, 124, 129
Bunsen, R. W. 194, 218-220, 225

Caldera, form of 235-240
    origin of 128-131, 151-152, 398, 425-433, 450-451
    relative age 130
Cantal 415, 416
Cape Akrotiri, rocks of 349, 366-370, 372, 381, 385-386, 395, 423
Cape Alonaki 380, 388
Cape Mavro 371-373, 386
Cape Mavrorachidi 344-345, 395
Cape Tripiti 388, 395
Cape Urcu 407
Carguairazo 407
Christiania 235
Christomanos 37, 84
    discovery of ruins on Therasia 95
Cigalla, Dr. de 36-37, 39-45, 55, 72-73, 83, 87, 90-91, 128
Civilization on Santorini 123-125
Classification of igneous rocks 310, 335-338, 344-345, 361-362, 446-450
Clay used in pottery 106, 113, 125
    sources of 126-127, 273
Cliff sections 249, 257, 260, 270, 277, 287
Columnar jointing 386
Commerce, prehistoric 128
Concretions 378
Cosiguina 407, 427-428

Cotopaxi 417
Craters of elevation, theory of 398-425, 450-451
Crops, prehistoric 128
Crystallization, rate and order of 302, 305-306, 313, 438
Cumulo-volcano, 441
Cyclones 52-53

Dacite, definition of 361
Dana, J. D. 410
Darwin, Charles 398
Daubrée, A. 380
Delesse, A. 319
Des Cloizeaux, A. L. 312, 319, 350
Dikes 137, 250, 252-254, 256-258, 271, 283, 284, 290-291, 298-340, 391, 395, 403-404, 414, 418, 424
    analyses of 333
    chilled margins 304
    compositional variations within 303-305
    numbering 339
    orientations 298, 335, 414, 424
Dion Cassius 6
Discoveries of prehistoric ruins 94-95, 104-105
Dissociation of water 254-255
Doelter 361
Domes, compositions of 417
Dufrénoy, A. 400, 402, 406, 408, 420, 422

Earthquakes
    preceding Minoan eruption 129
    recent 237, 259
Elba 388
Elevations, measurements of 245-246, 269, 276
Élie de Beaumont 2, 398-406, 410-411, 414, 417, 418, 422, 451
Erosion 397, 426
    responsible for calderas 425-426
Eruption
    Minoan 94, 129, 387-388
    of 46 AD 6, 8-9
    of 197 BC 8-9
    of 726 9-11
    of 1457 11-12
    of 1570 or 1573 12, 151
    of 1650 13-21, 30-31 35, 145
    of 1866-67 36-93, 158-174, 189-234, 440-441

    XXth century, see preface
    submarine 247, 263, 379-380, 445
Etna 384, 399, 400, 402-406, 409-410, 412, 414, 415, 421-422, 425
Eusebe 3, 5, 6, 8
Excavations, Akrotiri 104-123
    on Therasia 96-104
Explosive origin of caldera 426-430

Faults 19, 252, 256, 263-264, 273, 385, 417
Feldspars (see also plagioclase)
    analyses 197-198, 201, 202, 206, 212, 322-323, 326, 329, 347, 368, 371
    phenocrysts and groundmass 302, 318, 332-333, 336-337, 443
    properties 312, 327
    twinning 310, 312, 319, 341-342, 344-348, 351-353, 360-363, 373, 375
Filters, 112, 113, 122
Flames, 3-4, 6, 14-19, 23-29, 31-32, 38, 40-46, 48-50, 53, 59, 62-64, 66, 83-84, 86-88, 90-91, 225-226, 408, 434
    origin of 18-19, 225-226
Forbes, Edward 148-152
Form of island before Minoan eruption 298, 387-388, 392-398, 443
Fossils 125-126, 148-151, 280-282, 295, 382-385, 401, 446
Frescos 110-111, 123, 125
Fritsch, von, K. 282, 295, 395, 428
Fumaroles 43, 46, 48-51, 54-56, 60-61, 67, 82, 92
    temperatures of 54, 56, 70, 181, 435-436
    under water 29-30, 38, 138

Garnet, analyses 209
Gases, analytical methods 214-217, 224-225
    composition 180, 225-233, 434
    effects on humans 16-17, 24, 33
    effects on vegetation 35, 81- 84, 89, 157
    eruption of 1866-67 214-233
    sampling methods 57-58, 61, 214-224
    spectral analyses 87
    temperatures 56, 180, 435-436
Gelung-Gung 407
Germouny, map of 183
Giacomo di Toledo, account of Monte Nuovo 408, 420
Giorgios 143-145, 157-182, 186, 191, 195, 198, 201, 226-228, 230-233, 243, 246

mineral compositions 195-206
Glass, interstitial
    composition of 314, 316, 355, 438
Gold jewelry 105, 124, 127-128
Gorceix and Mamet, excavations by 107-109, 112-114, 116, 118-119, 122-125
Gorée, Father 12, 22
Grain found in ruins 98, 113-114, 120, 123, 128
Graves, map of 183
Grioux 400, 415-416
Ground movement 38-45, 47, 148-152, 170-173, 248-249, 263, 384, 440
    Banco 183
    Micra Kameni 141
    Nea Kameni 143-144

Hawaii 410, 429
Hiera 3-10
Historical accounts 3-35, 420-421
Humbolt, A. von 398, 400, 409, 411, 419-420, 422
Human skeleton, found on Therasia 99
Hunt, T. Sterry 319, 444
Hydrocarbons, origin of flames 225-226
Hydrogen 60, 77, 86-87
    origin of 254-255
Hypersthene, composition 212, 296
    found only in acidic rocks 313

Inclusions 262, 306, 313
    coarse-grained 210-212, 372-378, 387-388
    glassy 210, 307, 346, 374, 439
    metamorphic rocks 207-210, 377
    fluid 210, 308, 363, 371, 374, 377, 439
    in crystals 306-308, 439
    minerals of 75, 207-210, 286, 372-373, 376, 436, 438, 446
Isles of May 71-75, 87, 174, 201, 240, 442

Janssen 87
Jewelry, gold 105, 124, 127-128
Jorullo 408, 409, 411, 419-420
Justin 3, 5

Kallisty 379
Kircher 12
Kokkino Vouno 239. 248, 253, 388, 390, 393, 394
Kolombos 13-21

La Mottraye, Aubry de 31

Lanzarote 411, 421
Labbe de Bourges 6
Lava, analyses 335
    eruption of 1866-67 189-206
    fluidity of 166, 381
    form and flow mechanisms 85, 133, 136, 167-170, 391, 418, 441
    relations to craters of elevation 401-405, 411,
    submarine 263, 379, 442, 445
Leycester, Lieutenant 153-154
Lightning 15, 17
    origin of 18-19
Liparite, definition of 361
Loom weights 128-129
Lophiscos 147-153, 155
Loumaravi 279-282, 293, 345, 353, 361, 370-371, 381, 383-386, 396, 423
    amphibole andesites 279, 353
Lower pumice 126, 270, 285
Lyell, Charles 398, 405, 409, 412, 421

Magma chamber 387-388
    zoning of 304-305
Magnetic measurements 87
Magnetic separation 191, 193-194, 299
Mamet and Gorceix, excavations at Akrotiri 107-109, 112-114, 116, 118-119, 122-125
Map of Bory de Saint Vincent 153
Map of Germouny 183, 245
Map of Graves 56, 183, 244
Mauna Loa 410, 429
Megalo Vouno 247-248, 251-257, 293, 334, 388-391-394
Metamorphic rocks 269-274, 380, 423, 432
Metal tools 121, 124
Michel-Lévy, A. 197, 199, 200, 351-353, 447
    extinction angles of feldspars 197, 199, 210, 351-353
Micra Kameni 138-142, 144, 147, 149-151, 155, 158, 167, 172, 187
Microscope, importance of 317, 437
    study of pottery 125, 442, 443
Mikro Profitis Ilias 255-259, 261, 293, 334, 389-391, 394, 424
Milos 385
Mineral deposits 271
Minoan eruption 94, 152, 450
    dating of 21-31, 129-130

earthquakes preceding 129
effect on humans 94, 99
Mitscherlich, 325
Mt. Arkangelos 280-283, 293
Mt. Gavrillos 244, 247, 271, 274
Mt. Profitis Ilias 239, 243-244, 256-261, 269-274, 293-294, 379-381, 386, 396, 423, 432
Mont Dore 402, 406
Monte Nuovo 407, 408, 418-421

Name of prehistoric island 398
Nea Kameni 142-157
formation of 21-31
Nicephore 9
Nodules, see xenoliths
Nomicos, excavation by 96, 97-99, 101-104, 107, 128

Obelisk 280, 286, 370-371
Obsidian tools 105, 112, 115, 116, 121, 122-124, 128
Olivine, analyses 202, 203, 205, 330
incompatible with tridymite 310, 338
vs. hypersthene 327
Opal 126, 357-360
Oxidation by acidic gases 254-255
Oxidation from water 254

Palaea Kameni 7-12, 28, 132-138, 180, 187, 227
Palma 400, 406, 412, 418, 425
Pausanias 3, 5
Pègues 12-13, 15-16, 18-19, 21-22, 30, 140-141, 149, 154
Perlite 354, 368-369
Pisolites, see accretionary lapilli
Philostratus 11
Phlegrean Fields 406, 421
Plagioclase, see also feldspar
intermediate compositions 318-324, 444
Platanymos 274
Plaka 237, 244, 274
Pliny 3, 5, 6-8, 10, 16, 34, 379
Plutarch 3, 4
Popandayang 407, 427-428
Pottery 272
clay used in 106, 113, 125-127, 273
methods of making 106, 127
microscopic analysis 125, 443
Poulett Scrope, G. 398, 400, 409, 412

Pozzolana 94, 275
Pre-Minoan form of island 298, 387, 396-398, 442, 450
Pre-Minoan volcanism 334-335, 381, 386-395, see: vents, pre-caldera
Prehistoric culture
before the Minoan eruption 94, 123-125, 442
style of construction 97, 124, 129
Prehistoric dwellings, verticality of walls 129, 424
Prévost, Constant 398, 409
Profitis Ilias see Mt. Profitis Ilias
Pumice, relation to age of ruins 99, 100, 108, 123
quarrying of 94, 275, 293
use in cement 94
volume and distribution 248, 258-259, 266-267, 270, 275, 286, 292-297, 428
Puy de Dôme 400
Pyroxene, analyses 195-196, 202-205, 209, 212, 331
hypersthene 195, 205, 211, 212, 302-303, 313, 327, 332

Rain, effect of eruption on 83
Rames 421-422
Rammelsberg, C. 319
Rath, von 319
Reiss, W. and Stübel 37, 67, 69, 72-77, 150, 187, 233, 235, 245, 282, 292, 417
Reissite 371
Repose interval, relation to size of eruption 10
Reunion 409
Rhodes 3
Rhyolite, definition of 361
Richard, Father 12-13, 15-16, 18, 20-21
Richthofen, F. von 361
Roman ruins, Mt. Profitis Ilias 272
Rosenbuch, H., classification of 344-345, 361-362, 445-447, 449
Ruins, descriptions of
Mt. Profitis Ilias 272
Therasia 96-104
Akrotiri 104-123
Ross 8

Sainte-Claire Deville
Charles 2, 41, 48, 60, 64, 65, 92, 181, 184, 219, 220, 226, 232, 319
Henri 331
Schéerer 319

INDEX

Scrope, P. 398, 400, 409, 412
Seebach, von 37, 64-67
Seneca 3, 4-5, 9
Separation of minerals 189-195, 436
Silica, deposition from vapor 255
Silicification 314-315, 345, 348, 360, 364-365, 386
Sfoscioti, Nicolas 36, 38, 71
Soil under pumice 100, 105, 108, 112, 116
Solution, cause of collapse 432-433
Somma ridge 387, 392, 395-396, 399, 412, 422
Spectral analyses 87, 435
Spherulites 359, 366-368
Sterry Hunt 319, 444
Stone weights 118, 128
Stromatolites 248, 294-295
Strabo 3, 4, 9
Structures, prehistoric
    Akrotiri 104-123
    on Therasia 96-104
    style of construction 97, 102, 123-124, 129
Sublimates 56, 67, 87, 212-214, 436
Submarine cones 7-8, 10-11, 13-21, 185-187
Submarine fumaroles 55, 60, 154, 176
Submarine topography 188, 235-240
Submarine volcanism 13-21, 47, 263, 379-382, 386, 442, 445, 446
Suez Canal, pumice shipped to 94, 275, 293
Szabo, J. 234, 312, 327, 348, 349, 444

Tambora 427, 428
Tarillon, Father 21, 28-31
Temperature, basic magmas hotter that acidic 316
    of fumaroles 56, 435-436
    of water 38, 42-44, 46, 71, 77, 154, 176, 179-180, 274
Tenerife 400, 406, 411, 422, 424
Thera, shoreline 235-238
Thera, description of 247-286
Therasia, description of 286-291
Thermal springs 177, 256, 274
Theophanes 9

Thia 6-9
Thoulet 301, 351-352, 365, 367
Timor 407
Titus-Livy 3, 4
Tridymite, secondary origin of 255, 314
    incompatible with olivine 338
Tschermak, G. 318-325, 329, 336, 444
    classification of igneous rocks 336
    plagioclase series 318-325, 347
Tsunamis of 1650 15, 20

Uplift 22, 248-249, 263, 384-386, 399, 446, 450
    regional 380, 385
    see also ground movement

Vegetation on pre-caldera island 129, 397, 442
Vélain 338
Vents, pre-caldera 250-254, 256-260, 265-266, 269, 279, 283, 285-286, 290, 293, 294, 381, 387-395
Verneuil 54, 60, 62
Verticality of walls of buildings 419, 425
Vesicles, shapes of 356, 363, 369, 405
Vesuvius 305, 392, 399-400, 403, 409-410, 420, 426
    eruption of 1861 11, 18, 384
Virlet d'Aoust 398, 405, 417
Viscosity, factors governing 262-263
Vogelsang, H. 430
Volumes of calderas and ejecta 427-428

Wall decorations 110, 111, 123, 125
Waltershausen, S. von 319
Water, dissociation of 254-255, 434
Weights, measuring 98, 103, 118, 128
White Island 22-24, 26-28, 32-33, 146, 148-152, 385
Wollastonite analyses of 208

Xenoliths 207-212, 302, 372-377, 387-388, 436, 438, 445-446, 450

Zeolites 359, 371, 445